Quantum Field Theory

Quantum Field Theory

Editor: Henry Brosnan

NY RESEARCH PRESS
New York

Published by NY Research Press
118-35 Queens Blvd., Suite 400,
Forest Hills, NY 11375, USA
www.nyresearchpress.com

Quantum Field Theory
Edited by Henry Brosnan

International Standard Book Number: 978-1-63238-709-7 (Hardback)

Cataloging-in-Publication Data

Quantum field theory / edited by Henry Brosnan.
p. cm.
Includes bibliographical references and index.
ISBN 978-1-63238-709-7
1. Quantum field theory. 2. Quantum theory. 3. Field theory (Physics). I. Brosnan, Henry.
QC174.45 .Q36 2019
530.143--dc23

Contents

Preface ... IX

Chapter 1 **(1+1)-dimensional gauge symmetric gravity model and related exact black hole and cosmological solutions in string theory** ... 1
S. Hoseinzadeh and A. Rezaei-Aghdam

Chapter 2 **Broken boost invariance in the Glasma via finite nuclei thickness** ... 9
Andreas Ipp and David Müller

Chapter 3 **Calculable mass hierarchies and a light dilaton from gravity duals** ... 15
Daniel Elander and Maurizio Piai

Chapter 4 **Charged scalar quasi-normal modes for linearly charged dilaton-Lifshitz solutions** ... 20
M. Kord Zangeneh, B. Wang, A. Sheykhi and Z. Y. Tang

Chapter 5 **Compact Chern–Simons vortices** ... 27
D. Bazeia, L. Losano, M. A. Marques and R. Menezes

Chapter 6 **Constrained superfields from inflation to reheating** ... 32
Ioannis Dalianis and Fotis Farakos

Chapter 7 **Could the primordial radiation be responsible for vanishing of topological defects?** ... 38
Tomasz Romańczukiewicz

Chapter 8 **Coulomb effects in high-energy e^+e^- electroproduction by a heavy charged particles in an atomic field** ... 43
P. A. Krachkov and A. I. Milstein

Chapter 9 **Curved momentum spaces from quantum groups with cosmological constant** ... 47
Á. Ballesteros, G. Gubitosi, I. Gutiérrez-Sagredo and F. J. Herranz

Chapter 10 **Deuteron properties from muonic atom spectroscopy** ... 54
N. G. Kelkar and D. Bedoya Fierro

Chapter 11 **Diagrammar in an extended theory of gravity** ... 59
David C. Dunbar, John H. Godwin, Guy R. Jehu and Warren B. Perkins

Chapter 12 **Do the gravitational corrections to the beta functions of the quartic and Yukawa couplings have an intrinsic physical meaning?** ... 64
S. Gonzalez-Martin and C. P. Martin

Chapter 13 **Double and cyclic λ-deformations and their canonical equivalents** ... 70
George Georgiou, Konstantinos Sfetsos and Konstantinos Siampos

Chapter 14 **Emergent geometry, thermal CFT and surface/state correspondence** 77
Wen-Cong Gan, Fu-Wen Shu and Meng-He Wu

Chapter 15 **Entanglement entropy and complexity for one-dimensional holographic superconductors** 84
Mahdi Kord Zangeneh, Yen Chin Ong and Bin Wang

Chapter 16 **Entanglement entropy in a non-conformal background** 91
M. Rahimi, M. Ali-Akbari and M. Lezgi

Chapter 17 **Exact microstate counting for dyonic black holes in AdS_4** 96
Francesco Benini, Kiril Hristov and Alberto Zaffaroni

Chapter 18 **Exponentiating Higgs** 101
Marco Matone

Chapter 19 **Axion mass bound in very special relativity** 108
R. Bufalo and S. Upadhyay

Chapter 20 **Glueball-baryon interactions in holographic QCD** 114
Si-Wen Li

Chapter 21 **Hadronic structure functions in the $e^+e^- \to \bar{\Lambda}\Lambda$ reaction** 122
Göran Fäldt and Andrzej Kupsc

Chapter 22 **Holographic corrections to the Veneziano amplitude** 127
Adi Armoni and Edwin Ireson

Chapter 23 **Integration of trace anomaly in *6D*** 132
Fabricio M. Ferreira and Ilya L. Shapiro

Chapter 24 **Linking structure and dynamics in (*p*, *pn*) reactions with Borromean nuclei: The ^{11}Li(*p*, *pn*)^{10}Li case** 137
M. Gómez-Ramos, J. Casal and A. M. Moro

Chapter 25 **No nonminimally coupled massless scalar hair for spherically symmetric neutral black holes** 143
Shahar Hod

Chapter 26 **Onset of η-nuclear binding in a pionless EFT approach** 146
N. Barnea, B. Bazak, E. Friedman and A. Gal

Chapter 27 **On the mass of the world-sheet 'axion' in *SU (N)* gauge theories in 3 + 1 dimensions** 152
Andreas Athenodorou and Michael Teper

Chapter 28 **On the stability of non-supersymmetric supergravity solutions** 159
Ali Imaanpur and Razieh Zameni

Chapter 29 **Overlaps after quantum quenches in the sine-Gordon model** 164
D. X. Horváth and G. Takács

Chapter 30 **Photon mass via current confinement** ... 171
Vivek M. Vyas and Prasanta K. Panigrahi

Chapter 31 **Quantum non-equilibrium effects in rigidly-rotating thermal states** ... 176
Victor E. Ambruş

Chapter 32 **Quantum quench and scaling of entanglement entropy** ... 182
Paweł Caputa, Sumit R. Das, Masahiro Nozaki and Akio Tomiya

Chapter 33 **Small-x asymptotics of the quark helicity distribution: Analytic results** ... 187
Yuri V. Kovchegov, Daniel Pitonyak and Matthew D. Sievert

Chapter 34 **Aspects of ultra-relativistic field theories via flat-space holography** ... 192
Reza Fareghbal, Ali Naseh and Shahin Rouhani

Chapter 35 **Fermionic continuous spin gauge field in (A)dS space** ... 197
R. R. Metsaev

Chapter 36 **Analytical shear viscosity in hyperscaling violating black brane** ... 204
Xiao- Mei Kuang and Jian-Pin Wu

Permissions

List of Contributors

Index

Preface

Quantum field theory (QFT) is a theoretical approach to the study of subatomic particles and quasiparticles integrating classical field theory, quantum mechanics and the theory of special relativity. In QFT, particles are treated as excited states or quanta of their fields. Interactions between these particles are described by the interaction terms in the corresponding Lagrangian and through the Feynman diagrams. The concepts of path integral formulation, canonical quantization, renormalization, gauge theory, supersymmetry, etc. are integral to the development of quantum field theory. This book provides comprehensive insights into the field of quantum field theory. It unfolds the innovative aspects of this discipline, which will be crucial for the progress of this field in the future. It will serve as a valuable source of reference for graduate and post graduate students as well as experts.

After months of intensive research and writing, this book is the end result of all who devoted their time and efforts in the initiation and progress of this book. It will surely be a source of reference in enhancing the required knowledge of the new developments in the area. During the course of developing this book, certain measures such as accuracy, authenticity and research focused analytical studies were given preference in order to produce a comprehensive book in the area of study.

This book would not have been possible without the efforts of the authors and the publisher. I extend my sincere thanks to them. Secondly, I express my gratitude to my family and well-wishers. And most importantly, I thank my students for constantly expressing their willingness and curiosity in enhancing their knowledge in the field, which encourages me to take up further research projects for the advancement of the area.

Editor

$(1+1)$-dimensional gauge symmetric gravity model and related exact black hole and cosmological solutions in string theory

S. Hoseinzadeh, A. Rezaei-Aghdam *

Department of Physics, Faculty of Science, Azarbaijan Shahid Madani University, 53714-161, Tabriz, Iran

ARTICLE INFO

Editor: N. Lambert

ABSTRACT

We introduce a four-dimensional extension of the Poincaré algebra ($\mathcal{N}$) in $(1+1)$-dimensional space-time and obtain a $(1+1)$-dimensional gauge symmetric gravity model using the algebra $\mathcal{N}$. We show that the obtained gravity model is dual (canonically transformed) to the $(1+1)$-dimensional anti de Sitter (*AdS*) gravity. We also obtain some black hole and Friedmann–Robertson–Walker (FRW) solutions by solving its classical equations of motion. Then, we study $\frac{A_{4,8}}{A_1 \otimes A_1}$ gauged Wess–Zumino–Witten (WZW) model and obtain some exact black hole and cosmological solutions in string theory. We show that some obtained black hole and cosmological metrics in string theory are same as the metrics obtained in solutions of our gauge symmetric gravity model.

1. Introduction

$(1+1)$-dimensional gravity has been extensively studied by proposing various models. Two of the gravitational theories of most interest are singled out by their simplicity and group theoretical properties. One of them is proposed by Jackiw [1] and Teitelboim [2] (Liouville gravity) which is equivalent to the gauge theory of gravity with (anti) de Sitter group [3–5]. The other one is the string-inspired gravity [6–8] which is equivalent to the gauge theory of the Poincaré group $ISO(1,1)$ [7] and its central extension [9–13].

Recently, two algebras namely the Maxwell algebra [14,15] and the semi-simple extension of the Poincaré algebra [16] have been applied to construct some gauge invariant theories of gravity in four [16–20] and three [21–23] dimensional space-times. These algebras have been also applied to string theory as an internal symmetry of the matter gauge fields [24]. The Maxwell algebra in $(1+1)$-dimensional space-time, is the well-known central extension of the Poincaré algebra which, as we discussed above, has been applied to construct a $(1+1)$-dimensional gauge symmetric gravity action [9,10]. In this paper, we introduce a new four-dimensional extension of the Poincaré algebra ($\mathcal{N}$) in $(1+1)$-dimensional space-time, which is obtained from the 16-dimensional semi-simple extension of Poincaré algebra in $(3+1)$-dimensional space-time [16], by reduction of the dimensions of the space. Then, we construct a $(1+1)$-dimensional gauge symmetric gravity model, using this algebra. We obtain some black hole and cosmological solutions by solving its equations of motion.

On the other hand, in string theory, two-dimensional exact black hole has been found by Witten [6]. Another black hole solution to the string theory has been presented in [25] both in Schwarzschild-like and target space conformal gauges. Exact three-dimensional black string and black hole solutions in string theory have also been found in [26,27]. Here, we study the string theory in $(1+1)$-dimensional space-time, and show that some obtained black hole and cosmological solutions of the gravity model, are exact solutions of the beta function equations (in all loops).

The outline of this paper is as follows: In section 2, we construct a $(1+1)$-dimensional gauge symmetric gravity model using a four-dimensional gauge group related to the algebra $\mathcal{N}$. Then, by presenting a canonical map, we show that the obtained gravity model is dual (canonically transformed) to the $(1+1)$-dimensional *AdS* gravity model. In section 3, we solve the equations of motion and obtain some black hole and Friedmann–Robertson–Walker (FRW) cosmological solutions. Finally, in section 4, we study $\frac{A_{4,8}}{A_1 \otimes A_1}$ gauged Wess–Zumino–Witten (WZW) model, and show that some of the resulting string backgrounds, which are exact $(1+1)$-dimensional solutions of the string theory, are the same as the black hole and cosmological solutions obtained for our gravity model. Section 5, contains some concluding remarks.

* Corresponding author.
E-mail addresses: hoseinzadeh@azaruniv.edu (S. Hoseinzadeh), rezaei-a@azaruniv.edu (A. Rezaei-Aghdam).

2. (1 + 1)-dimensional gravity from a non-semi-simple extension of the Poincaré gauge symmetric model

The Poincaré algebra $Iso(1,1)$ in $(1+1)$-dimensional space-time has the following form:

$$[J, P_a] = \epsilon_{ab} P^b, \qquad [P_a, P_b] = 0, \tag{1}$$

where $\epsilon_{01} = -\epsilon^{01} = +1$, and J and P_a $(a = 0, 1)$ are generators of the rotation and translations in space-time, and the algebra indices $a = 0, 1$ can be raised and lowered by the $(1+1)$-dimensional Minkowski metric η_{ab} $(\eta_{00} = -1,\ \eta_{11} = 1)$ such that $P_a = \eta_{ab} P^b$. In $D = 1+1$, a four-dimensional non-semi-simple extension of the Poincaré algebra[1] $\mathcal{N} = (P_a, J, Z)$ has the following form:

$$[J, P_a] = \epsilon_{ab} P^b, \quad [P_a, P_b] = k\epsilon_{ab} Z, \quad [Z, P_a] - -\frac{\Lambda}{k}\epsilon_{ab} P^b, \tag{2}$$

where Z is the new generator and k and Λ are constants.[2] For $\Lambda = 0$, which leads to $[Z, P_a] = 0$, the above algebra reduces to a solvable algebra which is called the centrally extended Poincaré algebra (or Maxwell algebra[3] in $1+1$ dimensions) [9–11]. We construct the $\mathcal{N}$-algebra valued one-form gauge field as follows:

$$h_i = h_i{}^B X_B = e_i{}^a P_a + \omega_i\, J + A_i\, Z, \qquad i, j = 0, 1 \tag{3}$$

where the indices $i, j = 0, 1$ are the space-time indices, and the one-form fields have the following forms:

$$e^a = e_i{}^a dx^i\ , \quad \omega = \omega_i dx^i\ , \quad A = A_i dx^i,$$

where $e_i{}^a$, ω_i, A_i are the vierbein, spin connection and the new gauge field, respectively. Using the following infinitesimal gauge parameter:

$$u = \rho^a P_a + \tau\, J + \lambda\, Z,$$

and the gauge transformation as follows:

$$h_i \to h'_i = U^{-1} h_i U + U^{-1} \partial_i U,$$

with $U = e^{-u} \simeq 1 - u$ and $U^{-1} = e^{u} \simeq 1 + u$, we obtain the following transformations of the gauge fields:

$$\begin{aligned}
\delta e_i{}^a &= -\partial_i \rho^a - \epsilon^{ab} e_{ib}\, (\tau - \frac{\Lambda}{k}\lambda) + \epsilon^{ab} \rho_b\, (\omega_i - \frac{\Lambda}{k} A_i)\ ,\\
\delta\omega_i &= -\partial_i \tau, \qquad\qquad\qquad\qquad\qquad\qquad\qquad\qquad (4)\\
\delta A_i &= -\partial_i \lambda - k\, \epsilon^{ab} e_{ia}\, \rho_b.
\end{aligned}$$

The generic Ricci curvature can be obtained as follows:

$$\begin{aligned}
\mathcal{R} &= \mathcal{R}_{ij} dx^i \wedge dx^j = \mathcal{R}^A X_A = \mathcal{R}_{ij}{}^A X_A dx^i \wedge dx^j,\\
\mathcal{R}_{ij} &= \partial_{[i} h_{j]} + [h_i, h_j] = \mathcal{R}_{ij}{}^A X_A = T_{ij}{}^a\, P_a + R_{ij}\, J + F_{ij}\, Z\ , \qquad (5)
\end{aligned}$$

where the torsion $T_{ij}{}^a$, standard Riemannian curvature R_{ij} and the new gauge field strength F_{ij} have the following forms:

$$\begin{aligned}
T_{ij}{}^a &= \partial_{[i}\, e_{j]}{}^a + \epsilon^{ab} (e_{ib}\, \omega_j - e_{jb}\, \omega_i) - \frac{\Lambda}{k}\, \epsilon^{ab} (e_{ib}\, A_j - e_{jb}\, A_i),\\
R_{ij} &= \partial_{[i}\, \omega_{j]}, \qquad\qquad\qquad\qquad\qquad\qquad (6)\\
F_{ij} &= \partial_{[i}\, A_{j]} + k\, \epsilon^{ab} e_{ia}\, e_{jb}.
\end{aligned}$$

Now, one can write the gauge invariant action as [10]

$$\begin{aligned}
I &= \frac{1}{2} \int \eta_A \mathcal{R}^A = \frac{1}{2} \int d^2x\, \epsilon^{ij}\, \eta_A\, \mathcal{R}_{ij}{}^A \qquad (7)\\
&= \frac{1}{2} \int d^2x\, \epsilon^{ij}\, (\eta_a\, T_{ij}{}^a + \eta_2\, R_{ij} + \eta_3\, F_{ij}), \qquad (8)
\end{aligned}$$

where $\eta_A = (\eta_a, \eta_2, \eta_3)$ are the Lagrange multiplier-like fields. Now, using (6), one can rewrite this action in the following form:

$$\begin{aligned}
I = \int d^2x\, \epsilon^{ij} \Big\{ &\eta_a \Big(\partial_i e_j{}^a + \epsilon^{ab}\, e_{ib}\, (\omega_j - \frac{\Lambda}{k}\, A_j)\Big) + \eta_2\, \partial_i \omega_j\\
&+ \eta_3 \Big(\partial_i A_j + \frac{1}{2} k\, \epsilon^{ab}\, e_{ia}\, e_{jb}\Big)\Big\}. \qquad (9)
\end{aligned}$$

This action is invariant under the gauge transformations (4) and the following transformations of the fields η_a, η_2 and η_3:

$$\begin{aligned}
\delta\eta_a &= k\, \epsilon_a{}^b\, \eta_3\, \rho_b - \epsilon_a{}^b\, \eta_b (\tau - \frac{\Lambda}{k}\, \lambda),\\
\delta\eta_2 &= -\epsilon^{ab} \eta_a\, \rho_b, \qquad\qquad\qquad (10)\\
\delta\eta_3 &= \frac{\Lambda}{k}\, \epsilon^{ab} \eta_a\, \rho_b.
\end{aligned}$$

Now, we will show that the model (9) is dual to the $(1+1)$-dimensional AdS gravity. We know that $SO(2,1)$ gauge symmetric gravity action can be obtained by use of the following algebra (anti de Sitter algebra for $k' \neq 0$):

$$[J, P_a] = \epsilon_{ab} P^b, \qquad [P_a, P_b] = k' \epsilon_{ab} J, \tag{11}$$

as follows [10]:

$$\begin{aligned}
\tilde{I} = \int d^2x\, \epsilon^{ij} \Big\{ &\tilde{\eta}_a \Big(\partial_i e_j{}^a + \epsilon^{ab}\, e_{ib}\, \omega_j\Big)\\
&+ \tilde{\eta}_2 \Big(\partial_i \omega_j + \frac{1}{2} k'\, \epsilon^{ab}\, e_{ia}\, e_{jb}\Big)\Big\}. \qquad (12)
\end{aligned}$$

An $SO(2,1)$ invariant action for two-dimensional gravity was first constructed in [3] where the aim was to reconstruct the proposed two-dimensional Einstein equation from a two-dimensional gauge theory of gravity. Although the notation adopted in [3] is different from our notation,[4] but it can be shown that the action constructed in [3] is equivalent to the $(1+1)$-dimensional AdS gravity model (12). Now, by selecting $\eta_3 = -\frac{\Lambda}{k}\eta_2$ in our model (9), it is dual (canonically transformed) to the AdS gravity (12); i.e. the following map:

$$\begin{aligned}
&\omega_i \longrightarrow \omega_i - \frac{\Lambda}{k} A_i, \qquad e_i{}^a \longrightarrow e_i{}^a, \qquad \tilde{\eta}_a \longrightarrow \eta_a,\\
&\tilde{\eta}_2 \longrightarrow \eta_2, \qquad k' = -\Lambda, \qquad\qquad (13)
\end{aligned}$$

transforms the AdS_2 gravity model (12) to our model (9). In the following, we will show that this map is a canonical one. The canonical Poisson-brackets and the Hamiltonian related to the AdS_2 gravity model (12) are as follows:

$$\begin{aligned}
&\{(\tilde{\Pi}_e)_i{}^a(x)\ ,\ e_j{}^b(y)\} = \epsilon_{ij} \eta^{ab} \delta(x - y),\\
&\{(\tilde{\Pi}_\omega)_i(x)\ ,\ \omega_j(y)\} = \epsilon_{ij} \delta(x - y),\\
&\{(\tilde{\Pi}_{\tilde{\eta}_a})^a(x)\ ,\ \tilde{\eta}^b(y)\} = \eta^{ab} \delta(x - y),\\
&\{(\tilde{\Pi}_{\tilde{\eta}_2})(x)\ ,\ \tilde{\eta}_2(y)\} = \delta(x - y),
\end{aligned}$$

---

[1] This algebra is isomorphic to the four-dimensional Lie algebra $\mathcal{A}_{3,8} \oplus \mathcal{A}_1$ [28].

[2] Note that the commutation relation $[J, Z] = 0$ can be obtained from the Jacobi identity $[J, [P_a, P_b]] +$ cyclic terms $= 0$.

[3] Centrally extended Poincaré algebra (or Maxwell algebra) in $1+1$ dimensions is isomorphic to the four-dimensional Lie algebra $\mathcal{A}_{4,8}$ [28].

[4] The $SO(2,1)$ invariant action for two-dimensional gravity constructed in [3] is $\frac{1}{2}\int \epsilon^{ABC} R_{AB}\, \phi_C$ where $R_{AB} = d\omega_{AB} - \omega_{AC}\, \omega^C{}_B$ and $A, B = 0, 1, 2$. The field ω_{AB} contains both the spin connection ω_{ab} and vierbein e_a where $a, b = 0, 1$ such that we have $\omega_{a2} = \ell^{-1} e_a$. Using the field redefinitions $\phi_0 \equiv \ell\tilde{\eta}_1$, $\phi_1 \equiv \ell\tilde{\eta}_0$, $\phi_2 \equiv \tilde{\eta}_2$ and $\omega_{01} \equiv \omega$, one can easily show that this action is equivalent to the AdS_2 gravity action (12) with $k' = \ell^{-2}$.

$$\tilde{H} = \int d^3x\Big((\tilde{\Pi}_e)^i{}_a\, \partial_t e_i{}^a + (\tilde{\Pi}_\omega)^i{}_a\, \partial_t \omega_i{}^a + (\tilde{\Pi}_{\tilde{\eta}_a})^a\, \partial_t \tilde{\eta}_a + (\tilde{\Pi}_{\tilde{\eta}_2})\, \partial_t \tilde{\eta}_2 \Big) - \tilde{I} = 2\int d^2x \epsilon^{0i}\Big(\tilde{\eta}_a \partial_t e_i{}^a + \tilde{\eta}_2 \partial_t \omega_i\Big) - \tilde{I}, \tag{14}$$

where the coordinates of the space-time are $\{x^i\} = \{x^0, x^1\} = \{t, r\}$ such that $\partial_t = \partial_0 = \frac{\partial}{\partial t}$, η^{ab} is the inverse Minkowski metric, and the conjugate momentums corresponding to the fields are as follows:

$$(\tilde{\Pi}_e)^i{}_a = \frac{\partial \tilde{\mathcal{L}}}{\partial(\partial_t e_i{}^a)} = \epsilon^{0i}\tilde{\eta}_a, \qquad (\tilde{\Pi}_\omega)^i = \frac{\partial \tilde{\mathcal{L}}}{\partial(\partial_t \omega_i)} = \epsilon^{0i}\tilde{\eta}_2,$$
$$(\tilde{\Pi}_{\tilde{\eta}_a})^a = \frac{\partial \tilde{\mathcal{L}}}{\partial(\partial_t \tilde{\eta}_a)} = -\epsilon^{0i} e_i{}^a, \qquad (\tilde{\Pi}_{\tilde{\eta}_2}) = \frac{\partial \tilde{\mathcal{L}}}{\partial(\partial_t \tilde{\eta}_2)} = -\epsilon^{0i}\omega_i. \tag{15}$$

The map (13) is a canonical transformation and easily it can be shown that, under this map, the canonical Poisson-brackets and the Hamiltonian (14) related to the AdS_2 gravity model (12) are transformed to the following Poisson-brackets and the Hamiltonian related to our model (9):

$$\begin{aligned}
&\{(\Pi_e)_i{}^a(x)\,,\, e_j{}^b(y)\} = \epsilon_{ij}\eta^{ab}\delta(x-y),\\
&\{(\Pi_\omega)_i(x)\,,\, \omega_j(y)\} = \epsilon_{ij}\delta(x-y),\\
&\{(\Pi_A)_i(x)\,,\, A_j(y)\} = \epsilon_{ij}\delta(x-y),\\
&\{(\Pi_{\eta_a})^a(x)\,,\, \eta^b(y)\} = \eta^{ab}\delta(x-y),\\
&\{(\Pi_{\eta_2})(x)\,,\, \eta_2(y)\} = \delta(x-y),\\
&\{(\Pi_{\eta_3})(x)\,,\, \eta_3(y)\} = \delta(x-y),
\end{aligned} \tag{16}$$

$$H = \int d^3x\Big((\Pi_e)^i{}_a\, \partial_t e_i{}^a + (\Pi_\omega)^i{}_a\, \partial_t \omega_i{}^a + (\Pi_A)^i{}_a\, \partial_t A_i{}^a + (\Pi_{\eta_a})^a\, \partial_t \eta_a + (\Pi_{\eta_2})\, \partial_t \eta_2 + (\Pi_{\eta_3})\, \partial_t \eta_3 \Big) - I = 2\int d^2x \epsilon^{0i}\Big(\eta_a \partial_t e_i{}^a + \eta_2 \partial_t \omega_i + \eta_3 \partial_t A_i\Big) - I,$$

where the conjugate momentums corresponding to the fields are given as follows:

$$(\Pi_e)^i{}_a = \frac{\partial \mathcal{L}}{\partial(\partial_t e_i{}^a)} = \epsilon^{0i}\eta_a, \qquad (\Pi_\omega)^i = \frac{\partial \mathcal{L}}{\partial(\partial_t \omega_i)} = \epsilon^{0i}\eta_2,$$
$$(\Pi_A)^i = \frac{\partial \mathcal{L}}{\partial(\partial_t A_i)} = \epsilon^{0i}\eta_3, \qquad (\Pi_{\eta_a})^a = \frac{\partial \mathcal{L}}{\partial(\partial_t \eta_a)} = -\epsilon^{0i} e_i{}^a,$$
$$(\Pi_{\eta_2}) = \frac{\partial \mathcal{L}}{\partial(\partial_t \eta_2)} = -\epsilon^{0i}\omega_i, \qquad (\Pi_{\eta_3}) = \frac{\partial \mathcal{L}}{\partial(\partial_t \eta_3)} = -\epsilon^{0i} A_i. \tag{17}$$

Note that the conjugate momentums are transformed under the map (13) as:

$$(\tilde{\Pi}_e)^i{}_a \longrightarrow (\Pi_e)^i{}_a, \qquad (\tilde{\Pi}_{\tilde{\eta}_a})^a \longrightarrow (\Pi_{\eta_a})^a,$$
$$(\tilde{\Pi}_\omega)^i \longrightarrow (\Pi_\omega)^i, \qquad \tilde{\Pi}_{\tilde{\eta}_2} \longrightarrow (\Pi_{\eta_2} - \frac{\Lambda}{k}\Pi_{\eta_3}).$$

Since the Poisson-brackets and Hamiltonian of the model are preserved under the map (13), then the *AdS* gauge gravity model (12) is dual to our gravity model (9), and each can be transformed to the other by the canonical transformation (13), of course with $\eta_3 = -\frac{\Lambda}{k}\eta_2$. Finally, under the map (13), the equations of motion for the AdS_2 gravity (12) also transform to the equations of motion related to our model (9).

The equations of motion with respect to the fields η_a, η_2, η_3 have the following forms, respectively:

$$\begin{aligned}
&\epsilon^{ij}\Big(\partial_i e_j{}^a + \epsilon^{ab}\, e_{ib}\,(\omega_j - \frac{\Lambda}{k}\, A_j)\Big) = 0,\\
&\epsilon^{ij}\partial_i\, \omega_j = 0,\\
&\epsilon^{ij}\Big(\partial_i A_j + \frac{1}{2} k\, \epsilon^{ab}\, e_{ia}\, e_{jb}\Big) = 0,
\end{aligned} \tag{18}$$

and the equations of motion with respect to the fields $e_i{}^a$, ω_i, A_i are obtained as follows, respectively:

$$\begin{aligned}
&\epsilon^{ij}\Big(-\partial_j \eta_a + \epsilon_a{}^b\, \eta_b\,(\omega_j - \frac{\Lambda}{k}\, A_j) - k\, \epsilon_{ab}\, \eta_3\, e_j{}^b\Big) = 0,\\
&\epsilon^{ij}\Big(\partial_j \eta_2 - \epsilon^{ab}\, \eta_a\, e_{jb}\Big) = 0,\\
&\epsilon^{ij}\Big(\partial_j \eta_3 + \frac{\Lambda}{k}\epsilon^{ab}\, \eta_a\, e_{jb}\Big) = 0.
\end{aligned} \tag{19}$$

In the next section, we will try to solve these equations and obtain different solutions of them.

3. Solutions of the equations of motion

3.1. Radial solutions for $\Lambda \neq 0$

Using the following Ansatz for the metric:

$$ds^2 = e_i{}^a e_j{}^b \eta_{ab} dx^i dx^j = -N^2(r)\, dt^2 + M^2(r) dr^2, \tag{20}$$

where $\{x^0, x^1\} = \{t, r\}$ are the coordinates of the space-time ($0 \le t < \infty$, $-\infty < r < \infty$), one can obtain the following solution for the equations of motion (18)–(19):

$$\begin{aligned}
&M^2(r) = \frac{1}{-\Lambda N^2(r) + C_4}\Big(\frac{dN(r)}{dr}\Big)^2,\\
&\eta_0(r) = C_2 N(r), \qquad \eta_1(r) = 0,\\
&\eta_2(r) = \frac{C_2}{\Lambda}\sqrt{-\Lambda N^2(r) + C_4} + C_1,\\
&\eta_3(r) = -\frac{C_2}{k}\sqrt{-\Lambda N^2(r) + C_4},\\
&\omega(r) = C_3\, dt + f(r)\, dr,\\
&A(r) = \frac{k}{\Lambda}\Big((C_3 - \sqrt{-\Lambda N^2(r) + C_4})\, dt + f(r)\, dr\Big),
\end{aligned} \tag{21}$$

where C_1, C_2, C_3 and C_4 are arbitrary constants, and $N(r)$, $f(r)$ are arbitrary functions of r. The solution (21) describes a space-time with a constant scalar curvature $\mathcal{R} = 2\Lambda$.

3.1.1. AdS black hole solution

For $N^2(r) = -\Lambda r^2 - b$ and $C_4 = -\Lambda b$, the solution (21) reduces to the following *AdS* black hole solution:

$$ds^2 = -N^2(r)\, dt^2 + \frac{dr^2}{N^2(r)}, \tag{22}$$

and

$$\begin{aligned}
&\eta_0(r) = C_2 N(r), \qquad \eta_1(r) = 0,\\
&\eta_2(r) = -C_2\, r + C_1, \qquad \eta_3(r) = \frac{\Lambda}{k} C_2\, r,\\
&\omega(r) = C_3\, dt + f(r)\, dr,\\
&A(r) = (kr + \frac{k}{\Lambda} C_3)\, dt + \frac{k}{\Lambda} f(r)\, dr,
\end{aligned} \tag{23}$$

where b is a constant. Now, we calculate the mass (energy) of solution (23) using the ADM definition of mass (energy) as discussed in [29]. Varying the action (9) produces a bulk term, which is zero using the equations of motion, plus a boundary term which can be cancelled by adding the following boundary term to the Lagrangian:

$$\mathcal{L}_B = -\,\partial_r\Big(\eta_a e_0{}^a + \eta_2\omega_0 + \eta_3 A_0\Big), \tag{24}$$

together with an appropriate boundary condition. This boundary term is identified as the mass (energy) of solution. Our boundary condition is using the obtained values for fields in the solution (23) at spatial infinity ($r \to \pm\infty$). Then, the mass of the solution is obtained as follows:

$$m = \int_{-\infty}^{+\infty} dr\mathcal{L}_B = -\Big(\eta_a e_0{}^a + \eta_2\omega_0 + \eta_3 A_0\Big)\Big|_{-\infty}^{+\infty}, \tag{25}$$

which using (22) and (23) turns out to be

$$m = C_2 b - C_1 C_3. \tag{26}$$

The Kretschmann scalar for this metric is

$$K = R_{\mu\nu\rho\sigma}R^{\mu\nu\rho\sigma} = 4\Lambda^2. \tag{27}$$

Consequently, this solution has two singularities at the following points:

$$r_\pm = \pm\sqrt{\frac{b}{-\Lambda}}, \tag{28}$$

which are not the curvature singularities, but the coordinate singularities, and can be removed by definition of a new coordinate system. Using the Ansatz (22), we obtain another solution for the equations of motion (18)–(19) as follows:

$$\begin{aligned}
&N^2(r) = -\Lambda r^2 - 2Dr + C_3, \qquad \eta_0(r) = C_2 N(r),\\
&\eta_1(r) = 0, \qquad \eta_2(r) = -C_2\, r + C_1, \qquad \eta_3(r) = \frac{C_2}{k}(\Lambda\, r + D),\\
&\omega(r) = C_4\, dt + f(r)\, dr, \qquad A(r) = (kr + C_5)\, dt + \frac{k}{\Lambda} f(r)\, dr,
\end{aligned} \tag{29}$$

where C_1, C_2, C_3, C_4, C_5 and $D = \frac{\Lambda}{k}C_5 - C_4$ are arbitrary constants, and $f(r)$ is a function of r. The value of the Kretschmann scalar of this solution is same as that of the previous one $K = 4\Lambda^2$, and it has two coordinate singularities at points

$$r_\pm = \frac{-D \pm \sqrt{D^2 + \Lambda C_3}}{\Lambda}. \tag{30}$$

Using new coordinate $\rho = r + \frac{D}{2\Lambda}$, the latter solution (29) transforms to the *AdS* black hole solution (23). This also can be achieved by choosing $D = 0$ and $C_3 = -b$.

3.1.2. Black hole solutions

For negative Λ, by assuming $N(r) = \sinh(\sqrt{-\Lambda}\ r - b)$ and $C_4 = -\Lambda$, the solution (21) reduces to the following black hole solution:

$$\begin{aligned}
&ds^2 = -\sinh^2(\sqrt{-\Lambda}\, r - b)dt^2 + dr^2, \qquad (31)\\
&\eta_0(r) = C_2\sinh(\sqrt{-\Lambda}\, r - b), \qquad \eta_1(r) = 0,\\
&\eta_2(r) = -\frac{C_2}{\sqrt{-\Lambda}}\cosh(\sqrt{-\Lambda}\, r - b) + C_1,\\
&\eta_3(r) = -\frac{C_2}{k}\sqrt{-\Lambda}\cosh(\sqrt{-\Lambda}\, r - b),\\
&\omega(r) = C_3\, dt + f(r)\, dr,\\
&A(r) = \frac{k}{\Lambda}\Big((C_3 - \sqrt{-\Lambda}\cosh(\sqrt{-\Lambda}\, r - b))\, dt + f(r)\, dr\Big), \qquad (32)
\end{aligned}$$

where b is an arbitrary constant.

For positive Λ, by assuming $N(r) = \sin(\sqrt{\Lambda}\ r - b)$ and $C_4 = \Lambda$, the solution (21) reduces to the following black hole solution:

$$\begin{aligned}
&ds^2 = -\sin^2(\sqrt{\Lambda}\, r - b)dt^2 + dr^2, \qquad (33)\\
&\eta_0(r) = C_2\sin(\sqrt{\Lambda}\, r - b), \qquad \eta_1(r) = 0,\\
&\eta_2(r) = \frac{C_2}{\sqrt{\Lambda}}\cos(\sqrt{\Lambda}\, r - b) + C_1,\\
&\eta_3(r) = -\frac{C_2}{k}\sqrt{\Lambda}\cos(\sqrt{\Lambda}\, r - b),\\
&\omega(r) = C_3\, dt + f(r)\, dr,\\
&A(r) = \frac{k}{\Lambda}\Big((C_3 - \sqrt{\Lambda}\cos(\sqrt{\Lambda}\, r - b))\, dt + f(r)\, dr\Big). \qquad (34)
\end{aligned}$$

Both of the solutions (31) and (33) have coordinate singularities at

$$r = \frac{b}{\sqrt{|\Lambda|}}, \tag{35}$$

which can be removed by suitable coordinate transformations, because of their finite Ricci and Kretschmann scalars.

3.2. Radial solutions for $\Lambda = 0$

As previously discussed, for $\Lambda = 0$, the non-semi-simple extension of the Poincaré algebra $\mathcal{N}$, reduces to the centrally extended Poincaré algebra. Then, the $(1+1)$-dimensional gauge symmetric gravity model (9) reduces to the following central extension of Poincaré gauge symmetric action [9–11]:

$$\begin{aligned}
I = \int d^2x\,\epsilon^{ij}\,\Big\{&\eta_a\left(\partial_i e_j{}^a + \epsilon^{ab}\, e_{ib}\,\omega_j\right) + \eta_2\,\partial_i\omega_j\\
&+ \eta_3\left(\partial_i A_j + \frac{1}{2}k\,\epsilon^{ab}\, e_{ia}\, e_{jb}\right)\Big\}.
\end{aligned} \tag{36}$$

We use the Ansatz (20) to solve the equations of motions (18)–(19) by inserting $\Lambda = 0$ in them, and obtain the following solution:

$$\begin{aligned}
&M^2(r) = \Big(\frac{1}{D_1}\frac{dN(r)}{dr}\Big)^2, \qquad \eta_0(r) = -\frac{kD_2}{D_1}N(r), \qquad \eta_1(r) = 0,\\
&\eta_2(r) = \frac{kD_2}{2D_1^2}N^2(r) + D_3, \qquad \eta_3(r) = D_2,\\
&\omega(r) = D_1 dt, \qquad A(r) = \Big(\frac{k}{2D_1}N^2(r) + D_4\Big)\, dt + g(r)\, dr,
\end{aligned} \tag{37}$$

where D_1, D_2, D_3 and D_4 are arbitrary constants, and $N(r)$, $g(r)$ are arbitrary functions of r. The solution (37) describes a Ricci-flat space-time with zero scalar curvature ($\mathcal{R} = 0$). For $N^2(r) = 2D_1 r - D_5$, the metric in solution (37) reduces to the following Schwarzschild-type metric:

$$ds^2 = -(2D_1 r - D_5)dt^2 + \frac{dr^2}{2D_1 r - D_5}, \tag{38}$$

where D_5 is an arbitrary constant. The metric (38) has a coordinate singularity at

$$r = \frac{D_5}{2D_1}. \tag{39}$$

For $N^2(r) = (D_1 r - D_6)^2$, the metric in solution (37) turns out to be as follows:

$$ds^2 = -(D_1 r - D_6)^2 dt^2 + dr^2, \tag{40}$$

and has a coordinate singularity at

$$r = \frac{D_6}{D_1}, \tag{41}$$

where D_6 is an arbitrary constant. In section 4, we study the gauged Wess–Zumino–Witten model, and show that the solution (37) is an exact solution to the string theory, and specially the latter metric solution (40) describes an exact $(1+1)$-dimensional Ricci-flat black hole in string theory.

3.3. Friedmann–Robertson–Walker (FRW) solutions

Cosmology of the two-dimensional Jackiw–Teitelboim gravity model is studied in Ref. [30]. Moreover, some cosmological solutions of the string-inspired gravity coupled to the matter field, both for dust-filled and radiation-filled space-times have been discussed in Ref. [31]. Here, to obtain some cosmological solutions for the equations of motions (18)–(19), we use Friedmann–Robertson–Walker metric as follows:

$$ds^2 = -dt^2 + a^2(t)\frac{dr^2}{1-\kappa r^2}, \tag{42}$$

where $a(t)$ is the scale factor and describes the expansion of the world, and κ is a constant which can be equal to -1, 0 or $+1$ only. In $1+1$ dimensions, one can use the following coordinate transformation:

$$r \to \frac{1}{\sqrt{-\kappa}}\sinh(\sqrt{-\kappa}\, r) \quad \text{for } \kappa < 0 \tag{43}$$

$$r \to \frac{1}{\sqrt{\kappa}}\sin(\sqrt{\kappa}\, r) \quad \text{for } \kappa > 0 \tag{44}$$

to write the metric (42) as follows:

$$ds^2 = -dt^2 + a^2(t) dr^2. \tag{45}$$

Using the Ansatz (45) for the metric in the equations (18)–(19) and after some calculations, one can obtain three different solutions corresponding to the negative, positive or zero values of Λ, as follows:

3.3.1. FRW solution for $\Lambda < 0$

We have the following solution for the negative Λ:

$$\begin{aligned}
&a(t) = \frac{\dot a(0)}{\sqrt{-\Lambda}}\sin(\sqrt{-\Lambda}\, t) + a(0)\cos(\sqrt{-\Lambda}\, t),\\
&\omega(t,r) = \frac{\Lambda}{k}\Big(h(t,r)dt + s(t,r)dr\Big),\\
&A(t,r) = h(t,r)dt + \Big(s(t,r) + \frac{k}{\sqrt{-\Lambda}}\xi_1(t)\Big)dr,\\
&\eta_0(r) = \frac{dg(r)}{dr}, \qquad \eta_1(t,r) = -\sqrt{-\Lambda}\,\xi_1(t)g(r),\\
&\eta_2(t,r) = a(t)g(r), \qquad \eta_3(t,r) = -\frac{\Lambda}{k}a(t)g(r),
\end{aligned} \tag{46}$$

where

$$\begin{aligned}
&\xi_1(t) = a(0)\sin(\sqrt{-\Lambda}\, t) - \frac{\dot a(0)}{\sqrt{-\Lambda}}\cos(\sqrt{-\Lambda}\, t),\\
&s(t,r) = \int dt\frac{\partial h(t,r)}{\partial r},\\
&g(r) = C_1\cosh(\sqrt{\hat\lambda}\, r) - C_2\sinh(\sqrt{\hat\lambda}\, r),\\
&\hat\lambda = \dot a^2(0) - \Lambda\, a^2(0),
\end{aligned} \tag{47}$$

C_1 and C_2 are constants, $h(t,r)$ is an arbitrary function of t and r, and dot denotes derivative with respect to the timelike coordinate t. $a(0)$ and $\dot a(0)$ in solution (46) are the initial values of scale factor $a(t)$ and its time derivative $\dot a(t)$ at $t = 0$, respectively. Because the fields in the solution (46) are functions of the radial coordinate r, this solution is not a homogenous solution. In order to obtain a homogeneous solution, $h(t,r)$ must be r-independent $h(t,r) = h_0(t)$ and $C_1 = C_2 = 0$, where $h_0(t)$ is a function of timelike coordinate t only. Then, by this choice, all of the fields will be functions of the coordinate t only, and spatial homogeneity will be achieved. This solution will collapse at

$$t = \frac{1}{\sqrt{-\Lambda}}\arctan\Big(-\sqrt{-\Lambda}\,\frac{a(0)}{\dot a(0)}\Big). \tag{48}$$

The Hubble parameter $H(t)$ for this solution is as follows:

$$\begin{aligned}
H(t) &\equiv \frac{\dot a(t)}{a(t)}\\
&= \sqrt{-\Lambda}\Big(\frac{\dot a(0)\cos(\sqrt{-\Lambda}\, t) - \sqrt{-\Lambda}\, a(0)\sin(\sqrt{-\Lambda}\, t)}{\dot a(0)\sin(\sqrt{-\Lambda}\, t) + \sqrt{-\Lambda}\, a(0)\cos(\sqrt{-\Lambda}\, t)}\Big).
\end{aligned} \tag{49}$$

Using $\ddot a(t) = \Lambda a(t)$, the deceleration parameter $q(t)$ can be obtained as follows:

$$\begin{aligned}
q(t) &\equiv -\frac{a(t)\ddot a(t)}{\dot a^2(t)}\\
&= \Big(\frac{\dot a(0)\sin(\sqrt{-\Lambda}\, t) + \sqrt{-\Lambda}\, a(0)\cos(\sqrt{-\Lambda}\, t)}{\dot a(0)\cos(\sqrt{-\Lambda}\, t) - \sqrt{-\Lambda}\, a(0)\sin(\sqrt{-\Lambda}\, t)}\Big)^2,
\end{aligned} \tag{50}$$

which is obviously positive, and shows that the expansion of the universe is decelerating.

3.3.2. FRW solution for $\Lambda > 0$

For positive Λ, one obtains the following solution:

$$\begin{aligned}
&a(t) = \frac{\dot a(0)}{\sqrt{\Lambda}}\sinh(\sqrt{\Lambda}\, t) + a(0)\cosh(\sqrt{\Lambda}\, t),\\
&\omega(t,r) = \frac{\Lambda}{k}\Big(\hat h(t,r)dt + \hat s(t,r)dr\Big),\\
&A(t,r) = \hat h(t,r)dt + \Big(\hat s(t,r) + \frac{k}{\sqrt{\Lambda}}\xi_2(t)\Big)dr,\\
&\eta_0(r) = \frac{d\hat g(r)}{dr},\\
&\eta_1(t,r) = \sqrt{\Lambda}\,\xi_2(t)\hat g(r),\\
&\eta_2(t,r) = a(t)\hat g(r), \qquad \eta_3(t,r) = -\frac{\Lambda}{k}a(t)\hat g(r),
\end{aligned} \tag{51}$$

where

$$\hat{s}(t,r) = \int dt \frac{\partial \hat{h}(t,r)}{\partial r},$$

$$\xi_2(t) = a(0)\sinh(\sqrt{\Lambda}\, t) + \frac{\dot{a}(0)}{\sqrt{\Lambda}}\cosh(\sqrt{\Lambda}\, t),$$

$$\hat{g}(r) = \left\{ \begin{array}{cc} D_1 r + D_2 & \hat{\lambda} = 0 \\ D_1 \cosh(\sqrt{\hat{\lambda}}\, r) + D_2 \sinh(\sqrt{\hat{\lambda}}\, r) & \hat{\lambda} > 0 \\ D_1 \cos(\sqrt{-\hat{\lambda}}\, r) + D_2 \sin(\sqrt{-\hat{\lambda}}\, r) & \hat{\lambda} < 0 \end{array} \right\}, \tag{52}$$

$$\hat{\lambda} = \dot{a}^2(0) - \Lambda\, a^2(0),$$

D_1 and D_2 are constants, and $\hat{h}(t,r)$ is an arbitrary function. This solution is not homogenous, and as the previous solution, in order to have a homogenous solution we must put $\hat{h}(t,r) = \hat{h}_0(t)$ and $D_1 = D_2 = 0$. This solution will collapse for $|\frac{\dot{a}(0)}{a(0)}| \geqslant \sqrt{\Lambda}$, at

$$t = \frac{1}{\sqrt{\Lambda}} \operatorname{arctanh}\Big(-\sqrt{\Lambda}\, \frac{a(0)}{\dot{a}(0)} \Big), \tag{53}$$

but for $|\frac{\dot{a}(0)}{a(0)}| < \sqrt{\Lambda}$, it does not collapse. The Hubble parameter $H(t)$ for this solution is as follows:

$$H(t) \equiv \frac{\dot{a}(t)}{a(t)} = \sqrt{\Lambda}\Big(\frac{\dot{a}(0)\cosh(\sqrt{\Lambda}\, t) + \sqrt{\Lambda}\, a(0)\sinh(\sqrt{\Lambda}\, t)}{\dot{a}(0)\sinh(\sqrt{\Lambda}\, t) + \sqrt{\Lambda}\, a(0)\cosh(\sqrt{\Lambda}\, t)}\Big). \tag{54}$$

Using $\ddot{a}(t) = \Lambda a(t)$, the deceleration parameter $q(t)$ can be obtained as follows:

$$q(t) \equiv -\frac{a(t)\ddot{a}(t)}{\dot{a}^2(t)} = -\Big(\frac{\dot{a}(0)\sinh(\sqrt{\Lambda}\, t) + \sqrt{\Lambda}\, a(0)\cosh(\sqrt{\Lambda}\, t)}{\dot{a}(0)\cosh(\sqrt{\Lambda}\, t) + \sqrt{\Lambda}\, a(0)\sinh(\sqrt{\Lambda}\, t)}\Big)^2, \tag{55}$$

which is obviously negative. Note that for $\dot{a}(0) = \sqrt{\Lambda}\; a(0)$, the scale factor of the solution (51) has the following exponential form:

$$a(t) = a(0) e^{\sqrt{\Lambda}\, t}, \tag{56}$$

and leads to a constant Hubble parameter $H = \sqrt{\Lambda}$ and negative deceleration parameter $q = -1$, which means that the expansion of the universe is accelerating.

3.3.3. FRW solution for $\Lambda = 0$

For $\Lambda = 0$, we obtain the following solution:

$$a(t) = \dot{a}(0)t + a(0), \qquad \omega(t,r) = -\dot{a}(0)dr,$$
$$A(t,r) = \bar{h}(t,r)dt + \Big\{\bar{s}(t,r) + k\Big(\frac{1}{2}\dot{a}(0)t^2 + a(0)t + E_3\Big)\Big\}dr,$$
$$\eta_0(r) = -\frac{d\bar{g}(r)}{dr}, \qquad \eta_1(t,r) = -\dot{a}(0)\bar{g}(r),$$
$$\eta_2(t,r) = -a(t)\bar{g}(r), \qquad \eta_3(t,r) = 0, \tag{57}$$

where

$$\bar{s}(t,r) = \int dt \frac{\partial \bar{h}(t,r)}{\partial r},$$
$$\bar{g}(r) = E_1 \cosh\Big(\dot{a}(0)\, r\Big) + E_2 \sinh\Big(\dot{a}(0)\, r\Big),$$

E_1, E_2 and E_3 are arbitrary constants, and $\bar{h}(t,r)$ is an arbitrary function of t and r. To obtain a homogenous solution, $\bar{h}(t,r)$ must be independent of coordinate r $\Big(\bar{h}(t,r) = \bar{h}_0(t)\Big)$, and also we must have $E_1 = E_2 = 0$. This solution obviously will collapse at

$$t = -\frac{a(0)}{\dot{a}(0)}. \tag{58}$$

The Hubble parameter $H(t)$ and the deceleration parameter $q(t)$ for this solution can be obtained as follows:

$$H(t) = \frac{\dot{a}(0)}{\dot{a}(0)t + a(0)}, \qquad q(t) = 0, \tag{59}$$

which show that expansion of the universe is without acceleration. In the next section, by studying the gauged Wess–Zumino–Witten model, we show that the cosmological metric solution (57) is also an exact solution to the string theory.

4. $\frac{\mathbf{A_{4,8}}}{\mathbf{A_1 \otimes A_1}}$ gauged Wess–Zumino–Witten model

As we have explained in the introduction of this paper, it has been shown that two-dimensional string-inspired gravity model [6–8] is equivalent to the gauge theory (36) of the centrally extended Poincaré algebra (the Maxwell algebra in 1 + 1 spacetime dimensions) [9–11]. So we anticipate that the gravity model (36) which has the Maxwell symmetry, and the string theory obtained by a gauged WZW model on the Maxwell algebra ($\cong \mathcal{A}_{4,8}$), both have some common properties such as a common solution to them. In this section, we try to find some exact solutions to the string theory (obtained by gauged Wess–Zumino–Witten model using $\mathcal{A}_{4,8}$) which coincide with the solutions of our (1 + 1)-dimensional gravity model. We have mentioned in the footnote 3 that the Maxwell algebra in 1 + 1 dimensions is isomorphic to the Lie algebra $\mathcal{A}_{4,8}$ [28]:

$$[X_2, X_3] = X_1, \qquad [X_2, X_4] = X_2, \qquad [X_3, X_4] = -X_3, \tag{60}$$

where $\{X_i\}$, $i = 1 \ldots 4$ are the bases of the Lie algebra. We display the corresponding Lie group with $\mathbf{A_{4,8}}$. Now, we study G/H gauged Wess–Zumino–Witten model to obtain an exact solution to the string theory. Let the Lie group G be $G = \mathbf{A_{4,8}}$, and $H = \mathbf{A}_1^{(\mathbf{I})} \otimes \mathbf{A}_1^{(\mathbf{II})}$ is its subgroup which is a direct product of two one-dimensional Abelian non-compact Lie groups $\mathbf{A}_1^{(\mathbf{I})}$ and $\mathbf{A}_1^{(\mathbf{II})}$ generated by the bases X_1 and X_4, respectively. If g is an element of the Lie group G, then the Wess–Zumino–Witten action can be written as follows [32]:

$$L(g) = \frac{k}{4\pi} \int_\Sigma d^2 z\, \langle g^{-1}\partial g, g^{-1}\bar{\partial} g\rangle - \frac{k}{24\pi} \int_B d^3\sigma\, \epsilon^{\alpha\beta\gamma} \langle g^{-1}\partial_\alpha g, [g^{-1}\partial_\beta g, g^{-1}\partial_\gamma g]\rangle, \tag{61}$$

where B is a three-dimensional manifold with the coordinates $\sigma^\mu = \{z, \bar{z}, y\}$, and Σ is its boundary with the local complex coordinates $\{z, \bar{z}\}$. Furthermore, we use the notations $\partial = \frac{\partial}{\partial z}$, $\bar{\partial} = \frac{\partial}{\partial \bar{z}}$ and $\partial_\mu = \frac{\partial}{\partial \sigma^\mu}$, and $d^2 z$ and $d^3\sigma$ denote the measures $|dz d\bar{z}|$ and $d^2 z dy$, respectively. By introducing the gauge fields $\mathbf{A}$, $\bar{\mathbf{A}}$ which takes their values in the Lie algebra of H, the gauged Wess–Zumino–Witten action having the local axial symmetry $g \longrightarrow hgh$, $\mathbf{A} \longrightarrow h(\mathbf{A} + \partial)h^{-1}$ and $\bar{\mathbf{A}} \longrightarrow h^{-1}(\bar{\mathbf{A}} - \bar{\partial})h$ $(h \in H)$ can be written as follows [33]:

$$L(g, \mathbf{A}) = L(g) + \frac{k}{2\pi} \int_\Sigma d^2 z \Big(\langle \mathbf{A}, \bar{\partial} g\, g^{-1}\rangle + \langle \bar{\mathbf{A}}, g^{-1}\partial g\rangle + \langle \mathbf{A}, \bar{\mathbf{A}}\rangle + \langle g^{-1}\mathbf{A} g, \bar{\mathbf{A}}\rangle \Big). \tag{62}$$

Here, we consider the Lie algebra $\mathcal{H} = \mathcal{A}_1^{(I)} \oplus \mathcal{A}_1^{(II)}$ valued gauge fields $\mathbf{A}$ and $\bar{\mathbf{A}}$ as follows:

$$\mathbf{A} = A_1 X_1 + A_2 X_4, \qquad \bar{\mathbf{A}} = \bar{A}_1 X_1 + \bar{A}_2 X_4. \tag{63}$$

We parameterize $G = \mathbf{A_{4,8}}$ group by the following group element:

$$g = e^{aX_1}\, e^{bX_2}\, e^{uX_3}\, e^{vX_4}, \tag{64}$$

where $a, b, u, v \in \mathbb{R}$. Then, using the group element (64), and the following non-degenerate ad-invariant bilinear quadratic form [34] on the Lie algebra $\mathcal{A}_{4,8}$ (60):

$$\langle X_1, X_4\rangle = \alpha, \quad \langle X_2, X_3\rangle = -\alpha, \quad \langle X_4, X_4\rangle = \beta, \tag{65}$$

one can rewrite the Wess–Zumino–Witten action (61) as follows:

$$L(g) = \frac{k}{4\pi}\int_\Sigma d^2z\Big(\alpha(\partial a\bar\partial v + \bar\partial a\partial v) + \alpha u(\partial b\bar\partial v + \bar\partial b\partial v) - \alpha(\partial b\bar\partial u + \bar\partial b\partial u) + \beta\partial v\bar\partial v\Big) - \frac{k\alpha}{4\pi}\int_\Sigma d^2z\, v(\partial b\bar\partial u - \bar\partial b\partial u), \tag{66}$$

where α and β are arbitrary constants (α should be nonzero in order that the ad-invariant bilinear quadratic form (65) be non-degenerate). We gauge $\mathbf{A}_1^{(\mathbf{I})} \otimes \mathbf{A}_1^{(\mathbf{II})}$ subgroup generated infinitesimally by[5] $\delta g = \epsilon_1(X_1 g + gX_1) + \epsilon_2(X_4 g + gX_4)$ which using (64) gives the following transformations of the group parameters:

$$\delta a = 2\epsilon_1, \quad \delta b = -b\epsilon_2, \quad \delta u = u\epsilon_2, \quad \delta v = 2\epsilon_2, \tag{67}$$

together with the following transformations of the gauge fields:

$$\delta A_i = -\partial\epsilon_i, \quad \delta\bar{A}_i = -\bar\partial\epsilon_i, \quad (i = 1, 2) \tag{68}$$

where ϵ_1 and ϵ_2 are gauge parameters. Then, the resultant gauged WZW action is as follows:

$$L(g, \mathbf{A}) = L(g) + \frac{k}{2\pi}\int_\Sigma d^2z\,\Big\{\alpha\bar\partial v A_1 + \Big(\alpha(\bar\partial a + b\bar\partial u - bu\bar\partial v) + \beta\bar\partial v\Big)A_2 + \alpha\partial v\bar{A}_1 + \Big(\alpha(\partial a + u\partial b) + \beta\partial v\Big)\bar{A}_2 + 2\alpha A_1\bar{A}_2 + 2\alpha A_2\bar{A}_1 + (2\beta - \alpha bu)A_2\bar{A}_2\Big\}. \tag{69}$$

Variations of the action (69) with respect to the gauge fields A_i and $\bar{A}_i$ $(i = 1, 2)$ gives the following equations:

$$A_1 = -\frac{1}{2}(\partial a + u\partial b + \frac{1}{2}bu\partial v), \qquad A_2 = -\frac{1}{2}\partial v,$$
$$\bar{A}_1 = -\frac{1}{2}(\bar\partial a + b\bar\partial u - \frac{1}{2}bu\bar\partial v), \qquad \bar{A}_2 = -\frac{1}{2}\bar\partial v, \tag{70}$$

using which one eliminates the gauge fields in (69), and obtains the following gauged Wess–Zumino–Witten action:

$$L(g, \mathbf{A}) = \frac{k}{4\pi}\int_\Sigma d^2z\Big(\frac{1}{2}\alpha bu\partial v\bar\partial v + \alpha\partial v(u\bar\partial b - b\bar\partial u) - \alpha(\partial b\bar\partial u + \bar\partial b\partial u)\Big) - \frac{k\alpha}{4\pi}\int_\Sigma d^2z\, v(\partial b\bar\partial u - \bar\partial b\partial u). \tag{71}$$

This action is independent of the parameter a, and now we can fix the gauge by setting $b = u$, such that the gauged WZW action (71) turns out to be

$$L(g, \mathbf{A}) = -\frac{k\alpha}{2\pi}\int_\Sigma d^2z\,\Big(-\frac{1}{4}u^2\partial v\bar\partial v + \partial u\bar\partial u\Big). \tag{72}$$

Using the following field redefinition:

$$u(r) = \frac{\hat{N}(r)}{\hat{D}}, \qquad v(t) = 2\hat{D}t, \tag{73}$$

the gauged WZW action (72) becomes

$$L(g, \mathbf{A}) = -\frac{k\alpha}{2\pi}\int_\Sigma d^2z\,\Big(-\hat{N}^2(r)\partial t\bar\partial t + \Big(\frac{1}{\hat{D}}\frac{d\hat{N}(r)}{dr}\Big)^2\partial r\bar\partial r\Big), \tag{74}$$

where $\hat{D}$ is a constant and $\hat{N}(r)$ is an arbitrary function of the spatial coordinate r. Then, the obtained gauged WZW action (74) describes a string propagating in a space-time with the following metric

$$ds^2 = -\hat{N}^2(r)dt^2 + \Big(\frac{1}{\hat{D}}\frac{d\hat{N}(r)}{dr}\Big)^2 dr^2. \tag{75}$$

By assuming $\hat{D} = D_1$ and $\hat{N}(r) = N(r)$, the metric solution (75) is precisely same as the metric in (37) obtained as a solution of our gravity model (9). As we have discussed before, by assuming $\hat{N}^2(r) = (D_1 r - D_6)^2$ the metric (75) reduces to the metric (40) which describes an exact Ricci-flat black hole in string theory. Now, for obtaining the dilaton field, we consider the following one-loop beta function equations in $1+1$ dimensions [35]:

$$R_{\mu\nu} + 2\nabla_\mu\nabla_\nu\phi = 0, \tag{76}$$

$$R + \frac{8}{k'} + 4\nabla^2\phi - 4(\nabla\phi)^2 = 0, \tag{77}$$

where R and $R_{\mu\nu}$ are the scalar curvature and Ricci tensor of the target space, ϕ is the dilaton field, and $\frac{8}{k'}$ is the cosmological constant term. By requiring that the metric (75) must obey the one-loop beta function equations (76) and (77), we obtain the following relation for the dilaton field using (76):

$$\phi(t, r) = \hat{N}(r)(c_1\cosh(\hat{D}t) + c_2\sinh(\hat{D}t)), \tag{78}$$

where c_1 and c_2 are some real constants. By substituting $\phi(t, r)$ in (77), one can obtain the following relation between the constants c_1 and c_2:

$$\hat{D}^2({c_1}^2 - {c_2}^2) = \frac{2}{k'}. \tag{79}$$

In the same way, using another field redefinition as follows:

$$u(t) = \frac{\alpha t + \beta}{\alpha}, \qquad v(r) = \alpha r, \tag{80}$$

where α and β are arbitrary real constants, the gauged WZW action (72) turns out to have the following form:

$$L(g, \mathbf{A}) = \frac{k\alpha}{2\pi}\int_\Sigma d^2z\,\Big(-\partial t\bar\partial t + (\alpha t + \beta)^2\partial r\bar\partial r\Big), \tag{81}$$

which describes a string propagating in a space-time with the following cosmological metric:

$$ds^2 = -dt^2 + (\alpha t + \beta)^2 dr^2. \tag{82}$$

By assuming $\alpha = \dot{a}(0)$ and $\beta = a(0)$, this metric is precisely same as the FRW metric (57) which is obtained by solving the equations of motion for our gravity model discussed in the previous section. In the same way for obtaining the previous dilaton field, we use (76) and (77) to obtain the following dilaton field corresponding to the metric (82):

[5] The infinitesimal gauge transformation $\delta g = \epsilon_1(X_1 g + gX_1) + \epsilon_2(X_4 g + gX_4)$ can be obtained by substituting $h = e^{\epsilon_1 X_1 + \epsilon_2 X_4}$ in the axial symmetry $g \longrightarrow hgh$.

$$\phi(t,r) = (\alpha t + \beta)(d_1 \cosh(\alpha r) + d_2 \sinh(\alpha r)),$$
$$\alpha^2(d_2{}^2 - d_1{}^2) = \frac{2}{k'}, \qquad (83)$$

where d_1 and d_2 are real constants. Note that the black hole metric (40) converts to the FRW metric (82), and vice versa, using the following coordinate transformation:

$$t \to \hat{r}, \qquad r \to \hat{t}. \qquad (84)$$

5. Conclusions

We have presented a four-dimensional extension of the Poincaré algebra in $(1+1)$-dimensional space-time. Using this algebra, we have constructed a gauge theory of gravity, which is dual (canonically transformed) to the *AdS* gauge theory of gravity, under special conditions. We have also obtained black hole and Friedmann–Robertson–Walker (FRW) cosmological solutions of this model. Then, using $\frac{\mathbf{A_{4,8}}}{\mathbf{A_1 \otimes A_1}}$ gauged Wess–Zumino–Witten action, we have shown that some of the black hole and cosmological solutions of our gravity model are exact $(1+1)$-dimensional solutions of string theory. In this paper, we have shown that the Ricci-flat ($\mathcal{R} = 0$) solutions of our gravity model are also exact solutions to the string theory, only. But, it remains an interesting question to be investigated that if our other solutions of the gravity model (with $\mathcal{R} = 2\Lambda$) are also exact solutions to the string theory? Analysis of the constraints of the gravity model (9) and its quantization are some of the interesting open problems which may lead to some desired results. Another interesting problem may be the possibility of obtaining the $(1+1)$-dimensional gravity model (9) by an appropriate dimensional reduction from a gauge invariant $(2+1)$-dimensional gravity model (see [12,13]).

Acknowledgements

We would like to express our heartfelt gratitude to M.M. Sheikh-Jabbari for his useful comments. This research was supported by a research fund No. 217D4310 from Azarbaijan Shahid Madani University.

References

[1] R. Jackiw, Lower dimensional gravity, Nucl. Phys. B 252 (1985) 343–356.
[2] C. Teitelboim, Gravitation and hamiltonian structure in two space-time dimensions, Phys. Lett. B 126 (1983) 41–45.
[3] Takeshi Fukuyama, Kiyoshi Kamimura, Gauge theory of two-dimensional gravity, Phys. Lett. B 160 (1985) 259–262.
[4] K. Isler, C.A. Trugenberger, Gauge theory of two-dimensional quantum gravity, Phys. Rev. Lett. 63 (1989) 834.
[5] A. Chamseddine, D. Wyler, Gauge theory of topological gravity in $1+1$ dimensions, Phys. Lett. B 228 (1989) 75.
[6] E. Witten, On string theory and black holes, Phys. Rev. D 44 (1991) 314.
[7] H. Verlinde, Black holes and strings in two-dimensions, in: The Sixth Marcel Grossman Meeting on General Relativity, 1991, pp. 813–831.
[8] C. Callan, S. Giddings, A. Harvey, A. Strominger, Evanescent black holes, Phys. Rev. D 45 (1992) 1005, arXiv:hep-th/9111056.
[9] D. Cangemi, R. Jackiw, Gauge invariant formulations of lineal gravity, Phys. Rev. Lett. 69 (1992) 233–236, arXiv:hep-th/9203056.
[10] R. Jackiw, Gauge theories for gravity on a line, Theor. Math. Phys. 92 (1992) 979–987, arXiv:hep-th/9206093.
[11] D. Cangemi, R. Jackiw, Poincaré gauge theory for gravitational forces in $(1+1)$-dimensions, Ann. Phys. 225 (1993) 229–263, arXiv:hep-th/9302026.
[12] A. Achúcarro, Lineal gravity from planar gravity, Phys. Rev. Lett. 70 (1993) 1037–1040, arXiv:hep-th/9207108.
[13] G. Grignani, G. Nardelli, Poincaré gauge theories for lineal gravity, Nucl. Phys. B 412 (1994) 320–344, arXiv:gr-qc/9209013.
[14] H. Bacry, P. Combe, J.L. Richard, Group-theoretical analysis of elementary particles in an external electromagnetic field. 1. The relativistic particle in a constant and uniform field, Nuovo Cimento A 67 (1970) 267–299;
H. Bacry, P. Combe, J.L. Richard, Nuovo Cimento A 70 (1970) 289–312.
[15] R. Schrader, The Maxwell group and the quantum theory of particles in classical homogeneous electromagnetic fields, Fortschr. Phys. 20 (1972) 701–734.
[16] D.V. Soroka, V.A. Soroka, Gauge semi-simple extension of the Poincaré group, Phys. Lett. B 707 (2012) 160–162, arXiv:1101.1591 [hep-th].
[17] J.A. de Azcarraga, K. Kamimura, J. Lukierski, Generalized cosmological term from Maxwell symmetries, Phys. Rev. D 83 (2011) 124036, arXiv:1012.4402 [hep-th].
[18] J.A. de Azcarraga, K. Kamimura, J. Lukierski, Maxwell symmetries and some applications, Int. J. Mod. Phys. Conf. Ser. 23 (2013) 01160, arXiv:1201.2850 [hep-th].
[19] O. Cebecioğlu, S. Kibaroğlu, Gauge theory of the Maxwell–Weyl group, Phys. Rev. D 90 (2014) 084053, arXiv:1404.3969 [hep-th].
[20] O. Cebecioğlu, S. Kibaroğlu, Maxwell-affine gauge theory of gravity, Phys. Lett. B 751 (2015) 131–134, arXiv:1503.09003 [hep-th].
[21] P. Salgado, R.J. Szabo, O. Valdivia, Topological gravity and transgression holography, Phys. Rev. D 89 (2014) 084077, arXiv:1401.3653 [hep-th].
[22] J. Díaz, O. Fierro, F. Izaurieta, N. Merino, E. Rodríguez, P. Salgado, O. Valdivia, A generalized action for $(2+1)$-dimensional Chern–Simons gravity, J. Phys. A 45 (2012) 255207, arXiv:1311.2215 [hep-th].
[23] S. Hoseinzadeh, A. Rezaei-Aghdam, $2+1$ dimensional gravity from Maxwell and semi-simple extension of the Poincaré gauge symmetric models, Phys. Rev. D 90 (2014) 084008, arXiv:1402.0320 [hep-th].
[24] S. Hoseinzadeh, A. Rezaei-Aghdam, Exact three dimensional black hole with gauge fields in string theory, Eur. Phys. J. C 75 (2015) 227, arXiv:1501.02451 [hep-th].
[25] G. Mandal, A.M. Sengupta, S.R. Wadia, Classical solutions of two-dimensional string theory, Mod. Phys. Lett. A 6 (1991) 1685–1692.
[26] J. Horne, G. Horowitz, Exact black string solutions in three dimensions, Nucl. Phys. B 368 (1992) 444, arXiv:hep-th/9108001.
[27] G.T. Horowitz, D.L. Welch, Exact three dimensional black holes in string theory, Phys. Rev. Lett. 71 (1993) 328–331, arXiv:hep-th/9302126.
[28] J. Patera, R.T. Sharp, P. Winternitz, H. Zassenhaus, Invariants of real low dimension Lie algebras, J. Math. Phys. 17 (1976) 986–994.
[29] Dongsu Bak, D. Cangemi, R. Jackiw, Energy momentum conservation in general relativity, Phys. Rev. D 49 (1994) 5173–5181, arXiv:hep-th/9310025; Erratum: Phys. Rev. D 52 (1995) 3753.
[30] M. Cadoni, S. Mignemi, Cosmology of the Jackiw–Teitelboim model, Gen. Relativ. Gravit. 34 (2002) 2101–2109, arXiv:gr-qc/0202066.
[31] R.B. Mann, S.F. Ross, Gravitation and cosmology in $(1+1)$-dimensional dilaton gravity, Phys. Rev. D 47 (1993) 3312–3318, arXiv:hep-th/9206022.
[32] E. Witten, Nonabelian bosonization in two dimensions, Commun. Math. Phys. 92 (1984) 455–472.
[33] K. Gawedzki, A. Kupianen, G/H conformal field theory from gauged WZW model, Phys. Lett. B 215 (1988) 119.
[34] C.R. Nappi, E. Witten, A WZW model based on a non-semi-simple group, Phys. Rev. Lett. 71 (1993) 3751–3753, arXiv:hep-th/9310112.
[35] C.G. Callan, E.J. Martinec, M.J. Perry, D. Friedan, Strings in background fields, Nucl. Phys. B 262 (1985) 593.

2

Broken boost invariance in the Glasma via finite nuclei thickness

Andreas Ipp, David Müller *

Institut für Theoretische Physik, Technische Universität Wien, Wiedner Hauptstr. 8-10, A-1040 Vienna, Austria

ARTICLE INFO

Editor: J.-P. Blaizot

Keywords:
Heavy-ion collisions
Color glass condensate
Glasma
Boost invariance

ABSTRACT

We simulate the creation and evolution of non-boost-invariant Glasma in the early stages of heavy ion collisions within the color glass condensate framework. This is accomplished by extending the McLerran–Venugopalan model to include a parameter for the Lorentz-contracted but finite width of the nucleus in the beam direction. We determine the rapidity profile of the Glasma energy density, which shows deviations from the boost-invariant result. Varying the parameters both broad and narrow profiles can be produced. We compare our results to experimental data from RHIC and find surprising agreement.

1. Introduction

Heavy ion collisions at the Relativistic Heavy Ion Collider (RHIC) and the Large Hadron Collider (LHC) provide insight into the properties of nuclear matter under extreme conditions. The evolution of the Quark–Gluon Plasma (QGP) that is created in such collisions is well described by relativistic viscous hydrodynamics [1,2]. A first principles description of the initial state of heavy-ion collisions is provided by the Color Glass Condensate (CGC) framework [3–5]. The CGC is a classical effective field theory for nuclear matter at ultrarelativistic energies. Models such as the IP-Glasma [6, 7] in combination with hydrodynamics are able to correctly reproduce azimuthal anisotropies and event-by-event multiplicity distributions [8,9]. Furthermore, the CGC can explain long-range rapidity correlations like the ridge [10,11].

A Gaussian shaped rapidity profile of particle multiplicity can be found in experiments covering various energy ranges, from LHC [12] to RHIC Beam Energy Scan [13,14]. This shape is well explained by the Landau model [15] up to RHIC energies [13], which assumes full stopping of the colliding nuclei. The Landau model is in contrast to the Bjorken model [16] which relies on approximate boost invariance. A Gaussian profile has also been found in holographic calculations of colliding shock waves [17–19].

In its original formulation collisions in the CGC picture are assumed to be boost-invariant [20–23] and were thus only understood as an approximation valid close to midrapidity. This approach implicitly assumes infinitely thin Lorentz-contracted nuclei and entails a classical, boost-invariant evolution of the Glasma at leading order.

Only at the next-to-leading order the boost invariance is broken by a change of the initial conditions through JIMWLK evolution [24–27], and non-boost-invariant rapidity profiles can be obtained. Recently such rapidity dependencies have been found to agree reasonably well with experimental data where observables like charged particle multiplicities show a Gaussian rapidity profile [28,29]. On the other hand it has been suggested that if one considers nuclei with finite extent in the beam direction, deviations from boost invariance may arise already at the classical level [30]. In the case of proton–nucleus collisions methods have been developed to systematically include finite width corrections of the nucleus [31–33]. However, so far there has been no consistent simulation of the subsequent three-dimensional evolution for heavy-ion collisions even at the classical level.

In this letter, we show that Gaussian rapidity profiles of energy density can arise already from 3+1 dimensional purely classical CGC simulations, if incoming nuclei have a finite extent in the beam direction. As one would expect, the Gaussian profiles become broader at higher collision energy. In principle, we can cover the wide range from very thin nuclei with almost boost-invariant behavior to thick nuclei at low collision energies and narrow Gaussian profiles. We simulate the collision in the laboratory frame, see Fig. 1, which makes it necessary to include the propagating nuclei already before and during the collision.

* Corresponding author.
E-mail addresses: ipp@hep.itp.tuwien.ac.at (A. Ipp), david.mueller@tuwien.ac.at (D. Müller).

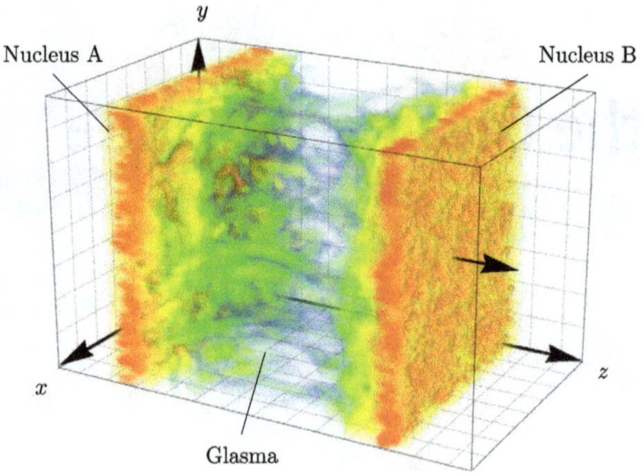

Fig. 1. A 3+1 dimensional colored particle-in-cell simulation of the collision of two thick sheets of relativistic nuclear matter. The simulation box covers a small part of the full transverse extent of the nuclei in the x, y plane. This figure shows a density plot of the energy density of both nuclei A and B, and the three-dimensional Glasma that is created in the collision. An animated version of this figure can be found at [34].

2. Theoretical framework

The nuclei in the CGC picture consist of hard partons that are surrounded by soft gluons. The hard partons can be described as classical color charges moving at the speed of light, while the soft gluons form a highly occupied coherent non-Abelian gauge field. The collision of two such infinitely thin condensates produces the Glasma whose evolution can be described classically by solving the Yang–Mills equations for early proper times. At finite nuclei width, the collision region is not pointlike anymore, and the nuclei, the collision, and the evolution of the Glasma can not be described separately and require one consistent simulation that covers all these steps.

A suitable numerical method was developed in our previous publication [35] based on the colored particle-in-cell method (CPIC) [36–39], which is a non-Abelian extension of the particle-in-cell method for the simulation of Abelian plasmas [40,41]. In contrast to the traditional approach of simulating the Glasma, where the field equations are solved in the forward light-cone parametrized by proper time τ and space–time rapidity η_s, we describe the collision in the laboratory frame using the lab-frame time t and the longitudinal beam direction z. The most striking difference of this approach is the explicit inclusion of the nuclei in the simulation, whereas in a boost-invariant simulation the information about the nuclei and their color currents is completely encoded in the initial conditions at the boundary of the light-cone, i.e. $\tau = 0$. To solve this problem numerically we simulate the continuous color charge densities of the nuclei with a large number of color-charged point-like particles, mimicking the dynamics of the continuous cloud of color charges on a lattice. This enables us to describe the full 3+1 dimensional collision and the subsequent evolution of the Glasma beyond the boost-invariant approximation. For a more detailed description we refer the reader to [35].

2.1. *Initial conditions*

The initial conditions in our simulation differ from the traditional approach as well. Instead of starting at $\tau = 0$, our simulation begins before the collision with the nuclei well-separated in the longitudinal direction. Here we quickly review how to solve the Yang–Mills equations in the covariant gauge and the transformation to the temporal gauge in the laboratory frame for a single nucleus. We base our model of the initial state on the McLerran–Venugopalan (MV) model [42,43], extended by a thickness parameter in longitudinal direction. The transverse charge density $\rho^a(x_T)$ as a function of the transverse coordinate x_T is a random variable following the usual gauge-invariant Gaussian probability functional $W[\rho]$ with the two-point correlation function

$$\left\langle \rho^a(x_T)\rho^b(y_T)\right\rangle = g^2\mu^2\delta^{ab}\delta^{(2)}(x_T - y_T), \tag{1}$$

where μ is the MV model parameter controlling average color charge density and g is the Yang–Mills coupling constant. For a single nucleus moving in the positive z direction, we embed this two-dimensional charge density into the three-dimensional laboratory frame via $\rho^a(x_T, x^-) = f(x^-)\rho^a(x_T)$ with a longitudinal profile function $f(x^-)$, where $x^\pm \equiv (t \pm z)/\sqrt{2}$ are the usual light-cone coordinates. For the longitudinal profile we choose a Gaussian

$$f(x^-) = \frac{1}{\sqrt{2\pi}L}\exp\left(-(x^-)^2/L^2\right), \tag{2}$$

where we introduce the thickness parameter L. In the limit of $L \to 0$ we have $f(x^-) \propto \delta(x^-)$ and restore the boost-invariant limit of the original MV model. Note that this model explicitly neglects non-trivial longitudinal color structure [44]. The only non-vanishing component of the light-like color current of the nucleus is then given by

$$J^+_{\text{cov}}(x_T, x^-) = \sqrt{2}f(x^-)\rho^a(x_T)t^a, \tag{3}$$

where t^a are the generators of the gauge group SU(N). The subscript "cov" denotes that this defines the color current in the covariant gauge $\partial_\mu A^{\mu,a} = 0$. Using this ansatz we can solve the Yang–Mills equations

$$D_\mu F^{\mu\nu} = J^\nu \tag{4}$$

in the covariant gauge by finding a solution to the two-dimensional Poisson equation

$$-\Delta_T A^+(x_T, x^-) = J^+_{\text{cov}}(x_T, x^-), \tag{5}$$

which is solved by

$$\phi^a(x_T) = \int_0^\Lambda \frac{d^2k_T}{(2\pi)^2}\frac{\tilde\rho^a(k_T)}{k_T^2 + m^2}e^{-ik_T\cdot x_T}, \tag{6}$$

$$A^+(x_T, x^-) = \sqrt{2}f(x^-)\phi^a(x_T)t^a, \tag{7}$$

where $\tilde\rho^a(k_T)$ is the Fourier transform of $\rho^a(x_T)$. We introduced an infrared regulator m and an ultraviolet cutoff Λ, since the MV model is both infrared and UV divergent. The regularization in (6) should be read as a modification of the charge densities $\rho^a(x_T)$ while the field equations remain unchanged.

Our numerical method requires the gauge fields to satisfy the temporal gauge condition $A^a_0 = 0$. Switching to this gauge from the covariant gauge renders the fields purely transverse. The transverse field components and the color current are given by

$$A_i(x_T, x^-) = \frac{i}{g}V(x_T, x^-)\partial_i V^\dagger(x_T, x^-), \tag{8}$$

$$J^+(x_T, x^-) = V(x_T, x^-)J^+_{\text{cov}}(x_T, x^-)V^\dagger(x_T, x^-), \tag{9}$$

with the temporal Wilson line

$$V(x_T, x^-) = \mathcal{T}\exp\left(-ig\int_{-\infty}^{t} dt' f(x'^-)\phi(x_T)\right). \tag{10}$$

Just like the gauge field $A^+(x_T, x^-)$ also the temporal Wilson line of the first nucleus is independent of x^+. The second nucleus moving in the negative z direction is placed initially in a safe distance where the gauge fields of the first nucleus vanish exponentially and vice versa. Having set up the initial conditions the system can be evolved forward in time via the equations of motion on the lattice.

In order to fix the other parameters of the initial conditions for a given collision energy $\sqrt{s_{NN}}$ we take the following approach: the longitudinal thickness L introduced in our implementation of the MV model somehow needs to be related to the Lorentz factor γ. It seems natural that L should be proportional to the Lorentz contracted length of the nucleus $\gamma^{-1}R$, where R is the nuclear radius. One possibility is to define [35]

$$\gamma = \frac{R}{2L}. \tag{11}$$

This way L is fixed by the geometry of the Lorentz-contracted nucleus. We determine the saturation momentum Q_s using the estimation $Q_s^2 \approx (\sqrt{s_{NN}})^{0.25}$ GeV^2 with $\sqrt{s_{NN}}$ given in GeV [45–47]. The coupling constant g is set by the one-loop beta function at the saturation scale Q_s, which gives values close to $g \approx 2$. The relation between the MV model parameter μ and Q_s is non-trivial [46] and for simplicity we choose $0.75\, g^2\mu \simeq Q_s$ as suggested in [7]. The Lorentz gamma factor is given by $\gamma = \sqrt{s_{NN}}/(2m_N)$ with the nucleon mass $m_N \approx 1$ GeV and consequently we can find L via Eq. (11), setting R to the radius of a Gold nucleus. Since our results depend on the IR regulator m we try out different values: $m = 0.2$ GeV, $m = 0.4$ GeV and $m = 0.8$ GeV. The UV modes are regulated by $\Lambda = 10$ GeV. The simulation is performed with the gauge group SU(2) instead of SU(3), which should give reasonable qualitative results [48].

2.2. Observables

During the simulation we record the components of the energy–momentum tensor $T^{\mu\nu}$ in the laboratory frame and average over a number of collision events to obtain the expectation value $\langle T^{\mu\nu}\rangle$. In the MV model most of the $T^{\mu\nu}$ components vanish after averaging over all initial conditions due to homogeneity and isotropy in the transverse plane. Therefore the energy–momentum tensor reduces to

$$\langle T^{\mu\nu}\rangle = \begin{pmatrix} \langle\varepsilon\rangle & 0 & 0 & \langle S_L\rangle \\ 0 & \langle p_T\rangle & 0 & 0 \\ 0 & 0 & \langle p_T\rangle & 0 \\ \langle S_L\rangle & 0 & 0 & \langle p_L\rangle \end{pmatrix}, \tag{12}$$

where $\langle\varepsilon\rangle$ is the energy density, $\langle p_L\rangle$ and $\langle p_T\rangle$ are the longitudinal and transverse pressure components and $\langle S_L\rangle$ is the longitudinal component of the Poynting vector. The local rest frame energy density is obtained by diagonalizing the energy–momentum tensor $\langle T^{\mu}_{\ \nu}\rangle$:

$$\langle\varepsilon_{\rm loc}\rangle = \frac{1}{2}\left(\langle\varepsilon\rangle - \langle p_L\rangle + \sqrt{(\langle\varepsilon\rangle + \langle p_L\rangle)^2 - 4\langle S_L\rangle^2}\right). \tag{13}$$

Given the electric and magnetic fields E_i^a and B_i^a we have

$$\langle\varepsilon\rangle = \frac{1}{2}\left\langle E_T^2 + B_T^2 + E_L^2 + B_L^2\right\rangle, \tag{14}$$

$$\langle p_T\rangle = \frac{1}{2}\left\langle E_L^2 + B_L^2\right\rangle, \tag{15}$$

$$\langle p_L\rangle = \frac{1}{2}\left\langle E_T^2 + B_T^2 - E_L^2 - B_L^2\right\rangle, \tag{16}$$

$$\langle S_L\rangle = \left\langle \left(\vec{E}^a \times \vec{B}^a\right)_L\right\rangle. \tag{17}$$

Using proper time $\tau = \sqrt{t^2 - z^2}$ and space–time rapidity $\eta_s = \frac{1}{2}\ln[(t-z)/(t+z)]$ we can plot the profile of the energy density as a function of η_s for various values of τ. Due to the extended collision region there is some ambiguity in setting the coordinate origin for the transformation to the comoving frame. We choose the space–time coordinate of the maximum of $\langle p_T(t,z)\rangle$, which is generated by the initially purely longitudinal Glasma fields. This coordinate origin is slightly later than the space–time coordinates defined by the maximum overlap of the nuclei.

3. Results and discussion

In Fig. 2 we see a calculation of the rapidity profile of the local rest frame energy density for a RHIC-like collision: choosing parameters $\sqrt{s_{NN}} = 200$ GeV and $m = 0.2$ GeV, we obtain an approximate Gaussian profile of the local rest frame energy density,

$$\varepsilon_{\rm loc}(\tau_0, \eta_s) \approx \varepsilon_{\rm loc}(\tau_0, 0)\exp\left(-\frac{\eta_s^2}{2\sigma_\eta^2}\right), \tag{18}$$

in the range $\eta_s \in (-1, 1)$. Using a Gaussian fit we compute σ_η and extrapolate to higher η_s. At higher values of the IR regulator m the rapidity profiles become more narrow, which shows that the transverse size of the correlated color structures $\sim m^{-1}$ has an effect on broken boost invariance. In the CGC literature it is well-known that quantities such as the initial energy density or the gluon multiplicity can be sensitive to the choice of the infrared regulator [44,46, 49,50]. Here we add another example of a strong dependency on the infrared regulation.

Even though the results of our model should be regarded as more qualitative than quantitative, we compare our findings to experimental results: it is an interesting observation that the rapidity profile of the energy density for $m = 0.2$ GeV agrees with the measured rapidity profile of pion multiplicities for the most central collisions at RHIC. Of course, a direct comparison of $\varepsilon_{\rm loc}$ and dN/dy profiles is not strictly valid: the gluon number distribution can be somewhat broader than the energy density [51]. The correct approach would be to use our results as initial conditions for hydrodynamic simulations in order to make a more direct connection with measured observables. The subsequent hydrodynamic expansion of the system likely increases the width of the profiles further as mentioned in [29], which would favor the more narrow curves (b) and (c) in Fig. 2. Under these assumptions, the width of the rapidity profile of measured charged particle multiplicities can be seen as an upper limit for realistic rapidity profiles computed from our simulation. We also compare to the Gaussian rapidity profile of the hydrodynamic Landau model [15] with $\sigma_{\rm Landau} = \sqrt{\ln\gamma}$.

The profiles have been computed at $\tau_0 = 1$ fm/c, which roughly corresponds to the transition from the Glasma to the QGP. In general we observe that the rapidity profiles quickly converge for $\tau \gtrsim 0.3$ fm/c and afterwards become independent of τ. We also observe free-streaming behavior signaled by $\varepsilon_{\rm loc} \propto 1/\tau$ for a wide range of η_s and longitudinal velocities $v_z \sim z/t$. This implies that there is only negligible flow of energy between different rapidity directions. Our results suggest that the rapidity dependence of the Glasma is fixed early on in the collision and remains unchanged thereafter. To see the effects of increased longitudinal thickness, we repeat the same calculation for $\sqrt{s_{NN}} = 130$ GeV in Fig. 3. We fit to a Gaussian shape and find that, as expected, the profiles become more narrow as compared to the RHIC-like case. This time, our results are also more narrow than Landau's prediction.

Investigating the reason behind the non-boost-invariant creation of the Glasma we look at the space–time distribution of the transverse pressure $\langle p_T(z,t)\rangle$ in and along the forward light-cone

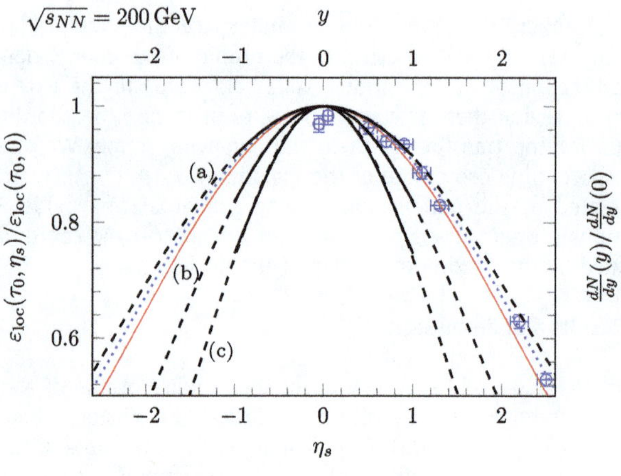

Fig. 2. Comparison of the space–time rapidity profile of the local rest frame energy density $\varepsilon_{loc}(\tau_0, \eta_s)$ for a RHIC-like collision (thick solid lines) at $\tau_0 = 1$ fm/c, a measured profile of π^+ multiplicity dN/dy at RHIC (data points) and Gaussian fits (dashed and dotted lines) for our simulation and experimental data ($\sigma_{exp} = 2.25$). The value of the infrared regulator m modifies the width of the profiles: (a) $m = 0.2$ GeV with $\sigma_\eta = 2.34$, (b) $m = 0.4$ GeV with $\sigma_\eta = 1.66$ and (c) $m = 0.8$ GeV with $\sigma_\eta = 1.28$. Data is taken from Ref. [13]. The thin, solid (red) line corresponds to the profile predicted by the Landau model with $\sigma_{Landau} = \sqrt{\ln \gamma} \approx 2.15$. (For interpretation of the references to color in this figure legend, the reader is referred to the web version of this article.)

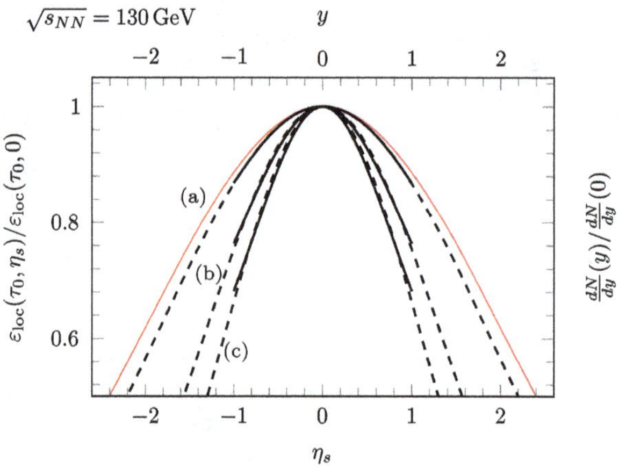

Fig. 3. Simulation results of collisions at $\sqrt{s_{NN}} = 130$ GeV at $\tau_0 = 1$ fm/c. The black solid lines are the computed profiles within $\eta_s \in (-1, 1)$. The dashed lines are fits to Gaussian profiles: (a) $m = 0.2$ GeV with $\sigma_\eta = 1.87$, (b) $m = 0.4$ GeV with $\sigma_\eta = 1.33$ and (c) $m = 0.8$ GeV with $\sigma_\eta = 1.10$. The thin, solid (red) line is the result predicted by the Landau model with $\sigma_{Landau} \approx 2.04$. Compared to Fig. 2 the profiles are more narrow than the Landau result. (For interpretation of the references to color in this figure legend, the reader is referred to the web version of this article.)

in Fig. 4. As can be seen from Eq. (15), the transverse pressure is solely due to the presence of longitudinal fields, i.e. $\langle p_T(z,t)\rangle$ is equivalent to the energy density generated by longitudinal fields $\langle \varepsilon_L(z,t)\rangle$. In the boost-invariant case the longitudinal fields would be constant along the boundary of the light-cone as determined by the boost-invariant Glasma initial conditions at $\tau = 0$. In our simulations we see very different behavior: the longitudinal fields are mostly centered around the maximum in the extended collision region $t \sim z \sim 0$ and decrease rather quickly along the $t = \pm z$ boundaries. Since the initially longitudinal fields are the starting point of the evolution of the Glasma, this sharp decrease means that for some fixed proper time τ there is less Glasma at higher values of rapidity η_s, leading to the observed Gaussian profiles of the energy density. Furthermore, we see a mostly constant spatial distribution of $\langle p_T(z,t)\rangle$ in Fig. 4 for later times t. A similar

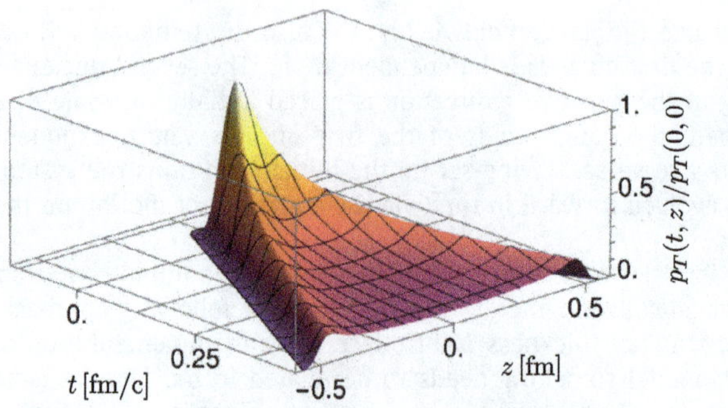

Fig. 4. Space–time distribution of the transverse pressure $\langle p_T(t,z)\rangle$ normalized to the maximum at the coordinate origin. The same parameters as in Fig. 2 (a) are used. In the Glasma the transverse pressure is generated by longitudinal magnetic and electric fields and equivalent to the longitudinal component of the energy density $\langle \varepsilon_L(t,z)\rangle$. The drop of $\langle \varepsilon_L(t,z)\rangle$ along the boundary of the light-cone is quite steep and becomes even steeper when decreasing the collision energy, resulting in more narrow rapidity profiles.

distribution has been found in holographic models of heavy-ion collisions [17,19].

The results in Figs. 2, 3 and 4 have been computed on a lattice with 2048 cells in the longitudinal direction with a length of 6 fm and 192^2 cells for the transverse plane with an area of $(6\,\text{fm})^2$ and a statistical average over 15 events. The tetragonal lattice (with much smaller longitudinal than transverse lattice spacing) is more suited for describing the collision of Lorentz-contracted nuclei. By varying the lattice spacings, the results were checked for discretization errors. Although accessing wider rapidity ranges is possible in principle, we restricted the results to $\eta_s \in (-1, 1)$ due to numerical issues: at high space–time rapidity η_s near the boundary of the light cone, the computation of ε_{loc} via Eq. (13) becomes very sensitive to cancellations in the square root term. Furthermore, the fields of the nuclei, being proportional to the longitudinal profile $f(x^-)$, become increasingly larger compared to the Glasma fields as $L \to 0$, which converge to the finite, boost-invariant result at mid-rapidity for high $\sqrt{s_{NN}}$. This leads to an unwanted modification of ε_{loc} by large transverse fields of the nuclei at larger η_s. We cannot strictly separate the Glasma from the nucleus fields, which is a clear disadvantage to simulations performed strictly in the forward light cone [29]. For these reasons we currently can not obtain reliable results for collision energies present at the LHC and further improvements in the numerical scheme are needed.

4. Conclusions

We found that Gaussian rapidity profiles in energy density can arise from CGC collisions of nuclei with finite longitudinal thickness in classical 3+1 dimensional Yang–Mills simulations. The width of the profiles is controlled by the energy of the incoming nuclei, but depends also crucially on the infrared modes of the fields. Presumably the infrared dependencies of our results could be fixed by more realistic initial conditions obtained from a JIMWLK evolution. It would be interesting to understand the mechanism behind the creation of Glasma fields in a non-boost-invariant setting with finite nucleus thickness, and in particular also the connection to infrared regulation. In any case, our result shows that there is a mechanism that breaks boost invariance at the classical level in the Glasma related to the finite thickness of the colliding nuclei, and further investigations are needed to see how this effect compares to other previously studied approaches like the rapidity dependence of JIMWLK. It is also exciting to see that by lifting the assumption of boost invariance, we have shown that our weak coupling results are in qualitative agreement with the strong coupling results exhibiting a Gaussian rapidity profile of the rest frame energy density and similar transverse pressure distributions in the laboratory frame [17]. That is to say, our result could be interpreted to indicate that the difference between the previous weak and strong coupling simulations is due to the initial conditions, not the weak or strong coupling dynamics.

This work has been supported by the Austrian Science Fund FWF, Project No. P 26582-N27 and Doctoral program No. W1252-N27. The computational results presented have been achieved using the Vienna Scientific Cluster. We would like to thank Aleksi Kurkela for useful discussions.

References

[1] U. Heinz, R. Snellings, Collective flow and viscosity in relativistic heavy-ion collisions, Annu. Rev. Nucl. Part. Sci. 63 (2013) 123–151, http://dx.doi.org/10.1146/annurev-nucl-102212-170540, arXiv:1301.2826.

[2] C. Gale, S. Jeon, B. Schenke, Hydrodynamic modeling of heavy-ion collisions, Int. J. Mod. Phys. A 28 (2013) 1340011, http://dx.doi.org/10.1142/S0217751X13400113, arXiv:1301.5893.

[3] F. Gelis, E. Iancu, J. Jalilian-Marian, R. Venugopalan, The color glass condensate, Annu. Rev. Nucl. Part. Sci. 60 (2010) 463–489, http://dx.doi.org/10.1146/annurev.nucl.010909.083629, arXiv:1002.0333.

[4] F. Gelis, Color glass condensate and glasma, Int. J. Mod. Phys. A 28 (2013) 1330001, http://dx.doi.org/10.1142/S0217751X13300019, arXiv:1211.3327.

[5] F. Gelis, Initial state and thermalization in the color glass condensate framework, Int. J. Mod. Phys. E 24 (10) (2015) 1530008, http://dx.doi.org/10.1142/S0218301315300088, arXiv:1508.07974.

[6] B. Schenke, P. Tribedy, R. Venugopalan, Fluctuating glasma initial conditions and flow in heavy ion collisions, Phys. Rev. Lett. 108 (2012) 252301, http://dx.doi.org/10.1103/PhysRevLett.108.252301, arXiv:1202.6646.

[7] B. Schenke, P. Tribedy, R. Venugopalan, Event-by-event gluon multiplicity, energy density, and eccentricities in ultrarelativistic heavy-ion collisions, Phys. Rev. C 86 (2012) 034908, http://dx.doi.org/10.1103/PhysRevC.86.034908, arXiv:1206.6805.

[8] C. Gale, S. Jeon, B. Schenke, P. Tribedy, R. Venugopalan, Event-by-event anisotropic flow in heavy-ion collisions from combined Yang–Mills and viscous fluid dynamics, Phys. Rev. Lett. 110 (1) (2013) 012302, http://dx.doi.org/10.1103/PhysRevLett.110.012302, arXiv:1209.6330.

[9] R. Snellings, Elliptic flow: a brief review, New J. Phys. 13 (2011) 055008, http://dx.doi.org/10.1088/1367-2630/13/5/055008, arXiv:1102.3010.

[10] A. Dumitru, F. Gelis, L. McLerran, R. Venugopalan, Glasma flux tubes and the near side ridge phenomenon at RHIC, Nucl. Phys. A 810 (2008) 91–108, http://dx.doi.org/10.1016/j.nuclphysa.2008.06.012, arXiv:0804.3858.

[11] E. Iancu, QCD in heavy ion collisions, in: Proceedings, 2011 European School of High-Energy Physics, ESHEP 2011, Cheile Gradistei, Romania, September 7–20, 2011, 2014, pp. 197–266, arXiv:1205.0579, https://inspirehep.net/record/1113441/files/1205.0579.pdf.

[12] E. Abbas, et al., Centrality dependence of the pseudorapidity density distribution for charged particles in Pb–Pb collisions at $\sqrt{s_{NN}} = 2.76$ TeV, Phys. Lett. B 726 (2013) 610–622, http://dx.doi.org/10.1016/j.physletb.2013.09.022, arXiv:1304.0347.

[13] I.G. Bearden, et al., Charged meson rapidity distributions in central Au+Au collisions at $\sqrt{s_{NN}} = 200$ GeV, Phys. Rev. Lett. 94 (2005) 162301, http://dx.doi.org/10.1103/PhysRevLett.94.162301, arXiv:nucl-ex/0403050.

[14] C.E. Flores, The rapidity density distributions and longitudinal expansion dynamics of identified pions from the STAR beam energy scan, Nucl. Phys. A 956 (2016) 280–283, http://dx.doi.org/10.1016/j.nuclphysa.2016.05.020.

[15] L.D. Landau, On the multiparticle production in high-energy collisions, Izv. Akad. Nauk Ser. Fiz. 17 (1953) 51–64.

[16] J.D. Bjorken, Highly relativistic nucleus–nucleus collisions: the central rapidity region, Phys. Rev. D 27 (1983) 140–151, http://dx.doi.org/10.1103/PhysRevD.27.140.

[17] J. Casalderrey-Solana, M.P. Heller, D. Mateos, W. van der Schee, From full stopping to transparency in a holographic model of heavy ion collisions, Phys. Rev. Lett. 111 (2013) 181601, http://dx.doi.org/10.1103/PhysRevLett.111.181601, arXiv:1305.4919.

[18] W. van der Schee, Gravitational Collisions and the Quark–Gluon Plasma, Ph.D. thesis, Utrecht Univ., 2014, arXiv:1407.1849.

[19] W. van der Schee, B. Schenke, Rapidity dependence in holographic heavy ion collisions, Phys. Rev. C 92 (6) (2015) 064907, http://dx.doi.org/10.1103/PhysRevC.92.064907, arXiv:1507.08195.

[20] A. Kovner, L.D. McLerran, H. Weigert, Gluon production from non-Abelian Weizsäcker–Williams fields in nucleus–nucleus collisions, Phys. Rev. D 52 (1995) 6231–6237, http://dx.doi.org/10.1103/PhysRevD.52.6231, arXiv:hep-ph/9502289.

[21] A. Kovner, L.D. McLerran, H. Weigert, Gluon production at high transverse momentum in the McLerran–Venugopalan model of nuclear structure functions, Phys. Rev. D 52 (1995) 3809–3814, http://dx.doi.org/10.1103/PhysRevD.52.3809, arXiv:hep-ph/9505320.

[22] A. Krasnitz, R. Venugopalan, Nonperturbative computation of gluon minijet production in nuclear collisions at very high-energies, Nucl. Phys. B 557 (1999) 237, http://dx.doi.org/10.1016/S0550-3213(99)00366-1, arXiv:hep-ph/9809433.

[23] T. Lappi, Gluon spectrum in the glasma from JIMWLK evolution, Phys. Lett. B 703 (2011) 325–330, http://dx.doi.org/10.1016/j.physletb.2011.08.011, arXiv:1105.5511.

[24] J. Jalilian-Marian, A. Kovner, L.D. McLerran, H. Weigert, The intrinsic glue distribution at very small x, Phys. Rev. D 55 (1997) 5414–5428, http://dx.doi.org/10.1103/PhysRevD.55.5414, arXiv:hep-ph/9606337.

[25] E. Iancu, A. Leonidov, L.D. McLerran, Nonlinear gluon evolution in the color glass condensate. 1, Nucl. Phys. A 692 (2001) 583–645, http://dx.doi.org/10.1016/S0375-9474(01)00642-X, arXiv:hep-ph/0011241.

[26] A.H. Mueller, A simple derivation of the JIMWLK equation, Phys. Lett. B 523 (2001) 243–248, http://dx.doi.org/10.1016/S0370-2693(01)01343-0, arXiv:hep-ph/0110169.

[27] E. Ferreiro, E. Iancu, A. Leonidov, L. McLerran, Nonlinear gluon evolution in the color glass condensate. 2, Nucl. Phys. A 703 (2002) 489–538, http://dx.doi.org/10.1016/S0375-9474(01)01329-X, arXiv:hep-ph/0109115.

[28] B. Schenke, P. Tribedy, R. Venugopalan, Multiplicity distributions in p+p, p+A and A+A collisions from Yang–Mills dynamics, Phys. Rev. C 89 (2) (2014) 024901, http://dx.doi.org/10.1103/PhysRevC.89.024901, arXiv:1311.3636.

[29] B. Schenke, S. Schlichting, 3D glasma initial state for relativistic heavy ion collisions, Phys. Rev. C 94 (4) (2016) 044907, http://dx.doi.org/10.1103/PhysRevC.94.044907, arXiv:1605.07158.

[30] Ş. Özönder, R.J. Fries, Rapidity profile of the initial energy density in heavy-ion collisions, Phys. Rev. C 89 (3) (2014) 034902, http://dx.doi.org/10.1103/PhysRevC.89.034902, arXiv:1311.3390.

[31] T. Altinoluk, N. Armesto, G. Beuf, M. Martínez, C.A. Salgado, Next-to-eikonal corrections in the CGC: gluon production and spin asymmetries in pA collisions, J. High Energy Phys. 07 (2014) 068, http://dx.doi.org/10.1007/JHEP07(2014)068, arXiv:1404.2219.

[32] T. Altinoluk, N. Armesto, G. Beuf, A. Moscoso, Next-to-next-to-eikonal corrections in the CGC, J. High Energy Phys. 01 (2016) 114, http://dx.doi.org/10.1007/JHEP01(2016)114, arXiv:1505.01400.

[33] T. Altinoluk, A. Dumitru, Particle production in high-energy collisions beyond the shockwave limit, Phys. Rev. D 94 (7) (2016) 074032, http://dx.doi.org/10.1103/PhysRevD.94.074032, arXiv:1512.00279.

[34] See Supplemental Material at http://dx.doi.org/10.1016/j.physletb.2017.05.032 for a movie of Fig. 1.

[35] D. Gelfand, A. Ipp, D. Müller, Simulating collisions of thick nuclei in the color glass condensate framework, Phys. Rev. D 94 (1) (2016) 014020, http://dx.doi.org/10.1103/PhysRevD.94.014020, arXiv:1605.07184.

[36] A. Dumitru, Y. Nara, Numerical simulation of non-Abelian particle-field dynamics, Eur. Phys. J. A 29 (2006) 65–69, http://dx.doi.org/10.1140/epja/i2005-10300-3i, arXiv:hep-ph/0511242.

[37] G.D. Moore, C.-r. Hu, B. Müller, Chern–Simons number diffusion with hard thermal loops, Phys. Rev. D 58 (1998) 045001, http://dx.doi.org/10.1103/PhysRevD.58.045001, arXiv:hep-ph/9710436.

[38] C. Hu, B. Müller, Classical lattice gauge field with hard thermal loops, Phys. Lett. B 409 (1997) 377–381, http://dx.doi.org/10.1016/S0370-2693(97)00851-4, arXiv:hep-ph/9611292.

[39] A. Dumitru, Y. Nara, M. Strickland, Ultraviolet avalanche in anisotropic non-Abelian plasmas, Phys. Rev. D 75 (2007) 025016, http://dx.doi.org/10.1103/PhysRevD.75.025016, arXiv:hep-ph/0604149.

[40] T.Z. Esirkepov, Exact charge conservation scheme for particle-in-cell simulation with an arbitrary form-factor, Comput. Phys. Commun. 135 (2001) 144–153, http://dx.doi.org/10.1016/S0010-4655(00)00228-9.

[41] J.P. Verboncoeur, Particle simulation of plasmas: review and advances, Plasma Phys. Control. Fusion 47 (5A) (2005) A231, http://dx.doi.org/10.1088/0741-3335/47/5A/017.

[42] L.D. McLerran, R. Venugopalan, Gluon distribution functions for very large nuclei at small transverse momentum, Phys. Rev. D 49 (1994) 3352–3355, http://dx.doi.org/10.1103/PhysRevD.49.3352, arXiv:hep-ph/9311205.

[43] L. McLerran, R. Venugopalan, Computing quark and gluon distribution functions for very large nuclei, Phys. Rev. D 49 (1994) 2233–2241, http://dx.doi.org/10.1103/PhysRevD.49.2233, arXiv:hep-ph/9309289.

[44] K. Fukushima, Randomness in infinitesimal extent in the McLerran–Venugopalan model, Phys. Rev. D 77 (2008) 074005, http://dx.doi.org/10.1103/PhysRevD.77.074005, arXiv:0711.2364.

[45] T. Lappi, Energy density of the glasma, Phys. Lett. B 643 (2006) 11–16, http://dx.doi.org/10.1016/j.physletb.2006.10.017, arXiv:hep-ph/0606207.
[46] T. Lappi, Wilson line correlator in the MV model: relating the glasma to deep inelastic scattering, Eur. Phys. J. C 55 (2008) 285–292, http://dx.doi.org/10.1140/epjc/s10052-008-0588-4, arXiv:0711.3039.
[47] D. Kharzeev, E. Levin, Manifestations of high density QCD in the first RHIC data, Phys. Lett. B 523 (2001) 79–87, http://dx.doi.org/10.1016/S0370-2693(01)01309-0, arXiv:nucl-th/0108006.
[48] A. Ipp, A. Rebhan, M. Strickland, Non-Abelian plasma instabilities: SU(3) vs. SU(2), Phys. Rev. D 84 (2011) 056003, http://dx.doi.org/10.1103/PhysRevD.84.056003, arXiv:1012.0298.
[49] H. Fujii, K. Fukushima, Y. Hidaka, Initial energy density and gluon distribution from the glasma in heavy-ion collisions, Phys. Rev. C 79 (2009) 024909, http://dx.doi.org/10.1103/PhysRevC.79.024909, arXiv:0811.0437.
[50] K. Fukushima, F. Gelis, The evolving glasma, Nucl. Phys. A 874 (2012) 108–129, http://dx.doi.org/10.1016/j.nuclphysa.2011.11.003, arXiv:1106.1396.
[51] T. Hirano, Y. Nara, Hydrodynamic afterburner for the color glass condensate and the parton energy loss, Nucl. Phys. A 743 (2004) 305–328, http://dx.doi.org/10.1016/j.nuclphysa.2004.08.003, arXiv:nucl-th/0404039.

Calculable mass hierarchies and a light dilaton from gravity duals

Daniel Elander [a,*], Maurizio Piai [b]

[a] *Departament de Física Quàntica i Astrofísica & Institut de Ciències del Cosmos (ICC), Universitat de Barcelona, Martí Franquès 1, ES-08028, Barcelona, Spain*
[b] *Department of Physics, College of Science, Swansea University, Singleton Park, Swansea SA2 8PP, UK*

ARTICLE INFO

Editor: M. Cvetič

ABSTRACT

In the context of gauge/gravity dualities, we calculate the scalar and tensor mass spectrum of the boundary theory defined by a special 8-scalar sigma-model in five dimensions, the background solutions of which include the 1-parameter family dual to the baryonic branch of the Klebanov–Strassler field theory. This provides an example of a strongly-coupled, multi-scale system that yields a parametrically light mass for one of the composite scalar particles: the dilaton. We briefly discuss the implications of these findings towards identifying a satisfactory solution to both the big and little hierarchy problems of the electro-weak theory.

1. Introduction

Many extensions of the Standard Model (SM) of particle physics are motivated by the (big) hierarchy problem. New dynamics and symmetries stabilize the electroweak scale, leading to the expectation that new particles should appear just above it. But such particles have not been detected experimentally in direct nor indirect searches. The little hierarchy between the mass of the Higgs and the new particles demands an explanation.

QCD dynamically explains the insensitivity to high-energy scales of the pion decay constant. New strong dynamics might replicate such success in the electro-weak theory. However, besides the calculability limitations of a strongly-coupled theory, the discovery of the Higgs particle [1] exacerbates the little hierarchy problem in such scenario, as one would have expected a proliferation of bound states to appear above the electroweak scale.

Fig. 1 provides a pictorial representation of the little hierarchy problem, by showing the SM mass spectrum, the current range of bounds from direct searches for exotica from the ATLAS collaboration, and the mass spectrum of a generic, hypothetic strongly-coupled new theory that evades them. Current bounds range from 570 GeV (Higgs triplet) [2] to 6.58 TeV (Kaluza–Klein graviton) [3], and (coarse-grained over the details) this range represents the LHC reach. The spectrum is model dependent, but consists of infinitely many bound states of all spins. If the Higgs scalar and the new physics have a common strong-coupling origin, and if the lack of

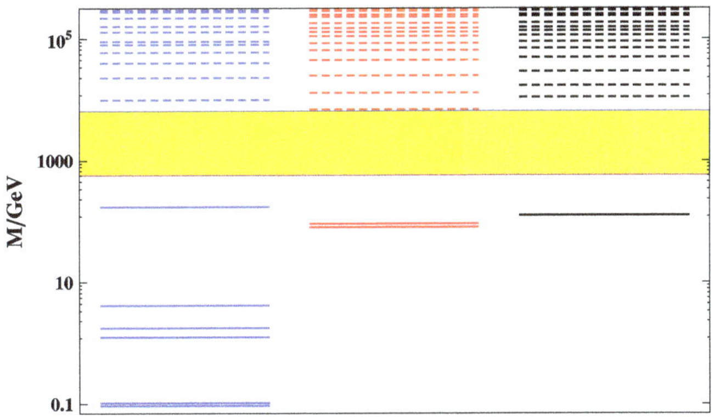

Fig. 1. The mass spectrum of SM particles (continuous lines) and of a generic strongly-coupled new theory (dashed lines) with new states heavy enough to evade the bounds from LHC direct searches (shaded region) [2,3]. Fermions are rendered in blue, vectors in red and scalars in black. (For interpretation of the references to color in this figure legend, the reader is referred to the web version of this article.)

evidence for new physics is confirmed, the anomalous suppression of the mass of the Higgs particle must also arise dynamically.

To make strongly-coupled models viable, it is imperative to find an example of a strongly-coupled, four-dimensional theory, no matter what the microscopic origin, that exhibits one scalar state parametrically lighter than the plethora of bound states. This possibility arose long ago within walking technicolor [4], and has been discussed at length in many contexts since [5–9], including Randall–Sundrum models stabilized à la Goldberger–Wise [10–12], suggesting that the gravity dual of a theory with a moduli space

* Corresponding author.
E-mail address: delander@ffn.ub.es (D. Elander).

Table 1
The field content, in terms of chiral superfields, and its classical symmetries [15]. $SU(M) \times SU(M+N)$ is the gauge group ($M = kN$). An additional Z_2 symmetry exchanges $A \leftrightarrow B$ and conjugates the gauge fields.

	$SU(M)$	$SU(M+N)$	$SU(2)_A$	$SU(2)_B$	$U(1)_B$	$U(1)_R$
A_α	M	$\overline{M+N}$	2	1	+1	+1/2
B_α	$\overline{M}$	$M+N$	1	2	−1	+1/2

might provide a concrete realisation of a system in which enhanced condensates and hierarchies of scales emerge. The underlying dynamics is approximately scale invariant, and condensates induce spontaneous symmetry breaking, yielding a light scalar particle in the spectrum: the dilaton.

In this paper, we provide a calculable example of a strongly-coupled theory that realizes this scenario, although it does not implement electro-weak symmetry breaking. Calculability is provided by the regular background in dual (super-)gravity [13,14]: the baryonic branch of the Klebanov–Strassler (KS) system [15, 16]. We compute the spectrum via the gauge-invariant fluctuations of the background in its 5-dimensional sigma-model description [17–20]. We refer to [21] for technical details. We report the results, discuss their origin, potential applications, and limitations.

2. The baryonic branch of KS: field theory

The four-dimensional $\mathcal{N} = 1$ supersymmetric theory is discussed for example in [22–26]. It has gauge group $SU(M) \times SU(M+N)$, with $M = kN$ (for k integer), and bifundamental matter fields (see Table 1).

The superpotential contains the nearly-marginal[1]:

$$W = h\mathrm{Tr}[A_1B_1A_2B_2 - A_1B_2A_2B_1]\,. \tag{1}$$

The moduli space of the theory contains a baryonic branch. Following [23,25], we illustrate some of its properties with tree-level arguments. For a fixed choice of $k = q$, the F-term equations are solved by $B_i = 0$. We define

$$\Phi_1 = \begin{pmatrix} \sqrt{q} & 0 & \cdots & 0 & 0 & 0 \\ 0 & \sqrt{q-1} & \cdots & 0 & 0 & 0 \\ \cdots & \cdots & \cdots & \cdots & \cdots & \cdots \\ 0 & 0 & \cdots & \sqrt{2} & 0 & 0 \\ 0 & 0 & \cdots & 0 & 1 & 0 \end{pmatrix}, \tag{2}$$

$$\Phi_2 = \begin{pmatrix} 0 & 1 & \cdots & 0 & 0 & 0 \\ 0 & 0 & \cdots & 0 & 0 & 0 \\ \cdots & \cdots & \cdots & \cdots & \cdots & \cdots \\ 0 & 0 & \cdots & 0 & \sqrt{q-1} & 0 \\ 0 & 0 & \cdots & 0 & 0 & \sqrt{q} \end{pmatrix}, \tag{3}$$

where each element represents a block proportional to the identity matrix $\mathbb{I}_N$. The D-term equations are

$$0 = -g_q \sum_i \mathrm{Tr}_{q+1}\left[A_i^\dagger T_q^A A_i + B_i T_q^A B_i^\dagger\right], \tag{4}$$

$$0 = -g_{q+1} \sum_i \mathrm{Tr}_q\left[A_i T_{q+1}^A A_i^\dagger + B_i^\dagger T_{q+1}^A B_i\right], \tag{5}$$

where the labels q and $q+1$ refer to the $SU(qN)$ and $SU((q+1)N)$ groups, respectively, g_j are the gauge couplings, and T_j^A the generators. Taking $A_i = c\ \Phi_i$:

$$\mathrm{Tr}\left[T_q^A\left(A_1A_1^\dagger + A_2A_2^\dagger\right)\right] = (q+1)|c|^2\mathrm{Tr}\,T_q^A = 0\,. \tag{6}$$

Another classical branch has $A_i \leftrightarrow B_i$. The operator

$$\mathcal{U} = \frac{1}{q(q+1)N}\mathrm{Tr}\left[A_i^\dagger A_i - B_iB_i^\dagger\right], \tag{7}$$

normalized so that for $A_i = c\Phi_i$ and $B_i = 0$ one has $\mathcal{U} = |c|^2$, is the order parameter of Z_2 symmetry breaking, and of the Higgsing $SU(qN) \times SU((q+1)N) \to SU(N)$.

The matrices Φ_i obey the $SU(2)$ algebra [25], and indeed the perturbative calculation of the spectrum of gauge bosons yields $M^2 = g^2|c|^2\lambda_{\ell,\pm}$ (for $g_q = g_{q+1} \equiv g$) where

$$\lambda_{\ell,\pm} = q + \frac{1}{2} \pm \sqrt{\left(q+\frac{1}{2}\right)^2 - \ell(\ell+1)}\,, \tag{8}$$

where $\ell = 0, 1, \cdots, q-1$ [25], and where the eigenvalues have multiplicity $(2\ell+1)N^2$ for $\ell \neq 0$ and $N^2 - 1$ for $\ell = 0$. In addition there are $(2q+1)N^2$ states with mass $M^2 = g^2|c|^2q$. The $N^2 - 1$ massless vectors represent the unbroken gauge group $SU(N)$. The unbroken $SU(N)$ theory with adjoint matter field content can be obtained by twisted compactification on a 2-sphere of the CVMN six dimensional field theory [28], the degeneracies exactly match, while the numerical values of the masses agree for $\ell \ll q$ [29]. This theory has a dynamical (confinement) scale Λ.

The perturbative approach provides a lot of insight [25], but leaves many open questions.

- The two gauge couplings and the coupling h are not independent, but non-perturbatively related [22].
- The couplings run, because of the presence of the anomaly. The RG flow is best described in terms of a cascade of Seiberg dualities that progressively reduce the group as in

$$\begin{aligned} SU(kN) \times SU((k+1)N) &\to SU(kN) \times SU((k-1)N) \\ &\to SU((k-2)N) \times SU((k-1)N) \\ &\to \cdots\,. \end{aligned} \tag{9}$$

- The cascade stops at $k = q$ because of the Higgsing due to $\mathcal{U}$, but the constant c should be determined non-perturbatively.
- Supersymmetry allows to infer that the gaugino condensate forms, breaking $Z_{2N} \to Z_2$ at scale Λ, and to classify the quantum moduli space [23], but not to calculate the spectrum of bound states.
- The Kähler part of the supersymmetric action is not protected by non-renormalization theorems, hence the whole spectrum requires non-perturbative treatment.
- The constant c should be linked with q. We assume in the following that the position along the (quantum) baryonic branch be characterized by a non-perturbatively defined α, such that $\alpha \to -\infty$ corresponds to $\mathcal{U} = 0$, and $\alpha \to +\infty$ to $q \to +\infty$.
- There are two dynamical scales: Λ is the scale of explicit symmetry breaking given by dimensional transmutation of the scale anomaly (beta functions), but $\mathcal{U}$ is unconstrained, defines a scale that can be taken to be larger than Λ, and breaks scale invariance spontaneously, suggesting the presence of a dilaton in the spectrum, if the latter effect is larger than the former.

[1] The $SU(M) \times SU(M)$ theory is a CFT [27], as for $N = 0$ there is no anomalous breaking of $U(1)_R \to Z_{2N}$. The CFT can be obtained as the low-energy IR fixed point reached by a mass deformation of the $\mathcal{N} = 2$ supersymmetric gauge theory, itself obtained by Z_2-orbifold of the $\mathcal{N} = 4$ theory with gauge group $SU(2M)$. At the IR fixed point the anomalous dimension is non-perturbative, and hence W in Eq. (1) is not irrelevant.

For all of these reasons, we need a non-perturbative description of the theory at strong coupling, which is provided (at large N) by the known gravity dual [16].

3. The baryonic branch of KS: gravity

The baryonic branch is described in gravity by a family of type-IIB supergravity backgrounds [16] (see also [30–32]) within the PT ansatz [33], characterized by a compact five-dimensional manifold with the symmetries of $T^{1,1}$ [35], and a non-compact five-dimensional space with metric ansatz:

$$\mathrm{d}s_5^2 = e^{2A(r)}\mathrm{d}s_{1,3}^2 + \mathrm{d}r^2 \,. \tag{10}$$

The space can be foliated along r in Minkowski slices, related by a conformal factor e^{2A} dependent only on r, so that the radial direction is interpreted in field-theory terms as a renormalization scale.

The general problem of finding solutions and studying their fluctuations can be conveniently formulated in terms of a truncation to a five-dimensional sigma-model with 8 scalars Φ^a coupled to gravity, and Lagrangian

$$\mathcal{L} = \frac{R}{4} - \frac{1}{2}G_{ab}g^{MN}\partial_M\Phi^a\partial_N\Phi^b - V(\Phi^a) \,, \tag{11}$$

where R is the five-dimensional Ricci scalar, G_{ab} is the sigma-model metric, g_{MN} is the metric, and $V(\Phi^a)$ is the potential. The detailed form of potential and kinetic terms can be found elsewhere [17]. The full lift to 10 dimensions is known [33].[2]

The background solution can be found by using Eq. (10) and the assumption that the scalars have non-trivial profiles $\Phi^a = \bar{\Phi}^a(r)$, and looking for non-singular (in 10 dimensions) solutions of the system of coupled differential equations [16].

The baryonic branch solutions are characterized by two scales r_0 and $\bar{r}$, the separation of which is controlled by the integration constant α [21]. When $r > \bar{r}$, the background is approximated by the KS solution, and captures the cascade of Seiberg dualities for $k > q$. For $r < \bar{r}$ the background approaches the CVMN solution [28]: the quiver gauge theory is higgsed to the $SU(N)$ one. As $r \to r_0$, the space ends: the theory confines and the gauginos condense. The gravity background captures all the non-perturbative features expected from field theory.

In computing the spectrum, we restrict our attention to tensor and scalar modes by fluctuating the five-dimensional sigma-model around the background solutions. We employ the gauge-invariant formalism of [17–19]. The tensor $\mathfrak{e}^{\mu}{}_{\nu}$ is the transverse and traceless part of the fluctuations of the four-dimensional metric. The gauge-invariant scalars $\mathfrak{a}^a$ are written in terms of the fluctuations φ^a of the bulk scalars and h of the trace of the four-dimensional part of the metric:

$$\mathfrak{a}^a = \varphi^a - \frac{\partial_r\bar{\Phi}^a}{6\partial_r A}h \,. \tag{12}$$

The tensors obey the linearized differential equations

$$0 = \left[\partial_r^2 + 4\partial_r A\partial_r + e^{-2A}m^2\right]\mathfrak{e}^{\mu}{}_{\nu} \,, \tag{13}$$

where m is the four-dimensional mass, and for the scalars

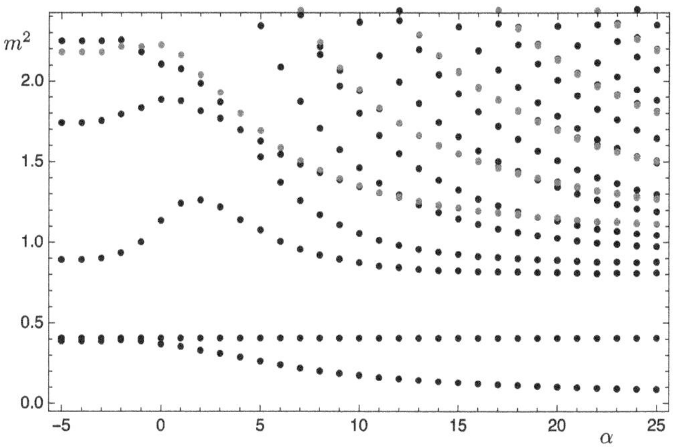

Fig. 2. The mass spectrum m^2 of scalar (black) and tensor (blue) states. (For interpretation of the references to color in this figure legend, the reader is referred to the web version of this article.)

$$0 = \left[\mathcal{D}_r^2 + 4\partial_r A\mathcal{D}_r + e^{-2A}m^2\right]\mathfrak{a}^a - \left[V^a{}_{|c} - \mathcal{R}^a{}_{bcd}\partial_r\bar{\Phi}^b\partial_r\bar{\Phi}^d + \frac{4(\partial_r\bar{\Phi}^a V^b + V^a\partial_r\bar{\Phi}^b)G_{bc}}{3\partial_r A} + \frac{16V\partial_r\bar{\Phi}^a\partial_r\bar{\Phi}^b G_{bc}}{9(\partial_r A)^2}\right]\mathfrak{a}^c \,. \tag{14}$$

In order to interpret the eigenstates as states in the dual theory, we impose the boundary conditions:

$$\partial_r\mathfrak{e}^{\mu}{}_{\nu}\Big|_{r=r_i} = 0 \,, \tag{15}$$

and [20]

$$\frac{2e^{2A}\partial_r\bar{\Phi}^a}{3m^2\partial_r A}\left[\partial_r\bar{\Phi}^b\mathcal{D}_r - \frac{4V\partial_r\bar{\Phi}^b}{3\partial_r A} - V^b\right]\mathfrak{a}_b + \mathfrak{a}^a\Big|_{r_i} = 0, \tag{16}$$

where $r_i = r_I, r_U$ are cutoffs, acting as regulators.

The two regulators have no physical meaning: they are needed in the numerical calculation for practical reasons, but the physical results are obtained by extrapolating to $r_I \to r_0$ and $r_U \to +\infty$ [21]. Here we report only the final physical results, contained in Fig. 2, where we show the spectrum of scalar and tensor modes as a function of the parameter α characterizing the position along the baryonic branch. We are interested only in ratios of masses, to facilitate the comparisons we chose the normalization so that for all values of α the next-to-lightest state agrees with the lightest state of the CVMN spectrum [18].

The main results that emerge are the following.

- Scalar and tensor spectra agree with the KS case for small α [17,36], and show evidence of deconstruction at the non-perturbative level: at large α, part of the spectrum consists of a dense sequence of states, above the continuum thresholds of the CVMN case ($\alpha \to +\infty$) [17,18].
- The spectrum of scalars contains one state the mass of which is suppressed going to large α. In the limit $\alpha \to +\infty$, we expect this state to become massless and decouple, as suggested by Fig. 2 and by the fact that this state is absent in the spectrum of the CVMN background, in which case it does not correspond to a normalizable, massive mode [21].

The latter is the main element of novelty of this paper.

[2] The five-dimensional system we study is obtained by imposing a set of constraints on the consistent truncation on $T^{1,1}$ in [34]. In particular, the reduction to eight scalars plus gravity is obtained by imposing a non-linear constraint that is in general not integrable. We treat the resulting sigma-model as a five-dimensional system on its own, and study its fluctuations. To study the full lift to type-IIB supergravity one would need to extend the five-dimensional sigma model along the lines of [34], to include additional scalar fields as well as vectors.

4. Discussion

All the qualitative expectations emerging from the study of the field theory are confirmed quantitatively by the gravity dual, including the fact that the spectrum of glueballs along the baryonic branch deconstructs a compact manifold, and interpolates between the known spectra of the KS and CVMN backgrounds.

In addition, we find a new result not accessible using field theory methods: the spectrum of scalars contains one parametrically light state, the mass of which is suppressed moving far from the origin of the baryonic branch. We interpret it as a dilaton, on the basis of the fact that all the solutions considered here are dual to the same field theory, with scale Λ controlled by the explicit breaking of scale invariance, but differ by the tunable choice of the vacuum value of the operator $\mathcal{U}$.

The dynamical scale Λ is controlled by explicit symmetry breaking (the scale anomaly), and hence is natural. There emerges in addition a tunable hierarchy between the mass of one isolated scalar (the dilaton) and the typical scale of the other states. This is hence an example of a strongly-coupled theory that naturally provides both a big and a little hierarchy of scales, thanks to the role of scale invariance and to the hierarchy between the scales of its spontaneous and explicit breaking.

A complete and rigorous understanding of the field theory requires extending the formalism for treating the fluctuations to adapt it to a more general truncation including vectors in the sigma model [34]. We leave this task for future work. Furthermore, the theory is supersymmetric, and the gravity calculation captures only the large-N limit. The latter provides a technical advantage, as it allows to perform a conceptually simple (although numerically challenging) calculation. The dilaton is parametrically light because the theory admits a classical moduli space along a flat direction that is lifted only by controllable quantum effects (the running of the couplings), with the quantum moduli space still non-trivial [23], and hence there are condensates that are parametrically larger than the scale introduced by the explicit breaking of scale invariance due to the anomaly. Whether such phenomena arise in non-supersymmetric theories is an open problem.

The results of this paper support the expectation that in a strongly-coupled theory in which condensates are enhanced, the mass spectrum of bound states (including the lightest scalar) would reproduce the qualitative features in Fig. 1. A phenomenologically viable solution of the hierarchy problem(s) of the electroweak theory would require to implement electro-weak symmetry and its breaking. One possible way to achieve this, within the specific context of the model studied here, might follow the lines of the Sakai–Sugimoto model [37], by embedding in the background extended objects [38,39] that implement the $SU(2) \times SU(2)$ global symmetry of the SM Higgs sector, and its breaking. The gauging of $SU(2)_L \times U(1)_Y$ must be reinstated via holographic renormalization, by including boundary-localised terms that cancel the divergence of the gauge boson wave function, and hence retain a finite gauge coupling in the limit in which the UV cutoff is taken to infinity. While finding the embedding is technically challenging [40], given the great potential of this or alternative model-building approaches, we think this line of reasoning deserves further future study.

Acknowledgements

We would like to thank D. Mateos and C. Nunez for useful discussions, and A. Faedo for important comments regarding the nonlinear constraint of the sigma-model. DE is supported by the ERC Starting Grant HoloLHC-306605 and by the grant MDM-2014-0369 of ICCUB. The work of MP is supported in part by the STFC grant ST/L000369/1.

References

[1] G. Aad, et al., ATLAS Collaboration, Phys. Lett. B 716 (2012) 1, arXiv:1207.7214 [hep-ex];
S. Chatrchyan, et al., CMS Collaboration, Phys. Lett. B 716 (2012) 30, arXiv:1207.7235 [hep-ex].
[2] The ATLAS collaboration, ATLAS Collaboration, ATLAS-CONF-2016-051.
[3] M. Aaboud, et al., ATLAS Collaboration, Phys. Rev. D 94 (3) (2016) 032005, arXiv:1604.07773 [hep-ex].
[4] B. Holdom, Phys. Lett. B 150 (1985) 301;
K. Yamawaki, M. Bando, K.i. Matumoto, Phys. Rev. Lett. 56 (1986) 1335;
T.W. Appelquist, D. Karabali, L.C.R. Wijewardhana, Phys. Rev. Lett. 57 (1986) 957.
[5] C.N. Leung, S.T. Love, W.A. Bardeen, Nucl. Phys. B 273 (1986) 649;
W.A. Bardeen, C.N. Leung, S.T. Love, Phys. Rev. Lett. 56 (1986) 1230;
D.K. Hong, S.D.H. Hsu, F. Sannino, Phys. Lett. B 597 (2004) 89, arXiv:hep-ph/0406200;
D.D. Dietrich, F. Sannino, K. Tuominen, Phys. Rev. D 72 (2005) 055001, arXiv:hep-ph/0505059;
M. Kurachi, R. Shrock, J. High Energy Phys. 0612 (2006) 034, arXiv:hep-ph/0605290;
M. Hashimoto, K. Yamawaki, Phys. Rev. D 83 (2011) 015008, arXiv:1009.5482 [hep-ph];
T. Appelquist, Y. Bai, Phys. Rev. D 82 (2010) 071701, arXiv:1006.4375 [hep-ph].
[6] W.D. Goldberger, B. Grinstein, W. Skiba, Phys. Rev. Lett. 100 (2008) 111802, arXiv:0708.1463 [hep-ph];
L. Vecchi, Phys. Rev. D 82 (2010) 076009;
Z. Chacko, R.K. Mishra, arXiv:1002.1721 [hep-ph], Phys. Rev. D 87 (11) (2013) 115006, arXiv:1209.3022 [hep-ph];
B. Bellazzini, C. Csaki, J. Hubisz, J. Serra, J. Terning, Eur. Phys. J. C 73 (2) (2013) 2333;
T. Abe, R. Kitano, Y. Konishi, K.y. Oda, J. Sato, S. Sugiyama, arXiv:1209.3299 [hep-ph], Phys. Rev. D 86 (2012) 115016, arXiv:1209.4544 [hep-ph];
E. Eichten, K. Lane, A. Martin, arXiv:1210.5462 [hep-ph];
M. Golterman, Y. Shamir, Phys. Rev. D 94 (5) (2016) 054502, arXiv:1603.04575 [hep-ph];
A. Kasai, K.i. Okumura, H. Suzuki, arXiv:1609.02264 [hep-lat];
S. Matsuzaki, K. Yamawaki, Phys. Rev. Lett. 113 (8) (2014) 082002, arXiv:1311.3784 [hep-lat];
M. Hansen, K. Langæble, F. Sannino, Phys. Rev. D 95 (3) (2017) 036005, arXiv:1610.02904 [hep-ph];
M. Golterman, Y. Shamir, arXiv:1610.01752 [hep-ph];
T. Appelquist, J. Ingoldby, M. Piai, arXiv:1702.04410 [hep-ph];
P. Hernandez-Leon, L. Merlo, arXiv:1703.02064 [hep-ph].
[7] D. Elander, C. Nunez, M. Piai, Phys. Lett. B 686 (2010) 64, arXiv:0908.2808 [hep-th];
D. Elander, M. Piai, Nucl. Phys. B 871 (2013) 164, arXiv:1212.2600 [hep-th];
D. Elander, Phys. Rev. D 91 (12) (2015) 126012, arXiv:1401.3412 [hep-th].
[8] S.P. Kumar, D. Mateos, A. Paredes, M. Piai, J. High Energy Phys. 1105 (2011) 008, arXiv:1012.4678 [hep-th];
D. Elander, M. Piai, Nucl. Phys. B 864 (2012) 241, arXiv:1112.2915 [hep-ph];
D. Kutasov, J. Lin, A. Parnachev, Nucl. Phys. B 863 (2012) 361, arXiv:1201.4123 [hep-th];
M. Goykhman, A. Parnachev, Phys. Rev. D 87 (2) (2013) 026007, arXiv:1211.0482 [hep-th];
D. Elander, M. Piai, Nucl. Phys. B 867 (2013) 779, arXiv:1208.0546 [hep-ph];
N. Evans, K. Tuominen, Phys. Rev. D 87 (8) (2013) 086003, arXiv:1302.4553 [hep-ph];
E. Megias, O. Pujolas, J. High Energy Phys. 1408 (2014) 081;
D. Elander, R. Lawrance, M. Piai, arXiv:1401.4998 [hep-th], Nucl. Phys. B 897 (2015) 583, arXiv:1504.07949 [hep-ph].
[9] Z. Fodor, K. Holland, J. Kuti, D. Nogradi, C. Schroeder, C.H. Wong, Phys. Lett. B 718 (2012) 657, arXiv:1209.0391 [hep-lat];
Y. Aoki, et al., LatKMI Collaboration, Phys. Rev. D 89 (2014) 111502, arXiv:1403.5000 [hep-lat];
T. Appelquist, et al., Phys. Rev. D 93 (11) (2016) 114514, arXiv:1601.04027 [hep-lat];
Z. Fodor, K. Holland, J. Kuti, S. Mondal, D. Nogradi, C.H. Wong, PoS LATTICE 2015 (2016) 219, arXiv:1605.08750 [hep-lat];
Y. Aoki, et al., LatKMI Collaboration, arXiv:1610.07011 [hep-lat];
A. Athenodorou, E. Bennett, G. Bergner, D. Elander, C.-J.D. Lin, B. Lucini, M. Piai, arXiv:1605.04258 [hep-th].
[10] L. Randall, R. Sundrum, Phys. Rev. Lett. 83 (1999) 3370, arXiv:hep-ph/9905221.
[11] W.D. Goldberger, M.B. Wise, Phys. Rev. Lett. 83 (1999) 4922, arXiv:hep-ph/9907447.

[12] O. DeWolfe, D.Z. Freedman, S.S. Gubser, A. Karch, Phys. Rev. D 62 (2000) 046008, arXiv:hep-th/9909134;
W.D. Goldberger, M.B. Wise, Phys. Lett. B 475 (2000) 275, arXiv:hep-ph/9911457;
C. Csaki, M.L. Graesser, G.D. Kribs, Phys. Rev. D 63 (2001) 065002, arXiv:hep-th/0008151;
N. Arkani-Hamed, M. Porrati, L. Randall, J. High Energy Phys. 0108 (2001) 017, arXiv:hep-th/0012148;
R. Rattazzi, A. Zaffaroni, J. High Energy Phys. 0104 (2001) 021, arXiv:hep-th/0012248;
L. Kofman, J. Martin, M. Peloso, Phys. Rev. D 70 (2004) 085015, arXiv:hep-ph/0401189.
[13] J.M. Maldacena, Int. J. Theor. Phys. 38 (1999) 1113, Adv. Theor. Math. Phys. 2 (1998) 231, arXiv:hep-th/9711200;
S.S. Gubser, I.R. Klebanov, A.M. Polyakov, Phys. Lett. B 428 (1998) 105, arXiv:hep-th/9802109;
E. Witten, Adv. Theor. Math. Phys. 2 (1998) 253, arXiv:hep-th/9802150.
[14] O. Aharony, S.S. Gubser, J.M. Maldacena, H. Ooguri, Y. Oz, Phys. Rep. 323 (2000) 183, arXiv:hep-th/9905111.
[15] I.R. Klebanov, M.J. Strassler, J. High Energy Phys. 0008 (2000) 052, arXiv:hep-th/0007191.
[16] A. Butti, M. Grana, R. Minasian, M. Petrini, A. Zaffaroni, J. High Energy Phys. 0503 (2005) 069, arXiv:hep-th/0412187.
[17] M. Berg, M. Haack, W. Mueck, Nucl. Phys. B 736 (2006) 82, arXiv:hep-th/0507285.
[18] M. Berg, M. Haack, W. Mueck, Nucl. Phys. B 789 (2008) 1, arXiv:hep-th/0612224.
[19] D. Elander, J. High Energy Phys. 1003 (2010) 114, arXiv:0912.1600 [hep-th].
[20] D. Elander, M. Piai, J. High Energy Phys. 1101 (2011) 026, arXiv:1010.1964 [hep-th].
[21] D. Elander, M. Piai, arXiv:1703.10158 [hep-th].
[22] M.J. Strassler, arXiv:hep-th/0505153.
[23] A. Dymarsky, I.R. Klebanov, N. Seiberg, J. High Energy Phys. 0601 (2006) 155, arXiv:hep-th/0511254.
[24] S.S. Gubser, C.P. Herzog, I.R. Klebanov, J. High Energy Phys. 0409 (2004) 036, arXiv:hep-th/0405282.
[25] J. Maldacena, D. Martelli, J. High Energy Phys. 1001 (2010) 104, arXiv:0906.0591 [hep-th].
[26] A. Ceresole, G. Dall'Agata, R. D'Auria, S. Ferrara, Phys. Rev. D 61 (2000) 066001, arXiv:hep-th/9905226;
F. Bigazzi, L. Girardello, A. Zaffaroni, Nucl. Phys. B 598 (2001) 530, arXiv:hep-th/0011041;
F. Bigazzi, A.L. Cotrone, M. Petrini, A. Zaffaroni, Riv. Nuovo Cimento 25 (12) (2002) 1, arXiv:hep-th/0303191;
D. Baumann, A. Dymarsky, S. Kachru, I.R. Klebanov, L. McAllister, J. High Energy Phys. 1006 (2010) 072, arXiv:1001.5028 [hep-th].
[27] I.R. Klebanov, E. Witten, Nucl. Phys. B 536 (1998) 199, arXiv:hep-th/9807080.
[28] J.M. Maldacena, C. Nunez, Phys. Rev. Lett. 86 (2001) 588, arXiv:hep-th/0008001;
See also A.H. Chamseddine, M.S. Volkov, Phys. Rev. Lett. 79 (1997) 3343, arXiv:hep-th/9707176.
[29] R.P. Andrews, N. Dorey, Phys. Lett. B 631 (2005) 74, arXiv:hep-th/0505107;
R.P. Andrews, N. Dorey, Nucl. Phys. B 751 (2006) 304, arXiv:hep-th/0601098.
[30] J. Gaillard, D. Martelli, C. Nunez, I. Papadimitriou, Nucl. Phys. B 843 (2011) 1, arXiv:1004.4638 [hep-th].
[31] E. Caceres, C. Nunez, L.A. Pando-Zayas, J. High Energy Phys. 1103 (2011) 054, arXiv:1101.4123 [hep-th].
[32] D. Elander, J. Gaillard, C. Nunez, M. Piai, J. High Energy Phys. 1107 (2011) 056, arXiv:1104.3963 [hep-th].
[33] G. Papadopoulos, A.A. Tseytlin, Class. Quantum Gravity 18 (2001) 1333, arXiv:hep-th/0012034.
[34] D. Cassani, A.F. Faedo, Nucl. Phys. B 843 (2011) 455, http://dx.doi.org/10.1016/j.nuclphysb.2010.10.010, arXiv:1008.0883 [hep-th];
I. Bena, G. Giecold, M. Grana, N. Halmagyi, F. Orsi, J. High Energy Phys. 1104 (2011) 021, http://dx.doi.org/10.1007/JHEP04(2011)021, arXiv:1008.0983 [hep-th].
[35] P. Candelas, X.C. de la Ossa, Nucl. Phys. B 342 (1990) 246.
[36] M. Krasnitz, arXiv:hep-th/0011179;
E. Caceres, R. Hernandez, Phys. Lett. B 504 (2001) 64, arXiv:hep-th/0011204;
A. Dymarsky, D. Melnikov, J. High Energy Phys. 0805 (2008) 035, arXiv:0710.4517 [hep-th];
M.K. Benna, A. Dymarsky, I.R. Klebanov, A. Solovyov, J. High Energy Phys. 0806 (2008) 070, arXiv:0712.4404 [hep-th];
A. Dymarsky, D. Melnikov, A. Solovyov, J. High Energy Phys. 0905 (2009) 105, arXiv:0810.5666 [hep-th];
I. Gordeli, D. Melnikov, J. High Energy Phys. 1108 (2011) 082, arXiv:0912.5517 [hep-th].
[37] T. Sakai, S. Sugimoto, Prog. Theor. Phys. 113 (2005) 843, arXiv:hep-th/0412141;
T. Sakai, S. Sugimoto, Prog. Theor. Phys. 114 (2005) 1083, arXiv:hep-th/0507073.
[38] C.D. Carone, J. Erlich, M. Sher, Phys. Rev. D 76 (2007) 015015, arXiv:0704.3084 [hep-th];
T. Hirayama, K. Yoshioka, J. High Energy Phys. 0710 (2007) 002, arXiv:0705.3533 [hep-ph];
O. Mintakevich, J. Sonnenschein, J. High Energy Phys. 0907 (2009) 032, arXiv:0905.3284 [hep-th].
[39] L. Anguelova, Nucl. Phys. B 843 (2011) 429;
L. Anguelova, P. Suranyi, L.C.R. Wijewardhana, arXiv:1006.3570 [hep-th];
L. Anguelova, P. Suranyi, L.C.R. Wijewardhana, Nucl. Phys. B 852 (2011) 39, arXiv:1105.4185 [hep-th];
T.E. Clark, S.T. Love, T. ter Veldhuis, Nucl. Phys. B 872 (2013) 1, arXiv:1208.0817 [hep-th].
[40] A. Dymarsky, S. Kuperstein, J. Sonnenschein, J. High Energy Phys. 0908 (2009) 005, arXiv:0904.0988 [hep-th].

4

Charged scalar quasi-normal modes for linearly charged dilaton-Lifshitz solutions

M. Kord Zangeneh [a,b,c,d], B. Wang [d,e], A. Sheykhi [c,b], Z.Y. Tang [d]

[a] Physics Department, Faculty of Science, Shahid Chamran University of Ahvaz 61357-43135, Iran
[b] Research Institute for Astronomy and Astrophysics of Maragha (RIAAM), Maragha, P. O. Box: 55134-441, Iran
[c] Physics Department and Biruni Observatory, Shiraz University, Shiraz 71454, Iran
[d] Center of Astronomy and Astrophysics, Department of Physics and Astronomy, Shanghai Jiao Tong University, Shanghai 200240, China
[e] Center for Gravitation and Cosmology, College of Physical Science and Technology, Yangzhou University, Yangzhou 225009, China

ARTICLE INFO

Editor: N. Lambert

ABSTRACT

Most available studies of quasi-normal modes for Lifshitz black solutions are limited to the neutral scalar perturbations. In this letter, we investigate the wave dynamics of massive charged scalar perturbation in the background of $(3+1)$-dimensional charged dilaton Lifshitz black branes/holes. We disclose the dependence of the quasi-normal modes on the model parameters, such as the Lifshitz exponent z, the mass and charge of the scalar perturbation field and the charge of the Lifshitz configuration. In contrast with neutral perturbations, we observe the possibility to destroy the original Lifshitz background near the extreme value of charge where the temperature is low. We find out that when the Lifshitz exponent deviates more from unity, it is more difficult to break the stability of the configuration. We also study the behavior of the real part of the quasi-normal frequencies. Unlike the neutral scalar perturbation around uncharged black branes where an overdamping was observed to start at $z=2$ and independent of the value of scalar mass, our observation discloses that the overdamping starting point is no longer at $z=2$ and depends on the mass of scalar field for charged Lifshitz black branes. For charged scalar perturbations, fixing m_s, the asymptotic value of ω_R for high z is more away from zero when the charge of scalar perturbation q_s increases. There does not appear the overdamping.

1. Introduction

In black hole physics, quasi normal mode (QNM) is a powerful tool to study the evolution of perturbations in the exterior of black holes [1–4]. The behavior of QNM can be used to identify the black hole existence and disclose dynamical stability of black hole configurations. Besides, QNM can serve as a testing ground of fundamental physics. It is widely believed that QNM can give deeper understandings of the AdS/CFT [4–8], dS/CFT [9] correspondences, loop quantum gravity [10] and also phase transitions of black holes [11] etc.

In this letter we will examine the QNM of the linearly charged dilaton-Lifshitz black brane solutions and try to disclose deep influences of the model parameters on the perturbation wave dynamics and examine the stability of the background configurations.

E-mail addresses: mkzangeneh@shirazu.ac.ir (M. Kord Zangeneh), wang_b@sjtu.edu.cn (B. Wang), asheykhi@shirazu.ac.ir (A. Sheykhi), tangziyu@sjtu.edu.cn (Z.Y. Tang).

Asymptotic Lifshitz black solutions are interesting duals to many condensed matter systems [12]. They are duals to systems with Schrodinger-like scaling symmetries i.e. $t \to \lambda^z t$, $\vec{\mathbf{x}} \to \lambda \vec{\mathbf{x}}$, where z is dynamical critical exponent. Lifshitz spacetime is not a vacuum solution of Einstein gravity with or without cosmological constant and some Lifshitz supporting fields are needed. Different Lifshitz supporting fields have been considered in literatures, such as including higher curvature corrections [13–15], inserting massive [16] and massless [17–22] Abelian gauge fields coupled to dilaton and non-Abelian $SU(2)$ Yang–Mills fields coupled to dilaton [23] etc.

The behavior of neutral scalar perturbations has been extensively studied for Lifshitz solutions in the presence of different Lifshitz supporting fields. In [24,25], scalar and spinorial perturbations around $(2+1)$-dimensional Lifshitz black holes with $z=3$ in the context of New Massive Gravity (which includes higher curvature terms) have been explored and it has been shown that black holes are stable under both of these perturbations. Moreover, it has been shown that higher-dimensional Lifshitz black branes are stable under massive scalar perturbations in the presence of higher curvature corrections [26,27]. In [28], (3 + 1)-dimensional Lifshitz black holes with $z = 0$ and in the presence of higher curvature corrections have been considered. In the case of massive scalar perturbations minimally coupled to curvature, these black holes are unstable, whereas such black holes with massless perturbations conformally coupled to curvature are stable. In the presence of a massive gauge field including Proca term, scalar massive QNMs for uncharged Lifshitz black branes with $z = 2$ have been studied in [29]. It was shown that the QNMs are purely damping (the real part of QNM vanishes) and Lifshitz black branes are always stable. It was further shown that quasi-normal frequencies of scalar massive perturbation around topological Lifshitz black holes with $z = 2$ [30] are always purely imaginary and negative, supporting that these black holes are always stable [31]. In [32], scalar QNMs of 2- and 3-dimensional uncharged Lifshitz black branes with $z = 3$ in the context of New Massive Gravity and 4-dimensional uncharged Lifshitz black branes with $z = 2$ in the presence of massive and massless gauge fields coupled to dilaton have been studied. It has been shown that quasi-normal frequencies are purely negative imaginary, reflecting that these solutions are globally stable. Considering dilaton field and massless gauge fields coupled to dilaton as Lifshitz supporting matter fields, massive QNMs with zero momenta for higher-dimensional Lifshitz black branes have been examined [33] and it has been shown that black branes are stable under these perturbations. In this context, massive and massless QNMs for higher-dimensional Abelian [34] and non-Abelian [35] charged Lifshitz black branes with hyperscaling violation have been studied as well. They were also found stable under scalar perturbations. Massive and massless scalar QNMs around (3 + 1)-dimensional dilaton-Lifshitz black holes/branes in the presence of nonlinear power-law Maxwell field have been explored in [36]. It was shown that these Lifshitz solutions are always stable. It was further found that QNMs for black branes can become overdamping when the dynamical critical exponent z and angular momentum take some special values. In [37], retarded Green functions of the current and momentum operator of a Lifshitz field theory have been investigated and it was shown that there exists a massive QNMs with an effective mass linearly proportional to temperature in non-vanishing momentum case.

Most available studies of QNMs for Lifshitz black solutions are limited to the neutral scalar perturbation. Some extensions to the Fermionic and electromagnetic perturbations for Lifshitz solutions have been reported recently [38,39]. It is of great interest to generalize the discussion to the charged scalar perturbation in the background of Lifshitz solutions. The charged scalar perturbation has been studied extensively in other contexts [40–45]. There has been examples showing that charged scalar perturbation can result in the spacetime instabilities in the Reissner–Nordström–de Sitter configuration and the anti-de Sitter charged black holes in [41,42] and [43–45], respectively. In the AdS space, the charged scalar field can condensate onto the AdS black hole to form a new hairy black hole. In the present letter, we will examine the dynamics of massive charged scalar perturbation in the charged Lifshitz black solution backgrounds. We will disclose the dependence of the QNMs on the model parameters, such as the mass and charge of the scalar perturbation field, the charge of the background spacetime etc. Especially we will examine in the Lifshitz configuration, whether the instability of the background configuration can appear and how the instability will depend on model parameters including the Lifshitz exponent z. For simplicity in our discussion, we will focus on the (3+1)-dimensional Lifshitz spacetime and leave the study on the higher dimensional influence on QNMs for our future work.

The letter is organized as follows. In the next section, we will review solutions of the four-dimensional linearly charged Lifshitz configurations. Then we will write out the wave equations for the charged scalar perturbation and explain the method of the numerical computation we are going to employ. In Section 4, we will report results of QNMs. In the last section we will conclude our results.

2. Review on 4-dimensional linearly charged Lifshitz solutions

In this section we will review the 4-dimensional charged dilaton-Lifshitz solutions. The line element of 4-dimensional Lifshitz black solutions can be written as [17]

$$ds^2 = -\frac{r^{2z}}{l^{2z}} f(r)dt^2 + \frac{l^2}{r^2}\frac{dr^2}{f(r)} + r^2 d\Omega_k^2, \tag{1}$$

where z is dynamical critical exponent and

$$d\Omega_k^2 = \begin{cases} d\theta^2 + \sin^2(\theta)d\phi^2 & k=1 \\ d\theta^2 + d\phi^2 & k=0 \\ d\theta^2 + \sinh^2(\theta)d\phi^2 & k=-1 \end{cases}, \tag{2}$$

represents a 2-dimensional hypersurface with constant curvature $2k$. $f(r)$ in the line element (1) has a solution in the context of Einstein-dilaton gravity in the presence of linear Maxwell electrodynamics and two Lifshitz supporting gauge fields

$$S = -\frac{1}{16\pi}\int_{\mathcal{M}} d^4x\sqrt{-g}\Big[\mathcal{R} - 2(\nabla\Phi)^2 - 2\Lambda + -e^{-2\lambda_1\Phi}F - \sum_{i=2}^{3} e^{-2\lambda_i\Phi}H_i\Big], \tag{3}$$

in which $\mathcal{R}$ is the Ricci scalar on manifold $\mathcal{M}$, Φ is the dilaton field and F and H_i's are the Maxwell invariants of electromagnetic fields $F_{\mu\nu} = \partial_{[\mu}A_{\nu]}$ and $(H_i)_{\mu\nu} = \partial_{[\mu}(B_i)_{\nu]}$, where A_μ and $(B_i)_\mu$'s are the electromagnetic potentials. Λ, λ_1 and λ_i's are constants. The solution for $f(r)$ in Einstein-dilaton gravity governed by the action (3), is [17,18]

$$f(r) = 1 + \frac{kl^2}{z^2r^2} - \frac{m}{r^{z+2}} + \frac{q^2l^{2z}b^{2(z-1)}}{zr^{2(z+1)}}, \tag{4}$$

where m and q are two constants which are, respectively, related to total mass and total charge of the black hole, the dilaton field is

$$\Phi(r) = \sqrt{z-1}\ln\left(\frac{r}{b}\right), \tag{5}$$

in which b is a constant and the gauge potentials are

$$A_t = \frac{qb^{2(z-1)}}{z}\left(\frac{1}{r_+^z} - \frac{1}{r^z}\right),$$

$$(B_2)_t = \frac{q_2 r^{z+2}}{(z+2)\,b^4}, \qquad (B_3)_t = \frac{q_3 r^z}{zb^2}, \tag{6}$$

where r_+ is the largest root of metric function $f(r)$, called event horizon. The constants of the model have been fixed as

$$\lambda_1 = -\sqrt{z-1}, \quad \lambda_2 = \frac{2}{\sqrt{z-1}}, \quad \lambda_3 = \frac{1}{\sqrt{z-1}},$$

$$q_2^2 = \frac{(z-1)(z+2)}{2b^{-4}l^{2z}}, \quad q_3^2 = \frac{kb^2(z-1)}{l^{2(z-1)}z},$$

$$\Lambda = -\frac{(z+1)(z+2)}{2l^2}. \tag{7}$$

It is clear from (4) that $f(r)$ tends to 1 at spatial infinity and therefore the metric (1) is asymptotically Lifshitz. Looking at q_3^2 in (7), we find that the $k=-1$ case where the constant curvature at horizon hypersurface is negative, causes an imaginary charge except for the AdS case with $z=1$ [17,18]. Since we will discuss the Lifshitz solutions with $z>1$, we will consider the cases $k=0$ (black brane) and $k=1$ (black hole) in following studies.

The Hawking temperature can be obtained as [17,18]

$$T=\frac{r_+^{z+1}f'(r_+)}{4\pi l^{z+1}}=\frac{1}{4\pi}\left(\frac{(z+2)m}{l^{z+1}r_+^2}-\frac{2r_+^{z-2}k}{z^2l^{z-1}}-\frac{2q^2(z+1)}{zb^{2(1-z)}l^{1-z}r^{z+2}}\right)=\frac{1}{4\pi}\left(\frac{kr_+^{z-2}}{zl^{z-1}}+\frac{(z+2)r_+^z}{l^{z+1}}-\frac{q^2l^{z-1}b^{2(z-1)}}{r_+^{z+2}}\right),\quad(8)$$

where m has been inserted by using the fact that $f(r_+)=0$. As the charge of black hole approaches the extreme value

$$q_{\text{ext}}^2=\frac{kr_+^{2z}}{zl^{2z-2}b^{2(z-1)}}+\frac{(z+2)r_+^{2z+2}}{l^{2z}b^{2(z-1)}},\quad(9)$$

the Hawking temperature tends to zero.

In the next section, we will consider a massive charged scalar perturbation around our Lifshitz solutions and discuss the numerical method to obtain the corresponding QNM frequencies.

3. Wave equations for charged scalar perturbation around Lifshitz solutions

Here, we intend to consider a massive charged scalar perturbation around our Lifshitz solutions. The dynamical wave equation for this scalar field perturbation is

$$D^\nu D_\nu\Psi=m_s^2\Psi,\quad(10)$$

where $D^\nu=\nabla^\nu-iq_sA^\nu$. The wave function can be separated into

$$\Psi=e^{-i\omega t}R(r)Y(\theta,\phi),\quad(11)$$

and the differential equation (10) can be written into angular part and radial part by using the metric (1)

$$\nabla^2Y(\theta,\phi)=-QY(\theta,\phi),\quad(12)$$

$$fR''+\left(f'+\frac{(3+z)f}{r}\right)R'+\left(\frac{\omega+q_sA_t}{r^{z+1}}\right)^2\frac{R}{f}-\left(m_s^2+\frac{Q}{r^2}\right)\frac{R}{r^2}=0,\quad(13)$$

where $Q=\ell(\ell+1)$ and $\ell=0,1,2,\cdots$. Hereafter we will fix the spacetime radius l to 1. Defining

$$R(r)=\frac{K(r)}{r},\quad(14)$$

and the tortoise coordinate

$$\frac{dr_*}{dr}=\left[r^{z+1}f(r)\right]^{-1},\quad(15)$$

we can rewrite (13) into the Schrödinger form

$$\frac{d^2K(r_*)}{dr_*^2}+\left[(\omega+q_sA_t(r))^2-V(r)\right]K(r_*)=0,\quad(16)$$

where the effective potential

$$V(r)=r^{2z}f(r)\left[rf'(r)+\frac{Q}{r^2}+m_s^2+(z+1)^2\right].\quad(17)$$

At spatial infinity $f(r)\to1$ and $f'(r)\to0$, it is easy to see that $V(r)\to\infty$ (note that $z\geq1$) and therefore we need to impose the Dirichlet boundary condition i.e. $R(r)\to0$ at the boundary of spatial infinity.

We are going to employ the improved asymptotic iteration method (AIM) [46] to solve (13) numerically. In order to do so, we rewrite (13) in terms of $u=1-r_+/r$

$$\frac{r_+^{z+1}f(u)}{(1-u)^{z-1}}R''(u)+\frac{r_+^{z+1}R'(u)}{(1-u)^z}\left[(1-u)f'(u)+(z+1)f(u)\right]+\frac{R(u)(1-u)^{z-1}}{r_+^{z-1}f(u)}(q_sA_t(u)+\omega)^2-\frac{R(u)r_+^{z+1}}{(1-u)^{z+1}}\left(m_s^2+\frac{Q(1-u)^2}{r_+^2}\right)=0.\quad(18)$$

Considering the asymptotic behaviors of $R(u)$ satisfying (18), at horizon $(u=0)$ we have $f(0)\approx uf'(0)$ and $A_t(0)=0$, so that (18) reduces to

$$R''(u)+\frac{R'(u)}{u}+\frac{\omega^2R(u)}{u^2r_+^{2z}f'(0)^2}=0,\quad(19)$$

which has the solution

$$R(u\to0)\sim C_1u^{-\xi}+C_2u^{\xi},\quad\xi=\frac{i\omega}{r_+^zf'(0)},\quad(20)$$

in which we have to set $C_2=0$ to respect the ingoing condition at horizon.

Near infinity $(u=1)$, we have

$$R''(u)+\frac{(z+1)R'(u)}{1-u}-\frac{m_s^2R(u)}{(1-u)^2}=0,\quad(21)$$

as asymptotic form of (18) (note that $z\geq1$). The above differential equation has the solution

$$R(u\to1)\sim D_1(1-u)^{\frac{1}{2}(z+2+\Pi)}+D_2(1-u)^{\frac{1}{2}(z+2-\Pi)},\quad(22)$$

where

$$\Pi=\sqrt{(z+2)^2+4m_s^2}.\quad(23)$$

In this case, we should set $D_2=0$ in order to satisfy Dirichlet boundary condition $R(r\to\infty)\to0$.

Now, the desired ansatz for (18) can be defined as

$$R(u)=u^{-\xi}(1-u)^{\frac{1}{2}(z+2+\Pi)}\chi(u).\quad(24)$$

Putting (24) into (18), we have

$$\chi''=\lambda_0(u)\chi'+s_0(u)\chi,\quad(25)$$

where

$$\lambda_0(u)=\frac{2i\omega}{ur_+^zf'(0)}-\frac{f'(u)}{f(u)}+\frac{1+\Pi}{1-u},\quad(26)$$

and

$$s_0(u)=\frac{r_+^{-2(z+1)}f'(0)^{-2}}{2(u-1)^2u^2f(u)^2}\Big[-2f'(0)^2r_+^2u^2(1-u)^{2z}(q_sA_t(u)+\omega)^2+f'(0)uf(u)r_+^z\Big(2f'(0)ur_+^z\left(m_s^2r_+^2+Q(u-1)^2\right)-r_+^2(u-1)f'(u)\left(f'(0)u(\Pi+z+2)r_+^z-2i(u-1)\omega\right)\Big)+2r_+^2f(u)^2\Big(-f'(0)^2m_s^2u^2r_+^{2z}+if'(0)(u-1)\times\omega r_+^z(u\Pi+1)+(u-1)^2\omega^2\Big)\Big].\quad(27)$$

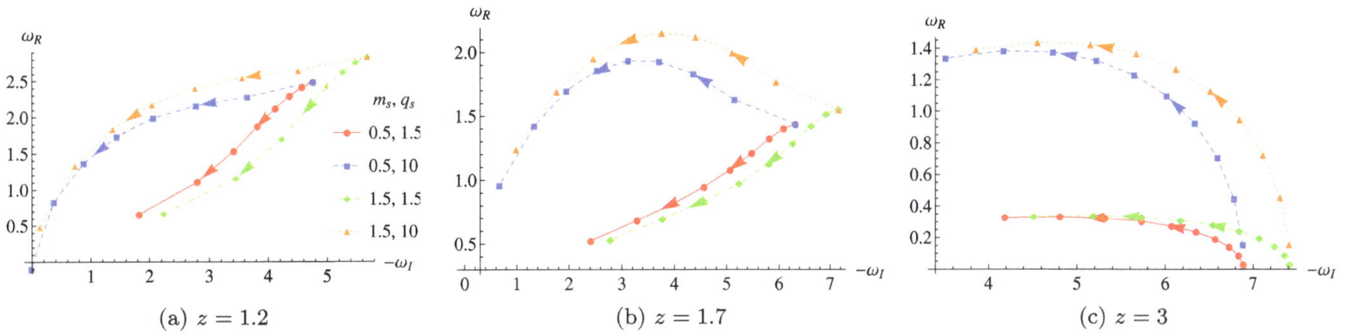

Fig. 1. The behaviors of real and imaginary parts of QNFs for different z's vs q where $k = 0$ and $m = 3$. The arrows show the direction of the increase of black branes charge q from 0 to extreme value q_{ext}. The curves with the same colors correspond to same values of m_s and q_s. (For interpretation of the references to color in this figure legend, the reader is referred to the web version of this article.)

(25) can be solved numerically by employing the improved AIM. In the following, we will set $l = b = 1$ and $Q = 0$. In [33–36], QNMs corresponding to neutral massive scalar perturbations around dilaton-Lifshitz solutions have been obtained by the improved AIM. Here, we will calculate the charged scalar perturbations and examine the influences of different model parameters on the real and imaginary parts of quasinormal frequencies around Lifshitz black solutions.

4. Numerical results

In this section, we report our numerical results of the QNMs of the charged scalar perturbation around Lifshitz black brane solutions.

Let us start with the imaginary part of QNMs shown in Fig. 1. Fixing the mass of the scalar field m_s and increasing the charge q of the background configuration, we observe that the absolute value of the imaginary part of quasi-normal frequency $|\omega_I|$ decreases. This behavior holds for all chosen values of the charge of the scalar field q_s. This property is shown in Fig. 1. Furthermore we find that fixing the charge of the black hole q and the mass of the scalar field m_s, the bigger value of q_s leads to the smaller $|\omega_I|$. But when $q = 0$, the q_s influence on the imaginary part of the quasi-normal frequency disappears. The increase of the mass of the scalar field can enhance $|\omega_I|$ when the charges of the scalar field q_s and the background q are fixed.

The decrease of the absolute value of the imaginary part of the quasinormal frequency $|\omega_I|$ shows that the perturbation outside black hole will persist longer period of time to decay completely. From holographic point of view, it means that it takes longer time for the dual system to go back to equilibrium [6]. When the charge of the black holes is increased, the $|\omega_I|$ is even smaller which shows that the perturbation can last even longer to finally vanish outside the black hole. This result is consistent with the Reissner–Nordström anti-de Sitter (RN–AdS) black hole case where it was found in [7] that the absolute value of the imaginary part of quasi-normal frequency $|\omega_I|$ decreases when the black hole charge approaches to the extreme value. In RN–AdS black hole background when the black hole becomes extreme, $|\omega_I|$ tends to zero, which makes the extreme RN–AdS black hole marginally unstable. But in the dilaton-Lifshitz black hole background, the imaginary part of the quasinormal frequency will not tend to zero except for some big (small) enough value of the scalar field charge (mass) of the perturbation field. This exhibits that the compared with the usual RN–AdS black hole, the stability of the dilaton-Lifshitz black hole is more easily to be protected. It is important to stress that the black hole stability found here is only stability against scalar perturbations in a certain region of parameters.

Now, we turn to discuss the behavior of the real part of the quasi-normal frequency (ω_R) exhibited in Fig. 1. When the Lifshitz exponent z is close to the unity, we find that for fixed m_s and q_s, ω_R monotonically decreases with the increase of the charge q in the background configuration. Besides, fixing m_s and q, we observe that ω_R is smaller when q_s becomes bigger, as shown in Fig. 1(a). It is worth mentioning that the smaller value of the real part of the quasinormal frequency ω_R shows that the scalar perturbation have less energy. When $q = 0$, the q_s influence disappears as well for the real part of the frequency. When z deviates a bit away from the unity, for small q_s, ω_R still decreases with the increase of black hole charge q. But when the scalar field is more charged with bigger q_s, ω_R behaves differently and does not monotonically decrease with the increase of q. Instead, there is a barrier in the value of ω_R so that it increases when q increases from zero, but then decreases when q is over a critical value (Fig. 1(b)). When z is much bigger, from Fig. 1(c), we see that ω_R increases from zero and flattens later with the increase of q from zero when q_s is small. For bigger q_s, the real part of the frequency will increase from zero to a maximum value and then slowly decreases when black hole charge approaches to the extreme value. Comparing with the small z case, here for zero black hole charge, we observe a purely damping mode for charged scalar perturbation. The oscillation adds when q is nonzero. When the mass of the scalar field is bigger, the difference caused by the small and large q_s will be enlarged in ω_R when the black hole charge is small. When the black hole charge is big enough, the influence of the mass of scalar field on ω_R fades away.

In Fig. 1(a), we can see that for some choices of parameters, the imaginary frequency can approach zero and can even jump to be above zero which shows that the stability of the background configuration can be destroyed. This phenomenon happens when the charge of the background black brane solution is high enough, which corresponds to the low temperature. The instability is consistent with the description of the charged scalar field condensation to make the original Lifshitz black brane to be a new hairy configuration. In usual RN–AdS black hole, it was found in [8] that when the black hole becomes extreme, the imaginary quasinormal frequency tends to zero, indicating that the extreme RN–AdS black hole is marginally unstable, since the perturbation will not die out and always persist outside the black hole background. Then if there comes another stronger wave of perturbation, the original black hole background is more likely to be destroyed. This was confirmed in [7]. Here in the dilaton-Lifshitz black hole background, we observed that when the scalar field is highly charged, the imaginary quasinormal frequency can even change to be positive. This shows that instead of the decay of the perturbation outside the dilaton-Lifshitz black hole background, the highly charged scalar perturbation can even blow up outside the black hole and de-

Table 1
The threshold values of m_s and q_s for chosen z to keep the stability of the black brane with extreme charge value q and $m = 3$.

$m_s = 0.6$					
z	1	1.2	1.6	2	3
q_s	5.3	7.3	8.8	12.3	22

$q_s = 8.8$					
z	1	1.2	1.6	1.7	1.8
m_s^2	6.25	2.25	0.36	−3.42	-

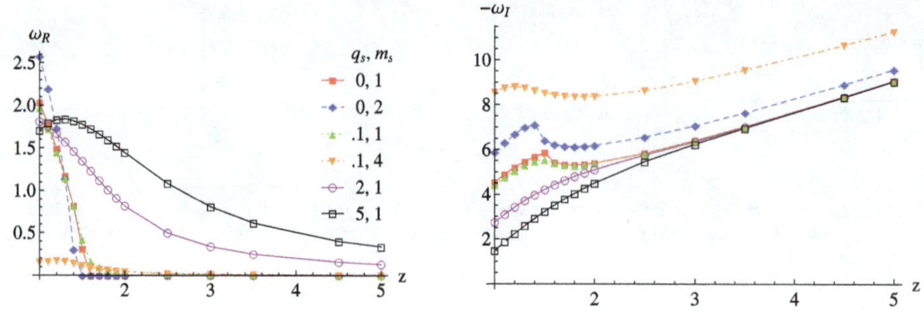

(a) ω_R vs z

(b) -ω_I vs z

Fig. 2. The behaviors of real and imaginary parts of QNMs vs z where $k = 0$, $m = 3$ and $q = 1$.

stroy the original black hole spacetime structure. This effect to make the spacetime unstable brought in by the highly charged perturbation field was also observed in the stability analysis in Reissner–Nordström black hole in de Sitter background [41].

As one can see from Fig. 1(a), when the Lifshitz exponent z is close to unity, and decreasing q_s or increasing m_s can keep the imaginary frequency to be negative to ensure the stability. When z is more away from unity (Figs. 1(b) and 1(a)) for given values of m_s and q_s, there is no possibility to destroy the original background configuration, since even when q takes the extreme value, the imaginary frequency is negative. For chosen value of the Lifshitz exponent z, we can find threshold values for the mass and charge of scalar perturbation field to ensure the stability of the original Lifshitz black brane when its charge q is nearly extreme. These threshold values are listed in Table 1. For fixing m_s, when q_s is below the corresponding threshold values for the chosen z, the black brane can keep to be stable. When we fix q_s, the mass of the scalar field is above the corresponding threshold value for the chosen z can ensure the stability. m_s^2 has a lower bound $-(z+2)^2/4$ according to Eq. (23). Thus m_s cannot be reduced infinitely so that in Table 1 at some z the decrease of m_s^2 stops.

It is clear from above discussions that increasing z makes the stability more easily to be protected. This result is physically reasonable from holographic point of view since it is in agreement with the behavior of Lifshitz superconductors where it was found that increasing dynamical critical exponent z makes superconductor more difficult to be formed [47].

In Fig. 2, the behaviors of real and imaginary parts of quasi-normal frequencies with the change of Lifshitz exponent z have been illustrated. In [33], it was shown that for $(3+1)$-dimensional uncharged dilaton-Lifshitz black branes, independent of the value of m_s, QNMs corresponding to neutral scalar perturbations start overdamping (with $\omega_R = 0$) at $z = 2$. Here, we find that for charged black branes, the point at which the real part of quasi-normal frequency of neutral scalar perturbation starts to vanish is no longer at $z = 2$, but depends on the value of m_s (see red and blue lines in Fig. 2(a)). For fixed m_s, when q_s is small, ω_R reduces and finally flattens when the Lifshitz exponent z becomes big. The value of ω_R for high z can approach to zero (see red and green lines in Fig. 2(a)). For bigger q_s, ω_R finally will be above zero (see purple line in Fig. 2(a)).

Now, we discuss the imaginary part of quasi-normal frequency as shown in Fig. 2. For small q_s, the absolute value of the imaginary part is always bigger than the corresponding value with bigger q_s. With the increase of the Lifshitz exponent z, the absolute value of ω_I increases. For large z, $|\omega_I|$ has little dependence on q_s and it is mainly influenced by the value of m_s.

We have mainly reported QNM behaviors of the charged scalar perturbation in the Lifshitz black brane background with $k = 0$. For the Lifshitz black hole case with $k = 1$ we have found similar results. We do not repeat explaining these similar results here for Lifshitz black holes.

5. Summary and conclusion

One of the important subject in black hole physics is to investigate the resonances for the scattering of incoming waves by black holes. Quasi-normal modes (QNMs) of a black hole spacetime are indeed the proper solutions of the perturbation equations. It is a general belief that QNMs carry unique footprints to directly identify the black hole existence.

Most available studies on QNMs corresponding to Lifshitz solutions are restricted to neutral scalar perturbations. In this letter, we have extended the study to the charged scalar perturbation in the background of $(3+1)$-dimensional charged dilaton Lifshitz branes/holes. Asymptotic Lifshitz solutions are interesting duals to many condensed matter systems.

We have used the improved asymptotic iteration method (AIM) to calculate the charged scalar perturbations numerically and examine the influences of different model parameters on the imaginary and real parts of quasi-normal frequencies of the charged scalar field perturbations around Lifshitz black solutions. In contrast to the case of neutral scalar perturbations, we found that it is possible to destroy the Lifshitz configuration. To be more clear, we observed that for suitable choices of model parameters, imaginary part of QNM frequencies can be positive near extreme charge value of Lifshitz configuration where temperature is low. For a chosen z, there are threshold values for mass and charge of scalar perturbation to guarantee the stability of the Lifshitz black configuration. Fixing q_s and taking the mass of the scalar field above the corresponding threshold value, or fixing m_s and keeping the charge of the scalar field below the threshold value, the black brane/hole stability can be protected. It is remarkable to note that here by stability of black hole, we mean only the stability against scalar perturbations in a certain region of parameters. In terms of z, it was shown that when it is more away from unity, it is more difficult for the Lifshitz background to become unstable. This result is consistent with the observation that Lifshitz superconductors are more difficult to be formed for greater values of Lifshitz exponent z [47]. We have also observed rich dependence of the quasinormal frequencies on the mass of the scalar field m_s, the charge q of the background configuration and the charge of the scalar field q_s.

We have also investigated the behaviors of the real and imaginary parts of quasi-normal frequencies in terms of the Lifshitz exponent z. We found out that for charged black branes, the point at which neutral QNMs start overdamping is no longer at $z=2$ as what was claimed in [33] for uncharged Lifshitz black branes, but it has dependence on the value of m_s. For fixed m_s, when q_s is small, ω_R reduces and finally flattens when the Lifshitz exponent z becomes big. The value of ω_R for high z can finally approach to zero, whereas for bigger q_s, this possibility does not exist and ω_R will be finally above zero. The rich dependence of the imaginary part $|\omega_I|$ on the Lifshitz exponent z, the scalar field charge q_s, mass m_s has also been investigated carefully. We have found that the QNMs properties of the Lifshitz black brane background with $k=0$ hold as well for the Lifshitz black hole case with $k=1$. To be concise, we do not repeat the discussion for the Lifshitz black hole here.

Finally, it is worth mentioning that in this letter we have only considered the gauge field as the linear Maxwell field. It is interesting to generalize this study to other nonlinear gauge fields such as power-law Maxwell, Born–Infeld, exponential and logarithmic nonlinear gauge fields in the background of Lifshitz spacetime. Besides, we studied the $(3+1)$-dimensional charged dilaton Lifshitz black solutions. It is also interesting if one could extend the study to higher dimensional Lifshitz spacetime. In addition, we investigated the scalar perturbations. It is worthwhile to extend this investigation to the vector and tensor perturbations in this context. These issues are now under investigation and the results will appear in our future works.

Acknowledgements

We thank the anonymous referee for constructive comments which helped us improve the letter significantly. MKZ is grateful to C. Y. Zhang for helpful discussions. MKZ and AS thank the research council of Shiraz University. The work of BW was partially supported by NNSF of China. This work has been supported financially by Research Institute for Astronomy and Astrophysics of Maragha (RIAAM).

References

[1] K.D. Kokkotas, B.G. Schmidt, Quasi-normal modes of stars and black holes, Living Rev. Relativ. 2 (1999) 2, arXiv:gr-qc/9909058.

[2] H.P. Nollert, Quasinormal modes: the characteristic sound of black holes and neutron stars, Class. Quantum Gravity 16 (1999) R159.

[3] R.A. Konoplya, A. Zhidenko, Quasinormal modes of black holes: from astrophysics to string theory, Rev. Mod. Phys. 83 (2011) 793, arXiv:1102.4014.

[4] B. Wang, Perturbations around black holes, Braz. J. Phys. 35 (2005) 1029, arXiv:gr-qc/0511133.

[5] B. Wang, C.Y. Lin, E. Abdalla, Quasinormal modes of Reissner–Nordström anti-de Sitter black holes, Phys. Lett. B 481 (2000) 79, arXiv:hep-th/0003295;
B. Wang, C. Molina, E. Abdalla, Evolution of a massless scalar field in Reissner–Nordström anti-de Sitter spacetimes, Phys. Rev. D 63 (2001) 084001, arXiv:hep-th/0005143;
J.M. Zhu, B. Wang, E. Abdalla, Object picture of quasinormal ringing on the background of small Schwarzschild anti-de Sitter black holes, Phys. Rev. D 63 (2001) 124004, arXiv:hep-th/0101133;
V. Cardoso, J.P.S. Lemos, Scalar, electromagnetic, and Weyl perturbations of BTZ black holes: quasinormal modes, Phys. Rev. D 63 (2001) 124015, arXiv:gr-qc/0101052;
V. Cardoso, J.P.S. Lemos, Quasinormal modes of Schwarzschild-anti-de Sitter black holes: electromagnetic and gravitational perturbations, Phys. Rev. D 64 (2001) 084017, arXiv:gr-qc/0105103;
V. Cardoso, J.P.S. Lemos, Quasi-normal modes of toroidal, cylindrical and planar black holes in anti-de Sitter spacetimes: scalar, electromagnetic and gravitational perturbations, Class. Quantum Gravity 18 (2001) 5257, arXiv:gr-qc/0107098;
E. Winstanley, Classical super-radiance in Kerr–Newman–anti-de Sitter black holes, Phys. Rev. D 64 (2001) 104010, arXiv:gr-qc/0106032;
J. Crisstomo, S. Lepe, J. Saavedra, Quasinormal modes of the extremal BTZ black hole, Class. Quantum Gravity 21 (2004) 2801, arXiv:hep-th/0402048;
S. Lepe, F. Mendez, J. Saavedra, L. Vergara, Fermions scattering in a three-dimensional extreme black-hole background, Class. Quantum Gravity 20 (2003) 2417, arXiv:hep-th/0302035;
D. Birmingham, I. Sachs, S.N. Solodukhin, Conformal field theory interpretation of black hole quasinormal modes, Phys. Rev. Lett. 88 (2002) 151301, arXiv:hep-th/0112055;
D. Birmingham, Choptuik scaling and quasinormal modes in the anti-de Sitter space/conformal-field theory correspondence, Phys. Rev. D 64 (2001) 064024, arXiv:hep-th/0101194;
B. Wang, E. Abdalla, R.B. Mann, Scalar wave propagation in topological black hole backgrounds, Phys. Rev. D 65 (2002) 084006, arXiv:hep-th/0107243;
J.S.F. Chan, R.B. Mann, Scalar wave falloff in topological black hole backgrounds, Phys. Rev. D 59 (1999) 064025;
S. Musiri, G. Siopsis, Asymptotic form of quasi-normal modes of large AdS black holes, Phys. Lett. B 576 (2003) 309, arXiv:hep-th/0308196;
R. Aros, C. Martinez, R. Troncoso, J. Zanelli, Quasinormal modes for massless topological black holes, Phys. Rev. D 67 (2003) 044014, arXiv:hep-th/0211024;
A. Nunez, A.O. Starinets, AdS/CFT correspondence, quasinormal modes, and thermal correlators in $N=4$ supersymmetric Yang–Mills theory, Phys. Rev. D 67 (2003) 124013, arXiv:hep-th/0302026.

[6] G.T. Horowitz, V.E. Hubeny, Quasinormal modes of AdS black holes and the approach to thermal equilibrium, Phys. Rev. D 62 (2000) 024027, arXiv:hep-th/9909056.

[7] B. Wang, C.Y. Lin, C. Molina, Quasinormal behavior of massless scalar field perturbation in Reissner–Nordström anti-de Sitter spacetimes, Phys. Rev. D 70 (2004) 064025, arXiv:hep-th/0407024.

[8] E. Berti, K.D. Kokkotas, Quasinormal modes of Reissner–Nordström–anti-de Sitter black holes: scalar, electromagnetic, and gravitational perturbations, Phys. Rev. D 67 (2003) 064020, arXiv:gr-qc/0301052.

[9] E. Abdalla, B. Wang, A. Lima-Santos, W.G. Qiu, Support of dS/CFT correspondence from perturbations of three-dimensional spacetime, Phys. Lett. B 538 (2002) 435, arXiv:hep-th/0204030;
E. Abdalla, K.H. CastelloBranco, A. Lima-Santos, Support of dS/CFT correspondence from space-time perturbations, Phys. Rev. D 66 (2002) 104018, arXiv:hep-th/0208065.

[10] S. Hod, Bohr's correspondence principle and the area spectrum of quantum black holes, Phys. Rev. Lett.81 (1998) 4293, arXiv:gr-qc/9812002;
A. Corichi, Quasinormal modes, black hole entropy, and quantum geometry, Phys. Rev. D 67 (2003) 087502, arXiv:gr-qc/0212126;
L. Motl, An analytical computation of asymptotic Schwarzschild quasinormal frequencies, Adv. Theor. Math. Phys. 6 (2003) 1135, arXiv:gr-qc/0212096;
L. Motl, A. Neitzke, Asymptotic black hole quasinormal frequencies, Adv. Theor. Math. Phys. 7 (2003) 307, arXiv:hep-th/0301173;
A. Maassen van den Brink, WKB analysis of the Regge–Wheeler equation down in the frequency plane, J. Math. Phys. 45 (2004) 327, arXiv:gr-qc/0303095;
O. Dreyer, Quasinormal modes, the area spectrum, and black hole entropy, Phys. Rev. Lett. 90 (2003) 08130, arXiv:gr-qc/0211076;
G. Kunstatter, d-dimensional black hole entropy spectrum from quasinormal modes, Phys. Rev. Lett. 90 (2003) 161301, arXiv:gr-qc/0212014;
N. Andersson, C.J. Howls, The asymptotic quasinormal mode spectrum of non-rotating black holes, Class. Quantum Gravity 21 (2004) 1623, arXiv:gr-qc/0307020;
V. Cardoso, J. Natario, R. Schiappa, Asymptotic quasinormal frequencies for black holes in nonasymptotically flat space-times, J. Math. Phys. 45 (2004) 4698, arXiv:hep-th/0403132;
J. Natario, R. Schiappa, On the classification of asymptotic quasinormal frequencies for d-dimensional black holes and quantum gravity, Adv. Theor. Math. Phys. 8 (2004) 1001, arXiv:hep-th/0411267;
V. Cardoso, J.P.S. Lemos, Quasinormal modes of the near extremal Schwarzschild–de Sitter black hole, Phys. Rev. D 67 (2003) 084020, arXiv:gr-qc/0301078;
K.H.C. CastelloBranco, E. Abdalla, Analytic determination of the asymptotic quasi-normal mode spectrum of small Schwarzschild–de Sitter black holes, arXiv:gr-qc/0309090.

[11] G. Koutsoumbas, S. Musiri, E. Papantonopoulos, G. Siopsis, Quasi-normal modes of electromagnetic perturbations of four-dimensional topological black holes with scalar hair, J. High Energy Phys. 0610 (2006) 006, arXiv:hep-th/0606096;
X.P. Rao, B. Wang, G.H. Yang, Quasinormal modes and phase transition of black holes, Phys. Lett. B 649 (2007) 472, arXiv:0712.0645;
R.G. Cai, Z.Y. Nie, B. Wang, H.Q. Zhang, Quasinormal modes of charged fermions and phase transition of black holes, arXiv:1005.1233;

Y. Liu, D.C. Zou, B. Wang, Signature of the Van der Waals like small-large charged AdS black hole phase transition in quasinormal modes, J. High Energy Phys. 1409 (2014) 179, arXiv:1405.2644;
J. Shen, B. Wang, C.Y. Lin, R.G. Cai, R.K. Su, The phase transition and the quasinormal modes of black holes, J. High Energy Phys. 0707 (2007) 037, arXiv:hep-th/0703102.

[12] S. Kachru, X. Liu, M. Mulligan, Gravity duals of Lifshitz-like fixed points, Phys. Rev. D 78 (2008) 106005, arXiv:0808.1725.

[13] E. Ayón-Beato, A. Garbarz, G. Giribet, M. Hassaïne, Lifshitz black hole in three dimensions, Phys. Rev. D 80 (2009) 104029, arXiv:0909.1347.

[14] E. Ayón-Beato, A. Garbarz, G. Giribet, M. Hassaïne, Analytic Lifshitz black holes in higher dimensions, J. High Energy Phys. 1004 (2010) 030, arXiv:1001.2361.

[15] H. Lu, Y. Pang, C.N. Pope, J.F. Vazquez-Poritz, AdS and Lifshitz black holes in conformal and Einstein–Weyl gravities, Phys. Rev. D 86 (2012) 044011, arXiv:1204.1062.

[16] K. Balasubramanian, J. McGreevy, An analytic Lifshitz black hole, Phys. Rev. D 80 (2009) 104039, arXiv:0909.0263.

[17] J. Tarrio, S. Vandoren, Black holes and black branes in Lifshitz spacetimes, J. High Energy Phys. 1109 (2011) 017, arXiv:1105.6335.

[18] M. Kord Zangeneh, A. Sheykhi, M.H. Dehghani, Thermodynamics of topological nonlinear charged Lifshitz black holes, Phys. Rev. D 92 (2015) 024050, arXiv:1506.01784.

[19] M. Kord Zangeneh, M.H. Dehghani, A. Sheykhi, Thermodynamics of Gauss–Bonnet–dilaton Lifshitz black branes, Phys. Rev. D 92 (2015) 064023, arXiv: 1506.07068.

[20] M. Kord Zangeneh, A. Dehyadegari, A. Sheykhi, M.H. Dehghani, Thermodynamics and gauge/gravity duality for Lifshitz black holes in the presence of exponential electrodynamics, J. High Energy Phys. 1603 (2016) 037, arXiv: 1601.04732.

[21] A. Dehyadegaria, A. Sheykhi, M. Kord Zangeneh, Holographic conductivity for logarithmic charged dilaton-Lifshitz solutions, Phys. Lett. B 758 (2016) 226, arXiv:1602.08476.

[22] M. Kord Zangeneh, A. Dehyadegari, M.R. Mehdizadeh, B. Wang, A. Sheykhi, Thermodynamics, phase transitions and Ruppeiner geometry for Einstein-dilaton Lifshitz black holes in the presence of Maxwell and Born–Infeld electrodynamics, arXiv:1610.06352.

[23] X.H. Feng, W.J. Geng, Non-abelian (hyperscaling violating) Lifshitz black holes in general dimensions, Phys. Lett. B 747 (2015) 395, arXiv:1502.00863.

[24] B. Cuadros-Melgar, J. de Oliveira, C.E. Pellicer, Stability analysis and area spectrum of 3-dimensional Lifshitz black holes, Phys. Rev. D 85 (2012) 024014, arXiv:1110.4856.

[25] B. Cuadros-Melgar, J. de Oliveira, C.E. Pellicer, Quasinormal modes and thermodynamical aspects of the 3d Lifshitz black hole, J. Phys. Conf. Ser. 453 (2013) 012025, arXiv:1302.6185.

[26] A. Giacomini, G. Giribet, M. Leston, J. Oliva, S. Ray, Scalar field perturbations in asymptotically Lifshitz black holes, Phys. Rev. D 85 (2012) 124001, arXiv: 1203.0582.

[27] E. Abdalla, O.P.F. Piedra, F.S. Nuñez, J. de Oliveira, Scalar field propagation in higher dimensional black holes at a Lifshitz point, Phys. Rev. D 88 (2013) 064035, arXiv:1211.3390.

[28] M. Catalan, E. Cisternas, P.A. Gonzalez, Y. Vasquez, Quasinormal modes and greybody factors of a four-dimensional Lifshitz black hole with $z = 0$, Astrophys. Space Sci. 361 (2016) 1, arXiv:1404.3172.

[29] P.A. Gonzalez, J. Saavedra, Y. Vasquez, Quasinormal modes and stability analysis for four-dimensional Lifshitz black hole, Int. J. Mod. Phys. D 21 (2012) 1250054, arXiv:1201.4521.

[30] R.B. Mann, Lifshitz topological black holes, J. High Energy Phys. 0906 (2009) 075, arXiv:0905.1136.

[31] P.A. Gonzalez, F. Moncada, Y. Vasquez, Quasinormal modes, stability analysis and absorption cross section for 4-dimensional topological Lifshitz black hole, Eur. Phys. J. C 72 (2012) 1, arXiv:1205.0582.

[32] Y.S. Myung, T. Moon, Quasinormal frequencies and thermodynamic quantities for the Lifshitz black holes, Phys. Rev. D 86 (2012) 024006, arXiv:1204.2116.

[33] W. Sybesma, S. Vandoren, Lifshitz quasinormal modes and relaxation from holography, J. High Energy Phys. 1505 (2015) 021, arXiv:1503.07457.

[34] P.A. González, Y. Vásquez, Scalar perturbations of nonlinear charged Lifshitz black branes with hyperscaling violation, Astrophys. Space Sci. 361 (2016) 224, arXiv:1509.00802.

[35] R. Becar, P.A. González, Y. Vásquez, Quasinormal modes of non-Abelian hyperscaling violating Lifshitz black holes, arXiv:1510.04605.

[36] R. Becar, P.A. González, Y. Vásquez, Quasinormal modes of four-dimensional topological nonlinear charged Lifshitz black holes, Eur. Phys. J. C 76 (2016) 78, arXiv:1510.06012.

[37] C. Park, A massive quasi-normal mode in the holographic Lifshitz theory, Phys. Rev. D 89 (2014) 066003, arXiv:1312.0826.

[38] M. Catalan, E. Cisternas, P.A. Gonzalez, Y. Vasquez, Dirac quasinormal modes for a 4-dimensional Lifshitz black hole, Eur. Phys. J. C 74 (2014) 1, arXiv:1312.6451.

[39] A. Lopez-Ortega, Electromagnetic quasinormal modes of an asymptotically Lifshitz black hole, Gen. Relativ. Gravit. 46 (2014) 1, arXiv:1406.0126.

[40] R.A. Konoplya, Decay of a charged scalar field around a black hole: quasinormal modes of RN, RNAdS, and dilaton black holes, Phys. Rev. D 66 (2002) 084007, arXiv:gr-qc/0207028;
R.A. Konoplya, A. Zhidenko, Decay of a charged scalar and Dirac fields in the Kerr–Newman–de Sitter background, Phys. Rev. D 76 (2007) 084018, Erratum: Phys. Rev. D 90 (2014) 029901, arXiv:0707.1890;
R.A. Konoplya, A. Zhidenko, Massive charged scalar field in the Kerr–Newman background I: quasinormal modes, latetime tails and stability, Phys. Rev. D 88 (2013) 024054, arXiv:1307.1812.

[41] Z. Zhu, S.J. Zhang, C.E. Pellicer, B. Wang, E. Abdalla, Stability of Reissner–Nordström black hole in de Sitter background under charged scalar perturbation, Phys. Rev. D 90 (2014) 044042, arXiv:1405.4931.

[42] R.A. Konoplya, A. Zhidenko, Charged scalar field instability between the event and cosmological horizons, Phys. Rev. D 90 (2014) 064048, arXiv:1406.0019.

[43] X. He, B. Wang, R.G. Cai, C.Y. Lin, Signature of the black hole phase transition in quasinormal modes, Phys. Lett. B 688 (2010) 230, arXiv:1002.2679.

[44] E. Abdalla, C.E. Pellicer, J. de Oliveira, A.B. Pavan, Phase transitions and regions of stability in Reissner–Nordström holographic superconductors, Phys. Rev. D 82 (2010) 124033, arXiv:1010.2806.

[45] Y. Liu, B. Wang, Perturbations around the AdS Born–Infeld black holes, Phys. Rev. D 85 (2012) 046011, arXiv:1111.6729.

[46] H.T. Cho, A.S. Cornell, J. Doukas, W. Naylor, Black hole quasinormal modes using the asymptotic iteration method, Class. Quantum Gravity 27 (2010) 155004, arXiv:0912.2740.

[47] J.W. Lu, Y.B. Wu, P. Qian, Y.Y. Zhao, X. Zhang, N. Zhang, Lifshitz scaling effects on holographic superconductors, Nucl. Phys. B 887 (2014) 112, arXiv: 1311.2699.

Compact Chern–Simons vortices

D. Bazeia [a,*], L. Losano [a], M.A. Marques [a], R. Menezes [b,c]

[a] Departamento de Física, Universidade Federal da Paraíba, 58051-970 João Pessoa, PB, Brazil
[b] Departamento de Ciências Exatas, Universidade Federal da Paraíba, 58297-000 Rio Tinto, PB, Brazil
[c] Departamento de Física, Universidade Federal de Campina Grande, 58109-970, Campina Grande, PB, Brazil

ARTICLE INFO

Editor: M. Cvetič

ABSTRACT

We introduce and investigate new models of the Chern–Simons type in the three-dimensional spacetime, focusing on the existence of compact vortices. The models are controlled by potentials driven by a single real parameter that can be used to change the profile of the vortex solutions as they approach their boundary values. One of the models unveils an interesting new behavior, the tendency to make the vortex compact, as the parameter increases to larger and larger values. We also investigate the behavior of the energy density and calculate the total energy numerically.

1. Introduction

Vortices are topological structures that appear in $(2, 1)$ spacetime dimensions, under the action of a complex scalar field coupled to a gauge field which evolves obeying the local $U(1)$ symmetry. The study of relativistic vortices started in Refs. [1,2], by considering a model governed by the Maxwell dynamics. However, one can also investigate vortices with the dynamics controlled by the Chern–Simons term, whose general properties where shown in Refs. [3–5]. The first studies considering Chern–Simons vortices appeared in Refs. [6–8]; for more on this see, e.g., Ref. [9].

The class of vortices that appears in models driven by the Chern–Simons term presents some interesting features, beyond the fact that they engender a magnetic flux that is quantized by its intrinsic vorticity. A peculiar behavior which deserves to be pointed out is the existence of an electric field and the fact that the magnetic field is related to the electric charge density associated to the Chern–Simons vortex solutions.

Soon after the first works on Chern–Simons vortex [6–8], other generalized models that also support vortex solutions appeared in Refs. [10,11]. On the other hand, over the past twenty years generalized models have been considered with other motivations. One of the first models with non-canonical kinetic term was presented in Ref. [12], in the context of inflation. Afterwards, Refs. [13,14] studied generalized models as a tentative to solve the cosmic coincidence problem [13,14]. It is worth mentioning that non-canonical models may present distinct features from the standard case. In the inflation scenario, for instance, these models may not require the presence of a potential to drive inflation. One can also study generalized models with defect structures in field theories, as in Refs. [15,16]. These models have been studied in several papers, in particular in [17–24]. An interesting fact here is that compact structures, firstly considered in Ref. [25] in models comprising nonlinearity and nonlinear dispersion, and later studied in Refs. [26–29], are also found in generalized models, as seen in Ref. [30]. Compact vortices in generalized Maxwell–Higgs models were found in Ref. [31]. They engender interesting features, such as the mapping of an infinitely long solenoid, if one includes a third spatial dimension.

In this work, we want to further investigate the presence of vortices in Chern–Simons models, but now focusing on generalized models. The idea is to follow Ref. [32] and propose new solutions of the vortex-like type in the case of non-canonical models. To do this, in Sec. 2 we first review the properties of the generalized model presented in Ref. [32], and then go on and investigate two distinct models in Sec. 3. The first model presents a parameter that controls the potential, which may be increased to make the model become the standard model described in Refs. [6,7]. The second model is also controlled by a real parameter, but now it starts at unit value with the standard model of Refs. [6,7] and makes the vortex compact if it is increased to larger and larger values. We conclude the work in Sec. 4.

* Corresponding author.
E-mail addresses: bazeia@fisica.ufpb.br (D. Bazeia), losano@fisica.ufpb.br (L. Losano), mam@fisica.ufpb.br (M.A. Marques), rmenezes@dce.ufpb.br (R. Menezes).

2. Generalized Chern–Simons vortices

We work with a complex scalar field and a gauge field governed by the pure Chern–Simons dynamics in (2, 1) spacetime dimensions. We use the class of generalized models presented in Ref. [32], which is described by the action $S = \int d^3x\mathcal{L}$, with the Lagrangian density $\mathcal{L}$ given by

$$\mathcal{L} = \frac{\kappa}{4}\epsilon^{\alpha\beta\gamma}A_\alpha F_{\beta\gamma} + K(|\varphi|)\overline{D_\mu\varphi}D^\mu\varphi - V(|\varphi|). \tag{1}$$

In the above expression, φ is the complex scalar field, $D_\mu = \partial_\mu + ieA_\mu$, $F_{\mu\nu} = \partial_\mu A_\nu - \partial_\nu A_\mu$ and $V(|\varphi|)$ is the potential. Also, κ has to be a constant, to keep gauge invariance of the action. The dimensionless function $K(|\varphi|)$ is in principle arbitrary, although it has to allow the existence of solutions with finite energy. The standard case studied in Ref. [7] is obtained for $K(|\varphi|) = 1$. We are using $A^\mu = (A^0, \vec{A})$, such that the electric and magnetic fields are defined by

$$E^i = F^{i0} = -\dot{A}^i - \partial_i A^0 \quad \text{and} \quad B = -F^{12}, \tag{2}$$

with $(E_x, E_y) \equiv E^i$ and $i = 1, 2$. The equations of motion for the fields φ and A_μ are given by

$$D_\mu(KD^\mu\varphi) = \frac{\varphi}{2|\varphi|}\left(K_{|\varphi|}\overline{D_\mu\varphi}D^\mu\varphi - V_{|\varphi|}\right), \tag{3a}$$

$$\frac{\kappa}{2}\epsilon^{\lambda\mu\nu}F_{\mu\nu} = J^\lambda, \tag{3b}$$

where the current is $J_\mu = ieK(|\varphi|)(\bar{\varphi}D_\mu\varphi - \varphi\overline{D_\mu\varphi})$. The energy momentum tensor $T_{\mu\nu}$ for the generalized model (1) is given by

$$T_{\mu\nu} = K(|\varphi|)\left(\overline{D_\mu\varphi}D_\nu\varphi + \overline{D_\nu\varphi}D_\mu\varphi\right) - \eta_{\mu\nu}\left(K(|\varphi|)\overline{D_\lambda\varphi}D^\lambda\varphi - V(|\varphi|)\right). \tag{4}$$

We now consider static configurations. In this case, the energy density is given by

$$\rho \equiv T_{00} = 2e^2K(|\varphi|)A_0^2|\varphi|^2 + V(|\varphi|) - K(|\varphi|)\overline{D_\lambda\varphi}D^\lambda\varphi. \tag{5}$$

The component A_0 that appears in the above equations is not an independent function. From the temporal component of the Eq. (3b), we get that A_0 is constrained to obey

$$A_0 = \frac{\kappa}{2e^2}\frac{B}{|\varphi|^2K(|\varphi|)}. \tag{6}$$

This leads to

$$\rho \equiv T_{00} = \frac{\kappa^2}{4e^2}\frac{B^2}{|\varphi|^2K(|\varphi|)} + K(|\varphi|)\overline{D_i\varphi}D_i\varphi + V(|\varphi|). \tag{7}$$

To search for vortexlike solutions, we take the usual ansatz

$$\varphi(r,\theta) = g(r)e^{in\theta}, \tag{8a}$$

$$A_0 = A_0(r), \tag{8b}$$

$$A_i = -\epsilon_{ij}\frac{x^j}{er^2}[a(r) - n] \tag{8c}$$

where r and θ are polar coordinates and n is an integer, the vortex winding number. The functions $g(r)$ and $a(r)$ must obey the boundary conditions

$$g(0) = 0, \quad a(0) = n, \quad \lim_{r\to\infty} g(r) = v, \quad \lim_{r\to\infty} a(r) = 0. \tag{9}$$

With the ansatz (8), we have $\overline{D_\mu\varphi}D^\mu\varphi = e^2g^2A_0^2 - (g'^2 + a^2g^2/r^2)$, where the prime denotes the derivative with respect to r. Furthermore, the electric and magnetic fields in Eq. (2) become

$$E^i = -A_0'\frac{x^i}{r} \quad \text{and} \quad B = -\frac{a'}{er}. \tag{10}$$

The magnetic flux $\Phi = 2\pi\int_0^\infty rdrB(r)$ is quantized:

$$\Phi = \frac{2\pi n}{e}. \tag{11}$$

The electric charge is given by $Q = 2\pi\int rdrJ^0$. By using Eq. (6), one can show that the charge can be written in terms of the magnetic flux (11) as $Q = -\kappa\Phi$, which makes the electric charge to be quantized. The equations of motion (3) with the ansatz (8) are given by

$$\frac{1}{r}(rKg')' + Kg\left(e^2A_0^2 - \frac{a^2}{r^2}\right) + \frac{1}{2}\left(\left(e^2g^2A_0^2 - g'^2 - \frac{a^2g^2}{r^2}\right)K_{|\varphi|} - V_{|\varphi|}\right) = 0, \tag{12a}$$

$$\frac{a'}{r} + \frac{2Ke^3g^2A_0}{\kappa} = 0, \tag{12b}$$

$$A_0' + \frac{2Keag^2}{\kappa r} = 0. \tag{12c}$$

Also, the energy density (7) becomes

$$\rho = \frac{\kappa^2}{4e^4}\frac{a'^2}{r^2g^2K(g)} + \left(g'^2 + \frac{a^2g^2}{r^2}\right)K(g) + V(g). \tag{13}$$

To find the energy density (13), one has to solve the equations of motion (12), which is a very complicated task, since they are coupled second order differential equations. However, in Ref. [32], it was shown that the first order equations

$$g' = \frac{ag}{r} \quad \text{and} \quad a' = -\frac{2e^2rg}{\kappa}\sqrt{KV} \tag{14}$$

solve the equations of motion (12) if the functions $K(|\varphi|)$ and $V(|\varphi|)$ are constrained to obey

$$\frac{d}{dg}\left(\sqrt{\frac{V}{g^2K}}\right) = -\frac{2e^2}{\kappa}gK. \tag{15}$$

Below we further explore generalized models of the class presented above, searching for vortex solutions that solve the first order equations (14) and obey the constraint (15). For simplicity, from now on we consider $e = \kappa = v = 1$ and work with dimensionless fields, setting $n = 1$ to explore the case of unit vorticity.

3. New models

We see from (1) that the model is specified once the two functions $K(|\varphi|)$ and $V(|\varphi|)$ are given explicitly. Below we study two distinct possibilities.

3.1. Model 1

Let us first consider the model that is specified by the functions

$$K(|\varphi|) = \left(1 - |\varphi|^{2l}\right)^2, \tag{16a}$$

$$V(|\varphi|) = |\varphi|^2\left(1 - |\varphi|^{2l}\right)^2 \times \left(1 - |\varphi|^2 + \frac{2}{l+1}|\varphi|^{2l+2} - \frac{1}{2l+1}|\varphi|^{4l+2}\right)^2, \tag{16b}$$

where l is a real parameter such that $l \geq 1$. The above pair of functions is compatible with Eq. (15). This potential has a minimum at $|\varphi| = 0$ and $|\varphi| = 1$, regardless the value of l. Whilst these

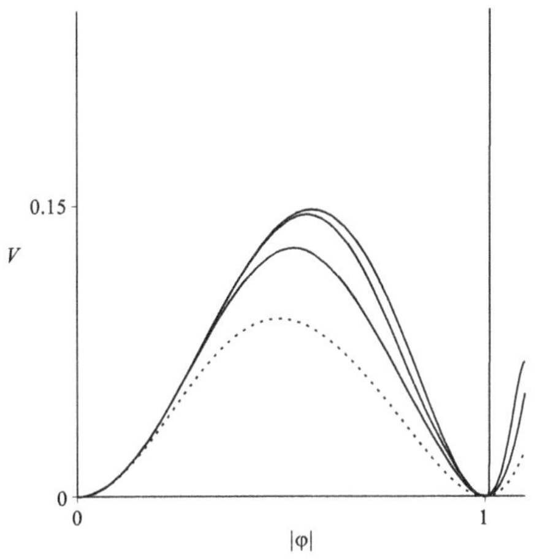

Fig. 1. The potential of Eq. (16b) is depicted for $l = 1, 2, 4$ and 64, with the case $l = 1$ as the dotted line.

minima are fixed, the maximum between them, which cannot be calculated analytically, changes with l. The above potential also has other minimum for $|\varphi| > 1$, which is solution of the equation

$$1 - |\varphi|^2 + \frac{2}{l+1}|\varphi|^{2l+2} - \frac{1}{2l+1}|\varphi|^{4l+2} = 0. \qquad (17)$$

However, here we are only interested to find solutions in the sector $0 \le |\varphi| \le 1$. To illustrate this, in Fig. 1 we depict the potential (16b) in the interval $0 \le |\varphi| \le 1 + \delta$, for δ small and for several values of l, to illustrate how the potential behaves around the minimum at $|\varphi| = 1$. One can see that the concavity of the potential in the minimum $|\varphi| = 1$ narrows as we increase l. However, for $\varphi \to 1$ from the left, the potential has almost the same behavior. Also, the minimum located outside of the aforementioned sector gets closer to $|\varphi| = 1$ as l gets larger. In the limit $l \to \infty$, this potential tends to become

$$V_\infty(r) = \begin{cases} |\varphi|^2\left(1 - |\varphi|^2\right)^2, & |\varphi| < 1 \\ \infty, & |\varphi| > 1. \end{cases} \qquad (18)$$

Therefore, inside the sector $0 \le |\varphi| \le 1$, the potential tends to behave as the one studied in Refs. [6,7].

In the general case, the first order equations (14) become

$$g' = \frac{ag}{r} \qquad (19a)$$

$$a' = -2rg^2\left(1 - g^{2l}\right)^2\left(1 - g^2 + \frac{2g^{2l+2}}{l+1} - \frac{g^{4l+2}}{2l+1}\right). \qquad (19b)$$

We can study the behavior of the above equations near the origin by considering $a(r) = 1 - a_0(r)$ and $g(r) = g_0(r)$ up to first order in $a_0(r)$ and $g_0(r)$. This leads to $a_0(r) \propto r^4$ and $g_0(r) \propto r$. The first order equations (19a) and (19b) are very hard to be solved analytically, so we conduct the investigation numerically. In Fig. 2, we display the solutions of Eqs. (19a) and (19b) for several values of l. Notice that, even though the solutions tend to approach their boundary condition as l increases, they do not compactify, due to the fact that the potential admits the limit (18), which does not give rise to compact solutions inside the sector that is being considered.

The electric and magnetic fields are obtained from Eq. (10), so we use the numerical solutions of Eqs. (19a) and (19b) and plot them in Fig. 3 for several values of l. We also display the temporal gauge component (6). The energy density can be calculated from Eq. (13). In this case, it is given by

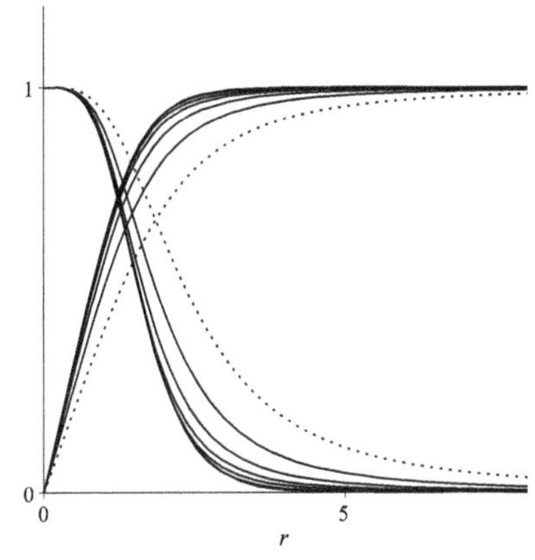

Fig. 2. The solutions $a(r)$ (descending lines) and $g(r)$ (ascending lines) of Eqs. (19a) and (19b), depicted for $l = 1, 2, 4, 8, 16, 64, 256$ and 1024, with the case $l = 1$ as the dotted line.

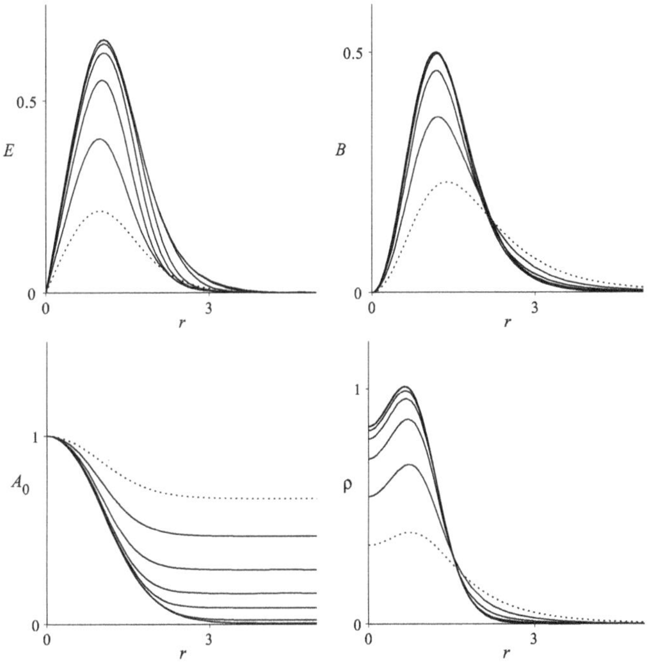

Fig. 3. The electric (top left) and magnetic (top right) fields (10), the temporal gauge field (6) (bottom left) and the energy density (20) (bottom right) plotted for $l = 1, 2, 4, 8, 16, 64, 256$ and 1024, with the case $l = 1$ as the dotted line.

$$\rho = \frac{a'^2}{4r^2g^2\left(1 - g^{2l}\right)^2} + \left(g'^2 + \frac{a^2g^2}{r^2}\right)\left(1 - g^{2l}\right)^2 + g^2\left(1 - g^{2l}\right)^2\left(1 - g^2 + \frac{2g^{2l+2}}{l+1} - \frac{g^{4l+2}}{2l+1}\right)^2. \qquad (20)$$

We have plotted the above energy density for the numerical solutions of Eqs. (19a) and (19b) in Fig. 3 for several values of l. As l gets larger and larger, the vortex energy density gets an internal structure. Numerical integration of these energy densities over all the space gives the energy $E \approx 2\pi$, which is independent of l.

3.2. Model 2

The second model which we consider is described by the generalized model (1), but is specified by the pair of functions

$$K(|\varphi|) = l|\varphi|^{2l-2} \quad \text{and} \quad V(|\varphi|) = l|\varphi|^{2l}\left(1 - |\varphi|^{2l}\right)^2, \qquad (21)$$

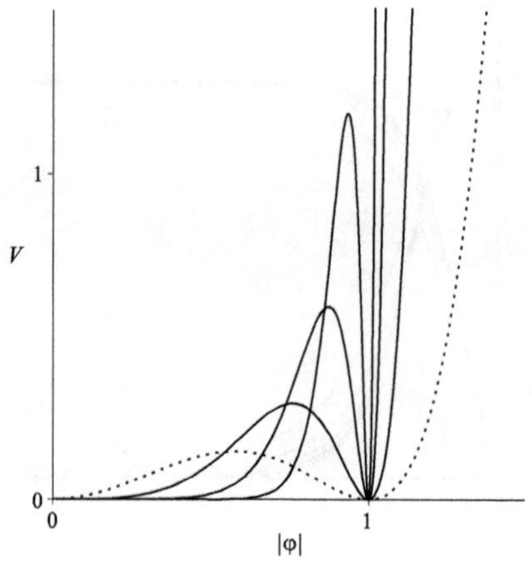

Fig. 4. The potential of Eq. (21), displayed for $l = 1, 2, 4$ and 8, with the case $l = 1$ as the dotted line.

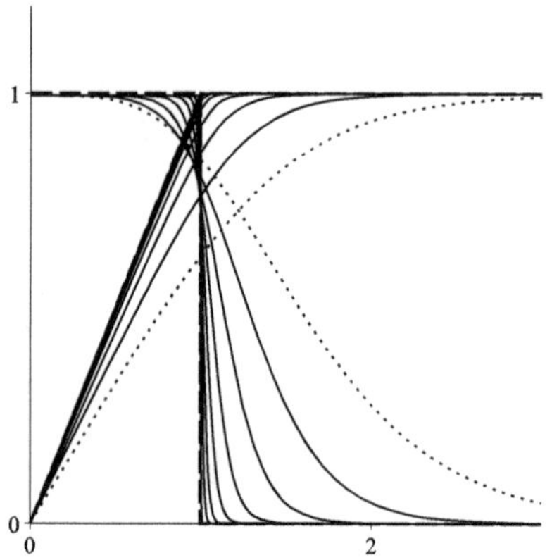

Fig. 5. The solutions $a(r)$ (descending lines) and $g(r)$ (ascending lines) of Eqs. (22) plotted for $l = 1$ and increasing to larger and larger values, with the case $l = 1$ as the dotted line. The dashed lines stand for the compact limit (23).

which are compatible with the constraint (15). In the above expressions, l is a real parameter such that $l \geq 1$. The model is simpler than the previous one. The potential has a minimum at $|\varphi| = 0$ and $|\varphi| = 1$. In between these two minima there is a local maximum at $|\varphi_{\max}| = 1/3^{1/(2l)}$ such that $V(|\varphi_{\max}|) = 4l/27$. As l increases, $|\varphi_{\max}|$ approaches more and more the minimum at $|\varphi| = 1$, and $V(|\varphi_{\max}|)$ increases to lager and larger values. The standard case, which was studied in Refs. [6,7], is obtained for $l = 1$. In Fig. 4, we display the potential (21) for several values of l to illustrate its general behavior.

In this case, the first order equations (14) become

$$g' = \frac{ag}{r} \quad \text{and} \quad a' = -2lrg^{2l}\left(1 - g^{2l}\right). \tag{22}$$

Before trying to solve them, we can study their behavior near the origin by taking $a(r) = 1 - a_0(r)$ and $g(r) = g_0(r)$ up to first order in $a_0(r)$ and $g_0(r)$. This shows that $a_0(r) \propto lr^{2(l+1)}/(l+1)$ and $g_0(r) \propto r$. A similar procedure can be used to find the asymptotic behavior of the functions by taking $a(r) = a_\infty(r)$ and $g(r) = 1 - g_\infty(r)$ in Eqs. (22). This leads to $a_\infty(r) \propto 2lrK_1(2lr)$ and $g_\infty(r) \propto K_0(2lr)$. In the latter expressions, $K_m(z)$ is the modified Bessel function of the second kind. The asymptotic behavior shows that the functions $a(r)$ and $g(r)$ approach their respective boundary condition faster and faster as l increases, a behavior that shows that the solutions tend to become compact for larger and larger values of l.

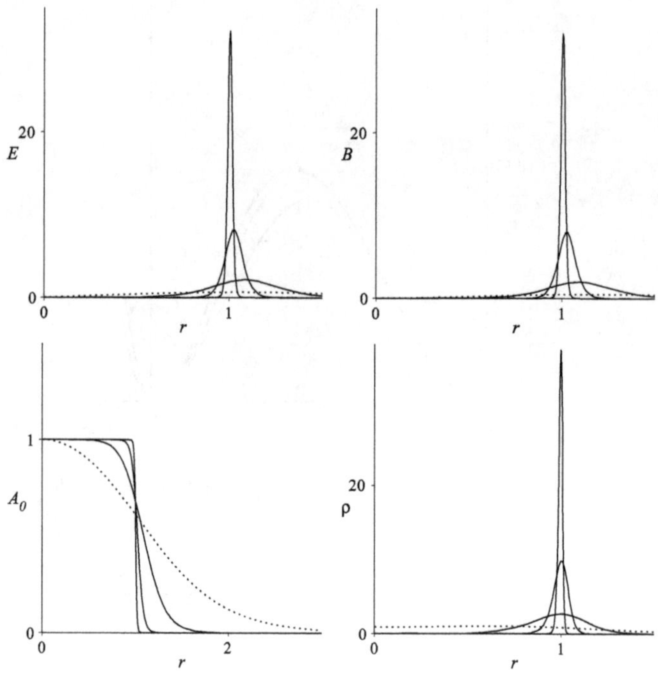

Fig. 6. The electric (top left) and magnetic (top right) fields (10), the temporal component of the gauge field (6) (bottom left) and the energy density (24) (bottom right), plotted for $l = 1, 4, 16$ and 64, with the case $l = 1$ as the dotted line.

The first order equations (22) are very hard to be solved analytically for a general l. However, we found that, as l increases, the solutions tend to get the compact profile

$$a_c(r) = \begin{cases} 1, & r \leq r_c \\ 0, & r > r_c, \end{cases} \tag{23a}$$

$$g_c(r) = \begin{cases} \frac{r}{r_c}, & r \leq r_c \\ 1, & r > r_c, \end{cases} \tag{23b}$$

where r_c is the compactification radius, which is calculated numerically for a very large value of l; it has the value $r_c \approx 0.995$. In Fig. 5, we depict the solutions of Eqs. (22) for several values of l, including the limit given by Eq. (23).

Also, we have to look at the electric and magnetic fields of Eq. (10) to study its behavior. In Fig. 6, we have sketched the electric and magnetic fields, as well as the temporal gauge field A_0 of Eq. (6), and the corresponding energy density. The uniform behavior of the functions $a_c(r)$ and $g_c(r)$ of Eq. (23) inside the compact sector makes the electric and magnetic fields tend to become a Dirac's delta as l increases. For l very large, these fields only exists as a very marrow ring around the unit value of r. Also, we study the energy density, given by Eq. (13), which becomes

$$\rho = \frac{a'^2}{4r^2lg^{2l}} + lg^{2l-2}g'^2 + \frac{la^2g^{2l}}{r^2} + g^{2l}\left(1 - g^{2l}\right)^2. \tag{24}$$

It is also displayed in Fig. 6 for several values of l.

We see that as l increases, the electric field, the magnetic field and the energy density become taller and thinner. They tend to become compact, living in a narrow area around $r = r_c$. In particular, the energy density behaves in a way such that the total energy obtained by numerical integration is always approximately equal to 2π, regardless the value of l. We then conclude that the vortex tends to become compact as l increases to larger and larger values, existing only inside a narrow ring around $r = r_c$.

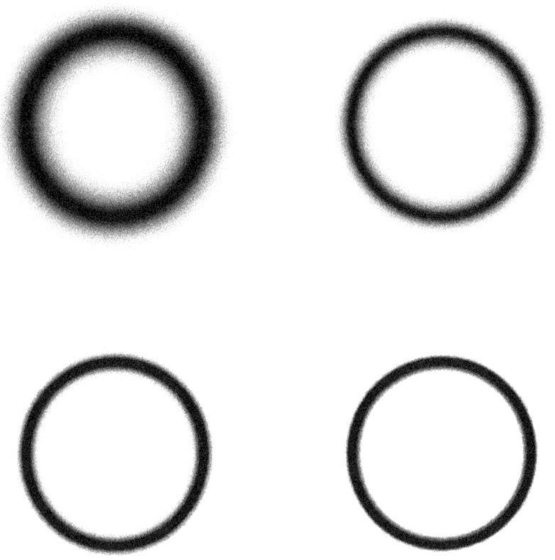

Fig. 7. The energy density (24) depicted in the polar coordinates (r, θ) for $l = 4$ (top left), $l = 8$ (top right), $l = 16$ (bottom left) and $l = 32$ (bottom right), illustrating that the vortex shrinks to become compact, living inside a narrower and narrower ring around $r = r_c$ as l increases to larger and larger values.

As far as we can see, this behavior was never seem before, so we illustrate it again in Fig. 7 where one depicts the energy density using polar coordinates, for some values of l.

4. Conclusions

In this work we studied two new models described by generalized Chern–Simons systems of the class (1). The two models are controlled by a single parameter that appears in the potential and in the function $K(|\varphi|)$ that modifies the scalar field dynamics. In the first model, we showed that although the solutions tend to go to their boundary conditions faster when the parameter gets larger, they do not shrink enough to make the vortices compact. This happens because the potential tends to acquire the very same form of the one that appears in the standard case presented before in Refs. [6,7].

The investigation continued with the study of another model, the second model which is described by a potential whose concavity around its unit minimum tends to become narrower as the parameter increases to larger and larger values. This behavior is illustrated in Fig. 4 and is very nice, since it makes the solutions shrink significantly, inside a narrow ring around $r = r_c$, as illustrated in Figs. 6 and 7. This model is very different from the previous one and we could see that it starts with the standard model described in Refs. [6,7] in the case $l = 1$. The effect of the compactification engendered by the second model is original and very interesting, and since it appears in a system that supports first-order differential equations, it may perhaps be extended to become supersymmetric and so of more general interest.

Other issues concern modification of the current work to consider the case of non-relativistic dynamics [33] and also, extensions of the Abelian $U(1)$ symmetry to the non-Abelian case.

Acknowledgements

We would like to thank the Brazilian agency CNPq for partial financial support. DB thanks support from fundings 455931/2014-3 and 306614/2014-6, LL thanks support from fundings 307111/2013-0 and 447643/2014-2, MAM thanks support from funding 140735/2015-1, and RM thanks support from fundings 455619/2014-0 and 306826/2015-1.

References

[1] H.B. Nielsen, P. Olesen, Nucl. Phys. B 61 (1973) 45.
[2] H.J. de Vega, F.A. Schaposnik, Phys. Rev. D 14 (1976) 1100.
[3] S.-S. Chern, J. Simons, Ann. Math. 99 (1974) 48.
[4] S. Deser, R. Jackiw, S. Templeton, Phys. Rev. Lett. 48 (1982) 975, Ann. Phys. 140 (1982) 372.
[5] C.R. Hagen, Ann. Phys. 157 (1984) 342, Phys. Rev. D 31 (1985) 2135.
[6] J. Hong, Y. Kim, P.Y. Pac, Phys. Rev. Lett. 64 (1990) 2230.
[7] R. Jackiw, E.J. Weinberg, Phys. Rev. Lett. 64 (1990) 2234.
[8] R. Jackiw, K. Lee, E.J. Weinberg, Phys. Rev. D 42 (1990) 3488.
[9] G. Dunne, Self-Dual Chern–Simons Theories, Springer-Verlag, 1995.
[10] J. Lee, S. Nam, Phys. Lett. B 261 (1991) 437.
[11] D. Bazeia, Phys. Rev. D 46 (1992) 1879.
[12] C. Armendariz-Picon, T. Damour, V. Mukhanov, Phys. Lett. B 458 (1999) 209.
[13] C. Armendariz-Picon, V. Mukhanov, Paul J. Steinhard, Phys. Rev. Lett. 85 (2000) 4438.
[14] C. Armendariz-Picon, V. Mukhanov, Paul J. Steinhardt, Phys. Rev. D 63 (2001) 103510.
[15] E. Babichev, Phys. Rev. D 74 (2006) 085004.
[16] E. Babichev, Phys. Rev. D 77 (2008) 065021.
[17] X. Jin, X. Li, D. Liu, Class. Quantum Gravity 24 (2007) 2773.
[18] S. Sarangi, J. High Energy Phys. 018 (2008) 0807.
[19] D. Bazeia, L. Losano, R. Menezes, J.C.R.E. Oliveira, Eur. Phys. J. C 51 (2007) 953.
[20] D. Bazeia, L. Losano, R. Menezes, Phys. Lett. B 668 (2008) 246.
[21] D. Bazeia, A.R. Gomes, L. Losano, R. Menezes, Phys. Lett. B 671 (2009) 402.
[22] R. Casana, M.M. Ferreira Jr, E. da Hora, Phys. Rev. D 86 (2012) 085034.
[23] Handhika S. Ramadhan, Phys. Lett. B 758 (2016) 149.
[24] R. Casana, M.L. Dias, E. da Hora, Phys. Lett. B 768 (2017) 254.
[25] P. Rosenau, J.M. Hyman, Phys. Rev. Lett. 70 (1993) 564.
[26] D. Bazeia, L. Losano, M.A. Marques, R. Menezes, Phys. Lett. B 736 (2014) 515.
[27] D. Bazeia, L. Losano, M.A. Marques, R. Menezes, Europhys. Lett. 107 (2014) 61001.
[28] D. Bazeia, M.A. Marques, R. Menezes, Europhys. Lett. 111 (2015) 61002.
[29] D. Bazeia, L. Losano, M.A. Marques, R. Menezes, R. da Rocha, Phys. Lett. B 758 (2016) 146.
[30] D. Bazeia, L. Losano, R. Menezes, Phys. Lett. B 731 (2014) 293.
[31] D. Bazeia, L. Losano, M.A. Marques, R. Menezes, I. Zafalan, Eur. Phys. J. C 77 (2017) 63.
[32] D. Bazeia, E. da Hora, C. dos Santos, R. Menezes, Phys. Rev. D 81 (2010) 125014.
[33] R. Jackiw, S.-Y. Pi, Phys. Rev. Lett. 64 (1990) 2969, Phys. Rev. D 42 (1990) 3500.

6

Constrained superfields from inflation to reheating

Ioannis Dalianis [a,*], Fotis Farakos [b,c]

[a] Physics Division, National Technical University of Athens, 15780 Zografou Campus, Athens, Greece
[b] Dipartimento di Fisica e Astronomia "Galileo Galilei", Università di Padova, Via Marzolo 8, 35131 Padova, Italy
[c] INFN, Sezione di Padova, Via Marzolo 8, 35131 Padova, Italy

ARTICLE INFO

Editor: M. Cvetič

ABSTRACT

We construct effective supergravity theories from customized constrained superfields which provide a setup consistent both for the description of inflation and the subsequent reheating processes. These theories contain the minimum degrees of freedom in the bosonic sector required for single-field inflation.

1. Introduction

Constrained superfields [1–6] provide an elegant, albeit effective, method for embedding inflation in supergravity [7–16]. Requiring the effective theory to be valid from inflation down to the present de Sitter phase of the universe results in important restrictions. For example, various consistency conditions were already pointed out in [10,17] for *sgoldstino-less* models. The utility of the constrained superfields extends to aspects of the supergravity cosmology besides inflation, see e.g. [18,19] for recent works on gravitino cosmology. The description of the gravitino production in this context is of particular interest and motivates this letter.

Gravitinos are cosmologically problematic when overabundant [20–22]. The preheating and reheating stages [23] are important sources of non-thermal gravitino production as has been demonstrated in notable works [24–35]. The number density of the $\pm 1/2$ helicity gravitinos, which are identified as the goldstinos at energies much larger than the gravitino mass, is not suppressed by inverse powers of $M_{\rm Pl}$ and it can be rather large [27–30]. In the standard picture, the goldstino is a linear combination of fermions, when several chiral superfields are present. Hence the composition of the helicity $\pm 1/2$ gravitino through the goldstino mixture varies due to the cosmic evolution. During the stage of the inflaton oscillations, one assumes the main contribution to the helicity $\pm 1/2$ gravitino to come from the fermionic superpartner of the inflaton, the inflatino, whereas at times after the reheating of the universe the longitudinal gravitino is given by the present vacuum golstino, which is the *true goldstino*. In [31,32] it was understood that this change with time of the definition of the longitudinal gravitino renders the production of gravitinos during the preheating stage cosmologically harmless.

In the framework of constrained superfields the gravitino production during preheating and reheating was examined recently [18] for the inflatino-less models constructed in [11–14]. The inflatino-less models of [11–14] have the property to contain the minimal bosonic sector essential for single-field inflation, but due to the constraints that are imposed it is also the inflatino which is eliminated from the spectrum. It was mentioned in [18] that gravitinos may be found to be cosmologically overabundant when the specific constrained superfields are considered as the inflationary sector. Indeed, when one eliminates all the spin-1/2 fermions in the theory except for the goldstino, then the longitudinal gravitino modes will be identified with the gravitino in the vacuum and therefore naturally lead to large gravitino number densities. Also, the perturbative decay of the inflaton may produce an excessive gravitino yield if the inflaton contributes to the vacuum supersymmetry breaking. Triggered by this novel approach to the gravitino cosmology we revisit the non-linear realizations of supersymmetric cosmology after inflation in order to build effective field theories that achieve agreement with the existing standard supergravity results.

Aiming at minimal effective theories which can describe both inflation and reheating processes, we introduce *customized* constrained superfields which contain minimum number of component fields. These models have a minimal bosonic sector which includes only the inflaton and the metric and deliver inflationary Lagrangians of the form

$$e^{-1}\mathcal{L} = -\frac{1}{2}R - \frac{1}{2}\,\partial_m\phi\,\partial^m\phi - V(\phi)\,. \tag{1}$$

* Corresponding author.
E-mail address: yiadalianis@gmail.com (I. Dalianis).

The fermion sector is left unconstrained, therefore will include the *vacuum goldstino* G_α, the gravitino ψ_m^α, and the fermions of the inflationary sector: *inflatino* χ_α, and *stabilino* λ_α when there is a stabilizer superfield. The number of constrained superfields in the models we construct is *not* essentially minimal, but this gives us the required flexibility to have a consistent effective description during the various stages of the evolution of the universe. In particular we make use of a constrained inflaton superfield Φ, a constrained stabilizer superfield S, and the nilpotent chiral superfield X which will break supersymmetry in the vacuum.

The organization of the article is the following. In the second section we revisit the inflatino-less models and discuss the origin of the gravitino overproduction cosmological problem during the non-thermal (p)reheating stage. In the third section we present the customized constrained superfields adequate to describe the Polonyi sector, the inflaton sector and the stabilizer sector. In the fourth section we present two working models, described in subsections (4.1) and (4.2), that realize our proposal and in this context we review some relevant (p)reheating results. In section five we conclude.

2. Inflatino-less models and the gravitino issue

The first supergravity inflationary models that contained the minimal bosonic sector (inflaton and gravitation) were constructed by two constrained superfields [11–14]: the chiral X satisfying $X^2 = 0$ and a second chiral superfield $\mathcal{A}$ satisfying [4]

$$X\mathcal{A} = X\overline{\mathcal{A}}\,. \tag{2}$$

The effect of the constraint (2) is to eliminate all the component fields of the superfield $\mathcal{A}$ leaving as independent only a real scalar residing in the lowest component field which is identified with the inflaton ϕ. Explicitly one finds: $\mathcal{A}| = \phi +$ terms with goldstino. The crucial side-effect of the constraint (2) is that the *inflatino* is also eliminated which would otherwise sit in the fermion component of the superspace expansion of $\mathcal{A}$.

After the end of inflation, gravitino production generically takes place during both phases of the reheating process: initially during preheating where the inflaton *oscillates* and excites the other fields in the theory, and subsequently during the perturbative *decay* of the inflaton. The gravitino production is cosmologically problematic when the corresponding yield $Y_{3/2} \equiv n_{3/2}/s$, where s the entropy density and $n_{3/2}$ the gravitino number density, violates either the cold dark matter density constraint [36]

$$\Omega_{3/2}h^2 \sim 10^8 \left(\frac{m_{3/2}}{\text{GeV}}\right) Y_{3/2}h^2 \lesssim 0.12\,, \tag{3}$$

for stable gravitinos, or the Big-Bang nucleosynthesis constraints

$$Y_{3/2} \lesssim 10^{-16} - 10^{-13}\,, \tag{4}$$

for unstable gravitinos with mass $m_{3/2} \lesssim 10^5$ GeV (see e.g. [34,37] for recent work and references therein). The $Y_{3/2}$ may be additionally constrained depending on the details of the LSP dark matter production mechanism.

Under general assumptions, it is known that during the preheating phase non-thermal production of longitudinal gravitinos takes place [27,29]. This issue came with a simple resolution [31]. Since both the kinetic and potential energies contribute to the supersymmetry breaking [29], the gravitino longitudinal component is dominated by the inflatino during inflation and preheating, and therefore the gravitino longitudinal components during these stages are not true goldstinos. They are rather harmless inflatinos which are produced during preheating and do not correspond to gravitinos in the vacuum. Essentially, the source of supersymmetry breaking during inflation and preheating has to be different from the source of supersymmetry breaking in the vacuum.

In the recent work [18], it has been pointed out that the inflatino-less models [11–14] generically suffer from a gravitino overproduction problem after inflation. Indeed, in contrast to the standard supergravity situation where inflatinos instead of gravitinos are produced during preheating, in this setup this is not possible. The reason is that when using the constraint (2), the inflatino is eliminated in terms of the goldstino. Therefore the longitudinal component of the gravitino is always aligned with the goldstino G_α during the cosmological phases from inflation to reheating and to the current de Sitter vacuum [38–43]. This means that any longitudinal gravitino production is essentially a *true goldstino* production and might easily lead to cosmological problems exactly as has been noted in the early literature [27,29].

We interpret the findings of [18] as an indication that the constrained superfield $\mathcal{A}$ might not always describe the inflaton in an appropriate way. In this work we construct effective theories with the appropriate inflaton superfield which can describe the interactions of the inflatino and still maintain a minimal bosonic sector. In this way minimal inflationary models with constrained superfields can produce the standard supergravity results regarding the gravitino production [31].

3. Customized constrained superfields

In this section we wish to introduce the constrained superfields which contain the minimal number of independent fields, while still serving the purpose they are introduced for in standard supergravity. The supergravity Lagrangians we will consider in this work are always built by the standard superspace couplings

$$\mathcal{L} = \int d^2\Theta\, 2\mathcal{E}\left[\frac{3}{8}\left(\overline{\mathcal{D}}^2 - 8\mathcal{R}\right) e^{-K/3} + W\right] + c.c.\,, \tag{5}$$

where we have set $M_{\text{Pl}} = 1$. The expressions for the chiral density $2\mathcal{E}$ and the curvature superfield $\mathcal{R}$ can be found in [44], where also the component form expression for (5) is given.

The Polonyi superfield. The Polonyi model provides the simplest realization of supersymmetry breaking in supergravity. In this setup one introduces a Kähler potential and superpotential of the form

$$K = X\overline{X}\,, \quad W = f_0 X + W_0\,. \tag{6}$$

By appropriately choosing the parameters f_0 and W_0 in the superpotential and introducing higher order corrections in the Kähler potential one obtains a vacuum with a small enough cosmological constant and a heavy scalar which essentially decouples. The effective description of this setup is provided by the nilpotent constraint superfield X which has found its way into various applications for supergravity cosmology [46]. Indeed, the decoupling of the heavy Polonyi scalar corresponds to the constraint

$$X^2 = 0\,, \tag{7}$$

which is solved for $X = \frac{G^2}{2F^X} + \sqrt{2}\Theta G + \Theta^2 F^X$, under the strict requirement

$$\langle F^X \rangle \neq 0\,. \tag{8}$$

This setup is tailor-made to describe the current de Sitter vacuum of the universe. The pure theories were constructed in [38–40], and the vacuum energy and gravitino mass are given by

$$\Lambda = f_0^2 - 3\,W_0^2\,, \quad m_{3/2} = W_0\,. \tag{9}$$

Throughout this work we will always require that the nilpotent Polonyi field X is the sole source responsible for supersymmetry breaking at the vacuum, therefore the G_α is the true vacuum goldstino. Any production of G_α corresponds to production of gravitino. Notice that since in the vacuum F^X is the only auxiliary field which will get a non-vanishing vacuum expectation value it is in fact singled out to be the one belonging to the superfield satisfying (7). Moreover, since the goldstino is a pure gauge and will be absorbed by the gravitino, it will be convenient to express our results directly in the gauge

$$G_\alpha = 0 \,. \tag{10}$$

In this gauge X becomes $X|_{G=0} = \Theta^2 F^X$.

The inflaton superfield. We will now present a new class of constrained superfields which are adequate for hosting the inflaton in their lowest component. The purpose of this constraint is to eliminate *only* the real scalar b residing in the imaginary part of $\Phi|$. Indeed, once we have the nilpotent X superfield, the method to construct customized superfields was outlined in [6]. Here we will use this method to construct the inflaton superfield. To this end we simply have to impose the constraint on Φ

$$X\overline{X}\left(\Phi - \overline{\Phi}\right) = 0 \,. \tag{11}$$

To solve this constraint, one first brings it to the form

$$\begin{aligned}\left(\Phi - \overline{\Phi}\right) =& 2\frac{\overline{\mathcal{D}}_{\dot\alpha}\overline{X}}{\overline{\mathcal{D}}^2\overline{X}}\overline{\mathcal{D}}^{\dot\alpha}\overline{\Phi} + \overline{X}\frac{\overline{\mathcal{D}}^2\overline{\Phi}}{\overline{\mathcal{D}}^2\overline{X}} \\ &- 2\frac{\mathcal{D}^\alpha X}{\mathcal{D}^2 X\,\overline{\mathcal{D}}^2\overline{X}}\mathcal{D}_\alpha\overline{\mathcal{D}}^2\left[\overline{X}\left(\Phi - \overline{\Phi}\right)\right] \\ &- \frac{X}{\mathcal{D}^2 X\overline{\mathcal{D}}^2\overline{X}}\mathcal{D}^2\overline{\mathcal{D}}^2\left[\overline{X}\left(\Phi - \overline{\Phi}\right)\right] ,\end{aligned} \tag{12}$$

and projects to $\theta = 0$ to find the component field expression. Then by performing recursive steps the real scalar component field b will be eliminated from the spectrum, in terms of the other component fields of Φ and X. The leading terms in the G expansion of $b = \left(\Phi - \overline{\Phi}\right)|/2i$ are

$$\begin{aligned}b =& -i\frac{G\chi}{\sqrt{2}F} + i\frac{\overline{G}\overline{\chi}}{\sqrt{2}\overline{F}} + i\frac{G^2}{4F^2}F^\Phi - i\frac{\overline{G}^2}{4\overline{F}^2}\overline{F}^\Phi \\ &- \frac{1}{2F\overline{F}}\left(G\sigma^c\overline{G}\right)e_c^m\partial_m\phi + \cdots\end{aligned} \tag{13}$$

where the $\cdots$ refer to terms with at least three goldstini. In (13) we used the component field definitions: $\chi_\alpha = \mathcal{D}_\alpha\Phi|/2$ and $F^\Phi = -\mathcal{D}^2\Phi|/4$. Note that (13) is the solution in curved superspace, but the gravitino and the auxiliary fields of the supergravity sector do not enter at this order. During inflation, the effect of (11) on Φ is more transparent once we write the inflaton superfield in the simplifying gauge (10). Indeed in this gauge we will have

$$\Phi|_{G=0} = \phi + 2\,\Theta\chi + \Theta^2 F^\Phi \,, \tag{14}$$

where ϕ is a real scalar (the inflaton), χ is the inflatino, and F^Φ is the auxiliary field of the inflaton multiplet. The constraint (11) can be imposed with a Lagrange multiplier in the spirit of [45].

The importance of introducing the inflatino and the auxiliary field of the inflaton multiplet has been stressed in the previous section. We recall that if the inflatino is eliminated then the longitudinal gravitino component production during preheating might lead to an overestimate of the gravitino number density. If the inflatino survives in the effective theory then it is in fact the inflatino which is dominantly produced during preheating [31]. We stress at this point that by just introducing the inflatino in the effective theory one does *not* automatically have a viable gravitino cosmology. Rather now the gravitino cosmology has the same status as it is in standard supergravity (see e.g. [34,37]). In the inflatino-less models this is not the case and one has to find new means of addressing this issue [18].

The stabilizer superfield. The third ingredient for the inflationary model building in standard supergravity is the stabilizer superfield S. This superfield has a dual role. First it has specific couplings which allow to construct a large class of inflationary potentials for the inflaton superfield [47]. Second, if it is the primary source of supersymmetry breaking during inflation it is used to strongly stabilize all scalars (during inflation), except for the inflaton. The important component field of the stabilizer superfield is its auxiliary field F^S and it should survive in the effective theory. Moreover, we wish to keep also the fermionic field of the stabilizer superfield, the *stabilino*. The reason is simple: during the preheating stage one has $F^S \neq 0$ therefore if the fermion of the stabilizer is eliminated from the spectrum its contribution to the longitudinal gravitino will be misinterpreted as true vacuum goldstinos G_α. On the other hand, the scalar of the stabilizer superfield can be very heavy and decouple. Therefore the appropriate constraint that should be imposed to describe the stabilizer superfield is

$$X\overline{X}\,S = 0 \,. \tag{15}$$

The solution to this constraint and minimal models can be found in [4,48,49]. For our purposes it suffices to present S in the gauge (10), where it has the form

$$S|_{G=0} = \sqrt{2}\Theta\,\lambda + \Theta^2 F^S \,. \tag{16}$$

The stabilizer will essentially couple to the inflaton superfield in the superpotential via terms of the form $S\,\Sigma(\Phi)$, which will reproduce the contribution to the scalar potential

$$|F^S|^2 \sim |\Sigma(\phi)|^2 \,. \tag{17}$$

In the subsequent section we will construct effective theories from these superfields which can describe single field inflation, the reheating phase and the current de Sitter phase of our universe. We will require that

$$\langle F^\Phi\rangle|_{\text{vacuum}} = 0 \,, \quad \langle F^S\rangle|_{\text{vacuum}} = 0 \,, \tag{18}$$

such that any inflatino or stabilino production will not correspond to true vacuum goldstino.

4. Effective theories

4.1. Models with the inflaton and the Polonyi superfields

In this section we will study models which contain only the inflaton superfield Φ and the Polonyi superfield X. We will not write the most general coupling, but we will rather construct simple inflationary models which however capture most of the possibilities for single field inflation. Since we present these supergravity effective theories here for the first time, we will study them in more detail. The Kähler potential and the superpotential we insert in (5) are

$$\begin{aligned}K =& X\overline{X} - \frac{1}{4}(\Phi - \overline{\Phi})^2 \,, \\ W =& f_0\,X + g(\Phi) \,,\end{aligned} \tag{19}$$

where f_0 is a real constant. The Kähler potential has the shift symmetry $\Phi \to \Phi + c$ (for c real constant) as proposed by [50] to address the supergravity η-problem. Notice that the X and Φ

sectors are separated and will therefore interact mostly gravitationally. We could have chosen instead of f_0 a function $f(\Phi)$, but we prefer to study a minimal setup here. In the $G_\alpha = 0$ gauge the full theory now reads

$$\begin{aligned} e^{-1}\mathcal{L} = &-\frac{1}{2}R + \frac{1}{2}\epsilon^{klmn}\left(\overline{\psi}_k\overline{\sigma}_l\mathcal{D}_m\psi_n - \psi_k\sigma_l\mathcal{D}_m\overline{\psi}_n\right) \\ &- i\overline{\chi}\overline{\sigma}^m\mathcal{D}_m\chi - \frac{1}{2}\partial_m\phi\,\chi\sigma^m\overline{\sigma}^n\psi_m - \frac{1}{2}\partial_n\phi\,\overline{\chi}\overline{\sigma}^m\sigma^n\overline{\psi}_m \\ &+ \frac{1}{4}\left[i\epsilon^{klmn}\psi_k\sigma_l\overline{\psi}_m + \psi_m\sigma^n\overline{\psi}^m\right]\chi\sigma_n\overline{\chi} \\ &- \frac{1}{8}\chi^2\overline{\chi}^2 - ig_\phi\chi\sigma^a\overline{\psi}_a - i\overline{g}_\phi\overline{\chi}\overline{\sigma}^a\psi_a \\ &- \left(g_{\phi\phi} - \frac{1}{2}g\right)\chi^2 - \left(\overline{g}_{\phi\phi} - \frac{1}{2}\overline{g}\right)\overline{\chi}^2 \\ &- \left(g(\phi)\overline{\psi}_a\overline{\sigma}^{ab}\overline{\psi}_b + \overline{g}(\phi)\psi_a\sigma^{ab}\psi_b\right) \\ &- \frac{1}{2}\partial^m\phi\,\partial_m\phi - V(\phi)\,. \end{aligned} \tag{20}$$

The scalar potential in (20) now has the form expected from standard supergravity

$$V(\phi) = f_0^2 + 2\left|\frac{\partial g(\phi)}{\partial\phi}\right|^2 - 3|g(\phi)|^2\,. \tag{21}$$

The term $+2\left|\frac{\partial g(\phi)}{\partial\varphi}\right|^2$ in particular comes exactly from integrating out the auxiliary field F^Φ. Notice that the bosonic sector in (20) is *minimal*: it contains only the inflaton and gravitation. We will show in a moment how to construct simple models for inflation from the expression (21).

The estimation of the gravitino abundance during preheating goes along the lines of the standard supergravity. The gravitino is not overproduced once a hierarchy between the inflationary scale and the supersymmetry breaking due to the Polonyi superfield is invoked. The inflationary scale is generically characterized by the inflaton mass in the vacuum m_ϕ, while the vacuum supersymmetry breaking by the gravitino mass. In this case one requires [31]

$$m_\phi|_{\text{vacuum}} \gg m_{3/2}|_{\text{vacuum}}\,, \tag{22}$$

and the conditions (8) and (18) to hold in the vacuum. Therefore in these models, during preheating supersymmetry breaking is dominated by the inflaton energy density and the longitudinal component of the gravitino should be identified with the inflatino χ_α instead of the *true goldstino* G_α [31].

Let us present a simple realization of inflation in these models. In fact our construction here is very similar to the *single superfield* inflationary models [51,52], and we require that the inflationary sector does not break supersymmetry in the vacuum. Consider the function entering the superpotential (19) to have the form

$$g = e^{2i\Phi}h(\Phi) + h_0\,, \tag{23}$$

where h_0 is a real constant. The scalar potential then reads

$$V = 2h_\phi^2 + 5h^2 + (f_0^2 - 3h_0^2) - 6hh_0\cos(2\phi)\,, \tag{24}$$

which can drive inflation by appropriately choosing the function $h(\phi)$. We also require to have a function h such that

$$\langle g_\phi\rangle = 0 = \langle h_\phi\rangle = \langle h\rangle\,, \tag{25}$$

which then automatically satisfies (18). The vacuum energy after the end of inflation will be $\Lambda_0 = f_0^2 - 3h_0^2$ which we assume to be very small.

An example for h is

$$h = \sqrt{V_0}\left[1 - (1+\phi)e^{-\phi}\right]\,, \tag{26}$$

with $\sqrt{V_0} \gg h_0$. The scalar potential becomes effectively

$$V \sim V_0\left\{2\phi^2 e^{-2\phi} + 5\left(1 - e^{-\phi} - \phi e^{-\phi}\right)^2\right\}\,, \tag{27}$$

which for large ϕ values approaches the constant value $5\times V_0$ and drives inflation with a plateau potential. In particular this model predicts for the tensor to scalar ratio: $r \sim 3\times 10^{-3}$ and for the spectral index: $1 - n_s \sim 0.035$, which are consistent with the latest data released by the Planck collaboration [53]. Moreover one can verify that the vacuum is at $\langle\phi\rangle = 0$, which then implies (25). Notice that during inflation $m_{3/2} > H^2$, therefore these models are within the perturbative unitarity bound discussed in [10]. For the inflaton and gravitino masses in the vacuum we have

$$m_\phi|_{\text{vacuum}} \sim \sqrt{V_0}\,, \quad m_{3/2}|_{\text{vacuum}} = h_0\,, \tag{28}$$

which realize (22).

4.2. Models including the stabilizer superfield

In this section we will construct inflationary models which on top of the inflaton superfield Φ and the Polonyi superfield X also include the stabilizer superfield S. The fact that there is now the stabilizer in the effective theory gives much more freedom in the inflationary model building. We can construct generic models by using the Kähler potential

$$K = X\overline{X} + S\overline{S} - \frac{1}{4}(\Phi - \overline{\Phi})^2\,, \tag{29}$$

and the superpotential

$$W = f_0 X + \Sigma(\Phi)S + g(\Phi) + g_0\,, \tag{30}$$

where g_0 is a complex constant. Once we insert (29) and (30) into (5) we find a theory with the minimal bosonic sector required for single field inflation, therefore the Lagrangian is (1), where the scalar potential reads

$$V(\phi) = f_0^2 + |\Sigma(\phi)|^2 + 2\left|\frac{\partial g(\phi)}{\partial\phi}\right|^2 - 3|g(\phi) + g_0|^2\,. \tag{31}$$

The gravitino mass is given by

$$m_{3/2} = |g(\phi) + g_0|\,. \tag{32}$$

Now the condition (22) can be easily satisfied because the gravitino mass (32) can be disentangled from the inflationary potential (31). We also require that in the vacuum $\langle\Sigma(\phi)\rangle = \langle g_\phi(\phi)\rangle = 0$, such that (18) are satisfied.

Simple models can be constructed by mimicking [10] and setting

$$\Sigma(\Phi) = \sqrt{3}\,g(\Phi)\,, \tag{33}$$

where now $g(\phi) = (g(\phi))^*$ but $g_0 = -g_0^*$ in (30), which brings the scalar potential to the form

$$V(\phi) = \left(f_0^2 - 3|g_0|^2\right) + 2\left|\frac{\partial g(\phi)}{\partial\phi}\right|^2\,. \tag{34}$$

We can now construct various inflationary potentials $V_{\text{infl.}}(\phi) = 2\mathcal{F}^2(\phi)$ by choosing

$$g(\Phi) = \int d\Phi\,\mathcal{F}(\Phi)\,, \tag{35}$$

and fixing the integration constant such that $\langle g_\phi \rangle = \langle g \rangle = 0$. For example, a realization of inflation with a plateau potential is achieved if we set

$$g(\Phi) = \sqrt{V_0}\left(-\sqrt{\frac{3}{2}} + \Phi + \sqrt{\frac{3}{2}} e^{-\sqrt{\frac{2}{3}}\Phi}\right), \quad (36)$$

with $\sqrt{V_0} \gg |g_0|$. The scalar potential then reads

$$V(\phi) = \left(f_0^2 - 3|g_0|^2\right) + 2V_0\left(1 - e^{-\sqrt{\frac{2}{3}}\phi}\right)^2. \quad (37)$$

The scalar potential (37) gives predictions for the r and $1-n_s$ similar to the Starobinsky model of inflation. The vacuum of this theory is at $\langle \phi \rangle = 0$, which gives

$$\langle \Sigma \rangle = \langle g_\phi \rangle = \langle g \rangle = 0. \quad (38)$$

During inflation one can see that $m_{3/2} > H^2$, therefore these models are also within the perturbative unitarity bound [10]. For the inflaton and gravitino masses in the vacuum we have

$$m_\phi|_{\text{vacuum}} \sim \sqrt{V_0}, \quad m_{3/2}|_{\text{vacuum}} = |g_0|, \quad (39)$$

which realize (22). Therefore during preheating the longitudinal gravitino components produced are to be interpreted as inflatinos and stabilinos, instead of true vacuum goldstinos that might be cosmologically hazardous.

4.3. Effective theories during the (p)reheating stage

In this section we outline some basics of the gravitino production during (p)reheating in standard supergravity considering the properties of the constrained superfield models discussed in the previous subsections. In the evolution of the universe subsequent to inflation, the inflaton field coherently oscillates about its vacuum and, among other particles, longitudinal gravitinos get produced due to the non-adiabatical change of the frequency of the gravitino $\pm 1/2$ helicity modes. The resulting number density of the longitudinal gravitinos is found to be [25,27–29]

$$n_l(t) = \langle 0|N/V|0\rangle = \frac{1}{\pi^2 a^3(t)} \int_0^{k_{\max}} dk\, k^2 |\beta_k|^2, \quad (40)$$

where $k_{\max}$ is the comoving momentum cut-off of the effective theory and β_k the corresponding Bogolyubov coefficient. If the physical momentum cut-off is of the order of the inflaton mass scale or somewhat lower, the resulting longitudinal gravitino number density might be cosmologically problematic. Namely, if $n_l(t_{\text{preh}}) \sim k_{\max}^3/a^3(t_{\text{preh}}) \sim m_\phi^3$ the extrapolation of the longitudinal gravitino number density at the time of the inflaton perturbative decay, t_{rh}, gives the yield

$$Y_l|_{\text{preh}} = \frac{n_l}{s} \simeq \frac{n_l(t_{\text{preh}})(t_{\text{preh}}/t_{\text{rh}})^2}{0.44 g_* T_{\text{rh}}^3}, \quad (41)$$

which rewrites

$$Y_l|_{\text{preh}} \sim 10^{-14}\alpha^2\left(\frac{T_{\text{rh}}}{10^{10}\,\text{GeV}}\right). \quad (42)$$

We considered that $a(t) \sim t^{2/3}$ during inflaton oscillations, $t_{\text{preh}} \sim \alpha m_\phi^{-1}$ with $\alpha \gg 1$ the characteristic time scale of preheating production, $t_{\text{rh}}^{-1} \sim H_{\text{rh}} \sim g_*^{1/2} T_{\text{rh}}^2/M_{\text{Pl}}$ and T_{rh} is the reheating temperature. If the above yield is interpreted as true vacuum goldstinos it generally violates the BBN bound (4) and/or might generate an overabundant dark matter at the time of the gravitino decay. Otherwise the estimation (41)–(42) is misleading. Evidently, the identification of the fermions that comprise the longitudinal gravitino component during the universe evolution is of central importance. To this end the field content of the theory and the corresponding dynamics should be specified.

As an illustrative example we choose a simple Kähler potential and superpotential for the inflaton–Polonyi sector

$$K = X\overline{X} + \Phi\overline{\Phi}, \quad W = f_0 X + g_0 + \frac{1}{2}m\Phi^2. \quad (43)$$

The constrained superfields X and Φ satisfy (7) and (11), but we have rescaled the fermion and scalar components of Φ such that

$$\Phi = \frac{\phi + ib}{\sqrt{2}} + \sqrt{2}\theta\chi + \theta^2 F^\Phi, \quad (44)$$

where b has the form (13) after appropriate rescaling of the component fields of Φ. As explained in [31,32], for models with Kähler potential and superpotential (43), the longitudinal gravitino component ψ^l is roughly given by

$$\psi^l \sim \sqrt{r_X}\, G + \sqrt{r_\Phi}\, \chi, \quad (45)$$

where

$$r_X = \frac{f_0^2}{f_0^2 + \frac{1}{2}\dot{\phi}^2 + \frac{1}{2}m^2\phi^2} = \frac{\rho_X}{\rho_X + \rho_\Phi}, \qquad r_\Phi = \frac{\frac{1}{2}\dot{\phi}^2 + \frac{1}{2}m^2\phi^2}{f_0^2 + \frac{1}{2}\dot{\phi}^2 + \frac{1}{2}m^2\phi^2} = \frac{\rho_\Phi}{\rho_X + \rho_\Phi}. \quad (46)$$

From (45) and (46) we see that if the inflaton superfield energy density ρ_Φ dominates over the Polonyi superfield energy density ρ_X during preheating, then the inflaton oscillation parametrically excites longitudinal gravitinos that should be identified as inflatinos [31–33]. In particular one has during preheating

$$\rho_X|_{\text{preh}} \ll \rho_\Phi|_{\text{preh}} \;\rightarrow\; \psi^l|_{\text{preh}} \sim \chi, \quad (47)$$

whereas at later times in the true vacuum, where the inflaton is strongly stabilized

$$\rho_X|_{\text{vacuum}} \gg \rho_\Phi|_{\text{vacuum}} \;\rightarrow\; \psi^l|_{\text{vacuum}} \sim G. \quad (48)$$

Numerical results can be found in the standard supergravity literature [31–33]. In models with the stabilizer superfield S, which has been studied recently [35], the description (45) for the longitudinal gravitino component will include an additional contribution given by $\sqrt{r_S}\lambda$ where $r_S = |F^S|^2/\rho_{\text{total}}$. These considerations imply that the yield (41)–(42) ought not be regarded as a realistic result.

In the working models studied in the previous subsections (4.1) and (4.2) one can easily check that the above results apply because the requirements (47) and (48) are met. In fact notice that if we expand the superpotentials from our previous examples in Φ (for $\Phi < M_{\text{Pl}}$) and keep only the leading terms we find the functions of the form (43), and then follow Planck-suppressed $\mathcal{O}(\Phi/M_{\text{Pl}})$ terms.

Finally we comment on the perturbative decay of the inflaton to gravitino. Here the standard results apply (see e.g. [34,54–56]) and we have a small gravitino production due to the small value of $W_\Phi|$. In particular for the examples we presented in the previous subsections, we have $\langle g_\phi \rangle = 0$, hence

$$\Gamma(\phi \rightarrow \psi^l\psi^l) \sim \frac{|g_\phi|^2 m_\phi^5}{m_{3/2}^4} \ll \Gamma(\phi \rightarrow \text{MSSM}), \quad (49)$$

that is the vacuum decay rate of the inflaton into two gravitinos can be sufficiently suppressed compared to the total inflaton decay rate in order that nonthermal gravitino overproduction problems are avoided.

5. Discussion

The main motivation of our work was to show that there do exist single-field inflationary models in supergravity constructed from constrained superfields which have two properties: a) They contain a minimal bosonic sector and b) They reproduce the results from standard supergravity for (p)reheating processes.

To achieve this we have presented a minimal setup for constructing effective theories for inflation in supergravity, such that the (p)reheating phase does not directly lead to gravitino overproduction. We have introduced a new customized constrained superfield Φ which includes the inflaton but also its fermion superpartner and the auxiliary field. We presented simple inflationary models which utilize this constrained superfield. It is now very interesting to couple these models to matter and study how the reheating works in a realistic setup.

We believe that our work offers a strong motivation in favor of using constrained superfields for constructing *effective theories*, against the compensator method for implementing non-linear realizations. Indeed, the component form methods where the Volkov–Akulov goldstino is used as a compensator to implement the local non-linear realization of supersymmetry will in principle not contain neither the required superpartner for the inflaton nor the inflaton multiplet auxiliary field. Therefore, unless additional fields with very specific couplings are introduced, supersymmetry breaking comes always from the same sector, the Volkov–Akulov sector, implying that the gravitino yield at the course of the cosmological evolution might be wrongly overestimated.

Acknowledgements

We thank N. Cribiori, G. Dall'Agata, A. Kehagias, A. Riotto and A. Van Proeyen. I.D. would like to thank the University of Padova for the hospitality offered to him during the preparation of this work. The work of I.D. is supported by IKY Scholarship Programs for Strengthening Post Doctoral Research, Co-financed by the European Social Fund – ESF and the Greek government. The work of F.F. is supported in parts by the Padova University Project CPDA119349.

References

[1] M. Rocek, Phys. Rev. Lett. 41 (1978) 451.

[2] U. Lindstrom, M. Rocek, Phys. Rev. D 19 (1979) 2300.

[3] R. Casalbuoni, S. De Curtis, D. Dominici, F. Feruglio, R. Gatto, Phys. Lett. B 220 (1989) 569.

[4] Z. Komargodski, N. Seiberg, J. High Energy Phys. 0909 (2009) 066, arXiv:0907.2441 [hep-th].

[5] N. Cribiori, G. Dall'Agata, F. Farakos, arXiv:1704.07387 [hep-th].

[6] G. Dall'Agata, E. Dudas, F. Farakos, J. High Energy Phys. 1605 (2016) 041, arXiv:1603.03416 [hep-th].

[7] I. Antoniadis, E. Dudas, S. Ferrara, A. Sagnotti, Phys. Lett. B 733 (2014) 32, arXiv:1403.3269 [hep-th].

[8] S. Ferrara, R. Kallosh, A. Linde, J. High Energy Phys. 1410 (2014) 143, arXiv:1408.4096 [hep-th].

[9] R. Kallosh, A. Linde, J. Cosmol. Astropart. Phys. 1501 (2015) 025, arXiv:1408.5950 [hep-th].

[10] G. Dall'Agata, F. Zwirner, J. High Energy Phys. 1412 (2014) 172, arXiv:1411.2605 [hep-th].

[11] Y. Kahn, D.A. Roberts, J. Thaler, J. High Energy Phys. 1510 (2015) 001, arXiv:1504.05958 [hep-th].

[12] S. Ferrara, R. Kallosh, J. Thaler, Phys. Rev. D 93 (4) (2016) 043516, arXiv:1512.00545 [hep-th].

[13] J.J.M. Carrasco, R. Kallosh, A. Linde, Phys. Rev. D 93 (6) (2016) 061301, arXiv:1512.00546 [hep-th].

[14] G. Dall'Agata, F. Farakos, J. High Energy Phys. 1602 (2016) 101, arXiv:1512.02158 [hep-th].

[15] R. Kallosh, A. Linde, T. Wrase, J. High Energy Phys. 1604 (2016) 027, arXiv:1602.07818 [hep-th].

[16] E. McDonough, M. Scalisi, J. Cosmol. Astropart. Phys. 1611 (11) (2016) 028, arXiv:1609.00364 [hep-th].

[17] E. Dudas, L. Heurtier, C. Wieck, M.W. Winkler, Phys. Lett. B 759 (2016) 121, arXiv:1601.03397 [hep-th].

[18] F. Hasegawa, K. Mukaida, K. Nakayama, T. Terada, Y. Yamada, Phys. Lett. B 767 (2017) 392, arXiv:1701.03106 [hep-ph].

[19] K. Benakli, Y. Chen, E. Dudas, Y. Mambrini, Phys. Rev. D 95 (9) (2017) 095002, arXiv:1701.06574 [hep-ph].

[20] J.R. Ellis, A.D. Linde, D.V. Nanopoulos, Phys. Lett. B 118 (1982) 59.

[21] D.V. Nanopoulos, K.A. Olive, M. Srednicki, Phys. Lett. B 127 (1983) 30.

[22] J.R. Ellis, J.E. Kim, D.V. Nanopoulos, Phys. Lett. B 145 (1984) 181.

[23] L. Kofman, A.D. Linde, A.A. Starobinsky, Phys. Rev. D 56 (1997) 3258, http://dx.doi.org/10.1103/PhysRevD.56.3258, arXiv:hep-ph/9704452.

[24] A.L. Maroto, A. Mazumdar, Phys. Rev. Lett. 84 (2000) 1655, arXiv:hep-ph/9904206.

[25] G.F. Giudice, M. Peloso, A. Riotto, I. Tkachev, J. High Energy Phys. 9908 (1999) 014, arXiv:hep-ph/9905242.

[26] M. Lemoine, Phys. Rev. D 60 (1999) 103522, arXiv:hep-ph/9908333.

[27] R. Kallosh, L. Kofman, A.D. Linde, A. Van Proeyen, Phys. Rev. D 61 (2000) 103503, arXiv:hep-th/9907124.

[28] G.F. Giudice, I. Tkachev, A. Riotto, J. High Energy Phys. 9908 (1999) 009, arXiv:hep-ph/9907510.

[29] G.F. Giudice, A. Riotto, I. Tkachev, J. High Energy Phys. 9911 (1999) 036, arXiv:hep-ph/9911302.

[30] R. Kallosh, L. Kofman, A.D. Linde, A. Van Proeyen, Class. Quantum Gravity 17 (2000) 4269, Erratum: Class. Quantum Gravity 21 (2004) 5017, arXiv:hep-th/0006179.

[31] H.P. Nilles, M. Peloso, L. Sorbo, Phys. Rev. Lett. 87 (2001) 051302, arXiv:hep-ph/0102264.

[32] H.P. Nilles, M. Peloso, L. Sorbo, J. High Energy Phys. 0104 (2001) 004, arXiv:hep-th/0103202.

[33] P.B. Greene, K. Kadota, H. Murayama, Phys. Rev. D 68 (2003) 043502, arXiv:hep-ph/0208276.

[34] M. Kawasaki, F. Takahashi, T.T. Yanagida, Phys. Rev. D 74 (2006) 043519, arXiv:hep-ph/0605297.

[35] Y. Ema, K. Mukaida, K. Nakayama, T. Terada, J. High Energy Phys. 1611 (2016) 184, http://dx.doi.org/10.1007/JHEP11(2016)184, arXiv:1609.04716 [hep-ph].

[36] P.A.R. Ade, et al., Planck Collaboration, Planck 2015 results. XIII. Cosmological parameters, arXiv:1502.01589 [astro-ph.CO].

[37] J. Ellis, M.A.G. Garcia, D.V. Nanopoulos, K.A. Olive, M. Peloso, J. Cosmol. Astropart. Phys. 1603 (03) (2016) 008, arXiv:1512.05701 [astro-ph.CO].

[38] E. Dudas, S. Ferrara, A. Kehagias, A. Sagnotti, J. High Energy Phys. 1509 (2015) 217, arXiv:1507.07842 [hep-th].

[39] E.A. Bergshoeff, D.Z. Freedman, R. Kallosh, A. Van Proeyen, Phys. Rev. D 92 (8) (2015) 085040, Erratum: Phys. Rev. D 93 (6) (2016) 069901, arXiv:1507.08264 [hep-th].

[40] F. Hasegawa, Y. Yamada, J. High Energy Phys. 1510 (2015) 106, arXiv:1507.08619 [hep-th].

[41] I. Bandos, L. Martucci, D. Sorokin, M. Tonin, J. High Energy Phys. 1602 (2016) 080, arXiv:1511.03024 [hep-th].

[42] I. Bandos, M. Heller, S.M. Kuzenko, L. Martucci, D. Sorokin, J. High Energy Phys. 1611 (2016) 109, arXiv:1608.05908 [hep-th].

[43] N. Cribiori, G. Dall'Agata, F. Farakos, M. Porrati, Phys. Lett. B 764 (2017) 228, arXiv:1611.01490 [hep-th].

[44] J. Wess, J. Bagger, Supersymmetry and Supergravity, Univ. Pr., Princeton, USA, 1992.

[45] S. Ferrara, R. Kallosh, A. Van Proeyen, T. Wrase, J. High Energy Phys. 1604 (2016) 065, arXiv:1603.02653 [hep-th].

[46] S. Ferrara, A. Kehagias, A. Sagnotti, Int. J. Mod. Phys. A 31 (25) (2016) 1630044, arXiv:1605.04791 [hep-th].

[47] R. Kallosh, A. Linde, T. Rube, Phys. Rev. D 83 (2011) 043507, arXiv:1011.5945 [hep-th].

[48] A. Brignole, F. Feruglio, F. Zwirner, J. High Energy Phys. 9711 (1997) 001, arXiv:hep-th/9709111.

[49] G. Dall'Agata, S. Ferrara, F. Zwirner, Phys. Lett. B 752 (2016) 263, arXiv:1509.06345 [hep-th].

[50] M. Kawasaki, M. Yamaguchi, T. Yanagida, Phys. Rev. Lett. 85 (2000) 3572, arXiv:hep-ph/0004243.

[51] S.V. Ketov, T. Terada, Phys. Lett. B 736 (2014) 272, arXiv:1406.0252 [hep-th].

[52] S.V. Ketov, T. Terada, Eur. Phys. J. C 76 (8) (2016) 438, arXiv:1606.02817 [hep-th].

[53] P.A.R. Ade, et al., Planck Collaboration, Planck 2015 results. XX. Constraints on inflation, Astron. Astrophys. 594 (2016) A20, arXiv:1502.02114 [astro-ph.CO].

[54] M. Endo, K. Hamaguchi, F. Takahashi, Phys. Rev. Lett. 96 (2006) 211301, arXiv:hep-ph/0602061.

[55] S. Nakamura, M. Yamaguchi, Phys. Lett. B 638 (2006) 389, arXiv:hep-ph/0602081.

[56] M. Dine, R. Kitano, A. Morisse, Y. Shirman, Phys. Rev. D 73 (2006) 123518, arXiv:hep-ph/0604140.

7

Could the primordial radiation be responsible for vanishing of topological defects?

Tomasz Romańczukiewicz

Institute of Physics, Jagiellonian University, Kraków, Poland

ARTICLE INFO

Editor: M. Trodden

Keywords:
Domain walls
Negative radiation pressure
Solitons

ABSTRACT

We study the motion of domain walls in 1+1, 2+1 and briefly 3+1 d relativistic ϕ^6 model with three equal vacua in the presence of radiation. We show that even small fluctuations can trigger a chain reaction leading to vanishing of the domain walls. Only one vacuum remains stable and domains containing other vacua vanish. We explain this phenomenon in terms of radiation pressure (both positive and negative). We construct an effective model which translates the fluctuations into additional term in the field theory potential. In case of two dimensional model we find a relation between the critical size of the bulk and amplitude of the perturbation.

1. Introduction

Topological defects arise in surprisingly many branches of physics. They can be found in liquid crystals [1], liquid helium [2], ferromagnets, superconductors, graphene [3] and many more important physical substances. It is also natural to expect that topological defects should have been created in large numbers in the early Universe via Kibble–Zurek mechanism during some symmetry breaking phase transitions [4,5]. Unfortunately there are no direct observation evidence proving such objects ever existed. However, it might be plausible that some linear defects (cosmic strings) could give origins to some large scale structures in the Universe. Some observed fluctuations in the cosmic microwave background referred to as "cold spot" could be explained as remnants of textures from early stages the Universe [6,7]. Topological defects are sometimes also considered as one of the dark matter candidates [8]. Surprisingly, there are no signs of other defects like monopoles and domain walls.

Topological defects can interact with each other as well as with some other objects like oscillons [9,10]. A very interesting interaction also can be observed between topological defects and radiation. In some cases the radiation can exert an ordinary radiation pressure proportional to the square of amplitude of incident wave. However, some defects, like kinks in a very popular ϕ^4 theory, are transparent to the radiation in the first order [11,12]. Sine-Gordon kinks are exactly transparent in all orders. Higher order analysis in ϕ^4 model revealed a surprising feature. The kinks undergo the negative radiation pressure (NRP) which accelerates the kinks towards the source of radiation. The acceleration of a kink in ϕ^4 model is proportional to the fourth power of amplitude of the wave. In models with two scalar fields with different masses it is possible that the radiation can exert both positive and negative radiation pressure depending on the composition of the wave [13,14]. In such a case the force exerted on the kink is proportional to the square of the amplitude. More recently some other examples were discussed in case of light and matter waves which, when scattered on a small objects, can bend in such a way that the object would feel the pulling force [15,16]. Mixing between different frequencies can cause the NRP in case of solitons with rotating phase as in Coupled Nonlinear Schrödinger Equation [17]. We want to emphasize that the NRP seen in case of solitons in the present and our previous papers is of a very different nature then the one described in optical physics. It can be exerted on flat and infinite surfaces where simple bending of the light or other wave trajectories is not an option.

In the present paper we consider a mechanism which could increase the rate of the domain wall collisions. In models with at least two equal minima of the potential but with different masses of small perturbations around those vacua. The interest in such models has increased recently [18–20] but they were considered many years ago as for example so called bag models of hadrons

E-mail address: trom@th.if.uj.edu.pl.

[21]. Static kinks (or domain walls in general) in such models have asymmetric profiles. We show that despite the fact that the vacua have the same energies (they are true vacua) the kinks usually accelerate in one direction no matter from where the radiation comes. Antikinks accelerate in opposite direction. Any small perturbation can therefore trigger a chain reaction during which defects collide and create more radiation accelerating other defects causing more collisions.

The present letter is organized as follows. First we define our example ϕ^6 model, than we show how kinks interact with monochromatic wave in case of two vacua with different masses of scalar field. We derive an analytic formula for the force with which such monochromatic wave acts on a kink. Next we show how generic perturbation can influence the stability of kink system (a lattice). In particular we study the effect of random fluctuations with Gaussian distribution filling the whole space. We show that the fluctuations are in some ways equivalent to the shift of the vacua. We also compare the results with other models. The last section concerns higher dimensional case. We find a critical size of a circular domain wall which could either grow or collapse depending on what type of vacuum is inside and how large the fluctuations are.

2. The model

In the present paper we consider one and two dimensional ϕ^6 theory, which can be defined by the rescaled Lagrangian density [22,18]

$$\mathcal{L} = \frac{1}{2}\partial_\mu\phi\partial^\mu\phi - U(\phi), \quad \text{where} \quad U(\phi) = \frac{1}{2}\phi^2\left(\phi^2-1\right)^2. \quad (1)$$

The model has three vacua $\phi_v \in \{-1, 0, 1\}$. Small perturbations around these vacua have different masses: $m_0 = 1$ and $m_1 = 2$ for $\phi_v = 0$ and $\phi_v = \pm 1$ respectively. In one dimensional case the kinks and antikinks can be found from a single solution

$$\phi_K(x) \equiv \phi_{(0,1)}(x) = \sqrt{\frac{1+\tanh x}{2}} \quad (2)$$

using the discrete symmetries of the model: The masses of all kinks are $M = 1/4$. Small perturbation added to the kink solution $\phi(x,t) = \phi_s(x) + \eta(x)e^{i\omega t}$ is governed by a linearized equation

$$-\eta_{xx} + V(x)\eta = \omega^2\eta \quad (3)$$

the potential $V(x)$ is

$$V(x) = U''(\phi_s) = 15\phi_s^4 - 12\phi_s^2 + 1. \quad (4)$$

Note that when $x \to -\infty$ the potential $V \to 1$ and $V(x \to \infty) \to 4$. Solutions to this linearized equation can be found in analytic form in [22]. Let us consider a wave traveling from the left side of kink *i.e.* from $\phi = 0$ vacuum. Asymptotic form of these solutions for frequencies above the two mass thresholds due to Lohe can be written as

$$\begin{cases} \eta_{+\infty}(x) \to e^{ikx}/A(q,k), \\ \eta_{-\infty}(x) \to e^{iqx} + \frac{A(-q,k)}{A(q,k)}e^{-iqx} \end{cases} \quad (5)$$

with

$$q = \sqrt{\omega^2-1}, \; k = \sqrt{\omega^2-4},$$
$$A(q,k) = \frac{\Gamma(1-ik)\Gamma(-iq)}{\Gamma(-\frac{1}{2}ik - \frac{1}{2}iq + \frac{5}{2})\Gamma(-\frac{1}{2}ik - \frac{1}{2}iq - \frac{3}{2})}. \quad (6)$$

This solution represents a wave traveling from $-\infty$ with amplitude 1. Amplitude of the reflected wave is equal to $\frac{A(-q,k)}{A(q,k)}$, and amplitude of the transient wave is $\frac{1}{A(q,k)}$. We can use this form to calculate the momentum and energy balance far away from the kink. From Noether's theorem, the conservation laws for energy and momentum density can be written as:

$$\partial_t\mathcal{E} = \partial_x\left(\phi'\dot{\phi}\right), \quad (7a)$$
$$\partial_t\mathcal{P} = -\frac{1}{2}\partial_x\left(\dot{\phi}^2 + \phi'^2 - 2U(\phi)\right). \quad (7b)$$

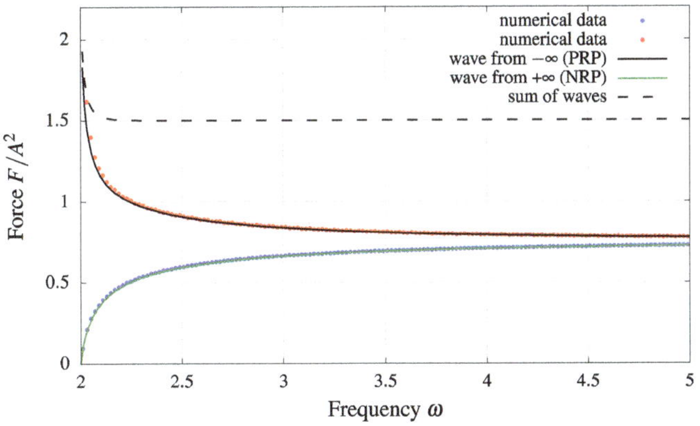

Fig. 1. Theoretical values of the force exerted on the kink. In both cases the force is positive. The color points are the results of numerical calculations for $\mathcal{A} = 0.05$.

Integrating the above expressions inside interval $[-L, L]$ and averaging over a period T we obtain energy and momentum balance just using asymptotic form of scattering solutions. If the kink does not move initially the conservation laws give (for $\phi = \phi_s + \mathcal{A}\,\mathrm{re}(e^{-i\omega t}\eta(x))$):

$$F_{+\infty}(q,k) = \frac{1}{2}\frac{\mathcal{A}^2}{|A(q,k)|^2}\left(2|A(-q,k)|^2q^2 + qk - k^2\right). \quad (8)$$

We can perform a very similar calculation for the second case when the wave is coming from $+\infty$. The force in this case can be expressed as:

$$F_{-\infty}(q,k) = -F_{+\infty}(k,q) \equiv \mathcal{A}^2(\omega)f(\omega). \quad (9)$$

Fig. 1 shows the force in both cases. Note that the force is positive for all frequencies. The kink will always accelerate towards $+\infty$ no matter which direction the wave comes from. In the first case the wave comes from $\phi = 0$ $(m = 1)$ and exerts positive radiation pressure. In the second case it comes from the second vacuum $\phi = 1$ with mass $m = 2$ and exerts negative radiation pressure.

3. General perturbations

Kinks interact very weakly with each other on large distances. Their profile vanish exponentially, and so do the interaction between them. For a pair of kinks initially separated by the distance L the estimated time to the collision is of order of $T \approx 2e^{L/2}/\sqrt{L}$. The value was numerically verified. System of static, separated kinks can last even longer, because the forces from neighboring kinks cancel each other (Fig. 2.a, here for the first three kinks $L = 20$ so the timescale to collision is about 10^4). However, adding a small perturbation causes the radiation which exerts pressure on those kinks. We have tested this idea by adding a localized Gaussian profile $\phi = \phi_{kinks} + ae^{-bx^2}$ or colliding two kinks (Fig. 2.b). The collapse of the system was evidently faster compared to the case when no perturbation was added. Because of the polarity in the direction of the radiation pressure, the kinks always accelerate in such a way that the regions with vacuum $\phi = 0$ grow and vacua $\phi = \pm 1$ shrink. Moreover, when kink and antikink collide

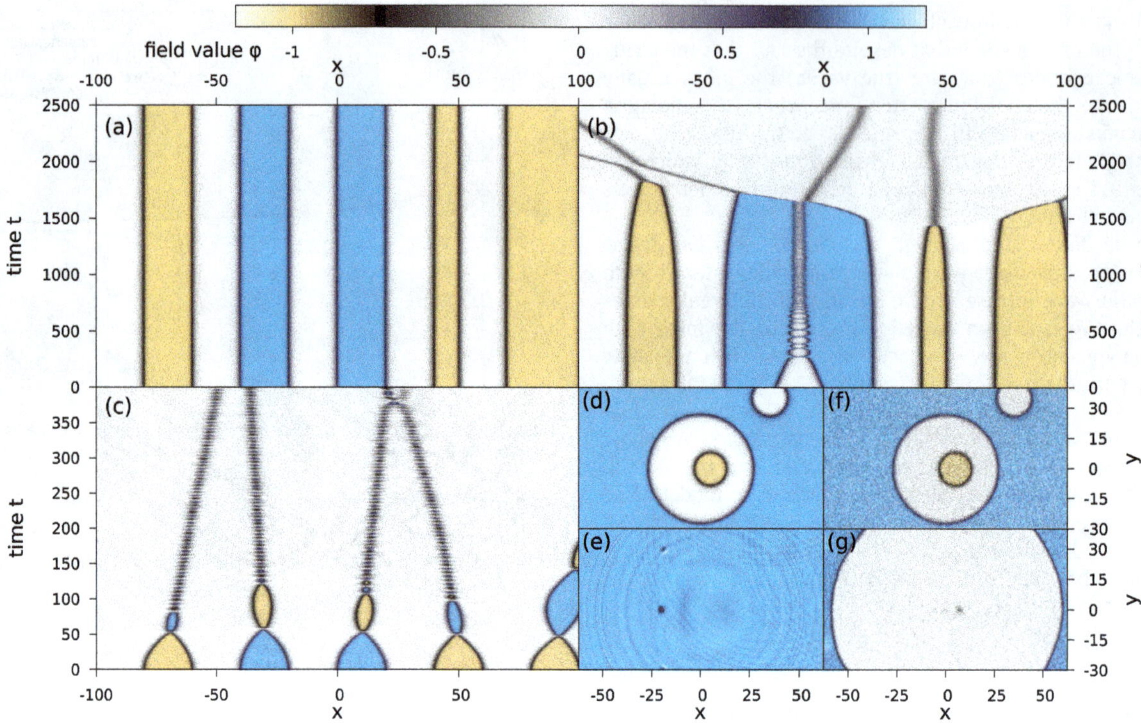

Fig. 2. Evolution of system of kinks: (a) nearly static configuration, (b) annihilation of close kink–antikink pair triggers the chain reaction of other annihilation, (c) with Gaussian noise with amplitude $A = 0.05$. Evolution of two dimensional circular domain walls: without perturbation (d) $t = 0$ (e) $t = 75$ and with Gaussian noise $A = 0.12$ (f) $t = 0$ (g) $t = 75$.

they form an oscillon. During such collisions more radiation is produced increasing the rate with which other kinks collide. Therefore a small perturbation can start a chain reaction leading to a collapse of the system of kinks. We have also simulated this radiation by adding random noise $\langle \mathcal{A} \rangle \zeta(x)$ to the initial conditions, where $\langle \zeta(x)\zeta(x') \rangle = \delta(x - x')$ is a random number with Gauss distribution. Adding such term is equivalent to introducing to the system background radiation which could be a relic of some phase transition (Fig. 2.c). Again, the system of kinks was quickly destabilized. We also solved Langevin equation but because of the damping term the dynamics was slowed down. After collisions of kink and antikink an oscillon was formed. The oscillons are not stable but they live for a very long time. However, perturbation usually increase the rate with which the oscillon radiates shortening its lifetime. Therefore, initially almost static system of kinks can end up as almost uniform state filled with radiation around one stable vacuum. The time when this state is reached is determined by the life span of the oscillons in given conditions. It is also worth mentioning that in some models collision of a kink and an antikink can lead to resonant windows and fractal structure. This happens usually when kinks have internal degrees of freedom (as in ϕ^4) or if some meson states can be trapped between the kinks as it could happen in our model (bag model). However, the presented mechanism leads to collisions of such configurations which do not trap mesons.

4. Other models

Similar simulations for other models like ϕ^4 revealed that the system of kinks can be destabilized by radiation, however, when the mass of small perturbation is the same around each vacuum there is no polarity in the direction of the acceleration of the kinks. The motion of kinks resembles a random walk. In ϕ^4 model the radiation from the collisions did not change the motion of the other kinks significantly. Kinks are transparent in the first order in the amplitude of the radiation. They undergo a negative radiation pressure which is proportional to the fourth power of the amplitude. Due to the nonlinear nature of this phenomenon, there is no simple superposition rule to sum all the effects coming from different frequencies. However we have found that the most dominant contribution to the motion of the kinks have collisions with oscillons created in earlier collisions. An oscillon can bounce back from the kink or go through it. However it can possess large amount of energy which can be transferred to the kink, significantly increasing its velocity.

In the sine-Gordon model the kinks are completely transparent to the radiation due to the integrability of the model. During collisions no radiation is lost. Therefore the evolution of system of kinks is completely determined by initial conditions. No annihilation or creation is possible within the sG model.

In general, if all the vacua have not only the same value but also the same mass of small perturbations there is no difference as to the direction from which the radiation came. Contributions coming from both sides should be equal and cancel each other. It does not matter whether the single, monochromatic wave traveling in one direction exerts a positive or negative radiation pressure. No vacuum is distinguished. On the other hand if the masses are different, as they are in the discussed ϕ^6 model, some excitations are more easily excited and can give nonvanishing contribution pushing the kink towards the vacuum around which the perturbations have larger mass.

It is also possible to construct more complicated theories including gauge fields or supersymmetry [23,31]. All the above considerations should hold only if there is a difference in masses of small fluctuations around different vacua.

5. Effective model

When the radiation fills uniformly the entire space we can integrate out small degrees of freedom and try to construct a more

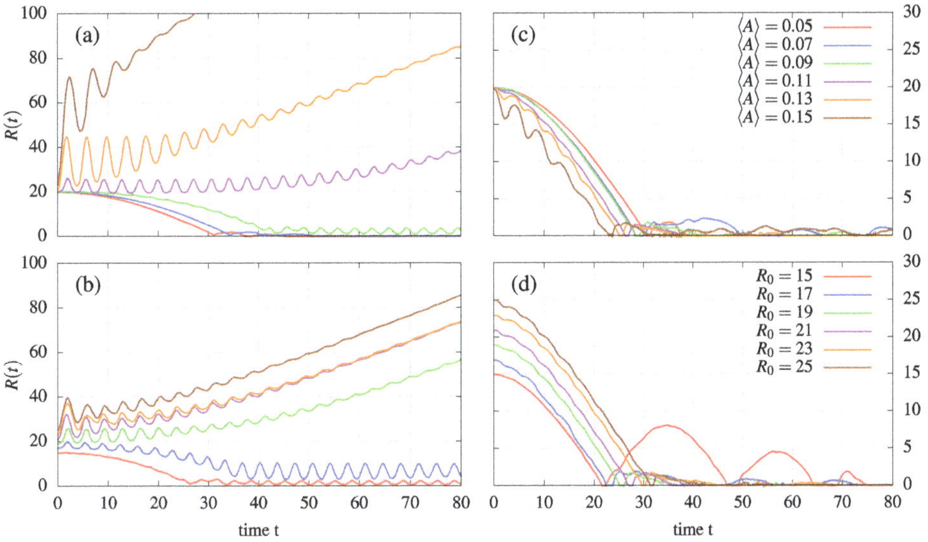

Fig. 3. A size of a bulk of vacuum in the presence of fluctuations. The upper plots (a) and (c) show the radius of the bulk for initial conditions $R_0 = 20$ and variable amplitude of fluctuations. The lower plots (b) and (d) show the evolution of the radius of bulks with different initial radii with the same fluctuation amplitude $\langle A \rangle = 0.1$. Plots (a) and (b) correspond to enclosed vacuum $\phi_{in} = 0$ and plots (c) and (d) to $\phi_{in} = 1$.

effective model describing only the motion of the kinks. Suppose that we know the spectral distribution of the radiation $\mathcal{A}(\omega)$ (which could be for instance the spectrum of the black body radiation). In ϕ^6 the radiation pressure comes from the first order perturbation series, therefore we can use the ordinary superposition rule. The total force exerted by the radiation could be written as an integral over all frequencies.

$$F_{tot} = \int_{-\infty}^{\infty} d\omega\, \mathcal{A}(\omega)^2 f(\omega) = \alpha \langle \mathcal{A} \rangle^2, \qquad (10)$$

where α is a value which in principle should depend on the distribution. In case of uniform distribution for all frequencies we have two contributions corresponding to two waves coming from opposite directions. They should have the same amplitude, so they should add. The sum of those functions is almost constant (within 1% for $\omega > 2.2$) and equal to 1.5 so we can assume that $\alpha = 0.75$. The discrepancy can be large only if low frequencies dominate. $\langle \mathcal{A} \rangle$ is the average amplitude of the perturbation. Note that all contributions $f(\omega)$ push the kink in the same direction (see eq. (9)). Kinks separated by the distance $L = 20$ as they are in Fig. 2.c should collide after time $T = \sqrt{ML/F_{tot}} \approx 52$ which is in very good agreement with the simulations.

Exactly the same force with just an opposite sign is exerted on the antikink. This can be effectively described as if one of the vacua is raised. It is quite easy to show that adding a term to the field theory potential, shifting vacua, results in external force acting on a kink, which in the first order can be written as $F = -\Delta U/M$, where ΔU is a gap between the vacua [24]. Therefore the total force exerted by the radiation can be effectively written as a shift in the potential of the vacua ± 1 by $\Delta U \sim \langle \mathcal{A} \rangle^2$.

It is known [26] that quantum radiative corrections can also change the effective potential via Weinberg–Coleman effect. The corrections depend on the second derivative of the potential around the vacuum. The quantum shift depends therefore also on the mass of the small perturbations. Another interesting results were recently presented in [27,28] where the author shows that quantum corrections lead to the polarization of the vacua which breaks the translational invariance and the moving kink gains energy.

6. Two dimensional case

Let us now consider domain walls in 2+1 dimensions. Effects described in the previous section are still present so the radiation pressure would try to close any bulks with vacua ± 1 and bulks enclosing vacuum $\phi = 0$ should grow. However, in two or more dimensions another phenomenon is present. The energy of a circular domain wall with large enough radius is equal to $E(R) = 2\pi MR$. This energy can be considered as the potential energy. The force acting on the unit length of the defect is equal to $F = -M/R$. More precise calculation of the evolution can be found for example in [29]. If the field inside the domain wall reaches the value $\phi_{in} = \pm 1$ and outside the wall $\phi_{out} = 0$ the forces of radiation pressure and tension act in the same direction. So any radiation would speed up the contraction of such circles. On the other hand when $\phi_{in} = 0$ and $\phi_{out} = \pm 1$ the forces act in opposite directions. The tension can be balanced with the radiation pressure $F = \alpha \langle \mathcal{A} \rangle^2$, where α contains all the information about the distribution of the perturbations and the geometry of the defect. $\langle \mathcal{A} \rangle$ is a amplitude of the noise. Therefore there is a critical radius of the bulk $R_{crit} = \frac{M}{\alpha \langle \mathcal{A} \rangle^2}$. We have performed numerical simulations with initial conditions of a circular domain wall with radius R_0 with additional random fluctuations of amplitude $\langle \mathcal{A} \rangle$. For given amplitude domain walls below certain radius shrank while domain walls which initially had a larger radius grew up (Fig. 3 (a),(b)). The simulations also confirmed our rough estimation that the critical radius indeed is proportional to $\langle \mathcal{A} \rangle^2$. From the critical values measured from simulations we can estimate that in the case of circular domain walls $\alpha \approx 1.3$ which is almost twice the value obtained in one-dimensional case.

Domain walls enclosing $\phi = \pm 1$ vacua shrank faster with larger amplitude of fluctuation (Fig. 3 (c),(d)). Moreover we have noticed that some radiation entering the bulk does not leave. After a while there is more radiative energy within the bulk. This creates additional pressure. Therefore initially static bulk starts to expand. However this effect is weak and does not influence much the above analysis. Another observation we have noticed is that a number of oscillons was created. The oscillons were created in more or less the same phase. This fact was earlier noticed by Gleiser [25].

For comparison we have performed analogous numerical simulations in case of ϕ^4 model. The radiation did not affect much the motion of the domain walls. Especially because the radiation coming from opposite sides of the wall exerts opposite effects which cancel each other out. In this case all the domain walls shrank with more or less the same rate.

In three dimensions the only difference is that the force shrinking the sphere is equal to $F = -2M/R$ which results in the critical radius being twice the critical radius in two dimensions. $R_{crit}^{(3D)} = \frac{2M}{\alpha \langle \mathcal{A} \rangle^2}$.

7. Conclusions

We have shown that in the ϕ^6 model the differences in masses of small perturbations around different vacua cause the polarization of radiation pressure. No matter what is the characteristic of perturbation, the kinks would always accelerate in such a way that the vacua ± 1 would disappear. The only stable vacuum is therefore $\phi = 0$. This behavior is very similar to the effect when the vacua ± 1 are raised. Therefore for uniformly distributed radiation one can introduce the effective model. The shift of the vacua can be expressed as an integral over all frequencies and is proportional to the square of the amplitude of the perturbation. Any radiation can force two kinks to accelerate towards each other. After the collision more radiation is created forcing other kinks to collide faster. Therefore even small, local perturbation can start a chain reaction leading to the collapse of a system of initially almost static kinks. In higher dimensions, closed domain walls in the absence of other perturbation, usually shrink. The tension is proportional to the inverse of a radius.

Radiation, or any type of fluctuations can speed up the process of domain wall decay, if vacuum with higher mass of small perturbation is closed inside. If the radius of the domain enclosing the vacuum with smaller mass, the radiation can stop the decay and even accelerate the growth of such domain. The critical radius is proportional to the inverse second power of the average of perturbation amplitude.

The radiation pressure can be a cause of vanishing of kinks and domain walls in some models. This process is very rapid compared with the interaction of static topological defects. No such phenomenon was found in case where masses of the small perturbations around vacua were the same.

We are sure that the similar phenomenon should exist in more complicated models widely discussed in the literature [30]. The only requirement is that the field theory potential is not entirely symmetric although the energies of the vacua are the same. Scalar field coupled to a gauge field can play a role of a Higgs field. Different expectation values of the Higgs field in different vacua can result in different masses of the gauge fields. The mechanism described in the present paper would favor the vacua with the smaller masses.

The aim of the paper is to present the mechanism in the simplest possible model. However, one of the possible future applications of the mechanism is the cosmological context. It still remains a mastery why there are no signs of existence of topological defect in the early Universe. Perhaps the presented mechanism can serve as a small step towards understanding of this problem, at least in the context of domain walls.

Although the mechanism presented in the letter is limited to the domain walls it is likely that some extension of it may be possible in case of higher co-dimensional topological defects such as vortices or monopoles. The existence of the defect is ensured by the vacuum manifold. However small perturbations around different points of this vacuum manifold can still have different masses. It is possible that such defects in a presence of radiation would accelerate more likely in certain directions pointing towards the smallest mass of the field. The collisions with other defects along these directions would be more likely.

Supplementary material. The letter is complemented with simulations showing some of the features discussed.

References

[1] M. Kleman, O.D. Lavrentovich, Philos. Mag. 86 (25–26) (2006) 4117.
[2] D. Vollhardt, P. Woelfle, Acta Phys. Pol. B 31 (2000) 2837.
[3] R.D. Yamaletdinov, V.A. Slipko, Y.V. Pershin, arXiv preprint, arXiv:1705.10684.
[4] T.W. Kibble, Acta Phys. Pol. B 13 (1982) 723.
[5] M. Hindmarsh, T.W. Kibble, Rep. Prog. Phys. 58 (1995) 477.
[6] N. Turok, D.N. Spergel, Phys. Rev. Lett. 64 (1990) 2736.
[7] M. Cruz, N. Turok, P. Vielva, E. Martínez-González, M. Hobson, Science 318 (5856) (2007) 1612.
[8] P. Forgács, A. Lukács, Phys. Rev. D 95 (2017) 035003.
[9] M. Hindmarsh, P. Salmi, Phys. Rev. D 77 (2008) 105025.
[10] T. Romańczukiewicz, Acta Phys. Pol. B, Proc. Suppl. 2 (2009) 611.
[11] T. Romańczukiewicz, Acta Phys. Pol. B 35 (2004) 523.
[12] P. Forgács, A. Lukács, T. Romańczukiewicz, Phys. Rev. D 77 (2008) 125012.
[13] T. Romańczukiewicz, Acta Phys. Pol. B 39 (2008) 3449.
[14] P. Forgács, A. Lukács, T. Romańczukiewicz, Phys. Rev. D 88 (2013) 125007.
[15] D.B. Ruffner, D.G. Grier, Phys. Rev. Lett. 109 (2012) 163903.
[16] A.A. Gorlach, M.A. Gorlach, A.V. Lavrinenko, A. Novitsky, Phys. Rev. Lett. 118 (2017) 180401.
[17] T. Romańczukiewicz, arXiv preprint, arXiv:1705.06955.
[18] P. Dorey, K. Mersh, T. Romańczukiewicz, Ya. Shnir, Phys. Rev. Lett. 107 (2011) 091602.
[19] A. Demirkaya, R. Decker, P.G. Kevrekidis, I.C. Christov, A. Saxena, arXiv preprint, arXiv:1706.01193.
[20] D. Bazeia, E.E.M. Lima, L. Losano, arXiv preprint, arXiv:1705.02839, 2017.
[21] N.H. Christ, T.D. Lee, Phys. Rev. D 12 (1975) 1606.
[22] M.A. Lohe, Phys. Rev. D 20 (1979) 3120.
[23] L. Pogosian, T. Vachaspati, Phys. Rev. D 67 (2003) 065012.
[24] V.G. Kiselev, Y.M. Shnir, Phys. Rev. D 57 (1998) 5174.
[25] M. Gleiser, Int. J. Mod. Phys. D 16 (2007) 219.
[26] S. Coleman, E.J. Weinberg, Phys. Rev. D 8 (1973) 1888.
[27] H. Weigel, Phys. Lett. B 766 (2017) 65.
[28] H. Weigel, Adv. High Energy Phys. 2017 (2017) 1486912.
[29] H. Arodź, R. Pełka, Phys. Rev. E 62 (2000) 6749.
[30] T. Vachaspati, Lectures at NATO ASI "Patterns of symmetry breaking", Cracow, September 2002.
[31] J.S. Rozowsky, R.R. Volkas, K.C. Wali, Phys. Lett. B 580 (2004) 249.

8

Coulomb effects in high-energy e^+e^- electro-production by a heavy charged particles in an atomic field

P.A. Krachkov*, A.I. Milstein

Budker Institute of Nuclear Physics, 630090 Novosibirsk, Russia

ARTICLE INFO

Editor: A. Ringwald

Keywords:
Electroproduction
Coulomb corrections

ABSTRACT

The cross section of high-energy e^+e^- pair production by a heavy charged particle in the atomic field is investigated in detail. We take into account the interaction with the atomic field of e^+e^- pair and a heavy particle as well. The calculation is performed exactly in the parameters of the atomic field. It is shown that, in contrast to the commonly accepted point of view, the cross section differential with respect to the final momentum of a heavy particle is strongly affected by the interaction of a heavy particle with the atomic field. However, the cross section integrated over the final momentum of a heavy particle is independent of this interaction.

1. Introduction

Production of e^+e^- pair by the ultra-relativistic heavy charged particle in the atomic field is very important because the cross section of this process is even larger than the cross section of bremsstrahlung of a heavy particle in the field. Thus, this process plays an important role in the energy losses of heavy particles in detectors. The cross section of the process under consideration in the leading Born approximation was derived many years ago [1,2]. In this approximation, the cross section depends on the atomic charge number Z and the charge number of a heavy particle Z_p as $Z^2Z_p^2$. In papers [1,2] the interaction of a heavy particle with the atomic field was not taken into account. Later, the Coulomb corrections with respect to the interaction of e^+e^- pair with the atomic field were obtained in Refs. [3,4] using the plane waves for the wave functions of a heavy particle and the Coulomb wave functions for electron and positron. Thus, the results in Refs. [3,4] were exact in the parameter $Z\alpha$ but still proportional to Z_p^2, where α is the fine-structure constant. The authors of Refs. [3,4] obtained the cross sections differential with respect to the final momentum of a heavy particle as well as that integrated over this momentum. For the latter case, the cross sections were obtained within another approach, see Refs. [5,6] and reviews [7,8]. In that approach, the cross sections were obtained for a fixed impact parameter ρ of a heavy particle with respect to the atomic center. Therefore, the interaction of a heavy particle with the atomic field was not taken into account. Then, the results were integrated over the impact parameter. Thus, the final results correspond to the cross sections integrated over the final momentum of a heavy particle. Note that the energy ω of created e^+e^- pair, which gives the main contribution to the cross section, is much smaller than the energy of a heavy particle. As a result, the cross sections of the process are independent of the spin and mass m_p of a heavy particle but depend on the relativistic factor $\gamma = \varepsilon_p/m_p$, where ε_p is the energy of a heavy particle, $\hbar = c = 1$. Thus, the formulas for the cross sections are the same for muons and light nuclei (with the corresponding substitutions of the charge numbers).

In the present paper, we investigate the differential cross section of high-energy e^+e^- electroproduction by heavy charged particles in the atomic field taking into account the interaction of a heavy particle with the atomic field. Our consideration is based on the quasiclassical approximation, see review in Ref. [9], developed in Ref. [10] for the problem of high-energy e^+e^- electroproduction by ultra-relativistic electron in the atomic field. Our results are exact in both parameters, $\eta = Z\alpha$ and $\eta_p = ZZ_p\alpha = Z_p\eta$. We show that the cross section differential over momentum of a heavy particle strongly depends on η_p, in contrast to the commonly accepted point of view. For light nuclei in the field of a heavy atom, this parameter can be large, $\eta_p \gtrsim 1$. However, the cross section integrated over the final momentum of a heavy particle is independent of the parameter η_p. It seems, the experimental investigation of a strong dependence of the differential cross section on η_p for moderate values of the relativistic factor γ is not a very difficult task.

* Corresponding author.
E-mail addresses: P.A.Krachkov@inp.nsk.su (P.A. Krachkov), A.I.Milstein@inp.nsk.su (A.I. Milstein).

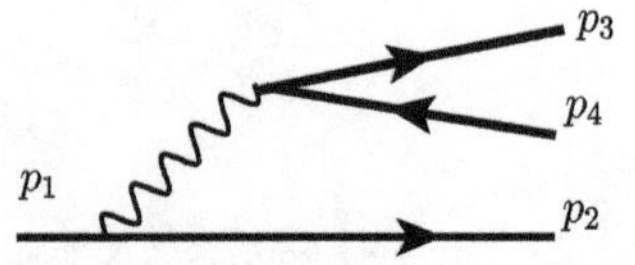

Fig. 1. Diagram for the amplitude of electroproduction by a heavy particle in the atomic field. Wavy line denotes the photon propagator, straight lines denote the wave functions in the atomic field.

2. General discussion

The differential cross section of the process under consideration reads [11], see Fig. 1 where the corresponding Feynman diagrams in the Furry representation is shown:

$$d\sigma = \frac{(Z_p\alpha)^2}{(2\pi)^8}\, d\varepsilon_3 d\varepsilon_4\, d\boldsymbol{p}_{2\perp}\, d\boldsymbol{p}_{3\perp} d\boldsymbol{p}_{4\perp}\, |T|^2\,, \tag{1}$$

where $\boldsymbol{p}_1$ and $\boldsymbol{p}_2$ are the initial and final momenta of a heavy particle, $\boldsymbol{p}_3$ and $\boldsymbol{p}_4$ are the momenta of electron and positron, $\varepsilon_1 = \varepsilon_2 + \omega$ is the energy of the incoming heavy particle, $\omega = \varepsilon_3 + \varepsilon_4$, $\varepsilon_{1,2} = \sqrt{p_{1,2}^2 + m_p^2}$, $\varepsilon_{3,4} = \sqrt{p_{3,4}^2 + m_e^2}$, m_e is the electron mass and m_p is the mass of a heavy particle. In Eq. (1) the notation $\boldsymbol{X}_\perp = \boldsymbol{X} - (\boldsymbol{X}\cdot\boldsymbol{v})\boldsymbol{v}$ for any vector $\boldsymbol{X}$ is used, $\boldsymbol{v} = \boldsymbol{p}_1/p_1$. Below we assume that $\varepsilon_{1,2} \gg m_p$, $\varepsilon_{3,4} \gg m_e$, and $m_p \gg m_e$. The main contribution to the cross section is given by the energy region $\varepsilon_{3,4} \lesssim \gamma m_e$, where $\gamma = \varepsilon_1/m_p \gg 1$, so that $\omega/\varepsilon_1 \lesssim m_e/m_p \ll 1$.

The straightforward application of the method developed in our recent paper [10] results in the expression for T exact in the parameters η and η_p for arbitrary atomic potential $V(r)$. We perform calculations of the matrix element T for definite helicities μ_3 and μ_4 of electron and positron, respectively, $\mu_i = \pm$ denotes a sign of the helicity.

We begin our consideration with the case of a pure Coulomb field and then discuss the effects of screening.

2.1. Coulomb field

It is convenient to write the amplitude T as a sum $T = T_\perp + T_\parallel$, where the corresponding contributions have the form (cf. Eq. (18) in Ref. [10]):

$$\begin{aligned}
T_\perp &= \frac{8i\eta}{\omega}|\Gamma(1-i\eta)|^2 \\
&\times \int \frac{d\boldsymbol{\Delta}_\perp\, A_{as}(\boldsymbol{\Delta}_\perp + \boldsymbol{p}_{2\perp})}{(Q^2+\Delta_{0\parallel}^2)M^2\,(\omega^2/\gamma^2+\Delta_\perp^2)}\left(\frac{\xi_2}{\xi_1}\right)^{i\eta}\mathcal{M}\,,\\
\mathcal{M} &= -\frac{\delta_{\mu_3\bar\mu_4}}{\omega}\left[\varepsilon_3(\boldsymbol{s}^*_{\mu_3}\cdot\boldsymbol{\Delta}_\perp)(\boldsymbol{s}_{\mu_3}\cdot\boldsymbol{I}_1) - \varepsilon_4(\boldsymbol{s}^*_{\mu_4}\cdot\boldsymbol{\Delta}_\perp)(\boldsymbol{s}_{\mu_4}\cdot\boldsymbol{I}_1)\right]\\
&+ \mu_3\delta_{\mu_3\mu_4}\frac{m_e}{\sqrt{2}}(\boldsymbol{s}^*_{\mu_3}\cdot\boldsymbol{\Delta}_\perp)I_0\,,\\
T_\parallel &= -\frac{8i\eta\varepsilon_3\varepsilon_4}{\omega^3}|\Gamma(1-i\eta)|^2\int\frac{d\boldsymbol{\Delta}_\perp\, A_{as}(\boldsymbol{\Delta}_\perp+\boldsymbol{p}_{2\perp})}{(Q^2+\Delta_{0\parallel}^2)M^2}\left(\frac{\xi_2}{\xi_1}\right)^{i\eta}\\
&\times I_0\delta_{\mu_3\bar\mu_4}\,,
\end{aligned} \tag{2}$$

where $\omega = \varepsilon_3+\varepsilon_4$, $\boldsymbol{s}_\lambda = (\boldsymbol{e}_x + i\lambda\boldsymbol{e}_y)/\sqrt{2}$, $\boldsymbol{e}_x$ and $\boldsymbol{e}_y$ are two orthogonal unit vectors perpendicular to $\boldsymbol{v} = \boldsymbol{p}_1/p_1$, $\Gamma(x)$ is the Euler Γ function, and the function $A_{as}(\boldsymbol{\Delta}_\perp)$ reads

$$A_{as}(\boldsymbol{\Delta}_\perp) = -\frac{4\pi\eta_p(L\Delta_\perp)^{2i\eta_p}\Gamma(1-i\eta_p)}{\Delta_\perp^2\Gamma(1+i\eta_p)}\,. \tag{3}$$

A specific value of L is irrelevant because the factor $L^{2i\eta_p}$ disappears in $|T|^2$. In Eq. (2) the following notations are used:

$$\begin{aligned}
&\boldsymbol{\Delta}_{0\perp} = \boldsymbol{p}_{2\perp} + \boldsymbol{p}_{3\perp} + \boldsymbol{p}_{4\perp}\,,\\
&\Delta_{0\parallel} = -\frac{1}{2}\left[\frac{\omega}{\gamma^2} + \frac{\omega(m_e^2+\zeta^2)}{\varepsilon_3\varepsilon_4} + \frac{\delta^2}{\omega} + \frac{p_{2\perp}^2}{\varepsilon_1}\right],\\
&M^2 = m_e^2 + \frac{\varepsilon_3\varepsilon_4}{\gamma^2} + \frac{\varepsilon_3\varepsilon_4}{\omega^2}\Delta_\perp^2\,,\\
&\boldsymbol{\zeta} = \frac{\varepsilon_4}{\omega}\boldsymbol{p}_{3\perp} - \frac{\varepsilon_3}{\omega}\boldsymbol{p}_{4\perp}\,,\quad \boldsymbol{\delta} = \boldsymbol{p}_{3\perp} + \boldsymbol{p}_{4\perp}\,,\\
&\boldsymbol{Q} = \boldsymbol{\Delta}_\perp - \boldsymbol{\delta}\,,\quad \boldsymbol{q}_1 = \frac{\varepsilon_3}{\omega}\boldsymbol{Q} - \boldsymbol{\zeta}\,,\quad \boldsymbol{q}_2 = \frac{\varepsilon_4}{\omega}\boldsymbol{Q} + \boldsymbol{\zeta}\,,\\
&I_0 = (\xi_1 - \xi_2)F(x) + (\xi_1+\xi_2-1)(1-x)\frac{F'(x)}{i\eta}\,,\\
&\boldsymbol{I}_1 = (\xi_1\boldsymbol{q}_1 + \xi_2\boldsymbol{q}_2)F(x) + (\xi_1\boldsymbol{q}_1 - \xi_2\boldsymbol{q}_2)(1-x)\frac{F'(x)}{i\eta}\,,\\
&\xi_1 = \frac{M^2}{M^2+q_1^2}\,,\quad \xi_2 = \frac{M^2}{M^2+q_2^2}\,,\quad x = 1 - \frac{Q^2\xi_1\xi_2}{M^2}\,,\\
&F(x) = F(i\eta, -i\eta, 1, x)\,,\quad F'(x) = \frac{\partial}{\partial x}F(x)\,,
\end{aligned} \tag{4}$$

where $F(a,b,c,x)$ is the hypergeometric function.

In terms of the variables $\boldsymbol{\zeta}$, $\boldsymbol{\delta}$, and $\boldsymbol{p}_{2\perp}$, see Eq. (4), the main contribution to the total cross section is given by the region ζ, δ, $p_{2\perp} \lesssim m_e$ and $\omega \lesssim m_e\gamma$. In this region $\omega/\varepsilon_1 \lesssim m_e/m_p$ and $p_{2\perp}/\varepsilon_1 \ll p_{3\perp}/\varepsilon_3$, $p_{4\perp}/\varepsilon_4$, i.e., the angle between the momenta $\boldsymbol{p}_2$ and $\boldsymbol{p}_1$ is much smaller than the angles between $\boldsymbol{p}_3$, $\boldsymbol{p}_4$ and $\boldsymbol{p}_1$.

Let us first consider the cross section integrated over $\boldsymbol{p}_{2\perp}$, which is one of the most interesting quantity from the experimental point of view. We show that this cross section is independent of the parameter η_p, so that it is not affected by the interaction of a heavy particle with the Coulomb center. First of all we note that the last term $p_{2\perp}^2/\varepsilon_1$ in $\Delta_{0\parallel}$ (see Eq. (4)) may be omitted because its contribution is small as compared with that of other terms (the relative contribution of this term to $\Delta_{0\parallel}$ is m_e/m_p). After that the variable $\boldsymbol{p}_{2\perp}$ presents in the amplitude T solely in the function $A_{as}(\boldsymbol{\Delta}_\perp + \boldsymbol{p}_{2\perp})$, see Eq. (2). Let us consider the integral,

$$R = \int d\boldsymbol{p}_{2\perp}\left|\int d\boldsymbol{\Delta}_\perp A_{as}(\boldsymbol{\Delta}_\perp + \boldsymbol{p}_{2\perp})G(\boldsymbol{\Delta}_\perp)\right|^2\,, \tag{5}$$

where $G(\boldsymbol{\Delta}_\perp)$ is some function and $A_{as}(\boldsymbol{\Delta}_\perp)$ is given by Eq. (3). The integral (5) is well-defined. However, to have a possibility to change the order of integration, we introduce the regularized function $A_{as}^{(r)}(\boldsymbol{\Delta}_\perp)$,

$$A_{as}^{(r)}(\boldsymbol{\Delta}) = -\frac{4\pi(\eta_p - i\epsilon)(L\Delta_\perp)^{2i\eta_p+2\epsilon}\Gamma(1-i\eta_p-\epsilon)}{\Delta_\perp^2\Gamma(1+i\eta_p+\epsilon)}\,, \tag{6}$$

where ϵ is a small positive parameter of regularization. Then we have

$$\begin{aligned}
R = \lim_{\epsilon\to 0}&\iint d\boldsymbol{x}d\boldsymbol{y}\, G(\boldsymbol{x})G^*(\boldsymbol{y})\\
&\times\int d\boldsymbol{p}_{2\perp}\, A_{as}^{(r)}(\boldsymbol{x}+\boldsymbol{p}_{2\perp})A_{as}^{(r)*}(\boldsymbol{y}+\boldsymbol{p}_{2\perp})\,.
\end{aligned} \tag{7}$$

The integral over $\boldsymbol{p}_{2\perp}$ can be easily take by means of the Feynman parametrization. We have

$$R = \lim_{\epsilon\to 0}\iint d\boldsymbol{x}d\boldsymbol{y}\, G(\boldsymbol{x})G^*(\boldsymbol{y})\frac{32\pi^3\epsilon}{|\boldsymbol{x}-\boldsymbol{y}|^{2-4\epsilon}} = (2\pi)^4\int d\boldsymbol{x}|G(\boldsymbol{x})|^2\,. \tag{8}$$

Thus, the final result is independent of the parameter η_p, so that it can be evaluated in the limit $\eta_p \to 0$ using the relation

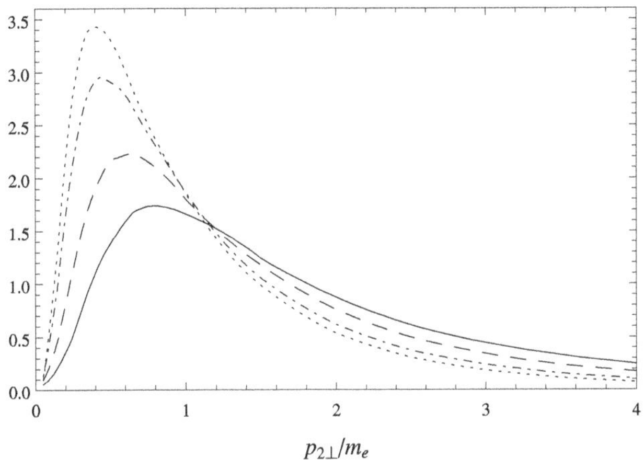

Fig. 2. The dependence of Σ, Eq. (11), on $p_{2\perp}/m_e$ for $\omega = m_e\gamma/4$, $\varepsilon_3 = \varepsilon_4 = \omega/2$, $\gamma = 100$, $Z = 79$ (gold), and a few values of Z_p; solid curve for $Z_p = 3$, dashed curve for $Z_p = 2$, dash-dotted curve for $Z_p = 1$, and dotted curve for $Z_p \to 0$ (without account for the interaction of a heavy particle with the Coulomb field).

$$\lim_{\eta_p \to 0} \eta_p \int d\boldsymbol{\Delta}_\perp |\boldsymbol{\Delta}_\perp + \boldsymbol{p}_{2\perp}|^{2i\eta_p - 2} G(\boldsymbol{\Delta}_\perp) = -i\pi G(-\boldsymbol{p}_{2\perp}). \quad (9)$$

The cross section $d\sigma_0$ in this limit is given by Eq. (1) with the replacement $T \to \mathcal{T} = \mathcal{T}_\perp + \mathcal{T}_\parallel$ with

$$\mathcal{T}_\perp = -\frac{32\pi^2\eta|\Gamma(1-i\eta)|^2\mathcal{M}_0}{\omega\Delta_0^2 M^2(\omega^2/\gamma^2 + p_{2\perp}^2)}\left(\frac{\xi_2}{\xi_1}\right)^{i\eta},$$
$$\mathcal{M}_0 = \frac{\delta_{\mu_3\bar{\mu}_4}}{\omega}\left[\varepsilon_3(\boldsymbol{s}^*_{\mu_3}\cdot\boldsymbol{p}_{2\perp})(\boldsymbol{s}_{\mu_3}\cdot\boldsymbol{I}_1) - \varepsilon_4(\boldsymbol{s}^*_{\mu_4}\cdot\boldsymbol{p}_{2\perp})(\boldsymbol{s}_{\mu_4}\cdot\boldsymbol{I}_1)\right],$$
$$-\mu_3\delta_{\mu_3\mu_4}\frac{m_e}{\sqrt{2}}(\boldsymbol{s}^*_{\mu_3}\cdot\boldsymbol{p}_{2\perp})I_0,$$
$$\mathcal{T}_\parallel = \frac{32\pi^2\eta\varepsilon_3\varepsilon_4|\Gamma(1-i\eta)|^2}{\omega^3\Delta_0^2M^2}\left(\frac{\xi_2}{\xi_1}\right)^{i\eta} I_0\delta_{\mu_3\bar{\mu}_4}, \quad (10)$$

where all notations are given in Eq. (4) with the replacement $\boldsymbol{\Delta}_\perp \to -\boldsymbol{p}_{2\perp}$. The result (10) agrees with that obtained in Ref. [3].

Though the cross section integrated over $\boldsymbol{p}_{2\perp}$ is independent of η_p, the cross section differential over $\boldsymbol{p}_{2\perp}$ strongly depends on this parameter. This statement is illustrated in Fig. 2, where the quantity Σ,

$$\Sigma = \frac{d\sigma}{S dp_{2\perp} d\varepsilon_3 d\varepsilon_4}, \quad S = \frac{(Z_p\alpha)^2}{\omega^2 m_e^3}, \quad (11)$$

which is the differential cross section in units S integrated over $\boldsymbol{p}_{3\perp}$ and $\boldsymbol{p}_{4\perp}$, is shown as the function of $p_{2\perp}$ for $\omega = m_e\gamma/4$, $\varepsilon_3 = \varepsilon_4 = \omega/2$, $\gamma = 100$, $Z = 79$ (gold), and a few values of Z_p.

It is seen that impact of interaction of a heavy particle with the Coulomb field on the cross section differential over $\boldsymbol{p}_{2\perp}$ is significant. At small value of $p_{2\perp}/m_e$, the cross section exact in η_p is essentially smaller than that obtained in the limit $\eta_p \to 0$. At large value of $p_{2\perp}/m_e$, the relation between these cross sections is opposite. As should be, $\int_0^\infty \Sigma\, dp_{2\perp}$ is independent of η_p.

In Fig. 3 the quantity Σ is shown as the function of $\omega/(m_e\gamma)$ for $p_{2\perp}/m_e = 2$ (left picture) and $p_{2\perp}/m_e = 0.5$ (right picture), $\varepsilon_3 = \varepsilon_4$, $\gamma = 100$, $Z = 79$ (gold), and a few values of Z_p. It is seen that the dependence of the function Σ on Z_p is very strong for all $\omega/(m_e\gamma)$.

In Fig. 4 we show the dependence of Σ on $x = \varepsilon_3/\omega$ for $\omega = m_e\gamma/4$, $\gamma = 100$, $Z = 79$ (gold), $p_{2\perp}/m_e = 2$ (left picture), $p_{2\perp}/m_e = 0.5$ (right picture), and a few values of Z_p. Again, it is seen that the account for the interaction of a heavy particle with the Coulomb field is very important for the differential cross section.

2.2. *Effect of screening*

An account for the effect of screening can be performed in the same way as it has been done in our paper [10]. In Eq. (2) one should replace $A_{as}(\boldsymbol{\Delta}_\perp) \to A(\boldsymbol{\Delta}_\perp)$,

$$A(\boldsymbol{\Delta}_\perp) = i\int d\boldsymbol{\rho}\, \exp[-i\boldsymbol{\Delta}_\perp\cdot\boldsymbol{\rho} - iZ_p\chi(\rho)],$$
$$\chi(\rho) = \int_{-\infty}^{\infty} dz\, V(\sqrt{z^2+\rho^2}), \quad (12)$$

where $V(r)$ is the atomic potential, and multiply the integrand in Eq. (2) by the function $F_a((\boldsymbol{\Delta}_\perp - \boldsymbol{\delta})^2 + \Delta_{0\parallel}^2)$, with $F_a(q^2)$ being the atomic form factor.

We show that the cross section integrated over $\boldsymbol{p}_{2\perp}$ is independent of η_p for any localized potential $V(r)$. Let us consider the function R_1,

$$R_1 = \int d\boldsymbol{p}_{2\perp}\left|\int d\boldsymbol{\Delta}_\perp A(\boldsymbol{\Delta}_\perp + \boldsymbol{p}_{2\perp})G(\boldsymbol{\Delta}_\perp)\right|^2. \quad (13)$$

Substituting Eq. (12) to Eq. (13) we obtain

$$R_1 = \iint d\boldsymbol{x}d\boldsymbol{y}G(\boldsymbol{x})G^*(\boldsymbol{y})\iint d\boldsymbol{\rho}_1 d\boldsymbol{\rho}_2 \exp\{iZ_p[\chi(\rho_2) - \chi(\rho_1)] + i\boldsymbol{y}\cdot\boldsymbol{\rho}_2 - i\boldsymbol{x}\cdot\boldsymbol{\rho}_1\} \times \int d\boldsymbol{p}_{2\perp}\exp[i\boldsymbol{p}_{2\perp}\cdot(\boldsymbol{\rho}_2 - \boldsymbol{\rho}_1)]. \quad (14)$$

Taking the integrals first over $\boldsymbol{p}_{2\perp}$ and then over $\boldsymbol{\rho}_1$, $\boldsymbol{\rho}_2$, and $\boldsymbol{y}$, we find the following result

$$R_1 = (2\pi)^4\int d\boldsymbol{p}_{2\perp}|G(\boldsymbol{p}_{2\perp})|^2, \quad (15)$$

which is independent of η_p. Thus, the cross section integrated over $\boldsymbol{p}_{2\perp}$ can be evaluated by means of Eq. (1), where $T = F_a(\Delta_0^2)(\mathcal{T}_\perp + \mathcal{T}_\parallel)$, $\mathcal{T}_\perp$ and $\mathcal{T}_\parallel$ are given in Eq. (10).

Let us discuss the impact of screening on the differential cross section. Without account for the interaction of a heavy particle with the atomic field, this question was investigated in Ref. [12]. In this case the effect of screening is important for $\gamma \gg m_e r_{scr} \sim Z^{-1/3}/\alpha$, where r_{scr} is the screening radius. Screening modifies the result of account for the interaction of a heavy particle with the atomic field in the cross section differential over $p_{2\perp}$ for $\gamma \gg (\omega/m_e)m_e r_{scr} \gg m_e r_{scr}$. Therefore, up to a very large γ one can use $A_{as}(\boldsymbol{\Delta})$, Eq. (3), instead of $A(\boldsymbol{\Delta})$, Eq. (12), but take into account the atomic form factor F_a.

3. Conclusion

Using the quasiclassical approximation, we have derived the differential cross section of high-energy e^+e^- electroproduction by heavy charged particles in the atomic field. The result is exact in the parameters η and η_p. It is shown that the cross section differential in $\boldsymbol{p}_{2\perp}$ strongly depends on the parameter η_p while the cross section integrated over $\boldsymbol{p}_{2\perp}$ is independent of this parameter. Though, at $\gamma \gtrsim m_e r_{scr} \gg 1$, screening is important for interaction of e^+e^- pair with the atomic field, it is not necessary to take screening into account for interaction of a heavy particle with the atomic field up to a very large γ.

Situation with the dependence of the cross sections of electroproduction on η_p reminds the situation with the Coulomb corrections to the cross sections of bremsstrahlung by high-energy muons in the atomic field, see Ref. [13]. The Coulomb corrections to the cross section of bremsstrahlung differential in both final

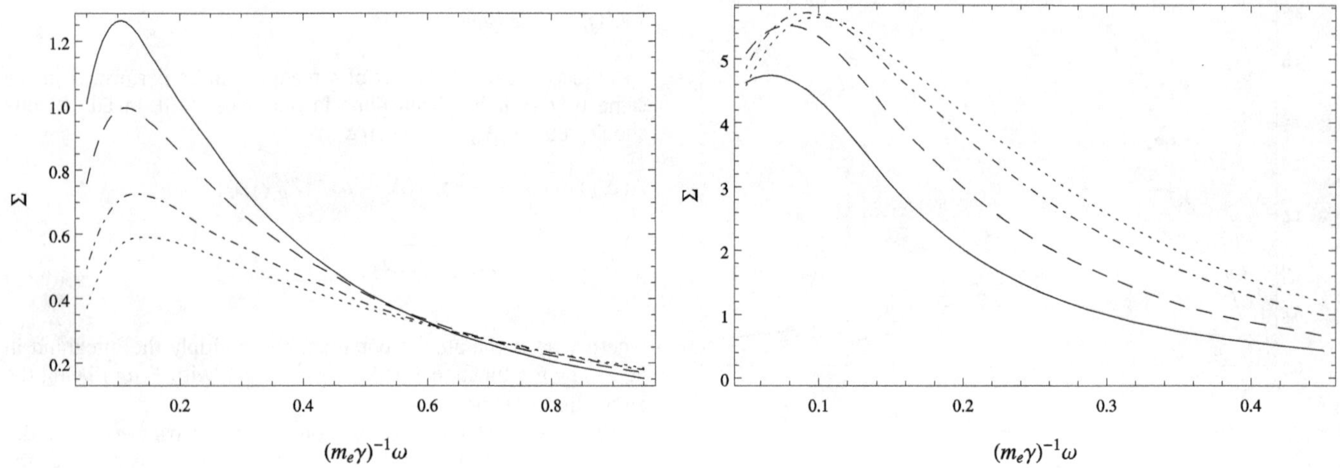

Fig. 3. The dependence of Σ, Eq. (11), on $\omega/(m_e\gamma)$ for $p_{2\perp}/m_e = 2$ (left picture) and $p_{2\perp}/m_e = 0.5$ (right picture), $\varepsilon_3 = \varepsilon_4$, $\gamma = 100$, $Z = 79$ (gold), and a few values of Z_p; solid curve for $Z_p = 3$, dashed curve for $Z_p = 2$, dash-dotted curve for $Z_p = 1$, and dotted curve for $Z_p \to 0$ (without account for the interaction of a heavy particle with the Coulomb field).

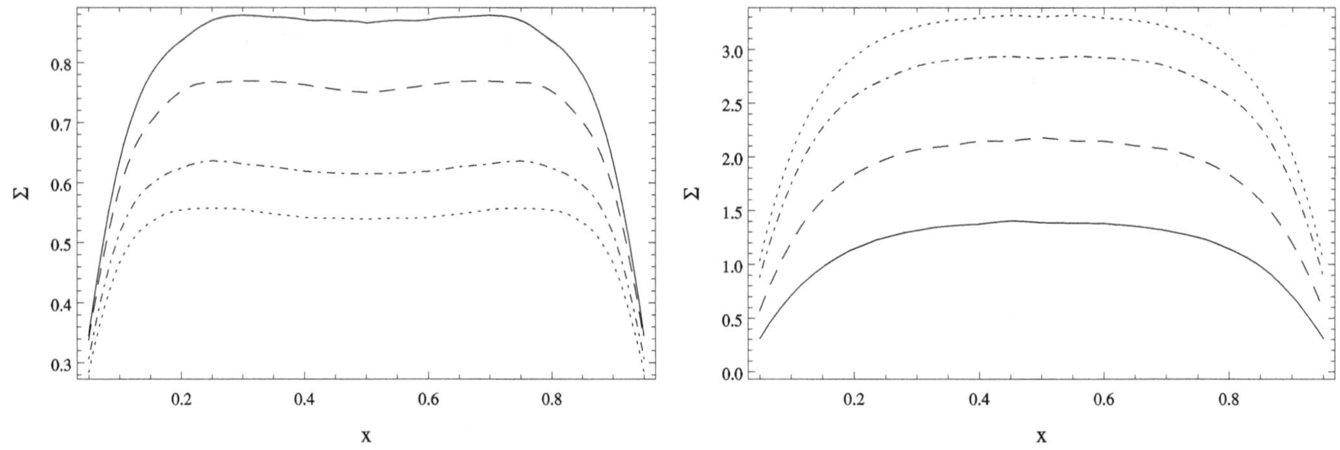

Fig. 4. The dependence of Σ, Eq. (11), on $x = \varepsilon_3/\omega$ for $\omega = m_e\gamma/4$, $\gamma = 100$, $Z = 79$ (gold), $p_{2\perp}/m_e = 2$ (left picture), $p_{2\perp}/m_e = 0.5$ (right picture), and a few values of Z_p; solid curve for $Z_p = 3$, dashed curve for $Z_p = 2$, dash-dotted curve for $Z_p = 1$, and dotted curve for $Z_p \to 0$ (without account for the interaction of a heavy particle with the Coulomb field).

muon and photon momenta modify essentially the result as compared with the Born result. However, the Coulomb corrections to the cross section integrated over the final muon momentum (or the photon momentum) vanish.

To observe experimentally a strong dependence of the cross section under discussion on the value of the parameter η_p, it is necessary to detect the final heavy particle. The angle between vectors $\boldsymbol{p}_2$ and $\boldsymbol{p}_1$ is essentially smaller than the angles between $\boldsymbol{p}_3$, $\boldsymbol{p}_4$, and $\boldsymbol{p}_1$. However, this experiment for moderate values of the relativistic factor γ seems to be not a very hard problem.

Acknowledgements

This work has been supported by Russian Science Foundation (Project No. 14-50-00080). It has been also supported in part by RFBR (Grant No. 16-02-00103).

References

[1] H. Bhabha, Proc. R. Soc. (London) A 152 (1935) 559.
[2] G. Racah, Nuovo Cimento 4 (1937) 112.
[3] A.I. Nikishov, N.V. Pichkurov, Sov. J. Nucl. Phys. 35 (1982) 561.
[4] D.Yu. Ivanov, E.A. Kuraev, A. Schiller, V.G. Serbo, Phys. Lett. B 442 (1998) 453.
[5] D.Yu. Ivanov, A. Schiller, V.G. Serbo, Phys. Lett. B 454 (1999) 155.
[6] R.N. Lee, A.I. Milstein, Phys. Rev. A 61 (2000) 032103.
[7] G. Baur, K. Hencken, D. Trautmann, Phys. Rep. 453 (2007) 1.
[8] K. Hencken, et al., Phys. Rep. 458 (2008) 1.
[9] P.A. Krachkov, R.N. Lee, A.I. Milstein, Usp. Fiz. Nauk 186 (2016) 689; Phys. Usp. 59 (2016) 619.
[10] P.A. Krachkov, A.I. Milstein, Phys. Rev. A 93 (2016) 062120.
[11] V.B. Berestetski, E.M. Lifshits, L.P. Pitayevsky, Quantum Electrodynamics, Pergamon, Oxford, 1982.
[12] T. Murota, A. Veda, H. Tanaka, Prog. Theor. Phys. (Kyoto) 16 (1956) 482.
[13] P.A. Krachkov, A.I. Milstein, Phys. Rev. A 91 (2015) 032106.

9

Curved momentum spaces from quantum groups with cosmological constant

Á. Ballesteros [a,*], G. Gubitosi [b], I. Gutiérrez-Sagredo [a], F.J. Herranz [a]

[a] *Departamento de Física, Universidad de Burgos, E-09001 Burgos, Spain*
[b] *Theoretical Physics, Blackett Laboratory, Imperial College, London SW7 2AZ, United Kingdom*

ARTICLE INFO

Editor: M. Cvetič

ABSTRACT

We bring the concept that quantum symmetries describe theories with nontrivial momentum space properties one step further, looking at quantum symmetries of spacetime in presence of a nonvanishing cosmological constant Λ. In particular, the momentum space associated to the κ-deformation of the de Sitter algebra in $(1+1)$ and $(2+1)$ dimensions is explicitly constructed as a dual Poisson–Lie group manifold parametrized by Λ. Such momentum space includes both the momenta associated to spacetime translations and the 'hyperbolic' momenta associated to boost transformations, and has the geometry of (half of) a de Sitter manifold. Known results for the momentum space of the κ-Poincaré algebra are smoothly recovered in the limit $\Lambda \to 0$, where hyperbolic momenta decouple from translational momenta. The approach here presented is general and can be applied to other quantum deformations of kinematical symmetries, including $(3+1)$-dimensional ones.

1. Introduction

Recent developments in quantum gravity research have revived and given new substance to the long-forgotten idea that momentum space should have a nontrivial geometry, an intuition originally due to Max Born [1]. After more than a decade since Deformed Special Relativity (DSR) was first proposed [2,3], it is now understood that a nontrivial geometry of momentum space is a general feature of DSR theories [4–8]. This is intimately related with the presence of the Planck energy as a second relativistic invariant (besides the speed of light), that can play the role of a curvature scale of the momentum manifold [9]. Nontrivial properties of momentum space emerge also in $(2+1)$-dimensional quantum gravity, where explicit computations show that the effective description of quantum gravity coupled to point particles is given by a theory with curved momentum space and noncommutative spacetime coordinates [10–13]. Of more direct interest for the results we are going to present here are models of noncommutative geometry, where the space of momenta that are dual to the noncommutative spacetime coordinates is curved [14–17].

Besides finding increasing theoretical support, Planck-scale modifications of the geometry of momentum space are extremely relevant from a phenomenological point of view. In fact, features due to curvature of momentum space are dual to those that in general relativity are ascribed to curvature of spacetime: in the same way as spacetime curvature induces redshift of energy, curvature of momentum space induces a dual redshift, that is, an energy-dependent correction to the time of flight of free particles [18]. Such effects open up a much needed observational window for Planck-scale physics, since they are testable with astrophysical observations [19].

Despite the recent significant theoretical and phenomenological progress just discussed, an important ingredient which is necessary to connect the properties of momentum space to observations is still missing. In fact, all of the models mentioned above are essentially deformations of special relativity: even though spacetime might be nontrivial (e.g. spacetime coordinates might not commute), still it has vanishing curvature. This is clearly a phenomenological shortcoming, since the most promising observations involve propagation of particles over cosmological distances, for which spacetime curvature cannot be neglected [20]. In the past few years several proposals aimed at extending relativistic models with curved momentum space were put forward in order to include nonvanishing spacetime curvature. The first concrete approach [21] focussed on constructing an extension of the Poincaré algebra that includes both the Planck scale and a (constant) space-

* Corresponding author.
E-mail addresses: angelb@ubu.es (Á. Ballesteros), g.gubitosi@imperial.ac.uk (G. Gubitosi), igsagredo@ubu.es (I. Gutiérrez-Sagredo), fjherranz@ubu.es (F.J. Herranz).

time curvature scale as relativistic invariants. The resulting algebra can be seen as a DSR version of the de Sitter (hereafter dS) algebra of symmetries, but the associated coalgebra was not investigated. Other proposals focussed on developing a unifying description of the whole phase space of free particles moving on a curved spacetime with deformed local Poincaré symmetries [22–26]. The general understanding coming from these approaches is that when both momentum space and spacetime have nonvanishing curvature they become so intertwined that it is not possible to give a neat geometrical description of the properties of momentum space on its own.

In this work we show that this is not necessarily the case. Indeed, we are able to explicitly construct the curved momentum space generated by quantum-deformed spacetime symmetries in presence of a nonvanishing cosmological constant. We achieve this result by enlarging the momentum space so that it is not only the manifold of momenta associated to translations on spacetime, but it also includes the 'hyperbolic' momenta associated to the boost transformations and the angular momenta associated to rotations. Within this construction we can also show that in the vanishing cosmological constant limit the Lorentz sector is not needed because it decouples from the energy–momentum sector, thus recovering previous results in the literature.

While we would like to argue that our results are general, we use the setting of Hopf algebras to present an explicit derivation. Hopf algebras have proved to be a very useful mathematical framework to model DSR effects. The most studied example is the κ-Poincaré Hopf algebra [27–29], the investigation of which provided inspiration and more precise understanding of several features of DSR models. For example, it can be explicitly shown that the manifold of momenta associated to the κ-Poincaré translation generators is a (portion of a) dS manifold, whose curvature is determined by the quantum deformation scale κ [17,30] and whose metric determines the free particle dispersion relation that is indeed compatible with the κ-Poincaré symmetries, thus showing that the phenomenology associated to the κ-Poincaré algebra fits very naturally within the framework of relative locality [17,31].

Here we present a generalization of all these results by working with the κ-deformation of the dS algebra (see [32–38]). The name is due to the fact that in the limit of vanishing cosmological constant Λ one recovers the κ-Poincaré algebra, while in the limit of vanishing quantum deformation parameter $z = 1/\kappa$ one recovers the algebra of symmetries of the dS spacetime. It is worth noticing that it was exactly using this Hopf algebra that the first pioneering investigations concerning the interplay between spacetime and momentum space curvature were undertaken [39].

The Poisson version of the κ-dS Hopf algebra in $(1+1)$ and in $(2+1)$ dimensions is defined in section 2, where it is shown that the main differences with respect to the corresponding κ-Poincaré structures fully arise in the $(2+1)$ setting: whilst in the vanishing cosmological constant limit the translation generators $\{P_0, P_1, P_2\}$ close a Hopf subalgebra, this is no longer the case for the κ-dS algebra, since the cosmological constant mixes the translation and Lorentz sectors within both the coproduct map and the deformed Casimir function. Thus, for nonvanishing Λ it seems natural to consider an enlarged momentum space including also the dual coordinates to the Lorentz generators. This idea allows us to construct the curved (generalized) momentum manifold in the nonvanishing cosmological constant setting as the full dual Poisson–Lie group manifold, whose explicit construction can be achieved through the Poisson version of the 'quantum duality principle' (see [40–43] and references therein).

The κ-dS dual Poisson–Lie groups are explicitly constructed in section 3. In $(1+1)$ dimensions the dual group coordinates are those associated to both the spacetime translations and boosts, and a certain linear action of the dual group on the origin of momentum space generates (half of) a $(2+1)$-dimensional dS manifold M_{dS_3}, spanned by the orbit of the group passing through the origin. In this case, the fact that boosts have the same role in the momentum space as translation generators can be understood since their coproducts have the same formal structure. In $(2+1)$ dimensions one spatial rotation comes into play and the structure of the κ-dS Hopf algebra is apparently much more involved. Nevertheless, the construction of the full dual Poisson–Lie group G^*_Λ gives the clue for the full geometrical description of the associated momentum space. The dual Lie algebra and its associated Poisson–Lie group are explicitly constructed in section 3.2, and the corresponding linear action on the enlarged momentum space can be defined in such a way that the dual rotation generates the isotropy subgroup of the origin of the momentum space. As a consequence, we find that a $(4+1)$-dimensional space of momenta associated to translations and boosts arises as a dual group orbit passing through the origin, and such a space again has the geometry of (half of) a dS manifold M_{dS_5}. Moreover, in the vanishing cosmological constant limit, the Lorentz sector completely decouples both in the dispersion relation and in the coproduct, thus recovering the well-known κ-Poincaré momentum space. The paper ends with a concluding section in which the applicability of the method here presented to the construction of the κ-AdS momentum space is shown, and the keystones for solving the corresponding $(3+1)$-dimensional problem are presented.

2. The κ-dS Poisson–Hopf algebra

Let us start by reviewing the structural properties of the κ-deformation of the $(1+1)$ and $(2+1)$ dS algebra, which will be presented by considering the cosmological constant $\Lambda > 0$ as an explicit parameter whose $\Lambda \to 0$ limit provides automatically the expressions for the κ-Poincaré algebra. In this way, the specific features of the construction leading to the κ-Poincaré momentum space will become transparent, and the proposed path to its nonvanishing cosmological constant generalization will arise in a natural way.

In the subsection on the $(1+1)$-dimensional case we just briefly present the essential formulas, postponing a more in-depth discussion of the relevant features of the κ-dS algebra to the following subsection focussing on the $(2+1)$-dimensional case.

2.1. The $(1+1)$ κ-dS algebra

The (undeformed) Poisson–Hopf dS algebra in $(1+1)$ dimensions is defined by the brackets

$$\{K, P_0\} = P_1, \qquad \{K, P_1\} = P_0, \qquad \{P_0, P_1\} = -\Lambda K, \tag{1}$$

where K is the generator of boost transformations, P_0 and P_1 are the time and space translation generators and the (undeformed) coproduct is given by $\Delta_0(X) = X \otimes 1 + 1 \otimes X$, with $X \in \{K, P_0, P_1\}$. The Poisson version of the $(1+1)$ κ-dS quantum algebra [34] is a Hopf algebra deformation of (1), given by

$$\{K, P_0\} = P_1, \qquad \{K, P_1\} = \frac{\sinh(zP_0)}{z}, \qquad \{P_0, P_1\} = -\Lambda K, \tag{2}$$

with deformed coproduct map

$$\begin{aligned} \Delta(P_0) &= P_0 \otimes 1 + 1 \otimes P_0, \\ \Delta(P_1) &= P_1 \otimes e^{\frac{z}{2}P_0} + e^{-\frac{z}{2}P_0} \otimes P_1, \\ \Delta(K) &= K \otimes e^{\frac{z}{2}P_0} + e^{-\frac{z}{2}P_0} \otimes K. \end{aligned} \tag{3}$$

The quantum deformation parameter is $z=1/\kappa$ and the deformed Casimir function for (2) is

$$\mathcal{C}_z=\left(\frac{\sinh(zP_0/2)}{z/2}\right)^2-P_1^2+\Lambda\,K^2. \tag{4}$$

The so-called bicrossproduct-type basis [27] for this algebra is given through the nonlinear change

$$P_0\to P_0,\qquad P_1\to e^{\frac{z}{2}P_0}\,P_1,\qquad K\to e^{\frac{z}{2}P_0}\,K, \tag{5}$$

so that the algebra becomes

$$\{K,P_0\}=P_1,\qquad \{K,P_1\}=\frac{1-\exp(-2zP_0)}{2z}-\frac{z}{2}(P_1^2-\Lambda\,K^2),$$
$$\{P_0,P_1\}=-\Lambda\,K, \tag{6}$$

with associated coproduct map

$$\begin{aligned}\Delta(P_0)&=P_0\otimes 1+1\otimes P_0,\\ \Delta(P_1)&=P_1\otimes 1+e^{-zP_0}\otimes P_1,\\ \Delta(K)&=K\otimes 1+e^{-zP_0}\otimes K.\end{aligned} \tag{7}$$

In this basis, the deformed Casimir reads

$$\mathcal{C}_z=\left(\frac{\sinh(zP_0/2)}{z/2}\right)^2-e^{zP_0}(P_1^2-\Lambda K^2). \tag{8}$$

We point out that for $\Lambda=0$ (the κ-Poincaré case), the momentum sector given by P_0 and P_1 generates an Abelian Hopf subalgebra, and the $\Lambda=0$ Casimir function provides the well-known $(1+1)$ κ-Poincaré deformed dispersion relation (see e.g. [17]). Note also that the coproduct (7) does not depend on Λ, although this property will not hold in higher dimensions.

2.2. The $(2+1)$ κ-dS algebra

In $(2+1)$ dimensions, the Poisson–Lie brackets of the (undeformed) dS algebra take the form

$$\begin{aligned}&\{J,P_i\}=\epsilon_{ij}P_j, &&\{J,K_i\}=\epsilon_{ij}K_j, &&\{J,P_0\}=0,\\ &\{P_i,K_j\}=-\delta_{ij}P_0, &&\{P_0,K_i\}=-P_i, &&\{K_1,K_2\}=-J,\\ &\{P_0,P_i\}=-\Lambda\,K_i, &&\{P_1,P_2\}=\Lambda\,J,\end{aligned} \tag{9}$$

where $i,j=1,2$, and ϵ_{ij} is a skew-symmetric tensor with $\epsilon_{12}=1$ (note that for negative values of Λ, this bracket defines the AdS Poisson–Lie algebra). The two quadratic Casimir functions for (9) are

$$\mathcal{C}=P_0^2-\mathbf{P}^2-\Lambda(J^2-\mathbf{K}^2),\qquad \mathcal{W}=-JP_0+K_1P_2-K_2P_1, \tag{10}$$

where $\mathbf{P}^2=P_1^2+P_2^2$ and $\mathbf{K}^2=K_1^2+K_2^2$. Recall that $\mathcal{C}$ comes from the Killing–Cartan form and is related to the energy of a point particle, while $\mathcal{W}$ is the Pauli–Lubanski vector. The undeformed Hopf algebra structure is given by Δ_0.

The $(2+1)$ κ-dS Poisson–Hopf algebra in the bicrossproduct basis is the Hopf algebra deformation with parameter $z=1/\kappa$ given by [35–37]

$$\begin{aligned}&\{J,P_0\}=0, &&\{J,P_1\}=P_2, &&\{J,P_2\}=-P_1,\\ &\{J,K_1\}=K_2, &&\{J,K_2\}=-K_1, &&\{K_1,K_2\}=-\frac{\sin(2z\sqrt{\Lambda}J)}{2z\sqrt{\Lambda}},\\ &\{P_0,P_1\}=-\Lambda\,K_1, &&\{P_0,P_2\}=-\Lambda\,K_2, &&\{P_1,P_2\}=\Lambda\,\frac{\sin(2z\sqrt{\Lambda}J)}{2z\sqrt{\Lambda}},\end{aligned}$$

$$\begin{aligned}&\{K_1,P_0\}=P_1, \qquad\qquad \{K_2,P_0\}=P_2,\\ &\{P_2,K_1\}=z(P_1P_2-\Lambda K_1K_2)\qquad \{P_1,K_2\}=z(P_1P_2-\Lambda K_1K_2),\\ &\{K_1,P_1\}=\frac{1}{2z}\left(\cos(2z\sqrt{\Lambda}J)-e^{-2zP_0}\right)\\ &\qquad+\frac{z}{2}\left(P_2^2-P_1^2\right)-\frac{z\Lambda}{2}\left(K_2^2-K_1^2\right),\\ &\{K_2,P_2\}=\frac{1}{2z}\left(\cos(2z\sqrt{\Lambda}J)-e^{-2zP_0}\right)+\frac{z}{2}\left(P_1^2-P_2^2\right)\\ &\qquad-\frac{z\Lambda}{2}\left(K_1^2-K_2^2\right),\end{aligned} \tag{11}$$

and with deformed coproduct map

$$\begin{aligned}&\Delta(P_0)=P_0\otimes 1+1\otimes P_0,\qquad \Delta(J)=J\otimes 1+1\otimes J,\\ &\Delta(P_1)=P_1\otimes\cos(z\sqrt{\Lambda}J)+e^{-zP_0}\otimes P_1+\Lambda K_2\otimes\frac{\sin(z\sqrt{\Lambda}J)}{\sqrt{\Lambda}},\\ &\Delta(P_2)=P_2\otimes\cos(z\sqrt{\Lambda}J)+e^{-zP_0}\otimes P_2-\Lambda K_1\otimes\frac{\sin(z\sqrt{\Lambda}J)}{\sqrt{\Lambda}},\\ &\Delta(K_1)=K_1\otimes\cos(z\sqrt{\Lambda}J)+e^{-zP_0}\otimes K_1+P_2\otimes\frac{\sin(z\sqrt{\Lambda}J)}{\sqrt{\Lambda}},\\ &\Delta(K_2)=K_2\otimes\cos(z\sqrt{\Lambda}J)+e^{-zP_0}\otimes K_2-P_1\otimes\frac{\sin(z\sqrt{\Lambda}J)}{\sqrt{\Lambda}},\end{aligned} \tag{12}$$

which explicitly depends on the cosmological constant Λ. The deformed Casimir function for this Poisson–Hopf algebra reads

$$\begin{aligned}\mathcal{C}_z=&\frac{2}{z^2}\left[\cosh(zP_0)\cos(z\sqrt{\Lambda}J)-1\right]\\ &-e^{zP_0}\left(\mathbf{P}^2-\Lambda\mathbf{K}^2\right)\cos(z\sqrt{\Lambda}\,J)-2\Lambda\,e^{zP_0}\frac{\sin(z\sqrt{\Lambda}J)}{\sqrt{\Lambda}}R_3,\end{aligned} \tag{13}$$

with $R_3=\epsilon_{3bc}K_bP_c$. Note that the projection to the κ-dS algebra in $(1+1)$ dimensions is obtained by setting to zero the generators $\{P_2,K_2,J\}$.

The $(2+1)$ κ-Poincaré Hopf algebra is smoothly recovered in the $\Lambda\to 0$ limit and in this 'flat' case the momentum sector $\{P_0,P_1,P_2\}$ generates an Abelian Hopf subalgebra with coproducts

$$\begin{aligned}\Delta(P_0)&=P_0\otimes 1+1\otimes P_0,\\ \Delta(P_1)&=P_1\otimes 1+e^{-zP_0}\otimes P_1,\\ \Delta(P_2)&=P_2\otimes 1+e^{-zP_0}\otimes P_2.\end{aligned} \tag{14}$$

Such a nonlinear superposition law for momenta is the essential footprint of a curved momentum space, which can be explicitly constructed by following the procedure presented in [9]. Essentially, the κ-Poincaré momentum space is a three-dimensional manifold generated by the action on a certain ambient space of the three-dimensional dual Lie group G^* whose Lie algebra g^*,

$$\left[X^0,X^1\right]=-z\,X^1,\qquad \left[X^0,X^2\right]=-z\,X^2,\qquad \left[X^1,X^2\right]=0, \tag{15}$$

is defined as the dual of the skew-symmetric part of the first order deformation in z of the coproducts (14). The Lie algebra (15) is the so-called $(2+1)$ κ-Minkowski noncommutative spacetime [27,44]. Moreover, when $\Lambda=0$ the deformed Casimir function

$$\mathcal{C}_z=\frac{2}{z^2}\left[\cosh(zP_0)-1\right]-e^{zP_0}\left(P_1^2+P_2^2\right), \tag{16}$$

provides the κ-Poincaré deformed dispersion relation in $(2+1)$ dimensions. The same construction can be straightforwardly generalized to the $(3+1)$ κ-Poincaré algebra (see [9] and references therein).

The main obstruction to a similar construction when $\Lambda \neq 0$ is readily seen by inspection of eq. (12). In fact, in the κ-dS case the momentum sector $\{P_0, P_1, P_2\}$ is no longer a Hopf subalgebra, since the coproduct of spatial momenta includes all the generators $\{J, K_1, K_2\}$ of the Lorentz sector (note that this is not the case in $(1+1)$ dimensions, where the coproduct does not depend on Λ). Moreover, the deformed Casimir C_z contains the Lorentz generators as well, and this feature is also present in the $(1+1)$ case (see eq. (8)). These two observations hold true also in the $(3+1)$ κ-dS Poisson–Hopf algebra that has been explicitly presented for the first time in [38].

We already mentioned that Hopf-algebraic deformations of spacetime symmetries can be endowed with a phenomenological interpretation. Specifically, the Casimir C_z of the algebra determines the dispersion relation of free particles, while the coproduct of the translation generators determines the rules of conservation of energy and spatial momentum in interactions [17]. Therefore, when $\Lambda \neq 0$ we can say that both the conservation rules in interactions and the deformed dispersion relation involve an enlarged set of 'momenta', including also the angular momentum and the 'hyperbolic' momenta corresponding, respectively, to the rotation and to boost transformations (hyperbolic rotations). In this framework, it seems natural to propose that when $\Lambda \neq 0$ the (curved) momentum space is defined by an enlarged space parametrized by the six coordinates that are dual to the generators of the full quantum algebra. Nevertheless, a simple inspection at the coproducts (12) shows that the role of the J generator is somewhat different from that of K_1 and K_2, since the latter have coproducts which are formally equivalent to those of P_1 and P_2. All these aspects will have a clear interpretation once the explicit construction of the κ-dS momentum space is performed in the following section.

3. Momentum space for the κ-dS Poisson–Hopf algebra

As anticipated above, in this section the momentum space for the κ-dS Poisson algebra with nonvanishing cosmological constant will be constructed as the full dual Poisson–Lie group G^*_Λ, whose Lie algebra g^*_Λ is provided by the dual of the cocommutator map δ generated by the coproduct of all the κ-dS generators in the bicrossproduct basis, including the Lorentz sector. This construction will be firstly illustrated in $(1+1)$ dimensions. While this case is simpler, it does not allow to appreciate the richness of structure characterising higher-dimensional models. The consistency and geometric features of our approach will be made fully explicit in the second subsection, where we demonstrate the full construction for the $(2+1)$-dimensional case.

3.1. *The* $(1+1)$ *case*

The cocommutator map for the full κ-dS algebra can be read from the skew-symmetric part of the first-order deformation in z of the coproduct (7), namely

$$\delta(P_0)=0, \qquad \delta(P_1)=z\,P_1\wedge P_0, \qquad \delta(K)=z\,K\wedge P_0. \tag{17}$$

If we denote by $\{X^0, X^1, L\}$ the generators dual to, respectively, $\{P_0, P_1, K\}$, the dual Lie algebra g^*_Λ is given by the Lie brackets

$$\left[X^0, X^1\right]=-z\,X^1, \qquad \left[X^0, L\right]=-z\,L, \qquad \left[X^1, L\right]=0. \tag{18}$$

A faithful representation ρ of this Lie algebra for $\Lambda \neq 0$ is given by the 4×4 matrices

$$\rho(X^0)=z\begin{pmatrix}0&0&0&1\\0&0&0&0\\0&0&0&0\\1&0&0&0\end{pmatrix} \qquad \rho(X^1)=z\begin{pmatrix}0&1&0&0\\1&0&0&1\\0&0&0&0\\0&-1&0&0\end{pmatrix}$$
$$\rho(L)=z\sqrt{\Lambda}\begin{pmatrix}0&0&1&0\\0&0&0&0\\1&0&0&1\\0&0&-1&0\end{pmatrix}. \tag{19}$$

If we denote as $\{p_0, p_1, \chi\}$ the local group coordinates which are dual, respectively, to $\{X^0, X^1, L\}$, then the group element of the dual Lie group G^*_Λ is given by:

$$G^*_\Lambda=\exp\left(p_1\rho(X^1)\right)\exp\left(\chi\rho(L)\right)\exp\left(p_0\rho(X^0)\right). \tag{20}$$

A straightforward computation leads to the following explicit matrix

$$G^*_\Lambda=\begin{pmatrix}\cosh(zp_0)+\frac{1}{2}e^{z\,p_0}z^2(p_1^2+\Lambda\chi^2) & zp_1 & z\sqrt{\Lambda}\chi & \sinh(zp_0)+\frac{1}{2}e^{z\,p_0}z^2(p_1^2+\Lambda\chi^2)\\ e^{z\,p_0}zp_1 & 1 & 0 & e^{z\,p_0}zp_1\\ e^{z\,p_0}z\sqrt{\Lambda}\chi & 0 & 1 & e^{z\,p_0}z\sqrt{\Lambda}\chi\\ \sinh(zp_0)-\frac{1}{2}e^{z\,p_0}z^2(p_1^2+\Lambda\chi^2) & -zp_1 & -z\sqrt{\Lambda}\chi & \cosh(zp_0)-\frac{1}{2}e^{z\,p_0}z^2(p_1^2+\Lambda\chi^2)\end{pmatrix}. \tag{21}$$

The multiplication law for the group G^*_Λ is obtained by multiplying two matrices of the form (21), and it can be written as a co-product (see [43]) in the form

$$\Delta(p_0)=p_0\otimes 1+1\otimes p_0, \qquad \Delta(p_1)=p_1\otimes 1+e^{-zp_0}\otimes p_1,$$
$$\Delta(\chi)=\chi\otimes 1+e^{-zp_0}\otimes\chi. \tag{22}$$

As the quantum duality principle indicates, this coproduct is just the one (7) for the κ-dS algebra once one identifies the dual group coordinates and the generators of the κ-dS Poisson–Hopf algebra as follows:

$$p_0\equiv P_0, \qquad p_1\equiv P_1, \qquad \chi\equiv K. \tag{23}$$

Moreover, by following the technique presented in [43] it can be shown that the unique Poisson–Lie structure on G^*_Λ that is compatible with the coproduct (22) and has the undeformed dS Lie algebra (1) as its linearization is given by the Poisson brackets

$$\{\chi, p_0\}=p_1, \qquad \{\chi, p_1\}=\frac{1-\exp(-2zp_0)}{2z}-\frac{z}{2}(p_1^2-\Lambda\,\chi^2),$$
$$\{p_0, p_1\}=-\Lambda\,\chi, \tag{24}$$

which is exactly the κ-dS algebra (6) under the identification (23). Evidently, the Casimir function for this Poisson bracket is

$$C_z=\left(\frac{\sinh(zp_0/2)}{z/2}\right)^2-e^{zp_0}(p_1^2-\Lambda\chi^2). \tag{25}$$

In this way, the composition law for the momenta with κ-dS symmetry (7) has been reobtained as the group law (22) for the coordinates of the dual Poisson–Lie group G^*_Λ, and the κ-dS Casimir function (8) can be interpreted as an on-shell relation (25) for these coordinates.

We stress that the main novelty with respect to the κ-Poincaré case described in [9] is the fact that the dual Lie group G^*_Λ is now three-dimensional, and the momentum space associated to κ-dS is parametrized by the three coordinates $\{p_0, p_1, \chi\}$, and not only by the momenta associated to spacetime translations. Moreover, both in the coproduct (22) and the Casimir function (25) the role of the parameters χ and p_1 turns out to be identical, which supports the role of the former as an additional 'hyperbolic' momentum for quantum symmetries with nonvanishing cosmological constant.

An explicit geometric interpretation of this enlarged momentum space can be obtained along the same lines of [9] by observing that the entries of the fourth column in G^*_Λ, given by

$$\begin{aligned} S_0 &= \sinh(zp_0) + \frac{1}{2} e^{z\,p_0} z^2 (p_1^2 + \Lambda \chi^2), \\ S_1 &= e^{z\,p_0} z\, p_1, \\ S_2 &= e^{z\,p_0} z \sqrt{\Lambda}\, \chi, \\ S_3 &= \cosh(zp_0) - \frac{1}{2} e^{z\,p_0} z^2 (p_1^2 + \Lambda \chi^2), \end{aligned} \tag{26}$$

satisfy the defining relation for the $(2+1)$-dimensional dS space,

$$-S_0^2 + S_1^2 + S_2^2 + S_3^2 = 1. \tag{27}$$

Moreover, if we consider a linear action of the Lie group G^*_Λ onto a four-dimensional ambient Minkowski space with coordinates (S_0, S_1, S_2, S_3), we have that

$$G^*_\Lambda \cdot (0,0,0,1)^T = (S_0, S_1, S_2, S_3)^T, \tag{28}$$

which means that the $(2+1)$-dimensional dS space is generated through G^*_Λ as the orbit that passes through the point $(0,0,0,1)$ in the ambient space, corresponding to the origin of the (generalized) momentum space. Note that the orbit passing through the point $(0,0,0,\alpha)$, with $\alpha \neq 0$, would satisfy $-S_0^2 + S_1^2 + S_2^2 + S_3^2 = \alpha^2$. Moreover, we have that the condition

$$S_0 + S_3 = e^{z\,p_0} > 0, \tag{29}$$

is automatically obeyed, so that only half of the $(2+1)$-dimensional dS space is generated as an orbit of the free action of G^*_Λ, and we will denote this manifold as M_{dS_3}. Finally, when $\Lambda = 0$ the ambient coordinate S_2 vanishes, as well as the realization $\rho(L)$ of the dual of the boost generator, thus recovering the well-known interpretation of the κ-Poincaré momentum space as (half of) a $(1+1)$-dimensional dS space, i.e., M_{dS_2}.

3.2. The $(2+1)$ case

The very same procedure described in the previous section can be applied to the construction of the momentum space associated to the $(2+1)$ κ-dS Poisson–Hopf algebra. The skew symmetrized first order in z of the coproduct (12) is given by the cocommutator map

$$\begin{aligned} \delta(P_0) &= \delta(J) = 0, \\ \delta(P_1) &= z(P_1 \wedge P_0 + \Lambda K_2 \wedge J), \\ \delta(P_2) &= z(P_2 \wedge P_0 - \Lambda K_1 \wedge J), \\ \delta(K_1) &= z(K_1 \wedge P_0 + P_2 \wedge J), \\ \delta(K_2) &= z(K_2 \wedge P_0 - P_1 \wedge J). \end{aligned} \tag{30}$$

Denoting by $\{X^0, X^1, X^2, L^1, L^2, R\}$ the generators dual to, respectively, $\{P_0, P_1, P_2, K_1, K_2, J\}$, the Lie brackets defining the Lie algebra g^* of the dual Poisson–Lie group G^*_Λ are

$$\begin{array}{lll} [X^0, X^1] = -z X^1, & [X^0, X^2] = -z X^2, & [X^1, X^2] = 0, \\ [X^0, L^1] = -z L^1, & [X^0, L^2] = -z L^2, & [L^1, L^2] = 0, \\ [R, X^2] = -z L^1, & [R, L^1] = z \Lambda X^2, & [L^1, X^2] = 0, \\ [R, X^1] = z L^2, & [R, L^2] = -z \Lambda X^1, & [L^2, X^1] = 0, \\ [R, X^0] = 0, & [L^1, X^1] = 0, & [L^2, X^2] = 0. \end{array} \tag{31}$$

A (faithful) representation ρ of this Lie algebra for $\Lambda \neq 0$ is given by the 6×6 matrices

$$\rho(X^0) = z \begin{pmatrix} 0&0&0&0&0&1 \\ 0&0&0&0&0&0 \\ 0&0&0&0&0&0 \\ 0&0&0&0&0&0 \\ 0&0&0&0&0&0 \\ 1&0&0&0&0&0 \end{pmatrix} \quad \rho(X^1) = z \begin{pmatrix} 0&1&0&0&0&0 \\ 1&0&0&0&0&1 \\ 0&0&0&0&0&0 \\ 0&0&0&0&0&0 \\ 0&0&0&0&0&0 \\ 0&-1&0&0&0&0 \end{pmatrix}$$

$$\rho(X^2) = z \begin{pmatrix} 0&0&1&0&0&0 \\ 0&0&0&0&0&0 \\ 1&0&0&0&0&1 \\ 0&0&0&0&0&0 \\ 0&0&0&0&0&0 \\ 0&0&-1&0&0&0 \end{pmatrix} \quad \rho(L^1) = z\sqrt{\Lambda} \begin{pmatrix} 0&0&0&1&0&0 \\ 0&0&0&0&0&0 \\ 0&0&0&0&0&0 \\ 1&0&0&0&0&1 \\ 0&0&0&0&0&0 \\ 0&0&0&-1&0&0 \end{pmatrix}$$

$$\rho(L^2) = z\sqrt{\Lambda} \begin{pmatrix} 0&0&0&0&1&0 \\ 0&0&0&0&0&0 \\ 0&0&0&0&0&0 \\ 0&0&0&0&0&0 \\ 1&0&0&0&0&1 \\ 0&0&0&0&-1&0 \end{pmatrix} \quad \rho(R) = z\sqrt{\Lambda} \begin{pmatrix} 0&0&0&0&0&0 \\ 0&0&0&0&-1&0 \\ 0&0&0&1&0&0 \\ 0&0&-1&0&0&0 \\ 0&1&0&0&0&0 \\ 0&0&0&0&0&0 \end{pmatrix}. \tag{32}$$

If we denote as $\{p_0, p_1, p_2, \chi_1, \chi_2, \theta\}$ the local group coordinates which are dual, respectively, to $\{X^0, X^1, X^2, L^1, L^2, R\}$, then the Lie group element G^*_Λ can be written as

$$\begin{aligned} G^*_\Lambda = {} & \exp(\theta \rho(R)) \exp\left(p_1 \rho(X^1)\right) \exp\left(p_2 \rho(X^2)\right) \\ & \times \exp\left(\chi_1 \rho(L^1)\right) \exp\left(\chi_2 \rho(L^2)\right) \exp\left(p_0 \rho(X^0)\right), \end{aligned} \tag{33}$$

and its explicit expression can be straightforwardly computed, although we omit it here for the sake of brevity. By multiplying two of these generic group elements, the group law for G^*_Λ can be directly derived and written as the following coproduct map for the six dual group coordinates:

$$\begin{aligned} \Delta(p_0) &= p_0 \otimes 1 + 1 \otimes p_0, \qquad \Delta(\theta) = \theta \otimes 1 + 1 \otimes \theta, \\ \Delta(p_1) &= p_1 \otimes \cos(z\sqrt{\Lambda}\theta) + e^{-zp_0} \otimes p_1 + \Lambda\, \chi_2 \otimes \frac{\sin(z\sqrt{\Lambda}\theta)}{\sqrt{\Lambda}}, \\ \Delta(p_2) &= p_2 \otimes \cos(z\sqrt{\Lambda}\theta) + e^{-zp_0} \otimes p_2 - \Lambda\, \chi_1 \otimes \frac{\sin(z\sqrt{\Lambda}\theta)}{\sqrt{\Lambda}}, \\ \Delta(\chi_1) &= \chi_1 \otimes \cos(z\sqrt{\Lambda}\theta) + e^{-zp_0} \otimes \chi_1 + p_2 \otimes \frac{\sin(z\sqrt{\Lambda}\theta)}{\sqrt{\Lambda}}, \\ \Delta(\chi_2) &= \chi_2 \otimes \cos(z\sqrt{\Lambda}\theta) + e^{-zp_0} \otimes \chi_2 - p_1 \otimes \frac{\sin(z\sqrt{\Lambda}\theta)}{\sqrt{\Lambda}}. \end{aligned} \tag{34}$$

Again, under the identification

$$p_0 \equiv P_0, \quad p_1 \equiv P_1, \quad p_2 \equiv P_2, \quad \chi_1 \equiv K_1, \quad \chi_2 \equiv K_2, \quad \theta \equiv J, \tag{35}$$

this is exactly the coproduct for the κ-dS Poisson–Hopf algebra given in (12), and the unique Poisson–Lie structure on G^*_Λ that is compatible with (34) and has the undeformed dS Lie algebra (9) as its linearization is the deformed Poisson algebra given by (11).

In order to provide a geometric interpretation of the six-dimensional generalized momentum space manifold, we proceed similarly to the $(1+1)$ case and consider the action of G^*_Λ onto an ambient space. The entries of the sixth column in the matrix realization (33) are

$$\begin{aligned} S_0 &= \sinh(zp_0) + \frac{1}{2} e^{z\,p_0} z^2 \left(p_1^2 + p_2^2 + \Lambda\left(\chi_1^2 + \chi_2^2\right)\right), \\ S_1 &= e^{z\,p_0} z\left(\cos(z\sqrt{\Lambda}\theta)\, p_1 - \sqrt{\Lambda} \sin(z\sqrt{\Lambda}\theta)\, \chi_2\right), \end{aligned}$$

$$S_2 = e^{z p_0} z\,(\cos(z\sqrt{\Lambda}\,\theta)\, p_2 + \sqrt{\Lambda}\, \sin(z\sqrt{\Lambda}\,\theta)\, \chi_1),$$
$$S_3 = e^{z p_0} z\,(-\sin(z\sqrt{\Lambda}\,\theta)\, p_2 + \sqrt{\Lambda}\, \cos(z\sqrt{\Lambda}\,\theta)\, \chi_1), \quad (36)$$
$$S_4 = e^{z p_0} z\,(\sin(z\sqrt{\Lambda}\,\theta)\, p_1 + \sqrt{\Lambda}\, \cos(z\sqrt{\Lambda}\,\theta)\, \chi_2),$$
$$S_5 = \cosh(zp_0) - \frac{1}{2} e^{z p_0} z^2 \left(p_1^2 + p_2^2 + \Lambda\left(\chi_1^2 + \chi_2^2\right)\right),$$

and satisfy the condition

$$-S_0^2 + S_1^2 + S_2^2 + S_3^2 + S_4^2 + S_5^2 = 1, \quad (37)$$

which is the defining relation for the $(4+1)$-dimensional dS space. Therefore, by assuming that the space of generalized momenta is the group manifold for the dual group G^*_Λ, we can conclude that a linear action of the Lie group G^*_Λ onto a six-dimensional ambient Minkowski space with coordinates $(S_0, S_1, S_2, S_3, S_4, S_5)$ allows us to obtain a $(4+1)$ dS space as the orbit that passes through the point in the ambient space with coordinates $(0, 0, 0, 0, 0, 1)$, which is the origin of the (generalized) momentum space. Moreover, we have that $S_0 + S_5 = e^{z p_0} > 0$, so only half of the dS space is generated in this way, and we will denote this manifold as M_{dS_5}. Therefore, the $(1+1)$ construction can be generalized to this $(2+1)$ setting, although some distinctive features of the latter are worth to be stressed.

Firstly, given that in the $(2+1)$ case one has six symmetry generators, one would naively expect that the generalized momentum space be a six dimensional manifold, given that in the $(1+1)$ case the dimensionality of the manifold corresponds to the number of symmetry generators. Instead, we demonstrated the emergence of a five-dimensional orbit under the action of G^*_Λ. The reason for this is the completely different role that the dual rotation (R, θ) plays with respect to the dual boosts (L^i, χ_i), both in the coproduct and in the action (36). In particular, it is immediate to check that the isotropy subgroup of the point $(0, 0, 0, 0, 0, 1)$ is just the one given by $G_0^* = \exp(\theta\rho(R))$. Therefore, the full momentum space for the κ-dS algebra in $(2+1)$ dimensions is the six-dimensional manifold $M_{dS_5} \times S^1$, where the rotation coordinate θ is the one parametrizing S^1 while (p_i, χ_i) parametrize M_{dS_5}.

Secondly, under the identification (35) the deformed Casimir is written as the following function on the generalized momentum space:

$$\begin{aligned} C_z = {} & \frac{2}{z^2}\left[\cosh(zp_0)\cos(z\sqrt{\Lambda}\theta) - 1\right] \\ & - e^{zp_0}\left(p_1^2 + p_2^2 - \Lambda(\chi_1^2 + \chi_2^2)\right)\cos(z\sqrt{\Lambda}\theta) \\ & - 2\Lambda\, e^{zp_0}\frac{\sin(z\sqrt{\Lambda}\theta)}{\sqrt{\Lambda}} R_3, \end{aligned} \quad (38)$$

which involves all the translation and Lorentz momenta. Nevertheless, if we specialize this function onto the five-dimensional orbit M_{dS_5} by taking the S^1 coordinate $\theta = 0$, we get

$$C_z = \frac{2}{z^2}\left[\cosh(zp_0) - 1\right] - e^{zp_0}\left(p_1^2 + p_2^2 - \Lambda(\chi_1^2 + \chi_2^2)\right), \quad (39)$$

which is an on-shell relation that is just a higher dimensional generalization of the one obtained in the $(1+1)$ κ-dS case, eq. (25). In this way, the striking equivalence between the role played by the momenta associated to space translations and boosts is manifestly shown.

Finally, the $(2+1)$ κ-Poincaré construction is again straightforwardly recovered in the limit $\Lambda \to 0$, where the action (36) provides $S_3 = S_4 = 0$ and the representation (32) is only defined for $\{X^0, X^1, X^2\}$, thus giving rise to (half of) a $(2+1)$ dS space as an orbit under the action of the corresponding three-dimensional dual group. Summarizing, in $(2+1)$ dimensions the momentum space for κ-dS is found to be the six-dimensional manifold $M_{dS_5} \times S^1$, while its κ-Poincaré limit was known to be the three-dimensional one M_{dS_3}.

4. Concluding remarks

Deformed special relativity (DSR) theories are characterized by the presence of an energy scale that plays the role of a second relativistic invariant besides the speed of light. Such an energy scale allows the geometry of momentum space to be nontrivial, and in fact it is a general feature of DSR models that the manifold of momenta has nonzero curvature.

In this paper we have shown that the curved momentum space construction can be extended to cases where also a nonvanishing spacetime cosmological constant is present. We explored in particular the momentum space of the κ-deformation of the dS algebra, called κ-dS, and we showed that one can construct a curved generalized-momentum space, that includes not only the momenta associated to spacetime translations but also the hyperbolic momenta associated to boosts. The procedure is an adaptation of the one that was successfully used to show that the momentum space of the κ-Poincaré algebra has the geometry of (half of) a dS manifold and is generated by the orbits of the dual Poisson–Lie group. The construction here presented can be applied to any other Hopf algebra deformation of kinematical symmetries with nonvanishing Λ, although the orbit structure of the momentum space so obtained will indeed depend on the chosen quantum deformation.

The construction in $(1+1)$ dimensions is quite straightforward once one realizes that the boosts play a very similar role to spatial translations in the structure of the algebra and coalgebra. We indeed found that the generalized-momentum manifold is a $(2+1)$-dimensional dS manifold, whose coordinates are the local group coordinates associated to spacetime translations and boosts.

In $(2+1)$ dimensions matters are complicated by the presence of a rotation generator in the algebra, that significantly complicates its structure. However the rotation generator has a peculiar role in the structure of the algebra and coalgebra, while boosts still behave similarly to spatial translations. We were indeed able to construct the generalized momentum space of the $(2+1)$ κ-dS algebra whose coordinates are the local group coordinates associated to spacetime translations and boosts, and we showed that this is half of a $(4+1)$-dimensional dS manifold, for which the dual rotation generator generates the isotropy subgroup of the origin.

It is worth mentioning that the formalism here presented, in which Λ is considered as an explicit 'classical' deformation parameter (and this fact is connected with the so-called 'semidualization' approaches in $(2+1)$ quantum gravity [45,46]), suggests the possibility of performing the same construction of the generalized momentum space for the κ-AdS (Anti de Sitter) algebra by taking $\Lambda < 0$. It turns out that one can indeed work out fully the κ-AdS counterpart of the results described above. The main difference between the κ-dS and κ-AdS cases arises from the dual group representation (32), which has to be modified in the $\Lambda < 0$ case in order to have a real representation of the corresponding dual Lie group G^*_Λ. The latter can be explicitly constructed and leads to an action on the point $(0, 0, 0, 0, 0, 1)$ that generates the quadric

$$-S_0^2 + S_1^2 + S_2^2 - S_3^2 - S_4^2 + S_5^2 = 1, \quad (40)$$

which is no longer the M_{dS_5} momentum space. Nevertheless, the $\Lambda \to 0$ limit of this action annihilates the S_3 and S_4 coordinates, thus giving rise to the same κ-Poincaré limit as the one previously obtained from the κ-dS algebra, as it should be.

While the point of this paper was clearly made limiting the analysis to lower-dimensional algebras, the application of the approach here presented to the construction of the momentum space for the $(3+1)$-dimensional κ-dS algebra seems to be feasible. In fact, the full κ-dS Poisson Hopf algebra in $(3+1)$ dimensions has been recently presented in [38], and its corresponding momentum space should be obtained as a 10-dimensional dual Poisson–Lie group manifold by mimicking the procedure here presented. In fact, by direct inspection of the expressions for the coproduct and cocommutator map in the $(3+1)$ κ-dS algebra [38], the formal similarity between boosts and spatial momenta is again evident, while the three rotations are composed in a completely different way. This is work in progress and will be presented elsewhere, including the analysis of the momentum space for the $(3+1)$ κ-AdS algebra. Also, it would be interesting to use this approach in order to construct the curved momentum spaces for other quantum deformations of kinematical symmetries with nonvanishing cosmological constant, like for instance the $(2+1)$ (A)dS quantum group recently introduced in [47] or other possible quantum (A)dS deformations (see [48,49]).

Acknowledgments

A.B., I.G-S and F.J.H. have been partially supported by Ministerio de Economía y Competitividad (MINECO, Spain) under grants MTM2013-43820-P and MTM2016-79639-P (AEI/FEDER, UE), by Junta de Castilla y León (Spain) under grants BU278U14 and VA057U16 and by the Action MP1405 QSPACE from the European Cooperation in Science and Technology (COST). G.G. acknowledges support from the John Templeton Foundation through grant Nr. 47633.

References

[1] M. Born, Proc. R. Soc. Lond. A 165 (1938) 291.
[2] G. Amelino-Camelia, Int. J. Mod. Phys. D 11 (2002) 35.
[3] G. Amelino-Camelia, Phys. Lett. B 510 (2001) 255.
[4] J. Kowalski-Glikman, S. Nowak, Class. Quantum Gravity 20 (2003) 4799.
[5] J. Kowalski-Glikman, Phys. Lett. B 547 (2002) 291.
[6] D. Raetzel, S. Rivera, F.P. Schuller, Phys. Rev. D 83 (2011) 044047.
[7] G. Amelino-Camelia, L. Freidel, J. Kowalski-Glikman, L. Smolin, Phys. Rev. D 84 (2011) 084010.
[8] G. Amelino-Camelia, L. Freidel, J. Kowalski-Glikman, L. Smolin, Gen. Relativ. Gravit. 43 (2011) 2547.
[9] J. Kowalski-Glikman, Int. J. Mod. Phys. A 28 (2013) 1330014.
[10] H.J. Matschull, M. Welling, Class. Quantum Gravity 15 (1998) 2981.
[11] L. Freidel, J. Kowalski-Glikman, L. Smolin, Phys. Rev. D 69 (2004) 044001.
[12] C. Meusburger, B.J. Schroers, Class. Quantum Gravity 20 (2003) 2193.
[13] L. Freidel, E.R. Livine, Phys. Rev. Lett. 96 (2006) 221301.
[14] S. Majid, Lect. Notes Phys. 541 (2000) 227.
[15] G. Amelino-Camelia, S. Majid, Int. J. Mod. Phys. A 15 (2000) 4301.
[16] J. Kowalski-Glikman, S. Nowak, Int. J. Mod. Phys. D 12 (2003) 299.
[17] G. Gubitosi, F. Mercati, Class. Quantum Gravity 30 (2013) 145002.
[18] G. Amelino-Camelia, L. Barcaroli, G. Gubitosi, N. Loret, Class. Quantum Gravity 30 (2013) 235002.
[19] G. Amelino-Camelia, Living Rev. Relativ. 16 (2013) 5.
[20] G. Amelino-Camelia, L. Smolin, Phys. Rev. D 80 (2009) 084017.
[21] G. Amelino-Camelia, A. Marciano, M. Matassa, G. Rosati, Phys. Rev. D 86 (2012) 124035.
[22] L. Barcaroli, L.K. Brunkhorst, G. Gubitosi, N. Loret, C. Pfeifer, Phys. Rev. D 92 (2015) 084053.
[23] L. Barcaroli, L.K. Brunkhorst, G. Gubitosi, N. Loret, C. Pfeifer, Phys. Rev. D 95 (2017) 024036.
[24] L. Barcaroli, L.K. Brunkhorst, G. Gubitosi, N. Loret, C. Pfeifer, arXiv:1703.02058.
[25] G. Amelino-Camelia, L. Barcaroli, G. Gubitosi, S. Liberati, N. Loret, Phys. Rev. D 90 (2014) 125030.
[26] F. Cianfrani, J. Kowalski-Glikman, G. Rosati, Phys. Rev. D 89 (2014) 044039.
[27] S. Majid, H. Ruegg, Phys. Lett. B 334 (1994) 348.
[28] J. Lukierski, A. Nowicki, H. Ruegg, Phys. Lett. B 293 (1992) 344.
[29] J. Lukierski, H. Ruegg, Phys. Lett. B 329 (1994) 189.
[30] J. Kowalski-Glikman, S. Nowak, arXiv:hep-th/0411154.
[31] G. Amelino-Camelia, M. Arzano, J. Kowalski-Glikman, G. Rosati, G. Trevisan, Class. Quantum Gravity 29 (2012) 075007.
[32] J. Lukierski, A. Nowicki, H. Ruegg, Phys. Lett. B 271 (1991) 321.
[33] J. Lukierski, H. Ruegg, A. Nowicki, V.N. Tolstoi, Phys. Lett. B 264 (1991) 331.
[34] A. Ballesteros, F.J. Herranz, M.A. del Olmo, M. Santander, J. Phys. A, Math. Gen. 26 (1993) 5801.
[35] A. Ballesteros, F.J. Herranz, M.A. del Olmo, M. Santander, J. Phys. A, Math. Gen. 27 (1994) 1283.
[36] A. Ballesteros, F.J. Herranz, N.R. Bruno, arXiv:hep-th/0401244, 2004.
[37] G. Amelino-Camelia, L. Smolin, A. Starodubtsev, Class. Quantum Gravity 21 (2004) 3095.
[38] A. Ballesteros, F.J. Herranz, F. Musso, P. Naranjo, Phys. Lett. B 766 (2017) 205.
[39] A. Marciano, G. Amelino-Camelia, N.R. Bruno, G. Gubitosi, G. Mandanici, A. Melchiorri, J. Cosmol. Astropart. Phys. 1006 (2010) 030.
[40] V.G. Drinfel'd, Quantum groups, in: A.V. Gleason (Ed.), Proc. Int. Cong. Math. Berkeley 1986, AMS, Providence, 1987, p. 798.
[41] M.A. Semenov-Tyan-Shanskii, Theor. Math. Phys. 93 (1992) 1292.
[42] N. Ciccoli, F. Gavarini, Adv. Math. 199 (2006) 104.
[43] A. Ballesteros, F. Musso, J. Phys. A, Math. Theor. 46 (2013) 195203.
[44] P. Maslanka, J. Phys. A, Math. Gen. 26 (1993) L1251.
[45] S. Majid, B. Schroers, J. Phys. A, Math. Theor. 42 (2009) 425402.
[46] P.K. Osei, B. Schroers, J. Math. Phys. 53 (2012) 073510.
[47] A. Ballesteros, F.J. Herranz, C. Meusburger, Phys. Lett. B 732 (2014) 201.
[48] A. Ballesteros, F.J. Herranz, F. Musso, J. Phys. Conf. Ser. 532 (2014) 012002.
[49] A. Borowiec, J. Lukierski, V.N. Tolstoy, Phys. Lett. B 754 (2016) 176.

10

Deuteron properties from muonic atom spectroscopy

N.G. Kelkar *, D. Bedoya Fierro

Dept. de Fisica, Universidad de los Andes, Cra. 1E No. 18A-10, Santafe de Bogotá, Colombia

ARTICLE INFO

Editor: W. Haxton

Keywords:
Deuteron radius
Electromagnetic form factors
Muonic atoms

ABSTRACT

Leading order (α^4) finite size corrections in muonic deuterium are evaluated within a few body formalism for the $\mu^- pn$ system in muonic deuterium and found to be sensitive to the input of the deuteron wave function. We show that this sensitivity, taken along with the precise deuteron charge radius determined from muonic atom spectroscopy can be used to determine the elusive deuteron D-state probability, P_D, for a given model of the nucleon–nucleon (NN) potential. The radius calculated with a P_D of 4.3% in the chiral NN models and about 5.7% in the high precision NN potentials is favoured most by the $\mu^- d$ data.

1. Introduction

The lightest nucleus, namely, the deuteron, has traditionally held an important place in nuclear physics as a testing ground for the nucleon–nucleon interaction. Determining the D-state probability in the deuteron wave function in particular has been a classic problem of nuclear physics [1–3]. Stating the problem in simple words, the deuteron has a quadrupole moment and hence cannot be in a pure S-state but rather a D-state admixture is required. However, as it was shown in [4] that the D-state probability, $P_D = \int_0^\infty w^2(r)dr$ (with $w(r)$ being the deuteron radial wave function with $l = 2$), is inaccessible directly to experiments, it is usually the asymptotic D-state to S-state wave function ratio, η [2, 5], which is determined. There do exist attempts to determine P_D from the measured magnetic moment of the deuteron, μ_D, with, $\mu_D = \mu_S - (3/2)P_D(\mu_S - 1/2) + \delta_R$, where, $\mu_S = \mu_P + \mu_N$ is the isoscalar nucleon magnetic moment. However, the term δ_R which includes mesonic exchange effects, relativistic corrections, dynamical effects and isobar configurations in the deuteron introduces uncertainties in the extraction of P_D [6]. This fact was noticed in one of the oldest works by Feshbach and Schwinger [1] on the theory of nuclear forces which gave the D-state probability, P_D, ranging between 2% to 6%. Much later, Ref. [7] listed values of P_D ranging from 0.28 to 6.47% for 9 different nucleon–nucleon (NN) potentials. However, earlier in [8] the possible minimum was shown to be 0.45%. With P_D not being a measurable quantity, Refs. [2] and [5] determined the asymptotic ratio $\eta = 0.0256 \pm 0.0004$ and 0.0268 $\pm$ 0.0013 from tensor analyzing powers in sub-Coulomb (d, p) reactions and dp elastic scattering respectively. In the absence of a "measured" D-state probability, theoretical models of the NN interaction also try to reproduce the asymptotic ratio η determined from experiments (in addition to other data) to confirm the reliability of the NN model [3].

The purpose of this work is to present a new method which provides a means to fix the percentage of the "elusive" [4] D-state probability, P_D, from experiments in an indirect manner. The method is particularly useful in view of the very high precision reported by recent muonic deuterium experiments [9]. It is based on a few body calculation of the leading order (α^4) finite size corrections (FSC) to the energy levels of muonic deuterium atoms. There exists extensive literature on corrections including the deuteron polarization [10–12], with detailed calculations of FSC at higher orders (α^5, α^6 etc) [13,10–12]. The sensitivity of the higher order FSC to the form of the nucleon–nucleon potential (and hence the deuteron wave function) is found to be small [10,12] or negligible [14]. The leading FSC at order α^4 in these works is written in terms of the deuteron charge radius. The few body formalism of the present work helps in revealing the dependence of the leading FSC term on the proton and neutron form factors as well as the deuteron wave function. We show that a comparison of the order α^4 FSC with those of Ref. [9] where the radius is precisely extracted from measurements in muonic deuterium provides a method to adjust the deuteron D-state probability. To be specific, we present calculations using different parametrizations of the deuteron wave function (with different amounts of the D-state probabilities) and compare the corrections with those given in [9] in a form dependent on the deuteron charge radius, r_d. Though

* Corresponding author.
E-mail address: nkelkar@uniandes.edu.co (N.G. Kelkar).

the general trend of the results is an increase in the radius for smaller values of P_D, the results are found to depend on the type of model used. In the class of chiral models [15], $P_D = 4.3\%$ is found to be favourable for the closest agreement with the precise value of $r_d = 2.12562(78)$ fm [9]. Using high precision NN potentials such as Nijmegen, Reid, Paris etc. [16], $P_D = 5.7\%$ to 5.8% is favoured by the μd data.

2. Finite size effects in muonic deuterium

Finite size corrections (FSC) to the energy levels in the hydrogen atom has been a topic of revived interest [17] in the past few years due to the increase in the precision achieved in atomic spectroscopy measurements. These effects are manifested more strongly in muonic atoms due to the fact that the muon is about 200 times heavier than the electron and hence has a Bohr radius which is much smaller. In view of the recent precise measurement of the Lamb shift in muonic deuterium [9], it seems timely to put forth the question as to what other impact (apart from the precise radius determination) does this measurement have on physics. In order to see this, we study the effects of deuteron structure on the energy levels in this atom. The present work considers the effects at leading order (α^4) and we refer the reader to [10–12] for higher order corrections.

2.1. Electromagnetic muon–deuteron potential

We investigate the finite size effects by calculating the energy correction, ΔE, using first order perturbation theory involving an electromagnetic muon–deuteron potential, $V_{\mu^- d}$. The latter is constructed using a three body approach to the muon–proton–neutron system with the proton and neutron being bound inside the deuteron. As we will see below, the $\mu^- p$ and $\mu^- n$ interactions are obtained using the proton and neutron electromagnetic form factors and the pn interaction is contained in the deuteron wave function. Such a potential can be constructed using standard techniques from scattering theory where we first write down the scattering amplitude to obtain the potential $V_{\mu^- d}(\boldsymbol{q})$ in momentum space and then evaluate its Fourier transform which enters the energy correction given by, $\Delta E = \int_0^\infty \Delta V(r) |\Psi_{nl}(\boldsymbol{r})|^2 d^3 r$. This procedure of obtaining potentials in coordinate space is also common in quantum field theory [18–20]. Here, ΔV is the difference of $V_{\mu^- d}(r)$ and the $\mu^- d$ electromagnetic potential assuming the deuteron to be point-like. Details of the few body formalism used here can be found in [21,22]. We shall repeat the relevant steps briefly below.

The Hamiltonian of the quantum system consisting of a muon and a nucleus (with A nucleons) is given as [21], $H = H_0 + V_{\mu^- A} + H_A$, where H_0 is the muon-nucleus kinetic energy operator (free Hamiltonian), $V_{\mu^- A} = \sum_{i=1}^{A} V_i$, the sum of muon-nucleon potentials, $V_i \equiv V_{\mu^- N}(|\boldsymbol{R} - \boldsymbol{r}_i|)$, where $\boldsymbol{R}$ and $\boldsymbol{r}_i$ are the coordinates of the muon and the ith nucleon with respect to the centre of mass of the nucleus and H_A is the total Hamiltonian of the nucleus containing the potential term, $\sum_{i \neq j} V_{NN}(|\boldsymbol{r}_i - \boldsymbol{r}_j|)$. We proceed with the assumption that the nucleus remains in its ground state during the scattering process, i.e., $H_A |\Phi\rangle = \epsilon |\Phi\rangle$ and that the nucleons occupy fixed positions inside the nucleus. The muon-nucleus elastic scattering amplitude can be expressed as [21] $f(\boldsymbol{k}', \boldsymbol{k}; E) = -(\mu/\pi) \langle \boldsymbol{k}', \Phi | T(E) | \boldsymbol{k}, \Phi \rangle$ in terms of the matrix elements of the operator T obeying the Lippmann–Schwinger (L-S) equation, $T = V + V(E - H_0 - H_A)^{-1} T$. $|\boldsymbol{k}, \Phi\rangle$ and $|\boldsymbol{k}', \Phi\rangle$ are the initial and final asymptotic states which differ only in the direction of the relative muon nucleus momenta $\boldsymbol{k}$ and $\boldsymbol{k}'$. Since the electromagnetic potential, $V_{\mu^- A}$, is proportional to the coupling constant $\alpha \sim 1/137$, it is reasonable to truncate the L-S equation at first order and approximate $T = V = \sum_i V_i$. Thus, $T(\boldsymbol{k}', \boldsymbol{k}) = V(\boldsymbol{k}', \boldsymbol{k})$ and denoting, $T(\boldsymbol{k}', \boldsymbol{k}) \equiv \langle \boldsymbol{k}', \Phi | T(E) | \boldsymbol{k}, \Phi \rangle$, we have $V(\boldsymbol{k}', \boldsymbol{k}) = \langle \boldsymbol{k}', \Phi | \sum_{i=1}^{A} V_i | \boldsymbol{k}, \Phi \rangle$. If the internal Jacobi coordinates are denoted by $\boldsymbol{x}_i$, then relating them with $\boldsymbol{r}_i = a_i \boldsymbol{x}_1 + b_i \boldsymbol{x}_2 + ... + g_i \boldsymbol{x}_{A-1}$, we can write,

$$V(\boldsymbol{k}', \boldsymbol{k}) = \int d\boldsymbol{x}_1 \, d\boldsymbol{x}_2 \ldots d\boldsymbol{x}_{A-1} \, |\Phi(\boldsymbol{x}_1, \boldsymbol{x}_2, \ldots)|^2 \sum_{i=1}^{A} V_i(\boldsymbol{k}', \boldsymbol{k}, \boldsymbol{r}_i), \tag{1}$$

where, $V_i(\boldsymbol{k}', \boldsymbol{k}, \boldsymbol{r}_i) = V_i(\boldsymbol{k}', \boldsymbol{k}) \exp[i(\boldsymbol{k} - \boldsymbol{k}') \cdot \boldsymbol{r}_i]$. The above discussion is valid for any nucleus with A nucleons. In case of the muon–deuteron system, this reduces to

$$V(\boldsymbol{k}', \boldsymbol{k}) = \int d\boldsymbol{x}_1 \, |\Phi_d(\boldsymbol{x}_1)|^2 \, [\, V_{\mu^- p}(\boldsymbol{k}', \boldsymbol{k}, \tfrac{1}{2}\boldsymbol{x}_1) + V_{\mu^- n}(\boldsymbol{k}', \boldsymbol{k}, -\tfrac{1}{2}\boldsymbol{x}_1)\,] \tag{2}$$

where we used, $\boldsymbol{x}_1 = \boldsymbol{r}_1 - \boldsymbol{r}_2$, $\boldsymbol{r}_1 = (1/2)\boldsymbol{x}_1$ and $\boldsymbol{r}_2 = -(1/2)\boldsymbol{x}_1$. We identify 1 and 2 with proton and neutron so that, $V_1 = V_{\mu^- p}$, $V_2 = V_{\mu^- n}$ and Φ_d is the deuteron wave function.

To evaluate (2), we need the μ^--nucleon electromagnetic potential, which, with the inclusion of the nucleon electromagnetic form factors $G_E^N(q^2)$ can be written using the formalism of the Breit equation [18] within the one-photon-exchange interaction. Since such a potential was explicitly derived in [17,18] by evaluating the elastic muon–nucleon amplitude expanded in powers of $1/c^2$, we shall not repeat the derivation here. This potential with form factors contains 23 terms [18] corresponding to the (i) Coulomb potential, (ii) Darwin terms, and (iii) spin dependent terms which give rise to fine and hyperfine structure. If we consider only the scalar parts of the Breit potential, they depend only on $\boldsymbol{q}^2$ and hence we can write, $V_{\mu^- N}(\boldsymbol{k}, \boldsymbol{k}') \equiv V_{\mu^- N}(\boldsymbol{q})$ [18,17], where, $\boldsymbol{q} = \boldsymbol{k} - \boldsymbol{k}'$ is the momentum transfer carried by the exchanged photon. Denoting $Q = |\boldsymbol{q}|$, the $\mu^- N$ potential is given as [17],

$$V_{\mu^- N}(Q) = -4\pi\alpha \frac{G_E^N(Q^2)}{Q^2} \left\{ 1 - \frac{Q^2}{8m_N^2 c^2} - \frac{Q^2}{8m_\mu^2 c^2} \right\}, \tag{3}$$

where m_N and m_μ are the nucleon and muon masses. $G_E^N(Q^2)$ is the nucleon electric form factor. A Fourier transform of the first term in the curly bracket leads to the $\mu^- N$ Coulomb potential for a finite sized nucleon. The next two terms in the curly brackets are relativistic corrections, the Darwin terms in the muon (spin 1/2)–nucleon (spin 1/2) $\mu^- N$ interaction Breit potential. The Darwin term $Q^2/8m_N^2 c^2$ is conventionally not considered as a part of the nucleon form factor $G_E^N(q^2)$ [23] and hence is kept explicitly in the muon–nucleon potential here. Putting together (2) and (3) we obtain the muon–deuteron electromagnetic potential, $V_{\mu^- d}(Q) = V_{\mu^- p}(Q) \int d\boldsymbol{x} |\Phi_d(\boldsymbol{x})|^2 e^{-i\boldsymbol{q}\cdot\boldsymbol{x}/2} + V_{\mu^- n}(Q) \int d\boldsymbol{x} |\Phi_d(\boldsymbol{x})|^2 e^{i\boldsymbol{q}\cdot\boldsymbol{x}/2}$, in momentum space. The integrals in this expression can be shown to reduce to [7] $G_0(Q) = \int_0^\infty [u^2(r) + w^2(r)] \, j_0(Qr/2) \, dr$, where, $u(r)$ and $w(r)$ are the radial parts of the deuteron S- and D-wave functions. Thus, $V_{\mu^- d}(Q) = (V_{\mu^- p}(Q) + V_{\mu^- n}(Q)) G_0(Q)$, so that,

$$V_{\mu^- d}(Q) = -4\pi\alpha \, \frac{G_0(Q)[G_E^p(Q^2) + G_E^n(Q^2)]}{Q^2} \left(1 - \frac{Q^2}{8m_N^2} - \frac{Q^2}{8m_\mu^2} \right), \tag{4}$$

where the proton and neutron masses have been written as $m_p \approx m_n \approx m_N$ for simplicity. We note here that the three body formalism allows us to include the relativistic corrections in the form of

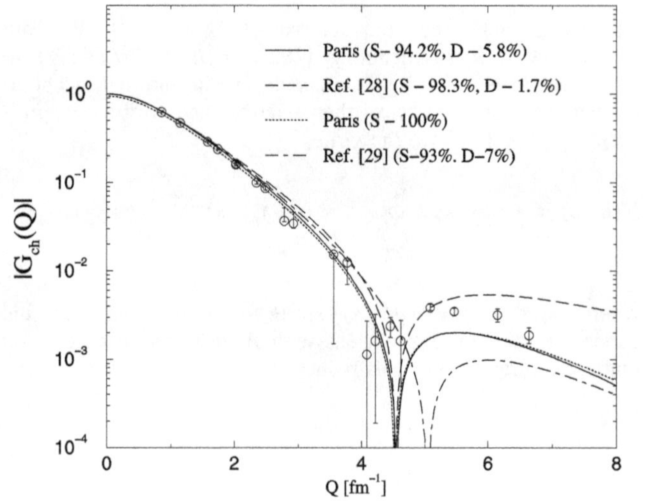

Fig. 1. Deuteron charge form factor using different D-state probabilities of the deuteron wave function. The data is from Ref. [30].

the Darwin terms since we are summing potentials between the muon and nucleons (both of which are spin - 1/2 objects) and folding them with the nuclear structure part. Including the relativistic corrections directly in a muon–deuteron potential is otherwise a formidable task since one has to work with an equation for spin 1/2–spin 1 elastic scattering with form factors. The above Darwin term is known as a recoil correction in atomic physics (see [24] for a detailed discussion).

The elementary potential (3) is calculated using the dipole proton form factor, $G_E^p(Q^2) = (1 + Q^2/0.71)^{-2}$ and a Galster form for the neutron, $G_E^n(Q^2) = [1.91\tau/(1 + 5.6\tau)](1 + Q^2/0.71)^{-2}$ (with $\tau = Q^2/4m_n^2$) as in [25]. These particular forms were chosen since using these forms of $G_E^{p,n}$ along with the matter distribution $G_0(Q)$ of the deuteron gives good agreement with the deuteron charge form factor defined by $G_{ch}(Q) = G_0(Q)[G_E^p(Q^2) + G_E^n(Q^2) - G_E^p(Q^2)Q^2/(8m_p^2)]$ in [25] (see Fig. 1). The proton radius corresponding to the dipole G_E^p is 0.81 fm and is smaller than that of the free proton radius. However, an input of the dipole form of G_E^p reproduces the *ed* data well as can also be found in [26].

2.2. Deuteron electric potential

This potential simply follows from the $\mu^- d$ interaction potential in (4) by noting that the deuteron electric potential should be associated with the Coulomb interaction with form factors but cannot depend on the mass of the probe, in this case the muon. Thus,

$$V_d(Q) = 4\pi e \frac{G_0(Q)[G_E^p(Q^2) + G_E^n(Q^2)]}{Q^2}\left(1 - \frac{Q^2}{8m_N^2}\right), \qquad (5)$$

where e is the positive charge of the deuteron. Denoting, $G_0(Q)(G_E^p(Q^2) + G_E^n(Q^2))(1 - Q^2/8m_N^2) = G_d(Q^2)$ so that, $V_d(Q) = 4\pi e(G_d(Q^2)/Q^2)$, its Fourier transform is the electric potential, $V_d(r) = 4\pi e \int e^{i\mathbf{Q}\cdot\mathbf{r}}(G_d(Q^2)/Q^2)d^3Q/(2\pi)^3$, the Laplacian of which gives the density $\rho_d(r)$.

Since we wish to study the sensitivity of the finite size corrections to the D-state probability in the deuteron wave function, we shall use different parametrizations of the deuteron wave function involving about 2 to 7% of D-state probabilities in order to calculate $G_0(Q)$. One choice involves a parametrization of the wave function obtained from the Paris NN potential [27] with $P_S = \int |u(r)|^2 dr = 0.942$ and $P_D = \int |w(r)|^2 dr = 0.058$. Our second choice is a phenomenological model [28] which uses similar forms as in [27] for parametrizing the wave functions but with different parameters, so that the probabilities are $P_S = 0.983$ and $P_D = 0.017$. Whereas the parameters in [27] were fitted to reproduce the numerical values of the Paris wave function, those in [28] were obtained by directly fitting the quadrupole moment and deuteron charge form factor data with $G_0(Q)$, assuming $G_E^p + G_E^n = 1$. In order to test the case with no D-wave component at all, we perform a calculation by normalizing the Paris $u(r)$ in [27] to 1 and not using the D-wave at all. We also use an older parametrization [29] of the Hamada–Johnston wave functions (once again having similar forms as in [27] and [28]) with $P_S = 0.93$ and $P_D = 0.7$. The charge form factor of the deuteron which is extracted from scattering experiments seems to be equally well produced (considering the error bars and the entire range shown) by all choices of the D-state probabilities (see Fig. 1).

2.3. Corrections to the 2S energy levels in muonic deuterium

Recent measurements of the 2S-2P transitions in muonic deuterium [9] have shown how precision spectroscopy of atomic energy levels can be used to determine the deuteron (and also the proton) radius more accurately than that extracted from any scattering experiment. The experiment was based on forming $\mu^- d$ atoms in an unstable 2S state and measuring the 2S-2P transitions by pulsed laser spectroscopy. The measured value of the 2S-2P Lamb shift is then compared with the theoretical calculations involving corrections from Quantum Electrodynamics (QED) and the finite size of the deuteron. The QED corrections can be calculated very accurately [24]. The finite size corrections (FSC) are incorporated as radius (r_d) dependent terms. The theoretical value of the Lamb shift thus calculated is given by, $\Delta E_{LS}^{theo} = 228.7766(10)$ meV $+ \Delta E^{TPE} - 6.1103(3)r_d^2$ meV/fm^2, where the second term is a deuteron polarizability contribution coming from two-photon exchange and is equal to 1.7096(200) meV. Comparing ΔE_{LS}^{theo} with the experimentally measured, $\Delta E_{LS}^{exp} = 202.8785(31)_{stat}(14)_{syst}$ meV, led to the precise value of the radius, $r_d = 2.12562(13)_{exp}(77)_{theo}$ fm. In order to compare the results of the present work with the above precision measurements, with the aim of extracting the D-state probability in deuteron, we first note that the finite size correction (FSC) term, 6.11019 r_d^2 meV/fm^2 is a sum of order α^4, α^5 and α^6 corrections given by 6.0731 r_d^2, 0.033804 r_d^2 and 0.003286 r_d^2 respectively. The order α^4 part given by 6.0731 r_d^2 meV/fm^2 will be derived below briefly.

2.4. Finite size Coulomb correction at order α^4

The effect of including the deuteron charge distribution, $\rho_d(r)$ in place of the point-like $1/r$ Coulomb potential can be incorporated by evaluating the energy correction using first order perturbation theory [31], as,

$$\Delta E_{FS} = \int |\Psi_{nl}(\mathbf{r})|^2 \left[e\, V_d(r) - \left[-\frac{4\pi\alpha}{r} \right] \right] d^3r, \qquad (6)$$

where, $\Psi_{nl}(\mathbf{r})$ is the unperturbed atomic wave function and $V_d(r)$ is the Fourier transform of the deuteron electric potential in Eq. (5). If we now approximate $\Psi_{nl}(\mathbf{r}) \approx \Psi_{nl}(0)$, it is easy to show that ΔE_{FS} reduces to [31],

$$\begin{aligned}\Delta E_{FS}^0 &= \frac{-e}{6}|\Psi_{nl}(0)|^2 \int d^3r\, r^2\, \nabla^2 V_d(r)\\ &= (2\pi\alpha/3)|\Psi_{nl}(0)|^2 \int d^3r\, r^2\, \rho_d(r)\\ &= (2\pi\alpha/3)|\Psi_{nl}(0)|^2 \langle r^2 \rangle, \end{aligned} \qquad (7)$$

Table 1
Finite size corrections to the 2S atomic level, ΔE_{FS} (Eq. (6)) and ΔE^0_{FS} (Eq. (7)).

% D-wave	7% [29]	5.8% [27]	1.7% [28]	0 [27]
ΔE_{FS} (meV)	26.2	26.72	27.03	27.57
ΔE^0_{FS} (meV)	26.53	27.01	27.35	27.87

Table 2
The estimated values of the radius for different D-state probabilities.

Model		% P_D	r_d (fm)
Chiral	EGM N3LO	3.28	2.1315
Ref. [15]	EMN N4LO	4.1	2.1277
	Jülich N4LO	4.29	2.1268
	EM N3LO	4.51	2.1296
High precision	CD-Bonn	4.85	2.1212
Ref. [16]	NjmNR	5.635	2.1222
	NjmR	5.664	2.1226
	Reid93	5.699	2.1236
	AV18	5.75	2.1221
	Paris [27]	5.8	2.126
Traditional	TRS	5.92	2.1297
Ref. [33]	RSC	6.47	2.1127
	RHC	6.497	2.1156
	HW	6.953	2.1223
	McGee [29]	7	2.107
Phenomenological [28]		1.7	2.14

since, $\nabla^2 V_d(r) = -4\pi e\rho_d$. For $n=2$, $l=0$, $(2\pi\alpha/3)|\Psi_{nl}(0)|^2 =$ 6.0731 meV/fm^2 and the right side of Eq. (7) is 6.0731 $\langle r_d^2\rangle$ as in [9,24]. The approximation $\Psi_{nl}(\mathbf{r}) \approx \Psi_{nl}(0)$ allows us to express the FSC in terms of the charge radius and thus opens the possibility of determining the charge radius of the proton or a nucleus from atomic spectroscopic data which would have otherwise been not possible. In Table 1, we show the tiny difference between the calculation of ΔE_{FS} using (6) or (7). The table also displays sensitivity of the corrections to the parametrization of the deuteron wave function. The magnitude of the corrections increases with the lowering of the D-state probability in the deuteron wave function. It is this sensitivity which leads us to the results shown in Table 2 which will be discussed in the next section. Note that even though there exists a tiny difference in the values of ΔE_{FS} and ΔE^0_{FS} in Table 1, for the comparison of the radius evaluated from $r_d = \Delta E^0_{FS}/6.0731$ with r_d^{exp} which has been fitted to data using a similar formula, this difference does not matter.

3. Deuteron charge radius and D-state probability

The electric potential $V_d(Q)$ in (5) can also be expressed as $V_d(Q) = 4\pi e G_{ch}(Q)/Q^2$ with, $G_{ch}(Q) = G_0(Q)[G_E^p(Q^2) + G_E^n(Q^2)][1 - (Q^2/8m_N^2)]$ being the Fourier transform of $\rho_d(r)$, so that using standard formulae for the expressions connecting radii and form factors [17], we obtain, $r_d^2 = r_p^2 + r_n^2 + (3/4m_p^2) + (1/4)\int_0^\infty [|u(r)|^2 + |w(r)|^2] r^2 dr$, where, the last term is the matter radius $r_m^2 = -6(dG_0/dQ^2)|_{Q^2=0}$. Thus, for a given parametrization of G_E^p and G_E^n which reproduce the data on $G_{ch}(Q)$ as defined above well (see Fig. 1), r_d can be seen to depend on the deuteron wave function $w(r)$. By choosing a certain $w(r)$, we choose also a certain P_D, since $P_D = \int |w(r)|^2 dr$. Knowing the values of ΔE^0_{FS} (see second line of Table 1), the radius can be determined from Eq. (7), namely, $\Delta E^0_{FS} = 6.0731\, r_d^2$. Since the fits in [9] assume a similar form of the α^4 FSC, it is appropriate to compare this r_d with the fitted value of $r_d^{exp} = 2.12562(78)$ fm in [9]. In studies of electron–deuteron scattering as in Ref. [25], data have been interpreted in terms of the plane wave Born approximation

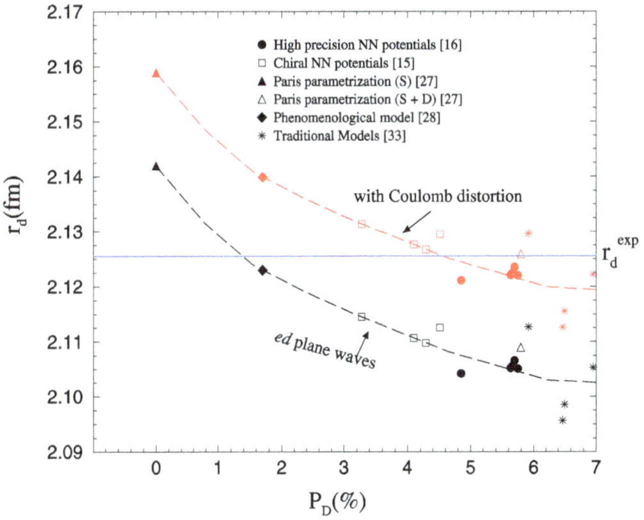

Fig. 2. Charge radius of the deuteron as a function of the D-state probability in the deuteron wave function evaluated using different models. The dashed lines are drawn to guide the eye with the red (black) one representing the numbers obtained with (without) the effects of Coulomb distortion included.

(PWBA). However, the effects of including distorted *ed* waves can become important for comparisons with precise data. Noting that the Coulomb distortion [32] changes the deuteron radius by 0.017 fm, the authors in [25] suggest an adjustment of the deuteron charge form factor by an amount $\delta G_C = -0.003 + 0.104\, Q^2$ which decreases the form factor at small Q^2 and increases the value of the radius. The results presented in Table 1, however, do not take these effects into account since it is not appropriate to evaluate (6) (which involves $V_d(r)$ obtained from a Fourier integral over all momenta) using the form factor corrected only at low Q^2. The correction at low Q^2 introduces a disagreement with data at large Q^2 as shown in [25]. The above correction is however important for the calculation of the radius defined by the derivative of the form factor at $Q^2 = 0$ and hence in Table 2, we present the deuteron charge radius, r_d, with the Coulomb distortion correction of 0.017 fm as in [32] for different choices of the nucleon–nucleon (NN) potentials. Fig. 2 displays the same with and without the Coulomb distortion included. From the figure we observe a general trend of increasing P_D for smaller radii. However, the results are model dependent with the chiral models indicating a value of $P_D = 4.3$ and the high precision NN models a value around 5.7 leading to a good agreement with the experimental $r_d =$ 2.12562(78) fm [9]. The choice of the proton and neutron form factor parametrization (which affects the values of r_p and r_n entering in $r_d^2 = r_p^2 + r_n^2 + (3/4m_p^2) + (1/4)\int_0^\infty [|u(r)|^2 + |w(r)|^2] r^2 dr$), can add a small uncertainty to the values deduced in Table 2. The magnitude of these uncertainties using different parametrizations for G_E^p and G_E^n which reproduce the deuteron charge form factor, $G_{ch}(Q) = G_0(Q)[G_E^p(Q^2) + G_E^n(Q^2)][1 - (Q^2/8m_N^2)]$ equally well, remains to be investigated in future.

4. Finite size Coulomb plus Darwin corrections at order α^4

For completeness, we also calculate the FSC with the Darwin terms in (4) within the few body formalism. The Fourier transform of the muon–deuteron interaction potential, $V_{\mu^- d}(Q)$, can be done numerically to obtain the potential in coordinate space which can then be used to evaluate the energy correction using first order time independent perturbation theory as,

$$\Delta E = \int |\Psi_{nl}(\mathbf{r})|^2 \left[V_{\mu^- d}(r) - V_{\mu^- d}(r)^{point} \right] d^3 r \qquad (8)$$

Table 3
Finite size corrections ΔE in meV (Eq. (8)), to the 1S and 2S atomic levels in $\mu^- d$, for different D-state probabilities of the deuteron wave function.

	7% [29]	5.8% [27]	1.7% [28]	0 [27]
1S	206.28	210.42	213.17	217.19
2S	25.78	26.31	26.65	27.15

where, $\Psi_{nl}(\mathbf{r})$ is the unperturbed atomic wave function. Note that we have subtracted the point-like contribution $-\alpha/r$ as well as the point-like Darwin terms from $V_{\mu-d}(r)$ so that the quantity in square brackets is the perturbative potential only due to deuteron structure. In Table 3, we list the finite size corrections (FSC) (to the Coulomb and Darwin terms) of order α^4 in muonic deuterium using Eq. (8) and different percentages of the D-state probabilities in the deuteron wave function for the energy levels with $l = 0$ and $n = 1, 2$. Since the numbers in Table 3 are not very different from those in Table 1 (compare the first line in Table 1 with the second line in Table 3), one can say that the FSC due to the Darwin terms are in general very small. Note that Eqs. (8) and (6) are different in the sense that (i) $V_{\mu-d}$ in (8) contains the additional muon Darwin term as compared to $e\,V_d$ and (ii) whereas (8) subtracts the point-like potential, $V_{\mu-d}(r)^{point}$, which contains the point-like Coulomb term, $-4\pi\alpha/r$ and two point-like Darwin terms $\delta^3(\mathbf{r})/8M_N^2$ and $\delta^3(\mathbf{r})/8M_\mu^2$, Eq. (6) subtracts only the point-like Coulomb, $-4\pi\alpha/r$.

To summarize, the leading order nuclear structure corrections in muonic deuterium have been evaluated within a few body formalism which reveals the dependence of the correction on the model of the deuteron wave function. Since scattering data do not have the high precision achieved by the muonic atom spectroscopy data, the deuteron charge form factor can be reproduced equally well (within error bars) by all the parametrizations of the deuteron wave function used, irrespective of the percentage of P_D in them. However, we notice that a comparison of the radius evaluated using these parametrizations with the precise radius value extracted from $\mu^- d$ spectroscopy provides a complementary tool to determine P_D. Though there do exist model dependent uncertainties in P_D (see Table 2 and Fig. 2), there seems to be a general trend of increasing values of P_D for smaller r_d. The few body formalism presented here can also be used to evaluate the nuclear structure corrections in muonic helium atoms which are expected to be studied in future.

Acknowledgements

One of the authors (D. B. F.) thanks the administrative department of science, technology and innovation of Colombia (COLCIENCIAS) grant number 0525-2013 and the Faculty of Science, Universidad de los Andes grant number P15.160322.009/01-02-FISI08 for the financial support provided. He is also grateful to IFIC, University of Valencia and Prof. Vicente Vacas for their kind hospitality and useful discussions.

References

[1] H. Feshbach, J. Schwinger, Phys. Rev. 84 (1951) 194.
[2] N.L. Rodning, L.D. Knutson, Phys. Rev. Lett. 57 (1986) 2248;
N.L. Rodning, L.D. Knutson, Phys. Rev. C 41 (1990) 898.
[3] J.A. Oller, Phys. Rev. C 93 (2016) 024002.
[4] R.D. Amado, Phys. Rev. C 19 (1979) 1473;
J.L. Friar, Phys. Rev. C 20 (1979) 325.
[5] H.E. Conzett, et al., Phys. Rev. Lett. 43 (1979) 572.
[6] E. Hadjimichael, Nucl. Phys. A 312 (1978) 341.
[7] L. Mathelitsch, H.F.K. Zingl, Nuovo Cimento 44 (1978) 81.
[8] J.S. Levinger, Phys. Lett. B 29 (1969) 216.
[9] R. Pohl, et al., Science 353 (2016) 669.
[10] W. Leidemann, R. Rosenfelder, Phys. Rev. C 51 (1995) 427.
[11] K. Pachucki, Phys. Rev. Lett. 106 (2011) 193007;
K. Pachucki, A. Wienczek, Phys. Rev. A 91 (2015) 040503(R).
[12] O.J. Hernandez, et al., Phys. Lett. B 736 (2014) 344.
[13] J.L. Friar, Ann. Phys. 122 (1979) 151.
[14] J.L. Friar, Phys. Rev. C 88 (2013) 034003.
[15] D.R. Entem, R. Machleidt, Phys. Rev. C 68 (2003) 041001;
E. Epelbaum, W. Göckle, U.-G. Meissner, Nucl. Phys. A 747 (2005) 362;
E. Epelbaum, H. Krebs, U.-G. Meissner, Phys. Rev. Lett. 115 (2015) 122301;
D.R. Entem, R. Machleidt, Y. Nosyk, preprint, arXiv:1703.05454, 2017.
[16] V.C.J. Stoks, R.A.M. Klomp, C.P.F. Terheggen, J.J. de Swart, Phys. Rev. C 49 (1994) 2950;
R. Machleidt, Phys. Rev. C 63 (2001) 024001;
R.B. Wiringa, V.C.J. Stoks, R. Schiavilla, Phys. Rev. C 51 (1995) 38.
[17] D. Bedoya Fierro, N.G. Kelkar, M. Nowakowski, J. High Energy Phys. 1509 (2015) 215;
N.G. Kelkar, F. Garcia Daza, M. Nowakowski, Nucl. Phys. B 864 (2012) 382.
[18] F. Garcia Daza, N.G. Kelkar, M. Nowakowski, J. Phys. G 39 (2012) 035103.
[19] H.B.G. Casimir, P. Polder, Phys. Rev. 73 (1948) 360;
E.M. Lifschitz, JETP Lett. 2 (1956) 73;
F. Ferrer, J.A. Grifols, Phys. Lett. B 460 (1999) 371.
[20] R. Machleidt, Adv. Nucl. Phys. 19 (1989) 181;
J.D. Walecka, Theoretical Nuclear and Subnuclear Physics, second edition, World Scientific, 2004.
[21] V.B. Belyaev, Lectures on the Theory of Few Body Systems, Springer-Verlag, Heidelberg, 1990;
S.A. Rakityansky, et al., Phys. Rev. C 53 (1996) R2043.
[22] N.G. Kelkar, Phys. Rev. Lett. 99 (2007) 210403.
[23] J.L. Friar, J.W. Negele, Advances in Nuclear Physics, vol. 8, Springer, New York, 1975, pp. 219–376;
J.L. Friar, J. Martorell, D.W.L. Sprung, Phys. Rev. A 56 (1997) 4579.
[24] J.J. Krauth, et al., Ann. Phys. 366 (2016) 168.
[25] R. Gilman, F. Gross, J. Phys. G 28 (2002) R37.
[26] H. Arenhövel, F. Ritz, T. Wilbois, Phys. Rev. C 61 (2000) 034002.
[27] M. Lacombe, et al., Phys. Lett. B 101 (1981) 139.
[28] Yu.A. Berezhnoy, V. Yu Korda, A.G. Gakh, Int. J. Mod. Phys. E 14 (2005) 1073.
[29] I.J. McGee, Phys. Rev. 151 (1966) 772.
[30] D. Abbott, et al., Eur. Phys. J. A 7 (2000) 421.
[31] C. Itzykson, J.-B. Zuber, Quantum Field Theory, Dover Publications, New York, 1980.
[32] I. Sick, D. Trautmann, Phys. Lett. B 375 (1996) 16;
I. Sick, D. Trautmann, Nucl. Phys. A 637 (1998) 559;
I. Sick, Prog. Part. Nucl. Phys. 47 (2001) 245.
[33] Roderick V. Reid Jr., Ann. Phys. 50 (1968) 411;
J.W. Humberston, J.B.G. Wallace, Nucl. Phys. A 141 (1970) 362;
R. De Tourreil, B. Rouben, D.W.L. Sprung, Nucl. Phys. A 242 (1975) 445.

11

Diagrammar in an extended theory of gravity

David C. Dunbar *, John H. Godwin, Guy R. Jehu, Warren B. Perkins

College of Science, Swansea University, Swansea, SA2 8PP, UK

ARTICLE INFO

Editor: M. Cvetič

ABSTRACT

We show how the S-matrix of an extended theory of gravity defined by its three-point amplitudes can be constructed by demanding factorisation. The resultant S-matrix has tree amplitudes obeying the same soft singularity theorems as Einstein gravity including the sub-sub-leading terms.

1. Introduction

Scattering amplitudes are traditionally defined from a quantum field theory and the resulting Feynman vertices and Feynman diagrams. Alternatively, the amplitudes can be regarded as the fundamental objects which define the theory perturbatively. It is not very useful to define a theory by specifying the entire S-matrix explicitly but it is an important question whether the S-matrix can be defined from a minimal set of data and rules i.e. a *"diagrammar"* [1]. Once a minimal set of amplitudes is specified we aim to construct all other amplitudes by demanding they have the correct symmetries and singularities. Defining the S-matrix using its singularities is a long-standing programme which is still active and fruitful [2–7].

In this letter we build an S-matrix from a set of three-point amplitudes using their singularity structure. The S-matrix corresponds to a theory of Einstein gravity extended by the addition of R^3 terms. We are working with massless theories and view the amplitude as a function of the twistor variables λ_i^a and $\bar{\lambda}_i^{\dot{a}}$, $M(\lambda_i, \bar{\lambda}_i)$. The spinor products $\langle i\,j\rangle$, $[i\,j]$ are $\langle i\,j\rangle = \epsilon_{ab}\lambda_i^a\lambda_j^b$, $[i\,j] = \epsilon_{\dot{a}\dot{b}}\bar{\lambda}_i^{\dot{a}}\bar{\lambda}_j^{\dot{b}}$. In this formalism amplitudes have a well-defined "spinor weight". Counting λ_i as weight $+1$ and $\bar{\lambda}_i$ as -1, then the amplitude has weight $+4$ for a negative helicity graviton and -4 for a positive helicity graviton.

We define the theory starting with the usual three-point amplitudes of Einstein gravity[1]:

$$V_3(1^-, 2^-, 3^+) = \frac{\langle 1\,2\rangle^6}{\langle 1\,3\rangle^2\langle 3\,2\rangle^2},$$

$$V_3(1^+, 2^+, 3^-) = \frac{[1\,2]^6}{[1\,3]^2[3\,2]^2},$$

$$V_3(1^+, 2^+, 3^+) = V_3(1^-, 2^-, 3^-) = 0. \tag{1}$$

These amplitudes have the correct spinor weight and are quadratic in the momenta. These amplitudes are only defined for complex momenta. For an on-shell three-point amplitude the condition $k_1 + k_2 + k_3 = 0$ demands $k_1 \cdot k_2 = 0$ etc. For real momenta this implies $\langle i\,j\rangle = [i\,j] = 0$ and the vertices are all zero. However if we consider complex momenta then we can have $\lambda_1 \sim \lambda_2 \sim \lambda_3$ but $[i\,j] \neq 0$.

The tree amplitudes for Einstein gravity can be computed recursively starting from these [8–10]. We show that a similar construction can be used for an extended theory.

We extend this theory by adding additional three-point amplitudes which are of higher power in momenta. To be non-trivial, these three-point amplitudes must either be functions of $\langle i\,j\rangle$ or $[i\,j]$ exclusively. The simplest polynomial amplitudes arise with six powers of momenta and are

$$V_3^\alpha(1^-, 2^-, 3^-) = \alpha\langle 1\,2\rangle^2\langle 2\,3\rangle^2\langle 3\,1\rangle^2,$$

$$V_3^\alpha(1^+, 2^+, 3^+) = \alpha[1\,2]^2[2\,3]^2[3\,1]^2 \tag{2}$$

where α is an arbitrary constant. We also have

$$V_3^\alpha(1^-, 2^-, 3^+) = V_3^\alpha(1^+, 2^+, 3^-) = 0, \tag{3}$$

there being no polynomial function with the correct spinor and momentum weight. These are essentially the unique choice for a three-point amplitude [11] (see Fig. 1).

The amplitudes in this theory can be expanded as a power series in α,

$$M_n(1, \cdots, n) = \sum_{r=0} \alpha^r M_n^{(r)}(1, \cdots, n) \tag{4}$$

* Corresponding author.
E-mail address: d.c.dunbar@swan.ac.uk (D.C. Dunbar).

[1] We remove a factor of $i(\kappa/2)^{n-2}$ from the n-point amplitude.

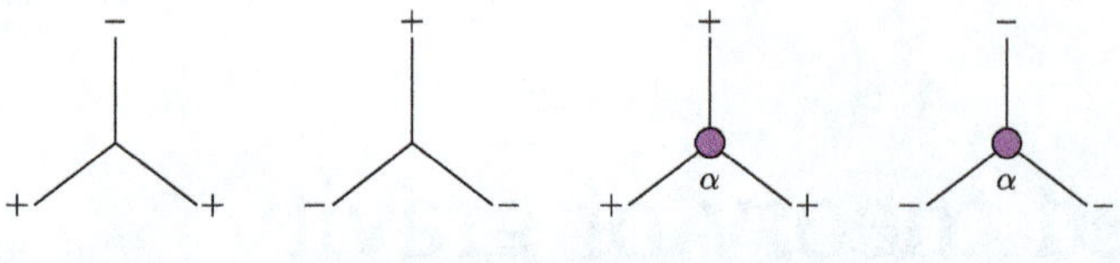

Fig. 1. The non-zero three-point amplitudes.

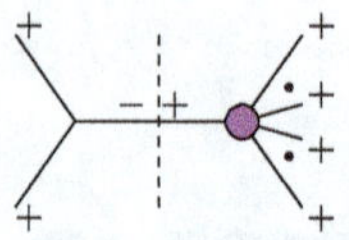

Fig. 2. Factorisations of the n-point all-plus.

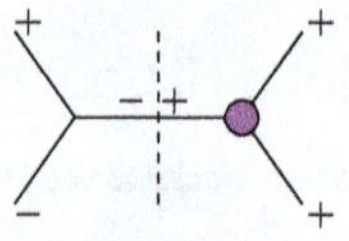

Fig. 3. Factorisations of the four-point single minus amplitude.

where $M_n^{(0)}$ is the Einstein gravity amplitude. Here we focus on the $r = 1$ part of the extended theory. This being the leading deformation of the theory from Einstein gravity.

The theory we are considering would arise using field theory methods from the Lagrangian

$$L = \int d^D x \sqrt{-g}(R + C_\alpha R_{abcd} R^{cdef} R_{ef}{}^{ab}) \tag{5}$$

where $C_\alpha = \alpha/60$. However we note that to do so would involve determining increasingly complicated n-point vertices as the Lagrangian is expanded in the graviton field. As we will see the three-point amplitudes are sufficient to completely determine the S-matrix.

The key element is that the entire S-matrix is determined from these vertices if we demand that the amplitudes factorise on simple poles. Specifically, for any partition of the external legs into two sets, $\{k_{L_1}, k_{L_2} \cdots, K_{L_l}\}$ and $\{k_{R_1}, k_{R_2} \cdots, k_{R_m}\}$ with $l + m = n$ and $l, m \geq 2$, if $K = \sum_{j=1}^{l} k_{L_j}$, then when $K^2 \longrightarrow 0$ the amplitude is singular with the simple pole being

$$M_n^{\text{tree}} \stackrel{K^2 \to 0}{\longrightarrow} \sum_{\lambda=\pm} \left[M_{l+1}^{\text{tree}}(k_{L_1}, \ldots, k_{L_l}, -K^\lambda) \frac{i}{K^2} \right. \\ \left. \times M_{m+1}^{\text{tree}}(K^{-\lambda}, k_{R_1}, \ldots, k_{R_m}) \right]. \tag{6}$$

We can excite the pole in K^2 by shifting to complex momenta and applying methods of complex analysis. There are two shifts which we use to generate the S-matrix. Firstly there is the original Britto–Cachazo–Feng–Witten (BCFW) shift [5],

$$\lambda_i \longrightarrow \lambda_i + z\lambda_j, \; \bar{\lambda}_j \longrightarrow \bar{\lambda}_j - z\bar{\lambda}_i. \tag{7}$$

For Einstein gravity this shift is sufficient to generate the tree level S-matrix [12]. Additionally we can use the Risager shift [13],

$$\begin{aligned} \lambda_i &\longrightarrow \lambda_i + z[jk]\lambda_\eta, \\ \lambda_j &\longrightarrow \lambda_j + z[ki]\lambda_\eta, \\ \lambda_k &\longrightarrow \lambda_k + z[ij]\lambda_\eta, \end{aligned} \tag{8}$$

where λ_η is an arbitrary spinor. Both shifts change the momenta to be functions of z whilst leaving all momenta null and preserving overall momentum conservation. We need both shifts to construct the S-matrix for the extended theory. By considering the integral

$$\int_\gamma \frac{M(z)}{z} \tag{9}$$

where γ is a closed contour, *provided* $M(z)$ vanishes at infinity the unshifted amplitude, $M(0)$, can be obtained from the singularities in the amplitude. These occur at points z_i where $K_i^2(z) = 0$. At these points,

$$K_i^2(z) = -\frac{(z - z_i)}{z_i} \times K_i^2(0) \tag{10}$$

and we obtain,

$$M_n^{\text{tree}}(0) = \sum_{i,\lambda} M_{l_i+1}^{\text{tree},\lambda}(z_i) \frac{i}{K_i^2(0)} M_{m_i+1}^{\text{tree},-\lambda}(z_i), \tag{11}$$

where the summation over i is only over factorisations where there are shifted legs on both sides of the pole. This is the on-shell recursive expression of [5]. Note that if $M(z)$ does not vanish at infinity this does not imply factorisation is insufficient to determine the amplitude but only that particular shift can not be used to engineer the amplitude.

Expressions obtained from (11) are not manifestly symmetric as the choice of shift legs breaks crossing symmetry, however symmetry is restored in the sum. This is a highly non-trivial check that the amplitude has been computed successfully.

2. Generating the amplitudes

In this section we give some of the details of the process of generating the leading α contribution to the S-matrix.

Four-point amplitudes: The three-point amplitudes are our inputs so the first outputs are the four-point amplitudes. There are three independent helicity configurations,

$$\begin{aligned} &M_4(1^+, 2^+, 3^+, 4^+), \; M_4(1^-, 2^+, 3^+, 4^+), \\ &M_4(1^-, 2^-, 3^+, 4^+). \end{aligned} \tag{12}$$

Of these the first two are vanishing in Einstein gravity with only the last being non-zero: which is consequently termed the "Maximally-Helicity-Violating" (MHV) amplitude. For $M_4^{(1)}$ the reverse is true: $M_4^{(1)}(1^-, 2^-, 3^+, 4^+) = 0$ since there are no possible factorisations, while $M_4^{(1)}(1^+, 2^+, 3^+, 4^+)$ and $M_4^{(1)}(1^-, 2^+, 3^+, 4^+)$ are non-zero.

The factorisations of the n-point all-plus amplitude are shown in Fig. 2, and the factorisations of the four-point single minus amplitude are shown on Fig. 3.

These factorisations can be excited using either of the shifts in (7) and (8). In the all-plus case only the second results in an amplitude with the correct symmetries. This in indication that (7) yields a shifted all-plus amplitude that does not vanish at infinity. Conversely, for the single minus amplitude we must use the BCFW shift. Performing the shifts and evaluating the amplitudes we obtain

$$\begin{aligned} M_4^{(1)}(1^+, 2^+, 3^+, 4^+) &= 10 \left(\frac{st}{\langle 12 \rangle \langle 23 \rangle \langle 34 \rangle \langle 41 \rangle} \right)^2 stu, \\ M_4^{(1)}(1^-, 2^+, 3^+, 4^+) &= \left(\frac{[24]^2}{[12] \langle 23 \rangle \langle 34 \rangle [41]} \right)^2 \frac{s^3 t^3}{u}. \end{aligned} \tag{13}$$

The other non-zero amplitudes are available by conjugation. For the all-plus amplitude the recursion generates terms that contain the arbitrary spinor λ_η, however the sum of terms is independent

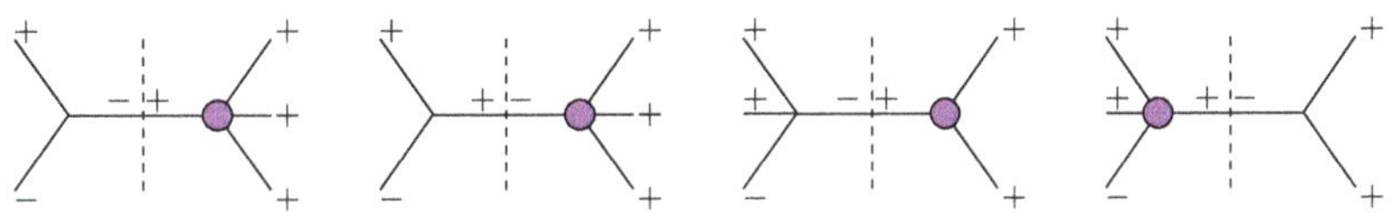

Fig. 4. Factorisations of the five-point single minus amplitude.

of λ_η and simplifies to the above. These four-point amplitudes due to a R^3 term have been computed using field theory methods long ago [14]. These amplitudes vanish to all orders in a supersymmetric theory: a fact used show supergravity was two-loop ultra-violet finite [15,16]. The above expressions are in a spinor helicity basis but agree once this is accounted for. In [17] these four-point amplitudes were also obtained using a "all-line recursion" technique where all legs have shifted momenta. These expressions also appear as the UV infinite pieces of both two-loop gravity in four dimensions [18,19] and one-loop gravity in six dimensions [20].

Five-point amplitudes: As before the shift (8) yields an all-plus amplitude that is independent of λ_η and has full crossing symmetry:

$$M_5^{(1)}(1^+,2^+,3^+,4^+,5^+) = \left(\sum_{P_6} T^A_{(1,2,3),(4,5)} + \sum_{P_3} T^B_{(1,2,3),4,5} \right) \tag{14}$$

where

$$T^A_{(1,2,3),(4,5)} = 10 \frac{[14]}{\langle 14\rangle} \frac{[53][52]}{\langle 1\eta\rangle^2 \langle 4\eta\rangle} \frac{[23]^2}{\langle 45\rangle} \times [5|K_{14}|\eta\rangle [2|K_{14}|\eta\rangle [3|K_{14}|\eta\rangle\,, \tag{15}$$

$$T^B_{(1,2,3),4,5} = -10 \frac{[14][15][23][1|K_{23}|\eta\rangle^2 [5|K_{23}|\eta\rangle [4|K_{23}|\eta\rangle}{\langle 23\rangle \langle 2\eta\rangle^2 \langle 3\eta\rangle^2} \frac{[45]}{\langle 45\rangle} \tag{16}$$

and P_3 denotes summation over the three cyclic permutations of legs 1,2 and 3. P_6 denotes the three permutations of P_3 together with interchange of legs 4 and 5. The λ_η independence of $M_5^{(1)}(1^+,2^+,3^+,4^+,5^+)$ is not manifest.

The factorisations of the five-point single minus amplitudes are more varied as shown on Fig. 4. Using the BCFW shift on $(\bar\lambda_1, \lambda_2)$ we obtain the amplitude

$$\begin{aligned} M_5^{(1)}&(1^-,2^+,3^+,4^+,5^+) \\ &= \frac{10}{[12]^2} \left(\prod_{i,j=2,3,4,5,i<j} [i\,j] \right) \\ &\times \left(\frac{\langle 15\rangle}{[15]} \frac{[25]^3}{\langle 34\rangle} + \frac{\langle 13\rangle}{[13]} \frac{[23]^3}{\langle 45\rangle} + \frac{\langle 14\rangle}{[14]} \frac{[24]^3}{\langle 53\rangle} \right) \\ &+ \frac{\langle 12\rangle^2}{\langle 34\rangle \langle 35\rangle \langle 45\rangle \prod_{i=3,4,5} \langle 1\,i\rangle} \left(\frac{[23]^5 [45] \langle 13\rangle^5}{\langle 23\rangle} \right. \\ &\left. + \frac{[24]^5 [53] \langle 14\rangle^5}{\langle 24\rangle} + \frac{[25]^5 [35] \langle 15\rangle^5}{\langle 25\rangle} \right) \\ &+ \frac{1}{\langle 12\rangle^2 \langle 34\rangle \langle 35\rangle \langle 45\rangle} \left(\frac{[23][45]^5 \langle 15\rangle^3 \langle 14\rangle^3}{\langle 23\rangle} \right. \\ &\left. + \frac{[24][53]^5 \langle 13\rangle^3 \langle 15\rangle^3}{\langle 24\rangle} + \frac{[25][34]^5 \langle 14\rangle^3 \langle 13\rangle^3}{\langle 25\rangle} \right). \end{aligned} \tag{17}$$

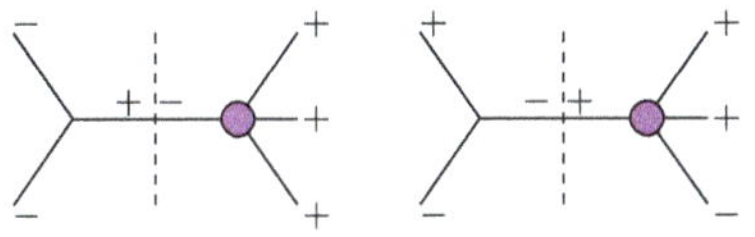

Fig. 5. Factorisations of the five-point MHV amplitude.

The five-point MHV amplitude is non-zero. The non-zero factorisations of the amplitude are shown in Fig. 5.

This amplitude can be obtained using a BCFW shift of either the two negative helicity legs or of a negative–positive pair. Shifting the two negative legs generates the expression (using only the second factorisation of Fig. 5),

$$\begin{aligned} M_5^{(1)}(1^-,2^-,3^+,4^+,5^+) &= -s_{34} \frac{\langle 15\rangle}{[15]} \frac{[34]^2 [35]^3 [45]^3}{[12]^2 [23][24]} \\ &- s_{45} \frac{\langle 13\rangle}{[13]} \frac{[45]^2 [43]^3 [53]^3}{[12]^2 [24][25]} - s_{53} \frac{\langle 14\rangle}{[14]} \frac{[53]^2 [54]^3 [34]^3}{[12]^2 [25][23]}\,. \end{aligned} \tag{18}$$

This completes the set of five-point amplitudes. We can continue in this way generating the tree-level S-matrix. We have made available $M_n^{(1)}$ for $n \le 7$ in `Mathematica` format at `http://pyweb.swan.ac.uk/~dunbar/Smatrix.html`. The amplitudes have been generated up to $n = 8$ and have the correct symmetries, are η-independent and have the correct leading soft-limits.

We have evaluated amplitudes in a $R + \alpha R^3$ theory. In ref. [21] amplitudes in Yang-Mills theory extended by F^3 terms were studied. Then using double copy techniques and the KLT relations [22] graviton scattering amplitudes were derived upto $n = 6$. As noted in [21] these correspond to amplitudes in a $R + \alpha R^3 + \sqrt{\alpha} R^2 \phi$ theory. The four-point amplitudes in the two theories are proportional [17,21] but beyond four-point the two sets of amplitudes are functionally different. The all-plus amplitude in the two theories remain proportional for $n > 4$ with

$$M_n^{(1),R^3+R^2\phi}(1^+,2^+,\cdots n^+) = \frac{5}{2} M_n^{(1),R^3}(1^+,2^+,\cdots n^+) \tag{19}$$

and we confirm this for $n \le 7$.

3. Soft limits

Graviton scattering amplitudes are singular as a leg becomes soft. Weinberg [23] many years ago presented the leading soft limit. If we parametrise the momentum of the n-th leg as $k_n^\mu = t \times k_s^\mu$ then in the limit $t \longrightarrow 0$ the singularity in the n-point amplitude is

$$M_n \longrightarrow \frac{1}{t} \times S^{(0)} \times M_{n-1} + O(t^0) \tag{20}$$

where M_{n-1} is the $n-1$-point amplitude. The soft-factor $S^{(0)}$ is universal and Weinberg showed that (20) does not receive corrections in loop amplitudes.

Recently it has also been proposed [24–26] that the sub-leading and sub-sub-leading terms are also universal. This can be best exposed, when a positive helicity leg becomes soft, by setting

$$\lambda_n = t \times \lambda_s\,, \quad \bar{\lambda}_n = \bar{\lambda}_s\,. \tag{21}$$

In the $t \longrightarrow 0$ limit the amplitude has t^{-3} singularities. At tree level the amplitudes satisfy soft-theorems [25] whereby their behaviour as $t \longrightarrow 0$ is

$$\begin{aligned} M_n^{\text{tree}} &= S_t M_{n-1}^{\text{tree}} + O(t^0) \\ &= \left(\frac{1}{t^3} S^{(0)} + \frac{1}{t^2} S^{(1)} + \frac{1}{t} S^{(2)}\right) M_{n-1}^{\text{tree}} + O(t^0) \end{aligned} \tag{22}$$

where, for a positive helicity-leg becoming soft [25,27,28]

$$S^{(0)} = -\sum_{i=1}^{n-1} \frac{[s\,i]\,\langle i\,\alpha\rangle\,\langle i\,\beta\rangle}{\langle s\,i\rangle\,\langle s\,\alpha\rangle\,\langle s\,\beta\rangle}\,, \tag{23}$$

$$S^{(1)} = -\frac{1}{2}\sum_{i=1}^{n-1} \frac{[s\,i]}{\langle s\,i\rangle}\left(\frac{\langle i\,\alpha\rangle}{\langle s\,\alpha\rangle} + \frac{\langle i\,\beta\rangle}{\langle s\,\beta\rangle}\right)\bar{\lambda}_s^{\dot{a}}\frac{\partial}{\partial\bar{\lambda}_i^{\dot{a}}}\,, \tag{24}$$

$$S^{(2)} = \frac{1}{2}\sum_{i=1}^{n-1} \frac{[i\,s]}{\langle i\,s\rangle}\bar{\lambda}_s^{\dot{a}}\bar{\lambda}_s^{\dot{b}}\frac{\partial}{\partial\bar{\lambda}_i^{\dot{a}}}\frac{\partial}{\partial\bar{\lambda}_i^{\dot{b}}}\,. \tag{25}$$

The proof of the soft theorems follows from Ward identities of extended Bondi, van der Burg, Metzner and Sachs (BMS) symmetry [29]. Although exact for tree level amplitudes these receive loop corrections [27,30,31].

Whether the soft theorems extend beyond Einstein gravity has been examined before. In particular the leading soft behaviour can often be used as a check upon amplitudes such, e.g. in [21]. The leading and sub-leading limits were shown to hold for a R^3 insertion in [32]. Here we examine the amplitudes and, in particular, test the sub-sub-leading soft behaviour.

We can summarise the behaviour of the leading amplitudes, $M_n^{(1)}$, simply by stating:

All the amplitudes calculated satisfy the soft limits of (22) up to and including the sub-sub-leading term.

We have verified this for all helicity amplitudes up to $n = 8$. Note: to check (22) one must implement momentum conservation consistently between the n-point amplitudes and the $n-1$-point amplitudes which in essence specifies how the point $t = 0$ is approached. These are several ways to do this. We have followed the prescription of [25] but alternative implementations are possible [27,28].

In principle we could have found a behaviour of the form

$$M_n^{(1)} \longrightarrow S_t M_{n-1}^{(1)} + S_t^{\alpha} M_{n-1}^{(0)} + R_n \tag{26}$$

where S_t^{α} would be an α correction to the soft functions and R_n is a non-factorising term. In terms of this we find $S_t^{\alpha} = R_n = 0$. Since the theory we are considering is higher derivative it is not surprising that the leading and sub-leading parts of S_t^{α} vanish however it is interesting that the vanishing continues for the sub-sub-leading – unlike the loop corrections to Einstein gravity.

Incidentally as a consequence of eq. (19) the amplitude $M_n^{(1),R^3+R^2\phi}(1^+,2^+,\cdots n^+)$ also satisfies the soft theorems to sub-sub leading level.

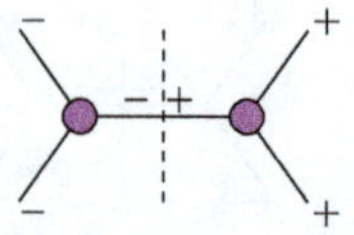

Fig. 6. Factorisations of the four-point MHV amplitude at α^2.

4. Other theories

We have chosen to extend gravity using a three-point vertex and use a diagrammar approach whereby we only consider the on-shell amplitudes. There is, of course, complementarity between this approach and that of Lagrangian based field theory. The single choice of three-point amplitude corresponds to the single R^3 field density that affects on-shell amplitudes. This makes the extended S-matrix simply depend upon the single parameter α.

If we were to deform Einstein gravity by an additional four-point amplitude then there are more choices consistent with symmetry and spinor weight, e.g. we could have

$$\begin{aligned} &M_4(1^+,2^+,3^+,4^+) \\ &\quad = \alpha_1(\langle 1\,2\rangle^4\langle 3\,4\rangle^4 + \langle 1\,3\rangle^4\langle 2\,4\rangle^4 + \langle 1\,4\rangle^4\langle 2\,3\rangle^4) \\ &\qquad + \alpha_2(\langle 1\,2\rangle\langle 2\,3\rangle\langle 3\,4\rangle\langle 4\,1\rangle + \text{permutations})^2 + \cdots \end{aligned} \tag{27}$$

From a field theory perspective this freedom corresponds to the observation that there are multiple R^4 tensors that contribute to on-shell amplitudes [33].

The same issue arises when we consider the further expansion in α. If we consider $M_4^{(2)}(1^-,2^-,3^+,4^+)$ there is a single factorisation as shown in Fig. 6. The amplitude

$$M_4^{(2)}(1^-,2^-,3^+,4^+) = \langle 1\,2\rangle^4\,[3\,4]^4\left(\frac{tu + \beta s^2}{s}\right) \tag{28}$$

has the correct factorisation for any choice of β. This ambiguity means we also have to specify the four-point amplitude to determine the S-matrix. In the diagrammar approach this ambiguity arises due to the existence of a polynomial function with the correct symmetries and spinor and momentum weight. From a field theory perspective, additional counterterms can contribute to this amplitude. Specifically, we could deform the theory via

$$R \longrightarrow R + C_\alpha R^3 + C_\beta D^2 R^4 \tag{29}$$

and the four-point amplitude is only specified once C_α and C_β are determined.

5. Conclusion

We have constructed the (leading part) of the S-matrix of an extended theory of gravity starting from three-point amplitudes and only demanding factorisation. The theory is extended by the addition of amplitudes which are polynomial in momentum, thus implicitly imposing locality and unitarity on the S-matrix. We also require the amplitudes to have the correct spinor helicity as appropriate for massless particles. The S-matrix is then generated entirely from on-shell amplitudes by demanding factorisation. Specifically, we have extended the theory by the addition of three-point amplitudes which, from a field theory perspective, corresponds to introducing R^3 terms. This S-matrix differs from that obtained by applying double copy or KLT techniques to a F^3 extension of Yang–Mills.

Beyond the leading part, polynomial amplitudes exist at higher point and these must be specified to fully determine the S-matrix. Consistency of this approach and a field theoretic approach beyond leading order requires a correspondence between these polynomial amplitudes and the counter terms contributing to on-shell amplitudes.

We find that these amplitudes satisfy the same soft theorems as the tree amplitudes of Einstein gravity up to and including the sub-sub leading terms. It is interesting that these theorems are robust to deformations of Einstein gravity even at the sub-sub-leading level particularly given the link to BMS symmetry which plays an important role in the recent understanding of black hole soft hair [34].

Acknowledgements

This work was supported by STFC grant ST/L000369/1. GRJ was supported by STFC grant ST/M503848/1. JHG was supported by the College of Science (CoS) Doctoral Training Centre (DTC) at Swansea University.

References

[1] G. 't Hooft, M.J.G. Veltman, NATO Sci. Ser. B 4 (1974) 177.
[2] R.J. Eden, P.V. Landshoff, D.I. Olive, J.C. Polkinghorne, The Analytic S Matrix, Cambridge University Press, 1966.
[3] Z. Bern, L.J. Dixon, D.C. Dunbar, D.A. Kosower, Nucl. Phys. B 425 (1994) 217, http://dx.doi.org/10.1016/0550-3213(94)90179-1, arXiv:hep-ph/9403226.
[4] Z. Bern, L.J. Dixon, D.C. Dunbar, D.A. Kosower, Nucl. Phys. B 435 (1995) 59, http://dx.doi.org/10.1016/0550-3213(94)00488-Z, arXiv:hep-ph/9409265.
[5] R. Britto, F. Cachazo, B. Feng, E. Witten, Phys. Rev. Lett. 94 (2005) 181602, arXiv:hep-th/0501052.
[6] N. Arkani-Hamed, J. Trnka, J. High Energy Phys. 1410 (2014) 030, http://dx.doi.org/10.1007/JHEP10(2014)030, arXiv:1312.2007 [hep-th].
[7] N. Arkani-Hamed, L. Rodina, J. Trnka, arXiv:1612.02797 [hep-th].
[8] J. Bedford, A. Brandhuber, B.J. Spence, G. Travaglini, Nucl. Phys. B 721 (2005) 98, http://dx.doi.org/10.1016/j.nuclphysb.2005.016, arXiv:hep-th/0502146.
[9] F. Cachazo, P. Svrcek, arXiv:hep-th/0502160.
[10] N.E.J. Bjerrum-Bohr, D.C. Dunbar, H. Ita, W.B. Perkins, K. Risager, J. High Energy Phys. 0601 (2006) 009, http://dx.doi.org/10.1088/1126-6708/2006/01/009, arXiv:hep-th/0509016.
[11] P. Benincasa, F. Cachazo, arXiv:0705.4305 [hep-th].
[12] P. Benincasa, C. Boucher-Veronneau, F. Cachazo, J. High Energy Phys. 0711 (2007) 057, http://dx.doi.org/10.1088/1126-6708/2007/11/057, arXiv:hep-th/0702032.
[13] K. Risager, J. High Energy Phys. 0512 (2005) 003, http://dx.doi.org/10.1088/1126-6708/2005/12/003, arXiv:hep-th/0508206.
[14] P. van Nieuwenhuizen, C.C. Wu, J. Math. Phys. 18 (1977) 182, http://dx.doi.org/10.1063/1.523128.
[15] M.T. Grisaru, Phys. Lett. B 66 (1977) 75, http://dx.doi.org/10.1016/0370-2693(77)90617-7.
[16] E. Tomboulis, Phys. Lett. B 67 (1977) 417, http://dx.doi.org/10.1016/0370-2693(77)90434-8.
[17] T. Cohen, H. Elvang, M. Kiermaier, J. High Energy Phys. 1104 (2011) 053, http://dx.doi.org/10.1007/JHEP04(2011)053, arXiv:1010.0257 [hep-th].
[18] D.C. Dunbar, G.R. Jehu, W.B. Perkins, arXiv:1701.02934 [hep-th].
[19] Z. Bern, H.H. Chi, L. Dixon, A. Edison, arXiv:1701.02422 [hep-th].
[20] D.C. Dunbar, N.W.P. Turner, Class. Quantum Gravity 20 (2003) 2293, http://dx.doi.org/10.1088/0264-9381/20/11/323, arXiv:hep-th/0212160.
[21] J. Broedel, L.J. Dixon, J. High Energy Phys. 1210 (2012) 091, http://dx.doi.org/10.1007/JHEP10(2012)091, arXiv:1208.0876 [hep-th].
[22] H. Kawai, D.C. Lewellen, S.H.H. Tye, Nucl. Phys. B 269 (1986) 1, http://dx.doi.org/10.1016/0550-3213(86)90362-7.
[23] Steven Weinberg, Phys. Rev. 140 (1965) B516–B524.
[24] C.D. White, J. High Energy Phys. 1105 (2011) 060, arXiv:1103.2981 [hep-th].
[25] F. Cachazo, A. Strominger, arXiv:1404.4091 [hep-th].
[26] T. He, V. Lysov, P. Mitra, A. Strominger, J. High Energy Phys. 1505 (2015) 151, http://dx.doi.org/10.1007/JHEP05(2015)151, arXiv:1401.7026 [hep-th].
[27] Z. Bern, S. Davies, J. Nohle, Phys. Rev. D 90 (8) (2014) 085015, arXiv:1405.1015 [hep-th].
[28] J. Broedel, M. de Leeuw, J. Plefka, M. Rosso, Phys. Rev. D 90 (6) (2014) 065024, arXiv:1406.6574 [hep-th].
[29] H. Bondi, M.G.J. van der Burg, A.W.K. Metzner, Proc. R. Soc. Lond. Ser. A 269 (1962) 21;
R.K. Sachs, Proc. R. Soc. Lond. Ser. A 270 (1962) 103.
[30] S. He, Y.t. Huang, C. Wen, J. High Energy Phys. 1412 (2014) 115, arXiv:1405.1410 [hep-th].
[31] S.D. Alston, D.C. Dunbar, W.B. Perkins, Phys. Rev. D 86 (2012) 085022, arXiv:1208.0190 [hep-th].
[32] M. Bianchi, S. He, Y.t. Huang, C. Wen, Phys. Rev. D 92 (6) (2015) 065022, http://dx.doi.org/10.1103/PhysRevD.92.065022, arXiv:1406.5155 [hep-th].
[33] S.A. Fulling, R.C. King, B.G. Wybourne, C.J. Cummins, Class. Quantum Gravity 9 (1992) 1151, http://dx.doi.org/10.1088/0264-9381/9/5/003.
[34] S.W. Hawking, M.J. Perry, A. Strominger, Phys. Rev. Lett. 116 (23) (2016) 231301, http://dx.doi.org/10.1103/PhysRevLett.116.231301, arXiv:1601.00921 [hep-th].

Do the gravitational corrections to the beta functions of the quartic and Yukawa couplings have an intrinsic physical meaning?

S. Gonzalez-Martin [a], C.P. Martin [b,*]

[a] Departamento de Física Teórica and Instituto de Física Teórica (IFT-UAM/CSIC), Universidad Autónoma de Madrid, Cantoblanco, 28049, Madrid, Spain
[b] Universidad Complutense de Madrid (UCM), Departamento de Física Teórica I, Facultad de Ciencias Físicas, Av. Complutense S/N (Ciudad Univ.), 28040 Madrid, Spain

ARTICLE INFO

Editor: M. Cvetič

Keywords:
Beta function
Renormalization
Gravitation

ABSTRACT

We study the beta functions of the quartic and Yukawa couplings of General Relativity and Unimodular Gravity coupled to the $\lambda\phi^4$ and Yukawa theories with masses. We show that the General Relativity corrections to those beta functions as obtained from the 1PI functional by using the standard MS multiplicative renormalization scheme of Dimensional Regularization are gauge dependent and, further, that they can be removed by a non-multiplicative, though local, field redefinition. An analogous analysis is carried out when General Relativity is replaced with Unimodular Gravity. Thus we show that any claim made about the change in the asymptotic behavior of the quartic and Yukawa couplings made by General Relativity and Unimodular Gravity lack intrinsic physical meaning.

1. Introduction

It is well known that perturbatively quantized general relativity is non-renormalizable due to the mass dimension of the coupling constant κ [1]. Moreover, the coupling to matter does not improve this behavior [2–5]. However, it can still be treated as an effective field theory well below the scale of the Planck mass $M_P \sim 10^{19}$ GeV [6,7]. On the other hand, unimodular gravity is known to be an alternative formulation to General Relativity; it yields the same classical predictions, and moreover it partially solves the so-called cosmological constant problem [8,9]. However, whether quantum corrections – putting aside the absence of such corrections to the Cosmological Constant in Unimodular Gravity – will turn Unimodular Gravity into a different theory from General Relativity is an open issue [10]. Despite this, it has the same problems of non-renormalizability that general relativity.

Then, it is clear that both theories should be regarded as effective field theories. It is in this sense, the asymptotic behavior of physical effects of quantum general relativity on other fields has been studied. Robinson and Wilczek suggested that when coupled to a Yang–Mills theory, it improves the behavior of the theory regarding asymptotic freedom [12]; but it was proved later that this result is gauge dependent [13,14]. Further, it is also known that a non-multiplicative renormalization can be used to eliminate some of the contributions to the beta functions in the Yang–Mills case [15].

The beta functions of the $\lambda\phi^4$ (quartic) and Yukawa couplings, and the logarithmic UV divergences that contribute to them, enter the quantitative analysis of a large number of High Energy Physics topics. Two chief topics among these are *a)* the study of the vacuum stability by using the renormalization-group-improved effective potential, with its implications in the study of the Physics of the Early Universe and the physics of the Standard Model and beyond – see [16] and references therein, and *b)* the construction of Asymptotically Safe theories of Quantum Gravity coupled to matter along the lines laid out in Reference [17]. Thus it is plain that the computation of the logarithmic UV divergent contributions which may lead to a change in the value of beta functions in question due to the interaction of the corresponding matter fields with gravitons is needed. In view of the fate of the corrections found in Reference [12], it is necessary to see whether or not these gravitational corrections are gauge independent and invariant under non-multiplicative field renormalization so that an intrinsic physical meaning can be ascribed to them.

In Reference [18], it was shown that the General Relativity contributions to beta functions of the quartic and Yukawa couplings obtained by using the multiplicative MS scheme of Dimensional Regularization applied to the 1PI functional do not vanish in the de Donder gauge of the graviton field. The contributions obtained

* Corresponding author.
E-mail addresses: sergio.gonzalezm@uam.es (S. Gonzalez-Martin), carmelop@fis.ucm.es (C.P. Martin).

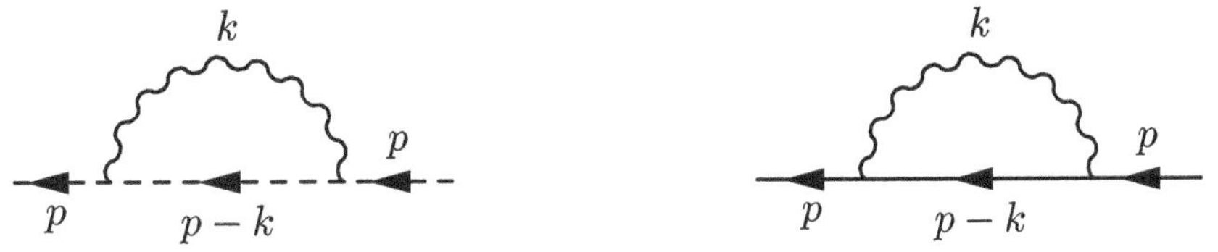

(a) P_S (b) P_Y

Fig. 1. Corrections to the scalar and fermion propagators.

lead to asymptotic freedom for appropriate values of the masses involved – among these values are masses of the real Higgs and top quark.

The first aim of the present paper is to show that the General Relativity corrections to the beta functions of the quartic and Yukawa couplings as computed in the de Donder gauge in [18] are gauge dependent artifacts and that, besides, they can be removed by appropriate non-multiplicative field redefinitions. Thus, we conclude that the General Relativity corrections to the beta functions in question obtained by using the multiplicative MS scheme of Dimensional Regularization applied to the 1PI functional have no intrinsic physical meaning and, that, therefore, any physical conclusion derived from them cannot be trusted. The second aim is to show that this same situation is reproduced when Unimodular Gravity is used instead of General Relativity. We shall actually see that in the gauge we shall use the Unimodular Gravity corrections to the beta function of the quartic coupling vanish in the Multiplicative MS scheme of Dimensional Regularization.

One word of caution: when, in this paper, we talk about gravity corrections – either from General Relativity or from Unimodular Gravity – we refer to corrections that are of order κ^2.

2. The setting

We start from the well known Einstein–Hilbert Lagrangian coupled to a massive real scalar ϕ via a ϕ^4 interaction and a Dirac fermion ψ via a Yukawa interaction. This is

$$\mathcal{L}_{\rm GR} = \sqrt{-g}\left\{-\frac{2}{\kappa^2}R + \bar\psi(i\not{D} - m_\psi)\psi + \frac{1}{2}g^{\mu\nu}\partial_\mu\phi\partial_\nu\phi + \right. \\ \left. -\frac{1}{2}m_\phi^2\phi^2 - g\phi\bar\psi\psi - \frac{\lambda}{4!}\phi^4\right\}, \qquad (1)$$

while for unimodular gravity

$$\mathcal{L}_{\rm UG} = -\frac{2}{\kappa^2}(-g)^{\frac{1}{4}}\left(R + \frac{3}{32}\frac{\nabla_\mu g\nabla^\mu g}{g^2}\right) + \bar\psi(i\not{D} - m_\psi)\psi + \\ + \frac{1}{2}g^{\mu\nu}\partial_\mu\phi\partial_\nu\phi - \frac{1}{2}m_\phi^2\phi^2 - g\phi\bar\psi\psi - \frac{\lambda}{4!}\phi^4, \qquad (2)$$

where $\kappa = 32\pi G$, and g, λ are – respectively – the Yukawa and the ϕ^4 coupling constants.

In order to keep explicit the gauge dependence, we use a generalized gauge condition for general relativity:

$$\mathcal{L}_{\rm GR} = \alpha\left(\partial^\mu h_{\mu\nu} - \frac{1}{2}\partial_\nu h\right)^2, \qquad (3)$$

where α is an arbitrary gauge parameter. This yields a propagator

$$\langle h_{\mu\nu}(k)h_{r\sigma}(-k)\rangle_{\rm GR} = \frac{i}{2k^2}\left(\eta_{\mu\sigma}\eta_{\nu\rho} + \eta_{\mu\rho}\eta_{\nu\sigma} - \eta_{\mu\nu}\eta_{\rho\sigma}\right) - \\ - i\left(\frac{1}{2} + \alpha\right)\left(\eta_{\mu\rho}k_\nu k_\sigma + \eta_{\mu\sigma}k_\nu k_\rho \right. \\ \left. + \eta_{\nu\rho}k_\mu k_\sigma + \eta_{\nu\sigma}k_\mu k_\rho\right). \qquad (4)$$

The gauge fixing and propagator of unimodular gravity are found in [11,10] and read

$$\langle h_{\mu\nu}(k)h_{r\sigma}(-k)\rangle_{\rm UG} = \frac{i}{2k^2}\left(\eta_{\mu\sigma}\eta_{\nu\rho} + \eta_{\mu\rho}\eta_{\nu\sigma}\right) \\ - \frac{i}{k^2}\frac{8\alpha^2 - 1}{16\alpha^2}\eta_{\mu\nu}\eta_{\rho\sigma} + \\ + i\left(\frac{k_\rho k_\sigma\eta_{\mu\nu}}{k^4} + \frac{k_\mu k_\nu\eta_{\rho\sigma}}{k^4}\right) \\ - 4i\frac{k_\mu k_\nu k_\rho k_\sigma}{k^6}. \qquad (5)$$

Let us remark that in the case of unimodular gravity the interaction comes from $h_{\mu\nu}\widehat{T}^{\mu\nu} = \widehat{h}_{\mu\nu}T^{\mu\nu}$ with $T^{\mu\nu}$ the energy-momentum tensor and the hat quantities the traceless ones. Therefore one can work with the traceless propagator $\langle\widehat{h}_{\mu\nu}(k)\widehat{h}_{r\sigma}(-k)\rangle$ (which can be trivially obtained from (5)) and the full energy-momentum tensor, therefore using the same Feynman rules for the vertices as in general relativity, or use (5) coupled to $\widehat{T}^{\mu\nu}$.

In order to compute the beta functions, the first step is to find the 1PI gravitational corrections to the scalar and fermion propagators. These are shown in Figs. 1a and 1b.

Using the propagators listed above, and computing divergences in dimensional regularization ($D = 4 + 2\epsilon$) these are,

$$P_S^{\rm GR} = \kappa^2\left(-\frac{i}{16\pi^2\epsilon}\right)m_\phi^2\left[1 + \left(\frac{1}{2} + \alpha\right)\right](p^2 - m_\phi^2), \qquad (6)$$

$$P_S^{\rm UG} = 0, \qquad (7)$$

$$P_Y^{\rm GR} = \kappa^2\left(-\frac{i}{16\pi^2\epsilon}\right)\left\{\frac{3}{8}m_\psi p^2 - \frac{1}{8}p^2\not{p} + \frac{1}{4}m_\psi^2(\not{p} - m_\psi) + \right. \\ \left. + \left(\frac{1}{2} + \alpha\right)\left[\frac{3}{4}m_\psi p^2 - \not{p}\left(\frac{15}{32}p^2 + \frac{29}{32}m_\psi^2\right) - \frac{19}{16}m_\psi^3\right]\right\}, \qquad (8)$$

$$P_Y^{\rm UG} = \kappa^2\left(-\frac{i}{16\pi^2\epsilon}\right)\left\{\not{p}\left(\frac{3}{16}m_\psi^2 - \frac{5}{16}p^2\right) + \frac{3}{8}m_\psi p^2\right\}. \qquad (9)$$

The corrections to the ϕ^4 (1PI) vertex (Fig. 2) read

$$V_\phi^{\rm GR} = \kappa^2\lambda\left(-\frac{i}{16\pi^2\epsilon}\right)\left(\frac{3}{2} + \alpha\right)\left[\frac{1}{2}\sum_{i=1}^{4}p_i^2 - 4m_\phi^2\right], \qquad (10)$$

$$V_\phi^{\rm UG} = 0. \qquad (11)$$

Finally, we compute the divergences of the (1PI) Yukawa vertices listed in Fig. 3. These ones read

$$V_\psi^{\rm GR} = g\kappa^2\left(-\frac{i}{16\pi^2\epsilon}\right)\left[-\frac{1}{4}m_\phi^2 - \frac{3}{4}m_\psi^2 + \frac{1}{16}(p_1 + p_2)^2 \right. \\ \left. + \frac{1}{4}m_\psi(\not{p}_1 + \not{p}_2) + \frac{1}{8}\not{p}_1\not{p}_2\right] + \\ + g\kappa^2\left(-\frac{i}{16\pi^2\epsilon}\right)\left(\frac{1}{2} + \alpha\right)\left[-m_\phi^2 - \frac{57}{16}m_\psi^2\right.$$

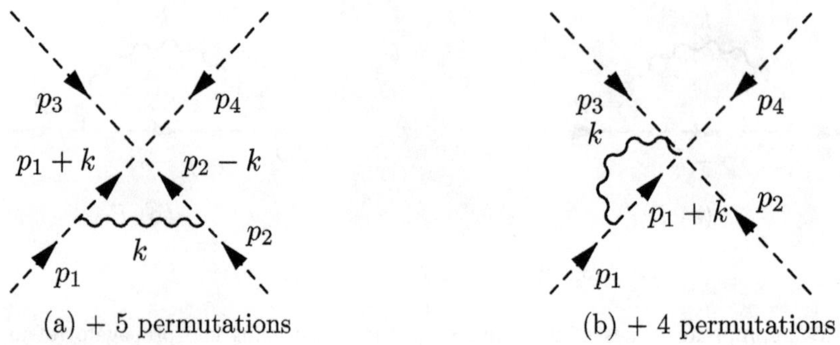

Fig. 2. ϕ^4 vertices.

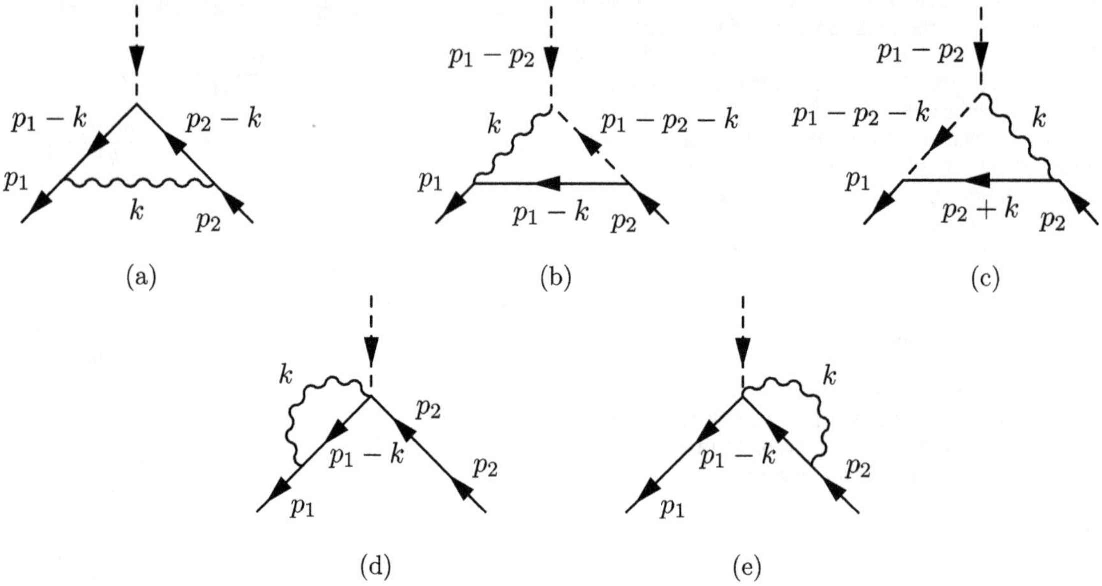

Fig. 3. Contributions to the Yukawa vertex.

$$+\frac{47}{32}(p_1^2+p_2^2)-\frac{13}{8}p_1\cdot p_2+m_\psi(\not{p}_1+\not{p}_2)-\frac{9}{16}\not{p}_1\not{p}_2\Big], \tag{12}$$

$$V_\psi^{\text{UG}}=g\kappa^2\left(-\frac{i}{16\pi^2\epsilon}\right)\Big[\frac{9}{16}(p_1^2+p_2^2)-\frac{3}{8}p_1\cdot p_2$$
$$+\frac{3}{16}m_\psi(\not{p}_1+\not{p}_2)-\frac{3}{8}\not{p}_1\not{p}_2\Big]. \tag{13}$$

3. Beta functions

We shall proceed now to the computation of the Yukawa and quartic coupling beta function gravitational corrections coming from General Relativity and Unimodular Gravity. To use the well known multiplicative MS renormalization scheme of Dimensional Regularization, we define

$$g_0=\mu^{-\epsilon}Z_gZ_\psi^{-1}Z_\phi^{-1/2}g, \qquad Z_g=1+\delta Z_g, \tag{14}$$
$$\Psi_0=Z_\Psi^{1/2}\Psi, \qquad Z_\Psi,=1+\delta Z_\Psi, \tag{15}$$
$$\bar{\Psi}_0=Z_\Psi^{1/2}\bar{\Psi}, \qquad Z_\phi=1+\delta Z_\phi, \tag{16}$$
$$m_{\Psi_0}=Z_mZ_\Psi^{-1}m_\Psi, \qquad Z_{m_\Psi}=1+\delta Z_{m_\Psi}, \tag{17}$$
$$m_{\phi_0}=Z_mZ_\phi^{-1}m_\phi, \qquad Z_{m_\phi}=1+\delta Z_{m_\phi}. \tag{18}$$

The counterterms obtained from the previous definitions are given in Fig. 4.

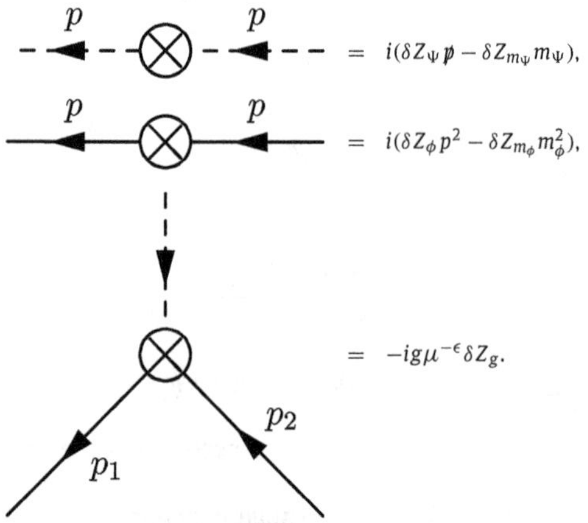

Fig. 4. Counterterms.

Following the standard MS procedure, the wave function renormalizations (δZ_Ψ and δZ_Ψ) are obtained by imposing that the contributions proportional to $\not{p}$ in the sum given in Fig. 5 are finite as $\epsilon \to 0$. This yields the values

$$\delta Z_\phi=\frac{1}{16\pi^2\epsilon}\kappa^2m_\phi^2\Big[1+\Big(\frac{1}{2}+\alpha\Big)\Big], \tag{19}$$

$$\delta Z_\Psi=\frac{1}{16\pi^2\epsilon}\kappa^2m_\Psi^2\Big[\frac{1}{4}+\Big(\frac{1}{2}+\alpha\Big)\frac{29}{32}\Big]. \tag{20}$$

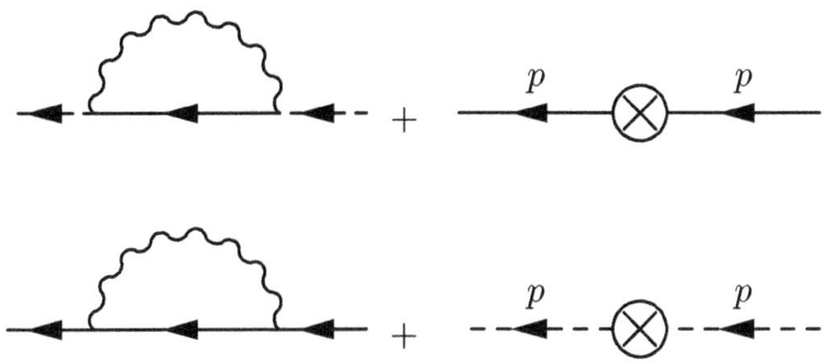

Fig. 5. Wave function renormalization.

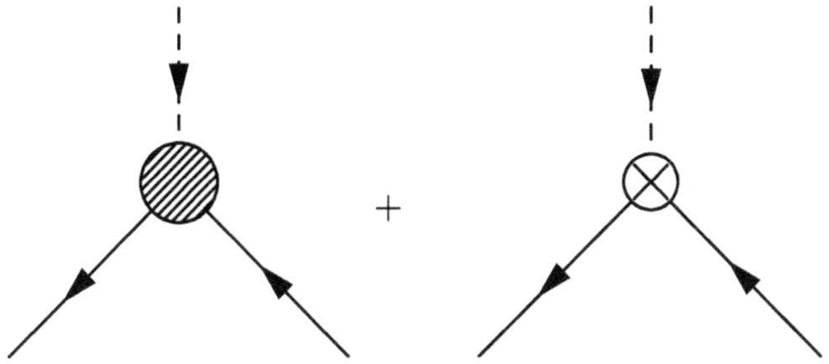

Fig. 6. Yukawa vertex renormalization.

For δZ_g, we demand that there is no singularity independent of the external momenta at $\epsilon \to 0$ in the sum of Fig. 6; hence

$$\delta Z_g = \frac{1}{16\pi^2\epsilon}\kappa^2\Big\{m_\phi^2\Big[\frac{1}{4}+\Big(\frac{1}{2}+\alpha\Big)\Big]+m_\Psi^2\Big[\frac{3}{4}\Big(\frac{1}{2}+\alpha\Big)\frac{57}{16}\Big]\Big\} \quad (21)$$

Defining $\beta_g = \mu\frac{dg(\mu)}{d\mu}$, and using standard techniques, one obtains the General Relativity contribution, $\beta_g^{\rm GR}$, to β_g, at order κ^2, from $\delta\widetilde{Z}_g = \delta Z_g - \delta Z_\Psi - \frac{1}{2}\delta Z_\phi$:

$$\beta_g^{\rm GR} = \frac{1}{16\pi^2}\kappa^2\Big\{m_\phi^2\Big[\frac{1}{2}-\Big(\frac{1}{2}+\alpha\Big)\Big]+m_\Psi^2\Big[-1-\Big(\frac{1}{2}+\alpha\Big)\frac{85}{16}\Big]. \quad (22)$$

The explicit dependence on the parameter α shows the gauge-dependent nature of this beta function in presence of gravity. Insofar as no physical observables can depend on the gauge, no physical consequences should be extracted from here.

We follow the same procedure for unimodular gravity to find

$$\delta Z_\Psi^{\rm UG} = \frac{1}{16\pi^2\epsilon}\kappa^2 m_\Psi^2\frac{3}{16}, \quad (23)$$

$$\delta Z_\phi^{\rm UG} = 0, \quad (24)$$

$$\delta Z_g^{\rm UG} = 0, \quad (25)$$

so that

$$\beta_g^{\rm UG} = \frac{1}{16\pi^2}\kappa^2 m_\Psi^2\frac{3}{16}. \quad (26)$$

We can see that we get a difference between general relativity and unimodular gravity by comparing (22) and (26). However we will see in the sequel that we can get rid of these beta functions by using a *non-multiplicative* renormalization, this is, by performing a field redefinition.

Let us now define

$$g_0 = \mu^{-\epsilon}Z_g Z_\psi^{-1}Z_\phi^{-1/2}g,$$
$$\phi_0 = \phi + \frac{1}{2}\delta Z_\phi\phi, \quad (27)$$

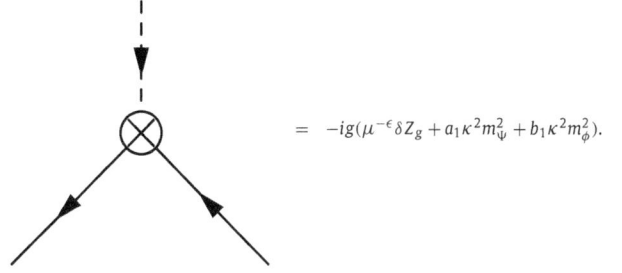

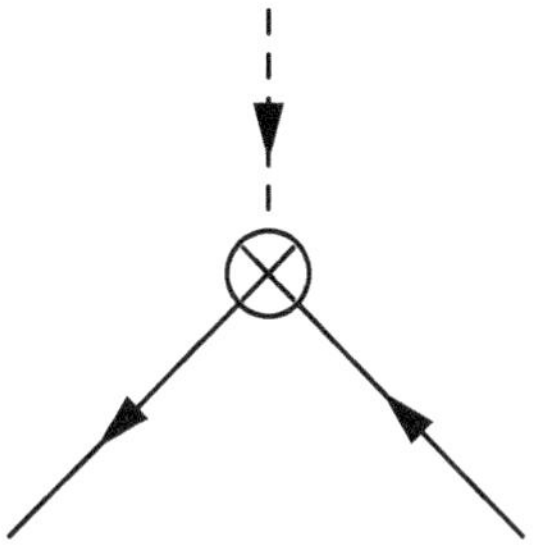

$$= -ig(\mu^{-\epsilon}\delta Z_g + a_1\kappa^2 m_\Psi^2 + b_1\kappa^2 m_\phi^2).$$

Fig. 7. Counterterm.

$$\Psi_0 = \Psi + \frac{1}{2}\delta Z_\Psi\Psi + \frac{1}{2}a_1\kappa^2 m_\Psi^2\phi\Psi + \frac{1}{2}b_1\kappa^2 m_\phi^2\phi\Psi,$$
$$m_{\Psi_0} = (1+\delta Z_{m_\Psi})m_\Psi, \quad (28)$$

$$\bar\Psi_0 = \bar\Psi + \frac{1}{2}\delta Z_\Psi\bar\Psi + \frac{1}{2}a_1\kappa^2 m_\Psi^2\bar\Psi\phi + \frac{1}{2}b_1\kappa^2 m_\phi^2\bar\Psi\phi,$$
$$m_{\phi_0} = (1+\delta Z_{m_\phi})m_\phi. \quad (29)$$

Therefore the matter Lagrangian can be written as

$$\begin{aligned}
&\bar\Psi_0(i\partial\!\!\!/ - m_{\Psi_0})\Psi_0 + \frac{1}{2}(\partial_\mu\phi_0\partial^\mu\phi_0 - m_{\phi_0}^2\phi_0^2) - g_0\bar\Psi_0\phi_0\Psi_0\\
&= \bar\Psi(i\partial\!\!\!/ - m_\Psi)\Psi + \frac{1}{2}(\partial_\mu\phi\partial^\mu\phi - m_\phi^2\phi^2) -\\
&\quad - g\mu^{-\epsilon}\bar\Psi\phi\Psi + \{\delta Z_\Psi\bar\Psi i\partial\!\!\!/\Psi + \delta Z_\phi\partial_\mu\phi\partial^\mu\phi\\
&\quad - m_\Psi\delta Z_{m_\Psi}\bar\Psi\Psi - \frac{1}{2}m_\phi^2\delta Z_{m_\phi}\} -\\
&\quad - g\mu^{-\epsilon}\{\delta Z_g + a_1\kappa^2 m_\Psi^2 + b_1\kappa^2 m_\phi^2\}\bar\Psi\phi\Psi\\
&\quad + \frac{1}{2}(a_1\kappa^2 m_\Psi^2 + b_1\kappa^2 m_\phi^2)[i\bar\Psi\phi\partial\!\!\!/\Psi + i\partial\!\!\!/(\phi\Psi)].
\end{aligned} \quad (30)$$

While the counterterms for the scalar and fermion field propagator remain unchanged with respect to the multiplicative renormalization, the counterterm for the vertex is now given by the expression in Fig. 7.

Imposing again that the sum in Fig. 6 is zero (plus terms depending on the external momenta) when $\epsilon \to 0$, we find

$$\delta Z_\phi = \frac{1}{16\pi^2\epsilon}\kappa^2 m_\phi^2\Big[1+\Big(\frac{1}{2}+\alpha\Big)\Big], \quad (31)$$

$$\delta Z_\Psi = \frac{1}{16\pi^2\epsilon}\kappa^2 m_\Psi^2\Big[\frac{1}{4}+\frac{29}{32}\Big(\frac{1}{2}+\alpha\Big)\Big], \quad (32)$$

$$\begin{aligned}
\delta\widetilde{Z}_g &= \delta Z_g - \delta Z_\Psi - \frac{1}{2}\delta Z_\phi =\\
&= \frac{1}{16\pi^2\epsilon}\kappa^2 m_\phi^2\Big[-\frac{1}{4}+\frac{1}{2}\Big(\frac{1}{2}+\alpha\Big)\Big]\\
&+ \frac{1}{16\pi^2\epsilon}\kappa^2 m_\Psi^2\Big[\frac{1}{2}+\frac{85}{32}\Big(\frac{1}{2}+\alpha\Big)\Big] - a_1\kappa^2 m_\Psi^2 - b_1\kappa^2 m_\phi^2.
\end{aligned} \quad (33)$$

It is clear that by choosing

$$a_1 = \frac{1}{16\pi^2\epsilon}\Big[\frac{1}{2}+\frac{85}{32}\Big(\frac{1}{2}+\alpha\Big)\Big], \quad (34)$$

$$b_1 = \frac{1}{16\pi^2\epsilon}\Big[-\frac{1}{4}+\frac{1}{2}\Big(\frac{1}{2}+\alpha\Big)\Big], \quad (35)$$

we shall wipe out the gravitational correction to $\delta\widetilde{Z}_g$ so the gravitational corrections to the beta function of the Yukawa coupling is now given by

$$\beta_g^{\rm GR}\Big|_{\rm gravitational} = 0. \quad (36)$$

We have seen here that the gravitational contribution to β_g can be brushed away by carrying out a field redefinition. Therefore, it is an *inessential* [17] contribution. However, notice that one cannot do the same with the contributions in absence of gravity, which show that they are *essential* contributions.

Finally, we can perform the same non-multiplicative renormalization for unimodular gravity finding

$$\delta Z_{\phi}^{\text{UG}} = 0, \tag{37}$$

$$\delta Z_{\Psi}^{\text{UG}} = \frac{1}{16\pi^2\epsilon}\kappa^2 m_{\Psi}^2 \frac{3}{16}, \tag{38}$$

$$\begin{aligned}\delta \widetilde{Z}_{g}^{\text{UG}} &= \delta Z_{g}^{\text{UG}} - \delta Z_{\Psi}^{\text{UG}} - \frac{1}{2}\delta Z_{\phi}^{\text{UG}} \\ &= -\frac{1}{16\pi^2\epsilon}\kappa^2 m_{\Psi}^2 \frac{3}{16} - a_1\kappa^2 m_{\Psi}^2 - b_1\kappa^2 m_{\phi}^2.\end{aligned} \tag{39}$$

Accordingly, we can set

$$a_1 = -\frac{1}{16\pi^2\epsilon}\frac{3}{16}, \tag{40}$$

$$b_1 = 0, \tag{41}$$

to make again $\delta\widetilde{Z}_g = 0$ and

$$\beta_g^{\text{UG}}\Big|_{\text{gravitational}} = 0. \tag{42}$$

The computation of the gravitational corrections of the beta function of the ϕ^4 interaction is done by following an akin the process. Defining

$$\lambda_0 = \lambda\mu^{-2\epsilon} Z_\lambda Z_\phi^{-2}, \quad Z_\lambda = 1 + \delta Z_\lambda, \tag{43}$$

we have obtained

$$\delta Z_\lambda^{\text{GR}} = \frac{4}{16\pi^2\epsilon}\left(\frac{3}{2} + \alpha\right), \tag{44}$$

$$\delta Z_\lambda^{\text{UG}} = 0. \tag{45}$$

Hence, one can compute the gravitational corrections to the beta functions of the quartic coupling, $\lambda\phi^4$, to be

$$\beta_\lambda^{\text{GR}} = -\frac{1}{4\pi^2}\kappa^2 m_\phi^2\left(\frac{3}{2} + \alpha\right)\lambda, \tag{46}$$

$$\beta_\lambda^{\text{UG}} = 0. \tag{47}$$

In this case the beta function of unimodular gravity is directly zero for this particular gauge. For general relativity, as we did with the Yukawa coupling, we can reabsorb this discrepancy by means of a non-multiplicative renormalization. In this case, we can carry out the following field redefinition

$$\phi_0 = \phi + \omega_1\phi + \omega_2\kappa^2\partial^2\phi + w_3\kappa^2\mu^{-2\epsilon}\phi^3\delta Z_\phi\phi, \tag{48}$$

and we can set $\beta_\lambda^{\text{GR}} = 0$ by choosing

$$\omega_1 = -\frac{1}{16\pi^2\epsilon}\kappa^2 m_\phi^2, \tag{49}$$

$$\omega_2 = 0, \tag{50}$$

$$\omega_3 = \frac{1}{16\pi^2\epsilon}\frac{1}{4!}2\lambda. \tag{51}$$

4. Summary and final discussion

The knowledge of beta functions of the $\lambda\phi^4$ (quartic) and Yukawa couplings, and the logarithmic UV divergences that contribute to them, is needed in the quantitative analysis of such important issues as the vacuum stability by using the renormalization-group-improved effective potential – see [16] and references therein – and the construction of Asymptotically Safe theories of Quantum Gravity coupled to matter as put forward in Reference [17]. Therefore, a computation of the logarithmic UV divergent contributions which may yield a change of the value of beta functions in question due to the interaction of the corresponding matter fields with gravitons is much needed. However, the fact that the corrections computed in Reference [12] for the gauge coupling constants turned out to lack any intrinsic physical meaning – see References [13,15] – makes necessary to ascertain whether or not these gravitational corrections to the quartic and Yukawa couplings are gauge independent and invariant under non-multiplicative field renormalizations so that an intrinsic physical meaning can be assigned to them.

In this paper, we have computed the General Relativity corrections to the beta functions of the Yukawa and $\lambda\phi^4$ theory as obtained from the 1PI functional by using the standard multiplicative MS dimensional regularization scheme. We have shown that they are gauge dependent and that, besides, they can be set to zero by appropriate, non-multiplicative, field redefinitions: they are *inessential* corrections [17]. We thus conclude that these corrections do not have any intrinsic physical meaning and, therefore, the statements about asymptotic freedom made in reference [18] are not physically meaningful. Of course, the gauge dependence of the gravitational corrections to the beta function can be avoided by using the DeWitt–Vilkovisky action instead of the 1PI functional – as done in reference [19] for the $\lambda\phi^4$ theory – but it is plain that those gauge-independent contributions can still be removed by appropriate non-multiplicative, but local, field redefinitions such as the ones – with different coefficients, of course – introduced in this paper. The use of the DeWitt–Vilkovisky effective action does not give the gravitational corrections in question any intrinsic physical meaning, so that any conclusion drawn from them also lack intrinsic physical content.

For the sake of comparison, we have carried out a similar computation for the case of Unimodular Gravity – for a gauge-fixing choice which yields no free parameters: the computations are hard enough already – and found that the corresponding gravitational corrections to the beta functions do not agree with those from General Relativity – curiously enough the corrections to the beta function of the $\lambda\phi^4$ vanish for Unimodular Gravity, and, that they can also be set to zero by appropriate local non-multiplicative field redefinitions. So one cannot use these gravitational corrections to the beta functions in question to distinguish between General Relativity and Unimodular Gravity. In fact, they behave in the same manner from the physical point of view: they are not *essential* in either case, for they correspond to field redefinitions.

Several final comments are in order. First, we would like to point out that our conclusions are quite in keeping with the conclusions – i.e., the inclusion of gravitational effects into the running coupling constants has not a universal meaning – in Reference [20] in the massless case, but our approach to the problem is not the same and, besides, our theories are massive. Notice that the contributions computed in the present paper – and in [18,19] – vanish if the masses are sent to zero. Secondly, that our analysis is in complete harmony with the discussion carried out in reference [15] for the Yang–Mills coupling constant. Thirdly, the results that we have presented are to be taken into account unavoidable when developing the asymptotic safety program as applied to Gravity interacting with matter, with the proviso that the UV divergences computed in this paper correspond to logarithmic divergences when a cutoff is used.

Acknowledgements

We are grateful to E. Alvarez for illuminating discussions. This work has received funding from the European Unions Horizon 2020 research and innovation programme under the Marie Sklodowska-Curie grants agreement No 674896 and No 690575. We also have been partially supported by the Spanish MINECO through grants FPA2014-54154-P and FPA2016-78645-P, COST actions MP1405 (Quantum Structure of Spacetime) and COST MP1210 (The string theory Universe). S.G-M acknowledges the support of the Spanish Research Agency (Agencia Estatal de Investigación) through the grant IFT Centro de Excelencia Severo Ochoa SEV-2016-0597.

References

[1] G. 't Hooft, M.J.G. Veltman, One loop divergencies in the theory of gravitation, Ann. Inst. Henri Poincaré A, Phys. Théor. 20 (1974) 69.

[2] S. Deser, P. van Nieuwenhuizen, Nonrenormalizability of the quantized Einstein–Maxwell system, Phys. Rev. Lett. 32 (1974) 245, http://dx.doi.org/10.1103/PhysRevLett.32.245.

[3] S. Deser, P. van Nieuwenhuizen, One loop divergences of quantized Einstein–Maxwell fields, Phys. Rev. D 10 (1974) 401, http://dx.doi.org/10.1103/PhysRevD.10.401.

[4] S. Deser, P. van Nieuwenhuizen, Nonrenormalizability of the quantized Dirac–Einstein system, Phys. Rev. D 10 (1974) 411, http://dx.doi.org/10.1103/PhysRevD.10.411.

[5] S. Deser, H.S. Tsao, P. van Nieuwenhuizen, Nonrenormalizability of Einstein Yang–Mills interactions at the one loop level, Phys. Lett. B 50 (1974) 491, http://dx.doi.org/10.1016/0370-2693(74)90268-8.

[6] J.F. Donoghue, Leading quantum correction to the Newtonian potential, Phys. Rev. Lett. 72 (1994) 2996, http://dx.doi.org/10.1103/PhysRevLett.72.2996, arXiv:gr-qc/9310024.

[7] J.F. Donoghue, General relativity as an effective field theory: the leading quantum corrections, Phys. Rev. D 50 (1994) 3874, http://dx.doi.org/10.1103/PhysRevD.50.3874, arXiv:gr-qc/9405057.

[8] G.F.R. Ellis, H. van Elst, J. Murugan, J.P. Uzan, On the trace-free Einstein equations as a viable alternative to general relativity, Class. Quantum Gravity 28 (2011) 225007, http://dx.doi.org/10.1088/0264-9381/28/22/225007, arXiv:1008.1196 [gr-qc].

[9] W.G. Unruh, A unimodular theory of canonical quantum gravity, Phys. Rev. D 40 (1989) 1048, http://dx.doi.org/10.1103/PhysRevD.40.1048.

[10] E. Alvarez, S. Gonzalez-Martin, M. Herrero-Valea, C.P. Martin, Quantum corrections to unimodular gravity, J. High Energy Phys. 1508 (2015) 078, http://dx.doi.org/10.1007/JHEP08(2015)078, arXiv:1505.01995 [hep-th].

[11] E. Alvarez, S. Gonzalez-Martin, C.P. Martin, Unimodular trees versus Einstein trees, Eur. Phys. J. C 76 (10) (2016) 554, http://dx.doi.org/10.1140/epjc/s10052-016-4384-2, arXiv:1605.02667 [hep-th].

[12] S.P. Robinson, F. Wilczek, Gravitational correction to running of gauge couplings, Phys. Rev. Lett. 96 (2006) 231601, http://dx.doi.org/10.1103/PhysRevLett.96.231601, arXiv:hep-th/0509050.

[13] A.R. Pietrykowski, Gauge dependence of gravitational correction to running of gauge couplings, Phys. Rev. Lett. 98 (2007) 061801, http://dx.doi.org/10.1103/PhysRevLett.98.061801, arXiv:hep-th/0606208.

[14] D. Ebert, J. Plefka, A. Rodigast, Absence of gravitational contributions to the running Yang–Mills coupling, Phys. Lett. B 660 (2008) 579, http://dx.doi.org/10.1016/j.physletb.2008.01.037, arXiv:0710.1002 [hep-th].

[15] J. Ellis, N.E. Mavromatos, On the interpretation of gravitational corrections to gauge couplings, Phys. Lett. B 711 (2012) 139, http://dx.doi.org/10.1016/j.physletb.2012.04.005, arXiv:1012.4353 [hep-th].

[16] M.B. Einhorn, D.R.T. Jones, J. High Energy Phys. 0704 (2007) 051, http://dx.doi.org/10.1088/1126-6708/2007/04/051, arXiv:hep-ph/0702295 [HEP-PH].

[17] S. Weinberg, Ultraviolet divergences in quantum theories of gravitation, in: General Relativity: An Einstein Centenary Survey, Cambridge University Press, 1979.

[18] A. Rodigast, T. Schuster, Gravitational corrections to Yukawa and φ^4 interactions, Phys. Rev. Lett. 104 (2010) 081301, http://dx.doi.org/10.1103/PhysRevLett.104.081301, arXiv:0908.2422 [hep-th].

[19] A.R. Pietrykowski, Phys. Rev. D 87 (2) (2013) 024026, http://dx.doi.org/10.1103/PhysRevD.87.024026, arXiv:1210.0507 [hep-th].

[20] M.M. Anber, J.F. Donoghue, M. El-Houssieny, Phys. Rev. D 83 (2011) 124003, http://dx.doi.org/10.1103/PhysRevD.83.124003, arXiv:1011.3229 [hep-th].

13

Double and cyclic λ-deformations and their canonical equivalents

George Georgiou [a], Konstantinos Sfetsos [b], Konstantinos Siampos [c,*]

[a] Institute of Nuclear and Particle Physics, National Center for Scientific Research Demokritos, Ag. Paraskevi, GR-15310 Athens, Greece
[b] Department of Nuclear and Particle Physics, Faculty of Physics, National and Kapodistrian University of Athens, Athens 15784, Greece
[c] Albert Einstein Center for Fundamental Physics, Institute for Theoretical Physics, Laboratory for High-Energy Physics, University of Bern, Sidlerstrasse 5, CH3012 Bern, Switzerland

ARTICLE INFO

Editor: N. Lambert

ABSTRACT

We prove that the doubly λ-deformed σ-models, which include integrable cases, are canonically equivalent to the sum of two single λ-deformed models. This explains the equality of the exact β-functions and current anomalous dimensions of the doubly λ-deformed σ-models to those of two single λ-deformed models. Our proof is based upon agreement of their Hamiltonian densities and of their canonical structure. Subsequently, we show that it is possible to take a well defined non-Abelian type limit of the doubly-deformed action. Last, but not least, by extending the above, we construct multi-matrix integrable deformations of an arbitrary number of WZW models.

0. Introduction and results

A new class of integrable theories based on current algebras for a semi-simple group was recently constructed [1]. The starting point was to consider two independent WZW models at the same positive integer level k and two distinct PCM models which were then left-right asymmetrically gauged with respect to a common global symmetry. The models are labeled by the level k and two general invertible matrices $\lambda_{1,2}$. For certain choices of $\lambda_{1,2}$ integrability is retained [1]. This idea can be generalized to include integrable deformations of exact CFTs on symmetric spaces. This construction is reminiscent to the one for single λ-deformations [2–4].

Subsequently, the quantum properties of the aforementioned multi-parameter integrable deformations were studied in [5], by employing a variety of techniques. One of the main results of that work was that the running of the couplings λ_1 and λ_2, as well as the anomalous dimensions of current operators depend only on one of the couplings, either λ_1 or λ_2 and are identical to those found for single λ-deformations [6–12]. These rather unexpected results seek for a simple explanation. The purpose of this work is to demonstrate that they are due to the fact that the doubly deformed models are canonically equivalent to the sum of two single λ-deformations, one with deformation matrix being λ_1 and the other with deformation matrix λ_2. Recall that all known forms of T-duality, i.e., Abelian, non-Abelian and Poisson–Lie T-duality can be formulated as canonical transformations in the phase space of the corresponding two-dimensional σ-models [13–17]. Moreover, it has been shown in various works that the running of couplings is preserved under these canonical transformations even though the corresponding σ-models fields are totally different [18–22]. All of the above strongly hint towards the validity of our assertion, which of course we will prove.

The plan of the paper is as follows: In section 1, after a brief review of the single and doubly λ-deformed models and of their *non-perturbative* symmetries, we will show that the doubly deformed models are canonically equivalent to the sum of two single λ-deformations. In section 2, we will present the type of non-Abelian T-duality that is based on the doubly deformed σ-models of [1]. Finally, in section 3, we will construct multi-matrix *integrable* deformations of an arbitrary number of independent WZW models by performing a left-right asymmetric gauging for each one of them but in such a way that the total classical gauge anomaly vanishes. This happens if these models are forced to obey the cyclic symmetry property or if they are infinitely many, resembling in structure either a closed or an infinitely open spin chain. Their action can be thought of as the all-loop effective action of several independent WZW models for G all at level k, perturbed by current bilinears mixing the different WZW models with nearest neighbour-type interactions. These models are also canonically equivalent to a sum of single λ-deformed models with appropriate couplings. Furthermore, we will argue that the Hamiltonian of

* Corresponding author.
E-mail addresses: georgiou@inp.demokritos.gr (G. Georgiou), ksfetsos@phys.uoa.gr (K. Sfetsos), siampos@itp.unibe.ch (K. Siampos).

these new models maps to itself under an inversion of all couplings $\lambda_i \mapsto \lambda_i^{-1}$, $i = 1, ..., n$ accompanied generically by non-local redefinitions of the group elements involved when $n = 3, 4, \ldots$. This symmetry, which in the special cases where $n = 1, 2$ simplifies to the one reviewed in section 1, is in accordance with the fact that the β-functions and anomalous dimensions of currents are again given by the same expressions as in the case of the single λ-deformed model.

1. Review and canonical equivalence

1.1. Single λ-deformed σ-models

The construction of the single λ-deformed σ-model starts by considering the sum of a gauged WZW and a PCM for a group G, defined with group elements g and $\tilde{g}$, respectively and next gauging the global symmetry [2]

$$g \mapsto \Lambda^{-1} g \Lambda \,, \quad \tilde{g} \mapsto \Lambda^{-1}\tilde{g} \,.$$

This is done by introducing gauge fields $A_\pm$ in the Lie-algebra of G transforming as

$$A_\pm \mapsto \Lambda^{-1} A_\pm \Lambda - \partial_\pm \Lambda \,.$$

The choice $\tilde{g} = \mathbb{I}$ completely fixes the gauge and the gauged fixed action reads

$$S_{k,\lambda}(g; A_\pm) = S_k(g) + \frac{k}{\pi}\int d^2\sigma \, \mathrm{Tr}\Big(A_-\partial_+ g g^{-1} - A_+ g^{-1}\partial_- g + A_- g A_+ g^{-1} - A_+ \lambda^{-1} A_-\Big) \,, \tag{1.1}$$

where $S_k(g)$ is the WZW model. The $A_\pm$'s are non-dynamical and their equations of motion read

$$\nabla_+ g \, g^{-1} = (\lambda^{-T} - \mathbb{I})A_+ \,, \quad g^{-1}\nabla_- g = -(\lambda^{-1} - \mathbb{I})A_- \,, \tag{1.2}$$

with $\nabla_\pm g = \partial_\pm g - [A_\pm, g]$. Solving them in terms of the gauge fields we find

$$A_+ = i\left(\lambda^{-T} - D\right)^{-1} J_+ \,, \quad A_- = -i\left(\lambda^{-1} - D^T\right)^{-1} J_- \,, \tag{1.3}$$

where

$$J_+^a = -i\,\mathrm{Tr}(t_a \partial_+ g g^{-1}) \,, \qquad J_-^a = -i\,\mathrm{Tr}(t_a g^{-1}\partial_- g) \,.$$
$$D_{ab} = \mathrm{Tr}(t_a g t_b g^{-1}) \,, \tag{1.4}$$

where t_a's are Hermitian representation matrices obeying $[t_a, t_b] = i f_{abc} t_c$, so that the structure constants f_{abc} are real. We choose the normalization such that $\mathrm{Tr}(t_a t_b) = \delta_{ab}$.

Using (1.3) into (1.1) one finds the action [2]

$$S_{k,\lambda}(g) = S_k(g) + \frac{k}{\pi}\int d^2\sigma \, \mathrm{Tr}\left(J_+(\lambda^{-1} - D^T)^{-1} J_-\right) \,. \tag{1.5}$$

For small elements of the matrix λ this action becomes

$$S_{k,\lambda}(g) = S_k(g) + \frac{k}{\pi}\int d^2\sigma \, \mathrm{Tr}\,(J_+ \lambda J_-) + \cdots \,.$$

Hence (1.5) represents the effective action of self-interacting current bilinears of a single WZW model. The action (1.5) has the remarkable *non-perturbative* symmetry [6,9]

$$k \mapsto -k \,, \quad \lambda \mapsto \lambda^{-1} \,, \quad g \mapsto g^{-1} \,. \tag{1.6}$$

As in the case of gauged WZW models [23], we define the currents $\mathcal{J}_\pm$

$$\mathcal{J}_+ = \nabla_+ g g^{-1} + A_+ - A_- \,, \quad \mathcal{J}_- = -g^{-1}\nabla_- g + A_- - A_+ \,, \tag{1.7}$$

The above form for the $\mathcal{J}_\pm^a$'s when rewritten in terms of phase space variables of the σ-model action, assumes the same form as the currents $J_\pm^a$ of the WZW action. Hence, they satisfy two commuting current algebras as in [24]

$$\{\mathcal{J}_\pm^a, \mathcal{J}_\pm^b\} = \frac{2}{k} f_{abc}\mathcal{J}_\pm^c \delta_{\sigma\sigma'} \pm \frac{2}{k}\delta_{ab}\delta'_{\sigma\sigma'} \,, \quad \delta_{\sigma\sigma'} = \delta(\sigma - \sigma') \,. \tag{1.8}$$

Moreover using (1.2) we can rewrite (1.7) as

$$\mathcal{J}_+ = \lambda^{-T} A_+ - A_- \,, \quad \mathcal{J}_- = \lambda^{-1} A_- - A_+ \,. \tag{1.9}$$

Inversely

$$A_+ = h^{-1}\lambda^T(\mathcal{J}_+ + \lambda \mathcal{J}_-) \,, \quad A_- = \tilde{h}^{-1}\lambda(\mathcal{J}_- + \lambda^T \mathcal{J}_+) \,, \quad h = \mathbb{I} - \lambda^T\lambda \,, \quad \tilde{h} = \mathbb{I} - \lambda\lambda^T \,, \tag{1.10}$$

assuming that the matrix λ is such that $h, \tilde{h}$ are positive-definite matrices. To obtain the Poisson algebra in the base of $A_\pm$ we use (1.8), (1.9) and (1.10).

To study the Hamiltonian structure of the problem we need to define its phase space [3,4]. This is given in terms of the currents $\mathcal{J}_\pm$, the gauge fields $A_\pm$ and the associated momenta $P_\pm$ to $A_\pm$. The $\mathcal{J}_\pm$ obey two commuting current algebras (1.8) and have vanishing Poisson brackets with $A_\pm$ and $P_\pm$

$$\{P_\pm^a(\sigma), A_\mp^b(\sigma')\} = \delta^{ab}\delta(\sigma - \sigma') \,.$$

Furthermore, since the $A_\pm$'s are non-dynamical their associated momenta $P_\pm$ vanish. This introduces two primary constraints

$$\varphi_1 = P_+ \approx 0 \,, \quad \varphi_2 = P_- \approx 0 \,.$$

Their time-evolution gives rise to the secondary constraints

$$\varphi_3 = \mathcal{J}_+ - \lambda^{-T} A_+ + A_- \approx 0 \,, \quad \varphi_4 = \mathcal{J}_- - \lambda^{-1} A_- + A_+ \approx 0 \,.$$

Time evolution generates no further constraints. The φ_i's with $i = 1, 2, 3, 4$, turn out to be second class constraints, since the matrix of their Poisson brackets is invertible in the deformed case. Finally, the Hamiltonian density of the single λ-deformed model before integrating out the gauge fields takes the form [2,23]

$$\mathcal{H}_{\text{single}} = \frac{k}{4\pi}\mathrm{Tr}\,\Big(\mathcal{J}_+\mathcal{J}_+ + \mathcal{J}_-\mathcal{J}_- + 4(\mathcal{J}_+ A_- + \mathcal{J}_- A_+) + 2(A_+ - A_-)(A_+ - A_-) - 4A_+(\lambda_1^{-1} - \mathbb{I})A_-\Big) \,,$$

or equivalently through (1.9), in terms of $A_\pm$'s

$$\boxed{\mathcal{H}_{\text{single}} = \frac{k}{4\pi}\,\mathrm{Tr}\left(A_+\left(\lambda^{-1}\tilde{h}\lambda^{-T}\right)A_+ + A_-\left(\lambda^{-T}h\lambda^{-1}\right)A_-\right)} \,. \tag{1.11}$$

1.2. Doubly λ-deformed σ-models

The action defining the doubly deformed models depends on two group elements $g_i \in G$, $i = 1, 2$ and is given by the deformation of the sum of two WZW models $S_k(g_1)$ and $S_k(g_2)$ as [1]

$$S_{k,\lambda_1,\lambda_2}(g_1, g_2) = S_k(g_1) + S_k(g_2) + \frac{k}{\pi}\int d^2\sigma \, \mathrm{Tr}\left\{(J_{1+} \;\; J_{2+})\begin{pmatrix}\Lambda_{21}\lambda_1 D_2^T\lambda_2 & \Lambda_{21}\lambda_1 \\ \Lambda_{12}\lambda_2 & \Lambda_{12}\lambda_2 D_1^T \lambda_1\end{pmatrix} \times \begin{pmatrix}J_{1-} \\ J_{2-}\end{pmatrix}\right\} \,, \tag{1.12}$$

where

$$\Lambda_{12}=(\mathbb{I}-\lambda_2 D_1^T\lambda_1 D_2^T)^{-1}\,,\quad \Lambda_{21}=(\mathbb{I}-\lambda_1 D_2^T\lambda_2 D_1^T)^{-1}\,. \quad (1.13)$$

The matrices D_{ab} and the currents $J_\pm^a$ are defined in (1.4). When a current or the matrix D has the extra index 1 or 2 this means that one should use the corresponding group element in its definition. The action (1.12) has the *non-perturbative* symmetry [1]

$$k\mapsto -k\,,\quad \lambda_1\mapsto\lambda_1^{-1}\,,\quad \lambda_2\mapsto\lambda_2^{-1}\,,\quad g_1\mapsto g_2^{-1}\,,\quad g_2\mapsto g_1^{-1}\,, \quad (1.14)$$

which is similar to (1.6). For small elements of the matrices λ_i's the action (1.12) becomes

$$S_{k,\lambda_1,\lambda_2}(g_1,g_2)=S_k(g_1)+S_k(g_2)+\frac{k}{\pi}\int \mathrm{d}^2\sigma\ \mathrm{Tr}(J_{1+}\lambda_1 J_{2-}+J_{2+}\lambda_2 J_{1-})+\cdots\,.$$

Hence (1.12) represents the effective action of two WZW models mutually interacting via current bilinears. Similarly to (1.7) we define the currents[1,2]

$$\begin{aligned}
\mathcal{J}_+^{(1)}&=\nabla_+ g_1 g_1^{-1}+A_+^{(1)}-A_-^{(1)}\,,\\
\mathcal{J}_-^{(1)}&=-g_1^{-1}\nabla_- g_1+A_-^{(2)}-A_+^{(2)}\,,\\
\mathcal{J}_+^{(2)}&=\nabla_+ g_2 g_2^{-1}+A_+^{(2)}-A_-^{(2)}\,,\\
\mathcal{J}_-^{(2)}&=-g_2^{-1}\nabla_- g_2+A_-^{(1)}-A_+^{(1)}\,.
\end{aligned} \quad (1.15)$$

These currents obey two commuting copies of current algebras [1]

$$\{\mathcal{J}_\pm^{(i)a},\mathcal{J}_\pm^{(i)b}\}=\frac{2}{k}f_{abc}\mathcal{J}_\pm^{(i)c}\delta_{\sigma\sigma'}\pm\frac{2}{k}\delta_{ab}\delta'_{\sigma\sigma'}\,,\quad i=1,2\,, \quad (1.16)$$

which encode the canonical structure of the theory. The action does not depend on derivatives of $A_\pm^{(i)}, i=1,2$, so that as in subsection 1.1, their equations of motion are second class constraints [1]

$$\begin{aligned}
\nabla_+ g_1\, g_1^{-1}&=(\lambda_1^{-T}-\mathbb{I})A_+^{(1)}\,,\quad g_1^{-1}\nabla_- g_1=-(\lambda_2^{-1}-\mathbb{I})A_-^{(2)}\,,\\
\nabla_+ g_2\, g_2^{-1}&=(\lambda_2^{-T}-\mathbb{I})A_+^{(2)}\,,\quad g_2^{-1}\nabla_- g_2=-(\lambda_1^{-1}-\mathbb{I})A_-^{(1)}\,,
\end{aligned} \quad (1.17)$$

determining the gauge fields in terms of the group elements similarly to (1.3) (for the precise expressions we refer to [1]). Then (1.15) rewrites as

$$\begin{aligned}
\mathcal{J}_+^{(1)}&=\lambda_1^{-T}A_+^{(1)}-A_-^{(1)}\,,\quad \mathcal{J}_-^{(1)}=\lambda_2^{-1}A_-^{(2)}-A_+^{(2)}\,,\\
\mathcal{J}_+^{(2)}&=\lambda_2^{-T}A_+^{(2)}-A_-^{(2)}\,,\quad \mathcal{J}_-^{(2)}=\lambda_1^{-1}A_-^{(1)}-A_+^{(1)}\,.
\end{aligned} \quad (1.18)$$

Equivalently the gauge fields in terms of the dressed currents are given by

$$\begin{aligned}
A_+^{(1)}&=h_1^{-1}\lambda_1^T(\mathcal{J}_+^{(1)}+\lambda_1\mathcal{J}_-^{(2)})\,,\quad A_-^{(1)}=\tilde h_1^{-1}\lambda_1(\mathcal{J}_-^{(2)}+\lambda_1^T\mathcal{J}_+^{(1)})\,,\\
A_+^{(2)}&=h_2^{-1}\lambda_2^T(\mathcal{J}_+^{(2)}+\lambda_2\mathcal{J}_-^{(1)})\,,\quad A_-^{(2)}=\tilde h_2^{-1}\lambda_2(\mathcal{J}_-^{(1)}+\lambda_2^T\mathcal{J}_+^{(2)})\,,\\
h_i&=\mathbb{I}-\lambda_i^T\lambda_i\,,\quad \tilde h_i=\mathbb{I}-\lambda_i\lambda_i^T\,,\quad i=1,2\,.
\end{aligned} \quad (1.19)$$

[1] To conform with the notation of the current work we have renamed the gauged fields $(A_\pm,B_\pm)$ of [1] by $(A_\pm^{(1)},A_\pm^{(2)})$.

[2] The various covariant derivatives are defined according to the transformation properties of the object they act on. For instance

$$\nabla_\pm g_1=\partial_\pm g_1-A_\pm^{(1)}g_i+g_iA_\pm^{(2)}\,,$$

$$\nabla_\pm(\nabla_\mp g_1 g_1^{-1})=\partial_\pm(\nabla_\mp g_1 g_1^{-1})-[A_\pm^{(1)},\nabla_\mp g_1 g_1^{-1}]\,.$$

To obtain the Poisson algebra in the base of $A_\pm^{(1)}$ and $A_\pm^{(2)}$ we use (1.16), (1.18) and (1.19). As a corollary one can easily show that $\{A_\pm^{(1)},A_\pm^{(2)}\}=0$, for all choices of signs and for generic coupling matrices $\lambda_{1,2}$. The Hamiltonian density of our system before integrating out the gauge fields takes the form [1]

$$\begin{aligned}
\mathcal{H}_{\text{doubly}}=\frac{k}{4\pi}\mathrm{Tr}\Big\{&\mathcal{J}_+^{(1)}\mathcal{J}_+^{(1)}+\mathcal{J}_-^{(1)}\mathcal{J}_-^{(1)}+\mathcal{J}_+^{(2)}\mathcal{J}_+^{(2)}+\mathcal{J}_-^{(2)}\mathcal{J}_-^{(2)}\\
&+4(\mathcal{J}_+^{(1)}A_-^{(1)}+\mathcal{J}_+^{(2)}A_-^{(2)}+\mathcal{J}_-^{(1)}A_+^{(2)}+\mathcal{J}_-^{(2)}A_+^{(1)})\\
&+2(A_+^{(1)}-A_-^{(1)})(A_+^{(1)}-A_-^{(1)})\\
&+2(A_+^{(2)}-A_-^{(2)})(A_+^{(2)}-A_-^{(2)})\\
&-4A_+^{(1)}(\lambda_1^{-1}-\mathbb{I})A_-^{(1)}-4A_+^{(2)}(\lambda_2^{-1}-\mathbb{I})A_-^{(2)}\Big\}
\end{aligned}$$

and can be rewritten through (1.18) in terms of $A_\pm^{(i)}$ and λ_i as

$$\boxed{\mathcal{H}_{\text{doubly}}=\frac{k}{4\pi}\sum_{i=1}^{2}\mathrm{Tr}\Big(A_+^{(i)}\left(\lambda_i^{-1}\tilde h_i\lambda_i^{-T}\right)A_+^{(i)}+A_-^{(i)}\left(\lambda_i^{-T}h_i\lambda_i^{-1}\right)A_-^{(i)}\Big)}\,. \quad (1.20)$$

The fact that the Hamiltonian density (1.20) is the sum of two terms one depending on $A_\pm^{(1)}$ and the other on $A_\pm^{(2)}$ combined with the fact that the currents $\mathcal{J}_\pm^{(i)}$, $i=1,2$, obey two commuting copies of the current algebra of the single λ-deformed model shows that the doubly deformed models are canonically equivalent to the sum of two single λ-deformed models, one with coupling λ_1 and the other with coupling λ_2. The relations defining the canonical transformation are given by

$$A_\pm^{(1)}=\tilde A_\pm^{(1)}\,,\quad A_\pm^{(2)}=\tilde A_\pm^{(2)}\,, \quad (1.21)$$

where the gauge fields without the tildes correspond to the doubly deformed models and depend on $(\lambda_1,\lambda_2;g_1,g_2)$, while the tilded gauge fields correspond to the canonically equivalent sum of two single λ-deformed models the first of which depends on $(\lambda_1;\tilde g_1)$ only while the second depends on $(\lambda_2;\tilde g_2)$.

Furthermore, the gauge fields of (1.21) should be considered as functions of the coordinates parametrising the group elements and their conjugate momenta. We may write relations involving worldsheet derivatives of the various group elements by using (1.3) and (1.17). As in all canonical transformation involving canonical variables as well as their momenta, the relation between the g_i's and the $\tilde g_i$'s is a non-local one.

A comment is in order concerning the η-deformed models [25–29] which are closely related to the single λ-deformed ones via Poisson–Lie T-duality [30] and an appropriate analytic continuation of the coordinates and the parameters [31–35]

$$\lambda\mapsto\frac{iE-\eta\mathbb{I}}{iE+\eta\mathbb{I}}\,,$$

where E is an arbitrary constant matrix. Poisson–Lie T-duality can also be formulated as a canonical transformation [16,17] and therefore there is a chain of canonical transformations from doubly λ-deformed, to two single λ-deformed and to η-deformed models. It would be interesting to formulate the canonical transformation (1.21) via a duality invariant action similarly perhaps to the case of Poisson–Lie T-duality in [36].

There is an important observation for further use in section 3. The Hamiltonian density (1.20) has the following *non-perturbative* symmetry

$$k \mapsto -k, \quad \lambda_i \mapsto \lambda_i^{-1}, \quad A_+^{(i)} \mapsto \lambda_i^{-T} A_+^{(i)},$$
$$A_-^{(i)} \mapsto \lambda_i^{-1} A_-^{(i)}, \quad i=1,2. \tag{1.22}$$

In other words $\mathcal{H}_{\text{doubly}}$ maps to itself under (1.22). By using (1.18) this implies the following transformation for the group elements g_1 and g_2

$$\mathcal{J}_+^{(1)} \mapsto -\mathcal{J}_-^{(2)}, \quad \mathcal{J}_+^{(2)} \mapsto -\mathcal{J}_-^{(1)},$$
$$\mathcal{J}_-^{(1)} \mapsto -\mathcal{J}_+^{(2)}, \quad \mathcal{J}_-^{(2)} \mapsto -\mathcal{J}_+^{(1)}. \tag{1.23}$$

Since the currents $\mathcal{J}_\pm^{(i)}$, $i=1,2$, depend both on the group elements and their derivatives, the transformation (1.23) can be viewed as a non-local transformation at the level of the group elements. In the special cases of the single and doubly λ-deformed theories the symmetry (1.22) and (1.23) can be realized locally simply by a mapping of group elements, i.e. (1.6) and (1.14). Indeed, it is not difficult to check that (1.6) and (1.14) imply for the gauge fields the transformation (1.22). The situation is slightly different for the generic cyclic models constructed below in section 3 which can have arbitrarily many group elements.

2. Doubly-deformed models and non-Abelian T-duality

It is has been known that the action (1.5) admits the non-Abelian T-dual limit that involves taking $k \to \infty$, whereas simultaneously taking the matrix λ and the group element g to the identity [2]. Specifically, if we let

$$\lambda = \mathbb{I} - \frac{E}{k}, \qquad g = \mathbb{I} + i\frac{v}{k}, \qquad k \to \infty,$$

where E is a constant matrix and $v = v_a t^a$, then the action (1.5) becomes

$$S(v,E) = \frac{1}{\pi}\int d^2\sigma \, \text{Tr}\left(\partial_+ v (E+f)^{-1} \partial_- v\right),$$

where f is a matrix with elements $f_{ab} = f_{abc} v^c$. This is the non-Abelian T-dual of the PCM action with general coupling matrix E

$$S_{\text{PCM}}(g,E) = -\frac{1}{\pi}\int d^2\sigma \, \text{Tr}\left(g^{-1}\partial_+ g \, E \, g^{-1}\partial_- g\right),$$

with respect to the global symmetry $g \mapsto \Lambda g$, $\Lambda \in G$. The above limit is well defined when is taken on the β-function for λ, as well as on the anomalous dimensions of various operators in the theory. In the case of doubly λ or even multiple/cyclic λ-deformations (see section 3) we have shown in particular that, the β-functions and current anomalous dimensions are the same with those of two or more simple λ-deformations. Hence, it is expected that it should be possible to take a well defined non-Abelian type limit in the action (1.12). This is not necessarily simple since a suitable limit involves the two group elements.

In the following we focus on the most interesting case in which the matrices λ_i, $i=1,2$ are isotropic, i.e. $(\lambda_i)_{ab} = \lambda_i \delta_{ab}$. It is convenient to use the group element $\mathcal{G} = g_1 g_2$ and also rename g_2 by g. Then employing the Polyakov–Wiegmann identity [44], the action (1.12), using also (1.13), takes the form

$$\begin{aligned}
&S_{k,\lambda_1,\lambda_2}(\mathcal{G},g) = S_k(\mathcal{G}) \\
&+\frac{k}{\pi}\int d^2\sigma \, \text{Tr}\Big((1-\lambda_2) g^{-1}\partial_+ g \,(\mathcal{D}-\lambda_1\mathbb{I})\,\Sigma\, g^{-1}\partial_- g \\
&-(1-\lambda_2) g^{-1}\partial_+ g \,\Sigma\, \partial_-\mathcal{G}\mathcal{G}^{-1} + \lambda_1(1-\lambda_2)\mathcal{G}^{-1}\partial_+\mathcal{G}\,\Sigma\, g^{-1}\partial_- g \\
&+\lambda_1\lambda_2\,\mathcal{G}^{-1}\partial_+\mathcal{G}\,\Sigma\,\mathcal{G}^{-1}\partial_-\mathcal{G}\Big),
\end{aligned} \tag{2.1}$$

where: $\Sigma = (\lambda_1\lambda_2\mathbb{I} - \mathcal{D})^{-1}$ and $\mathcal{D} = D(\mathcal{G}) = D(g_1)D(g_2)$. Next we take the limit

$$\lambda_i = 1 - \frac{\kappa_i^2}{k}, \quad i=1,2, \qquad \mathcal{G} = \mathbb{I} + i\frac{v}{k}, \qquad k \to \infty. \tag{2.2}$$

After some algebra we find that (2.1) becomes

$$\begin{aligned}
S_{\kappa_1^2,\kappa_2^2}(v,g) = &-\frac{1}{\pi}\int d^2\sigma \, \text{Tr}\Big(\kappa_2^2 g^{-1}\partial_+ g g^{-1}\partial_- g \\
&+\left(i\partial_+ v - \kappa_2^2 g^{-1}\partial_+ g\right)\left((\kappa_1^2+\kappa_2^2)\mathbb{I} + f\right)^{-1} \\
&\times \left(i\partial_- v + \kappa_2^2 g^{-1}\partial_- g\right)\Big).
\end{aligned} \tag{2.3}$$

It can be shown that this action is the non-Abelian T-dual of

$$\begin{aligned}
S = &-\frac{1}{\pi}\int d^2\sigma \, \text{Tr}\Big(\kappa_1^2 \tilde{g}^{-1}\partial_+\tilde{g}\tilde{g}^{-1}\partial_-\tilde{g} \\
&+\kappa_2^2 (g^{-1}\partial_+ g - \tilde{g}^{-1}\partial_+\tilde{g})(g^{-1}\partial_- g - \tilde{g}^{-1}\partial_-\tilde{g})\Big),
\end{aligned}$$

with respect to the global symmetry $\tilde{g} \mapsto \Lambda\tilde{g}$, $\Lambda \in G$. Note that, if we define the new group element $\widetilde{\mathcal{G}} = g\tilde{g}^{-1}$ one may write the previous action as

$$S = -\frac{1}{\pi}\int d^2\sigma \, \text{Tr}\Big(\kappa_1^2 \tilde{g}^{-1}\partial_+\tilde{g}\tilde{g}^{-1}\partial_-\tilde{g} + \kappa_2^2 \widetilde{\mathcal{G}}^{-1}\partial_+\widetilde{\mathcal{G}}\widetilde{\mathcal{G}}^{-1}\partial_-\widetilde{\mathcal{G}}\Big), \tag{2.4}$$

which is the sum of two independent PCM actions for a group G. The previous group element redefinition introduces interactions between them.

Finally consider a limit in which only λ_2 tends to one, whereas λ_1 stays inactive. Then, (2.2) has to be modified as

$$\lambda_2 = 1 - \frac{\kappa_2^2}{k}, \quad \mathcal{G} = \mathbb{I} + i\frac{v}{\sqrt{k}}, \quad k \to \infty,$$

in order for (2.1) to stay finite. In particular, this becomes

$$\begin{aligned}
S_{\kappa^2}(v,g) = &\frac{1}{2\pi}\frac{1+\lambda_1}{1-\lambda_1}\int d^2\sigma \, \text{Tr}(\partial_+ v \partial_- v) \\
&-\frac{\kappa_2^2}{\pi}\int d^2\sigma \, \text{Tr}(g^{-1}\partial_+ g g^{-1}\partial_- g),
\end{aligned} \tag{2.5}$$

representing $\dim G$ free bosons and a PCM model for a group G. This is consistent with the limit of the β-functions for λ_1 and λ_2 (see, eqs. (2.6) and (2.7) in [5]). In this limit, the constant λ_1 does not run since it can be absorbed into a redefinition of the v's. Also the coupling constant κ_2^2 obeys the same RG flow equation appropriate for the PCM model and its non-Abelian T-dual, since these models are canonically equivalent.

It would be very interesting to explore physical applications in an AdS/CFT context of this version of non-Abelian T-duality along the lines and developments of [37–43] (for a partial list of works in this direction). Prototype examples this can be applied are the backgrounds $AdS_3 \times S^3 \times S^3 \times S^1$ and $AdS_5 \times S^5$.

3. Cyclic λ-deformations

In this section we construct a class of multi-parameter deformations of conformal field theories of the WZW type Consider n WZW models and n PCMs for a group G, defined with group elements g_i and $\tilde{g}_i$, respectively. We would like to gauge the global symmetry

$$g_i \mapsto \Lambda_i^{-1} g_i \Lambda_{i+1}, \quad \tilde{g}_i \mapsto \Lambda_i^{-1}\tilde{g}_i, \quad i=1,2,\ldots,n,$$

with the periodicity condition $\Lambda_{n+1} = \Lambda_1$ implied. We introduce gauge fields $A_\pm^{(i)}$ in the Lie-algebra of G transforming as

$$A_\pm^{(i)} \mapsto \Lambda_i^{-1} A_\pm^{(i)} \Lambda_i - \Lambda_i^{-1} \partial_\pm \Lambda_i \,, \quad i = 1, 2, \ldots, n \,. \tag{3.1}$$

In this way we have a periodic chain of interacting models each one of which separately is gauge anomalous by a term independent of the group elements. The full model has no gauge anomaly since these cancel among themselves (the chain may be open as long as it is infinite long). The details are quite similar to those for the $n = 2$ case [1], so that we omit them here.

The choice $\tilde{g}_i = \mathbb{I}$, $i = 1, 2, \ldots, n$ completely fixes the gauge and is consistent with the equations of motion for the group elements $\tilde{g}_i$ of the PCMs which are automatically satisfied. Then, the gauged fixed action becomes

$$\begin{aligned} S_{k,\lambda_i}(\{g_i; A_\pm^{(i)}\}) &= \sum_{i=1}^{n} S_k(g_i) \\ &+ \frac{k}{\pi} \int \mathrm{d}^2\sigma \sum_{i=1}^{n} \mathrm{Tr}\Big(A_-^{(i)} \partial_+ g_i g_i^{-1} - A_+^{(i+1)} g_i^{-1} \partial_- g_i \\ &+ A_-^{(i)} g_i A_+^{(i+1)} g_i^{-1} - A_+^{(i)} \lambda_i^{-1} A_-^{(i)} \Big) \,, \end{aligned} \tag{3.2}$$

where the index i is defined modulo n. The equations of motion with respect to the $A_\pm^{(i)}$'s are given by

$$\lambda_i^T D_i A_+^{(i+1)} - A_+^{(i)} = -i\lambda_i^T J_+^{(i)} \,, \quad \lambda_{i+1} D_i^T A_-^{(i)} - A_-^{(i+1)} = i\lambda_{i+1} J_-^{(i)} \,.$$

Solving them we find that

$$A_+^{(1)} = i(\mathbb{I} - x_1 x_2 \cdots x_n)^{-1} \sum_{i=1}^{n} x_1 x_2 \cdots x_{i-1} \lambda_i^T J_+^{(i)} \,, \quad x_i = \lambda_i^T D_i \,. \tag{3.3}$$

The rest can be obtained by cyclic permutations. Plugging the latter into (3.2) we find that the on-shell action reads

$$\begin{aligned} S_{k,\lambda_i}(\{g_i\}) &= \frac{k}{12\pi} \int \mathrm{Tr}(g_1^{-1} \mathrm{d} g_1)^3 \\ &+ \frac{k}{\pi} \int \mathrm{d}^2\sigma \, \mathrm{Tr}\Bigg(\frac{1}{2} J_+^{(1)} D_1 \frac{\mathbb{I} + x_1^T x_n^T x_{n-1}^T \cdots x_2^T}{\mathbb{I} - x_1^T x_n^T x_{n-1}^T \cdots x_2^T} J_-^{(1)} \\ &+ \sum_{i=2}^{n} J_+^{(i)} \lambda_i x_{i-1}^T \cdots x_2^T (\mathbb{I} - x_1^T x_n^T x_{n-1}^T \cdots x_2^T)^{-1} J_-^{(1)} \Bigg) \\ &+ \text{cyclic in } 1, 2, \ldots, n \,, \end{aligned} \tag{3.4}$$

where we have separated the Wess–Zumino term from the WZW model action. For small values of the matrices we have that

$$\begin{aligned} S_{k,\lambda_i}(\{g_i\}) &= \sum_{i=1}^{n} S_k(g_i) + \frac{k}{\pi} \sum_{i=1}^{n} \int \mathrm{d}^2\sigma \, \mathrm{Tr}\left(J_+^{(i+1)} \lambda_{i+1} J_-^{(i)} \right) \\ &+ \mathcal{O}(\lambda^2) \,, \end{aligned} \tag{3.5}$$

representing n distinct WZW models interacting by mutual current bilinears, for which (3.4) is the all loop, in the λ_i's, effective action.

We would like to stress that the $n = 2$ is significantly different with respect to higher n's. Firstly, the *non-perturbative* symmetry $\lambda_i \mapsto \lambda_i^{-1}$ and $k \mapsto -k$, is seemingly realized at a local level for the group elements only when $n = 2$, see (1.14) (also for $n = 1$, see (1.6)). For higher values of n the group elements need to be transformed non-locally by using $\mathcal{J}_\pm^{(i)} \mapsto -\mathcal{J}_\mp^{(i+1)}$, with $n + 1 \equiv 1$. There are exceptions to this. In particular, if all λ_i are equal and isotropic, i.e. $\lambda_i = \lambda\mathbb{I}$, then this duality-type symmetry is

$$\begin{aligned} &k \mapsto -k \,, \quad \lambda \mapsto \frac{1}{\lambda} \,, \quad g_1 \leftrightarrow g_2^{-1} \,, \\ &g_n \leftrightarrow g_3^{-1} \,, \quad g_{n-1} \leftrightarrow g_4^{-1} \,, \quad \text{etc.} \,, \end{aligned} \tag{3.6}$$

that is the group elements are paired up as above. For odd n one group element simply gets inverted. Despite the fact that the symmetry can not be realized locally for the generic case it is still powerful enough to constrain the β-functions and current correlation functions of the cyclic model to have the same values as those of the single λ-deformations.

A second remark concerns the form of the action (3.4) when one of the coupling matrices vanishes. Consider this action for $n = 2$ and $n = 3$ when $\lambda_1 = 0$ while the other coupling matrices stay general

$$S_{k,0,\lambda_2}(g_1, g_2) = \sum_{i=1}^{2} S_k(g_i) + \frac{k}{\pi} \int \mathrm{d}^2\sigma \, \mathrm{Tr}\left(J_+^{(2)} \lambda_2 J_-^{(1)} \right) ,$$

$$\begin{aligned} S_{k,0,\lambda_2,\lambda_3}(g_1, g_2, g_3) &= \sum_{i=1}^{3} S_k(g_i) + \frac{k}{\pi} \int \mathrm{d}^2\sigma \, \mathrm{Tr}\Big(J_+^{(2)} \lambda_2 J_-^{(1)} \\ &+ J_+^{(3)} \lambda_3 J_-^{(2)} + J_+^{(2)} \lambda_3 D_2^T \lambda_2 J_-^{(1)} \Big) . \end{aligned}$$

When $n = 2$ the exact expression matches the approximate one in (3.5), while for $n = 3$ the last term couples the three WZW models and it is quadratic in the λ's.

3.1. Algebra and Hamiltonian

Here we provide the proof that the σ-model action (3.4) is integrable for specific choices of the matrices λ_i, $i = 1, 2, \ldots, n$. In particular, we will show that it is integrable for all choices of the deformation matrices λ_i which, separately, give an integrable λ-deformed model. These include the isotropic λ for semi-simple group and symmetric coset, the anisotropic $SU(2)$ and the λ-deformed Yang–Baxter model [2–4,33,45].

It is equivalent and more convenient to work with the gauged fixed action before integrating out the gauge fields. Varying the gauged fixed action with respect $A_-^{(i)}$ and $A_+^{(i+1)}$ we find the constraints

$$\nabla_+ g_i g_i^{-1} = (\lambda_i^{-T} - \mathbb{I}) A_+^{(i)} \,, \quad g_i^{-1} \nabla_- g_i = -(\lambda_{i+1}^{-1} - \mathbb{I}) A_-^{(i+1)} \,, \tag{3.7}$$

respectively. Varying with respect to g_i we obtain that

$$\nabla_-(\nabla_+ g_i g_i^{-1}) = F_{+-}^{(i)} \,, \quad \nabla_+(g_i^{-1} \nabla_- g_i) = F_{+-}^{(i+1)} \,, \tag{3.8}$$

which are in fact equivalent and where $F_{+-}^{(i)} = \partial_+ A_-^{(i)} - \partial_- A_+^{(i)} - [A_+^{(i)}, A_-^{(i)}]$.

Substituting (3.7) into (3.8) we obtain after some algebra that

$$\begin{aligned} &\partial_+ A_-^{(i)} - \lambda_i^{-T} \partial_- A_+^{(i)} = [\lambda_i^{-T} A_+^{(i)}, A_-^{(i)}] \,, \\ &\lambda_i^{-1} \partial_+ A_-^{(i)} - \partial_- A_+^{(i)} = [A_+^{(i)}, \lambda_i^{-1} A_-^{(i)}] \,. \end{aligned} \tag{3.9}$$

Hence the equations of motion split into n identical sets which are seemingly decoupled even though the $A_\pm^{(i)}$ depend on all group elements g_i and coupling matrices λ_i, $i = 1, 2, \ldots, n$. Moreover, each set is the same one that one would have obtained had we performed the corresponding analysis for the λ-deformed action (1.5). Working along the lines of subsection 1.2; Eqns. (1.15)–(1.20) we find (for $n = 2$ this was performed in detail in [1])

$$\begin{aligned} &\{\mathcal{J}_\pm^{(i)a}, \mathcal{J}_\pm^{(i)b}\} = \frac{2}{k} f_{abc}\mathcal{J}_\pm^{(i)c}\delta_{\sigma\sigma'} \pm \frac{2}{k}\delta_{ab}\delta'_{\sigma\sigma'}, \\ &\mathcal{J}_+^{(i)} = \lambda_i^{-T} A_+^{(i)} - A_-^{(i)}, \quad \mathcal{J}_-^{(i)} = \lambda_{i+1}^{-1} A_-^{(i+1)} - A_+^{(i+1)} \end{aligned} \tag{3.10}$$

and as a consequence $\{A_\pm^{(i)}, A_\pm^{(j)}\} = 0$, for $i \neq j$, for all choices of signs and for generic coupling matrices λ_i. Hence, all choices for matrices known to give rise to integrability for the λ-deformed models provide integrable models here as well with independent conserved changes. The Hamiltonian density of the system in terms of $A_\pm^{(i)}$ and λ_i is

$$\boxed{\begin{aligned} \mathcal{H}_{\text{cyclic}} = \frac{k}{4\pi}\sum_{i=1}^{n} \text{Tr}\Big(A_+^{(i)}\left(\lambda_i^{-1}\tilde{h}_i\lambda_i^{-T}\right)A_+^{(i)} \\ + A_-^{(i)}\left(\lambda_i^{-T}h_i\lambda_i^{-1}\right)A_-^{(i)}\Big) \end{aligned}}\,. \tag{3.11}$$

Using the above we generalize the result of subsection 1.2, that the cyclic λ-deformed models are canonically equivalent to n single λ-deformed σ-model. The relations which define the canonical transformation are given by: $A_\pm^{(i)} = \tilde{A}_\pm^{(i)}$, $i = 1, 2, \ldots, n$, where the gauge fields without the tildes correspond to the cyclic deformed models and depend on $(\lambda_1, \ldots, \lambda_n; g_1, \ldots, g_n)$, while those with tildes correspond to the canonically equivalent sum of n single λ-deformed models each one depending on $(\lambda_i; \tilde{g}_i)$.

3.2. RG flows and currents anomalous dimensions

Similar to the case with $n = 2$ considered in [5], the expression (3.5) can be used to argue that the RG flow equations of the n coupling matrices λ_i for the cyclic model (3.4) as well as the currents anomalous dimensions are the same with those obtained for the single λ-deformations model [6,9,46]. The basic reason is that the various interaction terms have regular OPE among themselves so that correlations functions involving currents factorize to those of n single λ-deformed models. This is also in agreement with the fact that the cyclic model is canonically equivalent to n single λ-deformations. Furthermore we mention without presenting any details that using the analysis performed in [5,46] we have explicitly checked the above claim for the cases of n isotropic couplings for general groups and symmetric spaces.

Acknowledgements

K. Sfetsos would like to thank the Physics Division, National Center for Theoretical Sciences of the National Tsing-Hua University in Taiwan and the Centre de Physique Théorique, École Polytechnique for hospitality and financial support during initial stages of this work. G. Georgiou and K. Siampos acknowledge the Physics Department of the National and Kapodistrian University of Athens for hospitality. Part of this work was developed during HEP 2017: Recent Developments in High Energy Physics and Cosmology in April 2017 at the U. of Ioannina.

References

[1] G. Georgiou, K. Sfetsos, A new class of integrable deformations of CFTs, JHEP 1703 (2017) 083, arXiv:1612.05012 [hep-th].
[2] K. Sfetsos, Integrable interpolations: from exact CFTs to non-Abelian T-duals, Nucl. Phys. B 880 (2014) 225, arXiv:1312.4560 [hep-th].
[3] T.J. Hollowood, J.L. Miramontes, D.M. Schmidtt, Integrable deformations of strings on symmetric spaces, JHEP 1411 (2014) 009, arXiv:1407.2840 [hep-th].
[4] T.J. Hollowood, J.L. Miramontes, D. Schmidtt, An integrable deformation of the $AdS_5 \times S^5$ superstring, J. Phys. A 47 (2014) 49, 495402, arXiv:1409.1538 [hep-th].
[5] G. Georgiou, E. Sagkrioti, K. Sfetsos, K. Siampos, Quantum aspects of doubly deformed CFTs, Nucl. Phys. B 919 (2017) 504, arXiv:1703.00462 [hep-th].
[6] K. Sfetsos, K. Siampos, Gauged WZW-type theories and the all-loop anisotropic non-Abelian Thirring model, Nucl. Phys. B 885 (2014) 583, arXiv:1405.7803 [hep-th].
[7] D. Kutasov, String theory and the nonabelian Thirring model, Phys. Lett. B 227 (1989) 68.
[8] B. Gerganov, A. LeClair, M. Moriconi, On the beta function for anisotropic current interactions in 2-D, Phys. Rev. Lett. 86 (2001) 4753, arXiv:hep-th/0011189.
[9] G. Itsios, K. Sfetsos, K. Siampos, The all-loop non-Abelian Thirring model and its RG flow, Phys. Lett. B 733 (2014) 265, arXiv:1404.3748 [hep-th].
[10] G. Georgiou, K. Sfetsos, K. Siampos, All-loop anomalous dimensions in integrable λ-deformed σ-models, Nucl. Phys. B 901 (2015) 40, arXiv:1509.02946 [hep-th].
[11] G. Georgiou, K. Sfetsos, K. Siampos, All-loop correlators of integrable λ-deformed σ-models, Nucl. Phys. B 909 (2016) 360, arXiv:1604.08212 [hep-th].
[12] G. Georgiou, K. Sfetsos, K. Siampos, λ-deformations of left-right asymmetric CFTs, Nucl. Phys. B 914 (2017) 623, arXiv:1610.05314 [hep-th].
[13] T. Curtright, C.K. Zachos, Currents, charges, and canonical structure of pseudodual chiral models, Phys. Rev. D 49 (1994) 5408, arXiv:hep-th/9401006.
[14] E. Alvarez, L. Alvarez-Gaume, Y. Lozano, A canonical approach to duality transformations, Phys. Lett. B 336 (1994) 183, arXiv:hep-th/9406206.
[15] Y. Lozano, Non-Abelian duality and canonical transformations, Phys. Lett. B 355 (1995) 165, arXiv:hep-th/9503045.
[16] K. Sfetsos, Poisson–Lie T duality and supersymmetry, Nucl. Phys. Proc. Suppl. B 56 (1997) 302, arXiv:hep-th/9611199.
[17] K. Sfetsos, Canonical equivalence of nonisometric sigma models and Poisson–Lie T duality, Nucl. Phys. B 517 (1998) 549, arXiv:hep-th/9710163.
[18] L.K. Balazs, J. Balog, P. Forgacs, N. Mohammedi, L. Palla, J. Schnittger, Quantum equivalence of sigma models related by non-Abelian duality transformations, Phys. Rev. D 57 (1998) 3585, arXiv:hep-th/9704137.
[19] J. Balog, P. Forgacs, N. Mohammedi, L. Palla, J. Schnittger, On quantum T duality in sigma models, Nucl. Phys. B 535 (1998) 461, arXiv:hep-th/9806068.
[20] K. Sfetsos, Poisson–Lie T duality beyond the classical level and the renormalization group, Phys. Lett. B 432 (1998) 365, arXiv:hep-th/9803019.
[21] K. Sfetsos, Duality invariant class of two-dimensional field theories, Nucl. Phys. B 561 (1999) 316, arXiv:hep-th/9904188.
[22] K. Sfetsos, K. Siampos, Quantum equivalence in Poisson–Lie T-duality, JHEP 0906 (2009) 082, arXiv:0904.4248 [hep-th].
[23] P. Bowcock, Canonical quantization of the gauged Wess–Zumino model, Nucl. Phys. B 316 (1989) 80.
[24] E. Witten, Nonabelian bosonization in two-dimensions, Commun. Math. Phys. 92 (1984) 455.
[25] C. Klimčík, YB sigma models and dS/AdS T-duality, JHEP 0212 (2002) 051, arXiv:hep-th/0210095.
[26] C. Klimčík, On integrability of the YB sigma-model, J. Math. Phys. 50 (2009) 043508, arXiv:0802.3518 [hep-th].
[27] F. Delduc, M. Magro, B. Vicedo, On classical q-deformations of integrable sigma-models, JHEP 1311 (2013) 192, arXiv:1308.3581 [hep-th].
[28] F. Delduc, M. Magro, B. Vicedo, An integrable deformation of the $AdS_5 \times S^5$ superstring action, Phys. Rev. Lett. 112 (5) (2014) 051601, arXiv:1309.5850 [hep-th].
[29] G. Arutyunov, R. Borsato, S. Frolov, S-matrix for strings on η-deformed $AdS_5 \times S^5$, JHEP 1404 (2014) 002, arXiv:1312.3542 [hep-th].
[30] C. Klimčík, P. Ševera, Dual non-Abelian duality and the Drinfeld double, Phys. Lett. B 351 (1995) 455, arXiv:hep-th/9502122.
[31] B. Vicedo, Deformed integrable σ-models, classical R-matrices and classical exchange algebra on Drinfel'd doubles, J. Phys. A, Math. Theor. 48 (2015) 355203, arXiv:1504.06303 [hep-th].
[32] B. Hoare, A.A. Tseytlin, On integrable deformations of superstring sigma models related to $AdS_n \times S^n$ supercosets, Nucl. Phys. B 897 (2015) 448, arXiv:1504.07213 [hep-th].
[33] K. Sfetsos, K. Siampos, D.C. Thompson, Generalised integrable λ- and η-deformations and their relation, Nucl. Phys. B 899 (2015) 489, arXiv:1506.05784 [hep-th].
[34] C. Klimčík, η and λ deformations as $\mathcal{E}$-models, Nucl. Phys. B 900 (2015) 259, arXiv:1508.05832 [hep-th].
[35] C. Klimčík, Poisson–Lie T-duals of the bi-Yang–Baxter models, Phys. Lett. B 760 (2016) 345, arXiv:1606.03016 [hep-th].
[36] C. Klimcik, P. Severa, Poisson–Lie T duality and loop groups of Drinfeld doubles, Phys. Lett. B 372 (1996) 65, arXiv:hep-th/9512040.
[37] K. Sfetsos, D.C. Thompson, On non-abelian T-dual geometries with Ramond fluxes, Nucl. Phys. B 846 (2011) 21, arXiv:1012.1320 [hep-th].
[38] G. Itsios, C. Nunez, K. Sfetsos, D.C. Thompson, On non-Abelian T-duality and new N = 1 backgrounds, Phys. Lett. B 721 (2013) 342, arXiv:1212.4840 [hep-th].
[39] G. Itsios, C. Nunez, K. Sfetsos, D.C. Thompson, Non-Abelian T-duality and the AdS/CFT correspondence: new N = 1 backgrounds, Nucl. Phys. B 873 (2013) 1, arXiv:1301.6755 [hep-th].

[40] Y. Lozano, C. Nunez, Field theory aspects of non-Abelian T-duality and $\mathcal{N} = 2$ linear quivers, JHEP 1605 (2016) 107, arXiv:1603.04440 [hep-th].
[41] Y. Lozano, N.T. Macpherson, J. Montero, C. Nunez, Three-dimensional $\mathcal{N} = 4$ linear quivers and non-Abelian T-duals, JHEP 1611 (2016) 133, arXiv:1609.09061 [hep-th].
[42] G. Itsios, C. Nunez, D. Zoakos, Mesons from (non) Abelian T-dual backgrounds, JHEP 1701 (2017) 011, arXiv:1611.03490 [hep-th].
[43] Y. Lozano, C. Nunez, S. Zacarias, BMN vacua, superstars and non-Abelian T-duality, arXiv:1703.00417 [hep-th].
[44] A.M. Polyakov, P.B. Wiegmann, Theory of nonabelian Goldstone bosons, Phys. Lett. B 131 (1983) 121.
[45] K. Sfetsos, K. Siampos, The anisotropic λ-deformed $SU(2)$ model is integrable, Phys. Lett. B 743 (2015) 160, arXiv:1412.5181 [hep-th].
[46] C. Appadu, T.J. Hollowood, Beta function of k deformed $AdS_5 \times \mathcal{S}^5$ string theory, JHEP 1511 (2015) 095, arXiv:1507.05420 [hep-th].

Emergent geometry, thermal CFT and surface/state correspondence

Wen-Cong Gan [a,b], Fu-Wen Shu [a,b,*], Meng-He Wu [a,b]

[a] Department of Physics, Nanchang University, Nanchang, 330031, China
[b] Center for Relativistic Astrophysics and High Energy Physics, Nanchang University, Nanchang 330031, China

ARTICLE INFO

Editor: N. Lambert

Keywords:
AdS/CFT correspondence
cMERA
Holography
BTZ black hole

ABSTRACT

We study a conjectured correspondence between any codimension-two convex surface and a quantum state (SS-duality for short). By applying thermofield double formalism to the SS-duality, we show that thermal geometries naturally emerge as a result of hidden quantum entanglement between two boundary CFTs. We therefore propose a general framework to emerge the thermal geometry from CFT at finite temperature, without knowing many details about the thermal CFT. As an example, the case of 2*d* CFT is considered. We calculate its information metric and show that it is either BTZ black hole or thermal AdS as expected.

1. Introduction

The fascinating idea that spacetimes might emerge from more fundamental degrees of freedom has attracted more and more attention in the past few years. This idea was revived recently by the discovery of the AdS/CFT correspondence [1–3]. Even though it has lead tremendous progresses in the past few years, fundamental mechanism of the AdS/CFT correspondence still remains a mystery. The situation became better not until the discovery of Ryu–Takayanagi formula [4–6], which states that the entanglement entropy of a subregion A of a $d+1$ dimensional CFT on the boundary of $d+2$ dimensional AdS is proportional to the area of a certain codimension-two extremal surface in the bulk:

$$S_A = \frac{Area(\gamma_A)}{4G_N^{d+2}}$$

where γ_A is the minimal surface whose boundary coincides the boundary of A: ∂A.

A recent step for our understanding of holography is made by Miyaji et al. in [7,8] where they proposed a duality called surface/state correspondence (SS-duality). It claims that any codimension two convex surface is dual to a quantum state of a QFT. With the help of the SS-duality, one can, in principle, find out the equivalent description of any spacetimes described by Einstein's gravity. In this way, we might encode the information of the boundary QFTs into the bulk geometry, and vice versa.

There is a very different way in mapping states and operators in the boundary Hilbert space to those in the bulk. This is known as the tensor networks. One specific example which is of particular significance is the multi-scale entanglement renormalization ansatz (MERA) [9–11] (see e.g. [12] for an introduction), and MERA of the ground state of a lattice model at critical point is naturally related to CFT [13]. The connection between the AdS/CFT and the MERA was first pointed out by Swingle in [14], where he noticed that the renormalization direction along the graph can be viewed as an emergent (discrete) radial direction of the AdS space (see e.g. [15–18] for further discussion on the resemblance between MERA and AdS geometry). This elegant method was latter generalized to continuous version (cMERA), which makes entanglement renormalization available for quantum fields in real space [19]. Equipped with this toolkit, the holographic (smooth) geometry can naturally emerge from QFTs [20].

Though it is very successful, the full investigation of AdS/cMERA is still very limited. Most past works paid their attention to zero-temperature systems. In this paper, we take a step forward and investigate how to emerge thermal spacetimes from boundary CFT at finite temperature, by making use of cMERA and SS-duality. At first glance the generalization is trivial and one can achieve this as long as the boundary CFT is replaced by a thermal one. However, there are two obstacles that prevent us from this generalization: First of all, the appearance of black hole (BH) horizon leads to a closed and topologically nontrivial surface in the bulk. This implies, according to the SS-duality, that the dual state in the boundary

* Corresponding author.
E-mail addresses: ganwencong@gmail.com (W.-C. Gan), shufuwen@ncu.edu.cn (F.-W. Shu), menghewu.physik@gmail.com (M.-H. Wu).

QFT is no longer a pure state. All calculations must be replaced by thermal mixed states. Secondly, for finite-temperature CFT, turning on a temperature introduces a scale which screens long-range correlations and the state have thermal correlations in addition to entanglement. One important effect is that the thermal correlations become more relevant as one runs the MERA. The MERA, therefore, truncates at a certain level, which is suggestive of a BH horizon [14]. We often call it the truncated MERA [21]. In our previous work [22], we discussed the emergent thermal geometry by generalizing the truncated MERA to continuous one.

An alternative way which is more natural is based on the thermofield double formalism [23] and the emergent tensor network is often called doubled MERA. This proposal [24] states that the eternal black hole is dual to two copies of the CFT, in the thermofield double state $|TFD\rangle$. Each asymptotic boundary of AdS is a copy of the original dual CFT. With the help of the SS-duality, we will find by this formulation that the thermal spacetimes naturally appear as a result of hidden quantum entanglement between two boundary CFTs. The road map is the following: we first propose a TFD-like state in CFT which is dual to bulk locally excited state in thermal spacetime background and then write down two TFD-like states which differ by infinitesimal parameters. The bulk coordinates can be naturally recognized as these parameters in CFT. Then similar to the proposal in [8], the Fisher information metric distance between two nearby TFD-like states which we propose to be dual to bulk local excitations is identified to the emergent bulk metric distance. Once the bulk metric is obtained, we get the emergent spacetime. In this way, we formulate a general framework by which thermal geometries emerge from dual CFTs, without knowing the details of the thermal correlations of the CFT.

2. Thermofield dynamics and surface/state correspondence

2.1. *Thermofield double formalism and doubled cMERA*

We start by introducing a new QFT H_{tot} which is two copies of the original QFT (with Hilbert spaces H_1 and H_2 respectively). The thermofield double formalism treat the thermal, mixed state $\rho = e^{-\beta H_i}$ $(i = 1, 2)$ as a pure state in the new double system $H_{tot} = H_1 \bigotimes H_2$. Thermofield double state (or Hartle–Hawking state in the dual bulk) in this doubled system is defined as

$$|TFD\rangle = \frac{1}{\sqrt{Z(\beta)}} \sum_n e^{-\beta E_n/2} |n\rangle_1 |n\rangle_2, \tag{1}$$

where $|n\rangle_1, |n\rangle_2$ are energy eigenstates of the two copies of QFT respectively. This is a particular (entangled) pure state in the doubled system. The density matrix of the doubled QFT in this state is

$$\rho_{tot} = |TFD\rangle\langle TFD|. \tag{2}$$

The thermofield double formalism can be applied to the case of the AdS eternal black hole. The Penrose diagram of an eternal black hole separates the whole spacetime into two asymptotically AdS regions as depicted in Fig. 1. Each asymptotic boundary of AdS is a copy of the original dual CFT. It is convenient to denote these two identical, non-interacting copies of CFT by CFT_1 and CFT_2, respectively. According to Maldacena [24], this eternal black hole which is described by the Hartle–Hawking state $|HH\rangle$, is dual to two copies of the CFT in the thermofield double state $|TFD\rangle$.

Due to the presence of the horizon, observers in one of those two asymptotically AdS regions (say, region I) cannot come in contact with the other one directly.[1] From the viewpoint of the dual CFTs, for CFT_1, information from CFT_2 must be traced out. As a consequence, the CFT_1 is in a thermal state described by

$$\rho_1 = \mathrm{Tr}_2 \rho_{tot} = e^{-\beta H_1}. \tag{3}$$

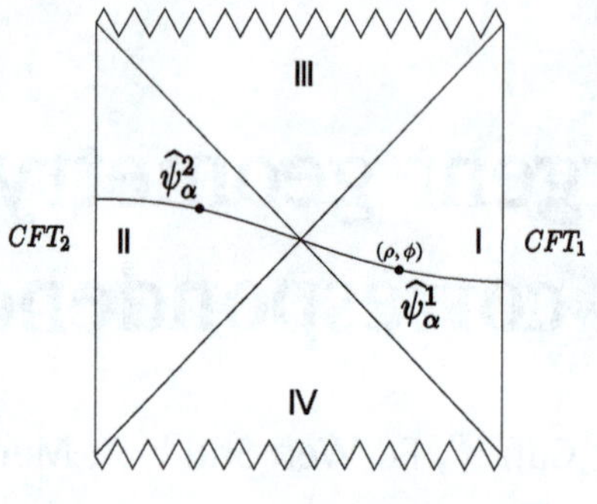

Fig. 1. Penrose diagram for an eternal black hole. There are two asymptotically AdS regions which are dual to two copies of CFT. The bulk quantized fields $\hat{\psi}^1_\alpha$ and $\hat{\psi}^2_\alpha$ are put in the bulk points (ρ, ϕ) in the two asymptotically AdS time slices, respectively.

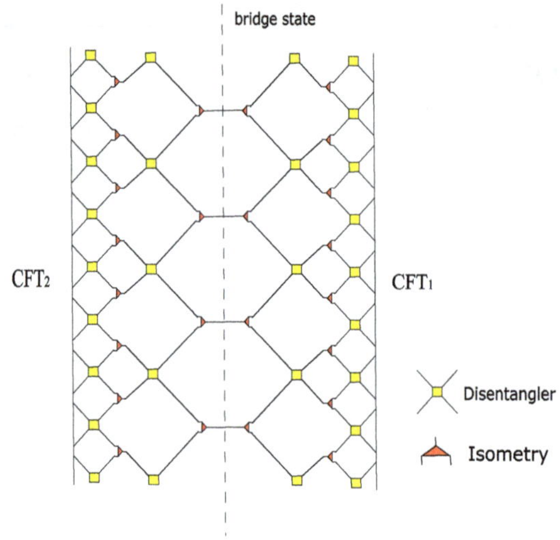

Fig. 2. Doubled MERA network. At the center there is a bridge state which glues two copies of the standard MERA. This state is usually viewed as a black hole horizon.

The above picture nicely agrees with the MERA at finite temperature as proposed in [25–28], which is known as the doubled MERA network. It is composed of two copies of the standard MERA for a pure state which are gluing together at infrared points by a "bridge" state. Fig. 2 shows a schematic representation of the doubled MERA network. The continuous version of MERA (cMERA) at finite temperature has already been considered in [29], where the authors found that, similar to the MERA, finite-temperature cMERA can be constructed by doubling two copies of the standard cMERA.

2.2. *SS-duality description of thermofield dynamics*

Now let us generalize the above picture to a description in terms of the SS-duality. The SS-duality argues a correspondence between any codimension-two convex surface Σ and a quantum state of a quantum theory which is dual to the Einstein's gravity. It can be applied to any spacetimes described by Einstein's gravity and therefore can be viewed as a generalization of the AdS/CFT

[1] However, they connect with each other indirectly through hidden quantum entanglement. The hidden quantum entanglement entropy of the thermal CFT can be viewed as the black hole entropy.

correspondence. More specifically, this duality states that a closed topological trivial convex surface is dual to a pure quantum state $|\Phi(\Sigma)\rangle$, while a closed topological non-trivial convex surface[2] Σ corresponds to a mixed quantum state $\rho(\Sigma)$, such as the surface which wraps a black hole. In particular, the zero-size closed surface (i.e. a point) is dual to a boundary state $|B\rangle$ [7,30,31]. When Σ_1 and Σ_2 are related by a smooth deformation which preserves convexity, the deformation can be expressed by a unitary transformation

$$\rho(\Sigma_1) = U(s_1, s_2)\rho(\Sigma_2)U^{-1}(s_1, s_2), \tag{4}$$

where $U = \mathcal{P}\exp\{-i\int_{u_2}^{u_1} \hat{M}(s)\,ds\}$ with $\mathcal{P}$ the path-ordering and $\hat{M}(s)$ a Hermitian operator.

To proceed, let us turn to a cMERA description of the CFT state. From [8], we learn that a CFT ground state can be expressed in terms of cMERA as follows:

$$|0\rangle_{CFT} = \mathcal{P}\exp\left(-i\int_{-\infty}^{0} du\hat{K}(u)\right)|I_0\rangle, \tag{5}$$

where $\hat{K}(u)$ is the disentangling operator of cMERA at scale u, and $|I_0\rangle$ is the Ishibashi state. The cMERA flow can be adjusted by a conformal transformation. Specifically, for the case of $1+1$ dimensions, it was shown in [8] that one has a transformation $g(\rho, \phi)$ which takes the origin $\rho = 0$ to any point (ρ, ϕ). After acting $g(\rho, \phi)$ transformation, Eq. (5) can be rewritten as

$$|0\rangle_{CFT} = \mathcal{P}\exp\left(-i\int_{-\infty}^{0} du\hat{K}_{(\rho,\phi)}(u)\right)|I_0\rangle \equiv U(\rho, \phi)|I_0\rangle, \tag{6}$$

where

$$\hat{K}_{(\rho,\phi)}(u) = g(\rho, \phi)\hat{K}(u)g(\rho, \phi)^{-1}.$$

Similarly, the CFT excited states $|\Psi_\alpha(\rho, \phi)\rangle_{CFT}$ can be expressed in terms of Ishibashi states $|I_\alpha\rangle$ for primary field Ψ_α

$$|\Psi_\alpha(\rho, \phi)\rangle_{CFT} = U_{(\rho,\phi)}|I_\alpha\rangle. \tag{7}$$

According to surface/state correspondence, there are dualities between quantum states in the CFT and states in the bulk gravity. In particular,

$$|0\rangle_{CFT} \Leftrightarrow |0\rangle_{bulk},$$
$$|\Psi_\alpha(\rho, \phi)\rangle_{CFT} \Leftrightarrow |\Psi_\alpha(\rho, \phi)\rangle_{bulk} \equiv \hat{\psi}_\alpha(\rho, \phi)|0\rangle_{bulk}, \tag{8}$$

where $|0\rangle_{bulk} \in \mathcal{H}_{bulk}$ is the vacuum state of the bulk gravity, and $|\Psi_\alpha(\rho, \phi)\rangle_{bulk} \in \mathcal{H}_{bulk}$ denotes the locally excited state in the bulk. The proposal [8] is to find the dual state of $|\Psi_\alpha\rangle_{bulk} \equiv |\Psi_\alpha(0,0)\rangle_{bulk}$ by noting that the $SL(2,R)$ subgroup of $SL(2,R) \times SL(2,R)$ (whose generators are (L_1, L_0, L_{-1}) and $(\tilde{L}_1, \tilde{L}_0, \tilde{L}_{-1})$) which preserves the point $\rho = t = 0$ impose constraints on $|\Psi_\alpha\rangle_{bulk}$. This is equivalent to impose the same constraints on $|\Psi_\alpha\rangle \equiv |\Psi_\alpha(0,0)\rangle_{CFT}$ in the dual CFT, i.e.,

$$(L_0 - \tilde{L}_0)|\Psi_\alpha\rangle = (L_1 + \tilde{L}_{-1})|\Psi_\alpha\rangle = (L_{-1} + \tilde{L}_1)|\Psi_\alpha\rangle = 0. \tag{9}$$

In the same footing, as we try to generalize it to the thermal case, we firstly insert a locally excited field $\hat{\psi}_\alpha(\rho, \phi)$ on the time slice of the thermal spacetimes, which is the Hartle–Hawking vacuum $|HH\rangle$[3] for an eternal black hole as shown in Fig. 1. This implies that the following dual relation which is similar to (8) holds

$$|\Psi_\alpha^\beta(\rho, \phi)\rangle_{CFT} \Leftrightarrow |\Psi_\alpha(\rho, \phi)\rangle_{bulk} \equiv \hat{\psi}_\alpha^1(\rho, \phi)\hat{\psi}_\alpha^2(\rho, \phi)|HH\rangle, \tag{10}$$

where two equivalent local bulk quantized fields $\hat{\psi}_\alpha^1$ and $\hat{\psi}_\alpha^2$ act on CFT_1 and CFT_2, respectively, and superscript β introduced in $|\Psi_\alpha^\beta(\rho, \phi)\rangle$ to distinguish them from the one at zero temperature. And the bulk coordinate (ρ, ϕ) can be naturally recognized as parameters in CFT.

As the second step, we would like to learn what is the explicit form of the states $|\Psi_\alpha^\beta(\rho, \phi)\rangle_{CFT}$. Without loss of the generality, we only need to focus on $|\Psi_\alpha^\beta\rangle_{CFT} \equiv |\Psi_\alpha^\beta(0,0)\rangle_{CFT}$. Other cases can be obtained by using $|\Psi_\alpha^\beta(\rho, \phi)\rangle_{CFT} = g(\rho, \phi)|\Psi_\alpha^\beta\rangle_{CFT}$. The dual relation (10) implies that $|\Psi_\alpha^\beta\rangle_{CFT}$ should satisfy the following conditions:

(i) The duality (10) strongly favors that these are TFD-like states;

(ii) $|\Psi_\alpha^\beta\rangle_{CFT}$ should be the energy eigenstates of the dual CFT;

(iii) Just like the zero-temperature case, constraints imposed by the generators which preserves the point $\rho = t = 0$ in the bulk will, through the duality, impose the same constraints on $|\Psi_\alpha^\beta\rangle_{CFT}$ in the dual doubled CFTs, equally. Each of them is similar to (9).

The above conditions force us to propose the following form of $|\Psi_\alpha^\beta\rangle_{CFT}$

$$|\Psi_\alpha^\beta\rangle_{CFT} \equiv |TFD - like\rangle_{CFT} = \frac{1}{\sqrt{Z(\beta)}}\sum_\alpha e^{-\beta\Delta_\alpha/2}|\tilde{\Psi}_\alpha\rangle_1|\tilde{\Psi}_\alpha\rangle_2,$$

where $|\tilde{\Psi}_\alpha\rangle_i$ are eigenstates of the dual CFTs and as well, should be solutions of the constraints similar to (9), which implies they should be generated by some primary states $|\alpha\rangle$ (and their descendants) with conformal dimensions Δ_α. In the next section we will show that in the case of BTZ black hole, $|\tilde{\Psi}_\alpha\rangle_i$ are nothing but $|\Psi_\alpha\rangle_{CFT}$ in (8) i.e. the zero-temperature solution satisfying the constraints (9).

The density matrix of the double CFTs which is dual to bulk locally excited state is given by

$$\rho_{tot} = |\Psi_\alpha^\beta\rangle_{CFT}\langle\Psi_\alpha^\beta| \equiv |TFD - like\rangle_{CFT}\langle TFD - like|. \tag{11}$$

Note that the TFD-like state though is not the usual TFD states, they are very similar. It can be viewed as a perturbative version of the usual TFD state. The reason is the following: the duality (8) shows that the vacuum $|0\rangle_{CFT}$ corresponds to a pure AdS configuration in the bulk, and the excitations $|\Psi_\alpha\rangle_{CFT}$ are dual to perturbative bulk states, which are perturbative configurations deviated slightly from the pure AdS. The TFD-like state here is a generalization of the above picture to the eternal black hole case, by putting two equivalent local bulk quantized fields into the bulk. The induced bulk configuration is a deviation from the unperturbed geometry.

Similar to the thermofield double, the thermal density matrix in one of the copies of the CFT is obtained by tracing out the contributions of the other copy of the CFT, as shown in (3). Explicitly, for density matrix of the double CFT given by (11), the reduced density matrix of one of the CFT is

$$\rho_{CFT_1} = \mathrm{Tr}_{CFT_2}\,\rho_{tot} = \frac{1}{Z(\beta)}\sum_\alpha e^{-\beta\Delta_\alpha}|\tilde{\Psi}_\alpha(\rho, \phi)\rangle_{CFT_1}\langle\tilde{\Psi}_\alpha(\rho, \phi)|, \tag{12}$$

where $Z(\beta) = \mathrm{Tr}\sum_\alpha e^{-\beta\Delta_\alpha}|\tilde{\Psi}_\alpha(\rho, \phi)\rangle\langle\tilde{\Psi}_\alpha(\rho, \phi)|$ is the partition function.

[2] Topologically trivial surface is the surface which can be smoothly deformed into a point, as a contrast, topologically non-trivial one fails to do so.

[3] To make it more readable, we use $|HH\rangle$ to denote the bulk TFD state, so as to distinguish it from TFD state in the CFT.

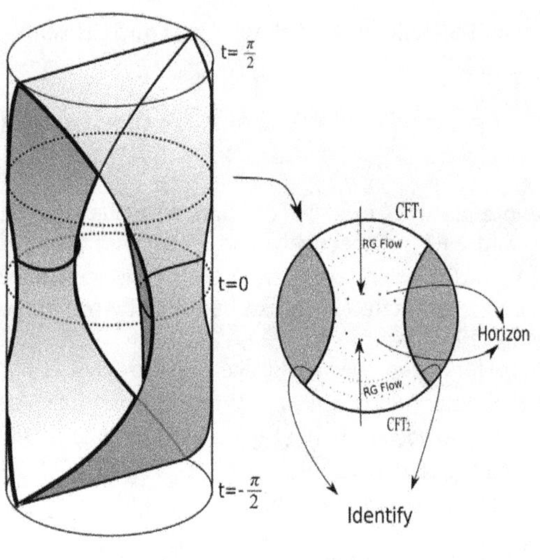

Fig. 3. The 2+1 dimensional (spinless) BTZ solution in coordinates (t, ρ, ϕ). All points inside the cylinder belong to anti-de Sitter space, its surface ($\rho = 1$) representing spatial infinity. The BTZ spacetime lies between the two surfaces inside the cylinder which are identified under an isometry generated by (16)-(18). The RG flow and the horizon (bridge states) are indicated by the dashed lines in the constant time slices to the right.

To proceed, let us employ the idea of Fisher information metric

$$ds^2 = \mathcal{D}_B = 1 - \mathrm{Tr}\sqrt{\rho_1^{1/2}\rho_2\rho_1^{1/2}}, \tag{13}$$

where

$$\rho_1 \equiv \rho_{CFT}(\lambda) = \frac{1}{Z}\sum_{\alpha} e^{-\beta\Delta_\alpha}|\tilde{\Psi}_\alpha(\lambda)\rangle_{CFT}\langle\tilde{\Psi}_\alpha(\lambda)|, \tag{14}$$

$$\rho_2 \equiv \rho_{CFT}(\lambda+\delta\lambda) = \frac{1}{Z}\sum_{\alpha'} e^{-\beta\Delta_{\alpha'}}|\tilde{\Psi}_{\alpha'}(\lambda+\delta\lambda)\rangle_{CFT}\langle\tilde{\Psi}_{\alpha'}(\lambda+\delta\lambda)|, \tag{15}$$

and it measures the distance between two infinitesimally close states ρ_1 and ρ_2 which parameterized by $\lambda = (\rho, \phi)$. As usual, we identify this information metric with the metric of the time slice of the emergent spacetime. In this way, the dual geometry of the eternal black hole can be obtained, according to the SS-duality, by considering the distance between two local excitations, which is given by (13).

3. Emergent BTZ black hole

As an explicit example, in this section we would like to employ our proposal to 2*d* CFT and to see how the expected BTZ black hole is emergent. The quantum distance (13) plays a significant role in the derivation of the geometry. The key to the derivation is to find out the form of the states $|\tilde{\Psi}_\alpha(\rho,\phi)\rangle_{CFT} = g(\rho,\phi)|\tilde{\Psi}_\alpha\rangle_{CFT}$ which is the building blocks of our proposal $|\Psi_\alpha^\beta\rangle_{CFT}$. This can be achieved by imposing conditions (ii) and (iii) as mentioned in the last section.

Let us start with a brief review on how to make a BTZ black hole (the AdS black hole in $2+1$ dimensions [32,33]). Fig. 3 shows a sketch of the way in obtaining a BTZ black hole. Roughly speaking, we first find out the "identification surfaces" in AdS. These are hypersurfaces that divide the whole spacetime into several regions, some of them have the timelike or null Killing vectors. These regions must be cut out from anti-de Sitter space to make the identifications permissible. This means that they should lie entirely within the region where the Killing vector field is space-like. Identifying corresponding points on these surfaces gives us the BTZ black hole. Before identification, it is a patch of the whole AdS space, whose isometry is given by $SL(2,R) \times SL(2,R)$ generated by the (global) Virasoro generators (L_1, L_0, L_{-1}) and $(\tilde{L}_1, \tilde{L}_0, \tilde{L}_{-1})$ of the dual 2*d* CFT. In the global coordinate, they are

$$L_0 = i\partial_+ = i\frac{\partial}{\partial x^+},\quad \tilde{L}_0 = i\partial_- = i\frac{\partial}{\partial x^-}, \tag{16}$$

$$L_{\pm1} = ie^{\pm ix^+}[\frac{\cosh 2\rho}{\sinh 2\rho}\partial_+ - \frac{1}{\sinh 2\rho}\partial_- \mp \frac{i}{2}\partial\rho], \tag{17}$$

$$\tilde{L}_{\pm1} = ie^{\pm ix^-}[\frac{\cosh 2\rho}{\sinh 2\rho}\partial_- - \frac{1}{\sinh 2\rho}\partial_+ \mp \frac{i}{2}\partial\rho], \tag{18}$$

where $x^\pm = t \pm \phi$. The identification breaks the symmetry group from $SL(2,R)\times SL(2,R)$ to $SL(2,R)\times U(1)$, however, the BTZ black hole (and its higher dimensional generalization [34]) remains locally AdS. Our procedures of deriving the thermal spacetimes in the last section only need local information, it is therefore safe enough to start with (16)-(18). For the same sake, the fact that the BTZ geometry has the same local isometry as the AdS_3 suggests the same constraints should be imposed on $|\tilde{\Psi}_\alpha\rangle_{CFT}$ as the one given in (9), which implies $|\tilde{\Psi}_\alpha\rangle_{CFT}$ has exactly the same solution as the one for AdS, i.e., $|\tilde{\Psi}_\alpha\rangle_{CFT} = |\Psi_\alpha\rangle_{CFT}$.

Following [8], the excited state $|\Psi_\alpha(\rho,\phi)\rangle_{CFT}$ can be obtained by acting the conformal transformation $g(\rho,\phi)$ to $|\Psi_\alpha\rangle_{CFT} \equiv |\Psi_\alpha(0,0)\rangle_{CFT}$, that is

$$|\Psi_\alpha(\rho,\phi)\rangle_{CFT} = g(\rho,\phi)|\Psi_\alpha\rangle_{CFT}, \tag{19}$$

where $g(\rho,\phi) = e^{i\phi l_0}e^{\frac{\rho}{2}(l_1 - l_{-1})}$ with $l_0 = L_0 - \tilde{L}_0$, $l_{-1} = \tilde{L}_1 - L_{-1}$, $l_1 = \tilde{L}_{-1} - L_1$. We will see later that the CFT parameters (ρ,ϕ) can be recognized as bulk coordinates. The state $|\Psi_\alpha\rangle_{CFT}$ turns out to be of the following form

$$|\Psi_\alpha\rangle_{CFT} \propto e^{-\delta(L_0+\tilde{L}_0)}e^{i\frac{\pi}{2}(L_0+\tilde{L}_0)}|J_\alpha\rangle, \tag{20}$$

where $\delta \sim 1/c$ is a UV cut off, $|J_\alpha\rangle = \sum_{k=0}^{\infty}|k\rangle_L \otimes |k\rangle_R$ are boundary states, and $|k\rangle_L \propto (L_{-1}^k)|\alpha\rangle$, $|k\rangle_R \propto (\tilde{L}_{-1}^k)|\alpha\rangle$ are descendants of the primary states $|\alpha\rangle$. In the following manuscript, we just denote $|\Psi_\alpha\rangle_{CFT}$ as $|\Psi_\alpha\rangle$ to simplify notation.

For later convenience, we first calculate the following inner product,

$$\begin{aligned}&|\langle\Psi_\alpha(\rho,\phi)|\Psi_{\alpha'}(\rho+d\rho,\phi+d\phi)\rangle|^2\\ =&[1-\frac{1}{8}(d\rho^2+\sinh^2\rho d\phi^2)\langle\Psi_\alpha|l_{-1}l_1+l_1l_{-1}|\Psi_{\alpha'}\rangle]^2\\ =&[1-\frac{1}{8\delta^2}(d\rho^2+\sinh^2\rho d\phi^2)\delta_{\alpha\alpha'}]^2,\end{aligned} \tag{21}$$

where in the second line the following relations have been employed [8]

$$\begin{aligned}&|\langle\Psi_\alpha(\rho,\phi)|\Psi_\alpha(\rho+d\rho,\phi+d\phi)\rangle|\\ =&1-\frac{1}{8}(d\rho^2+\sinh^2\rho d\phi^2)\langle\Psi_\alpha|l_{-1}l_1+l_1l_{-1}|\Psi_\alpha\rangle,\end{aligned} \tag{22}$$

and in the third line, we used (see Appendix A for more detail).

$$\begin{aligned}&{}_{CFT}\langle\Psi_\alpha|l_{-1}l_1+l_1l_{-1}|\Psi_{\alpha'}\rangle_{CFT}\\ =&{}_{CFT}\langle\Psi_\alpha|l_{-1}l_1+l_1l_{-1}|\Psi_\alpha\rangle_{CFT}\delta_{\alpha\alpha'} \simeq \frac{1}{\delta^2}\delta_{\alpha\alpha'}.\end{aligned} \tag{23}$$

With the above preparation, we are now in the situation to derive the Fisher information metric. In the limit $\beta^3\delta\lambda^3 \ll 1$, we have (see Appendix B for more detail)

$$
\begin{aligned}
ds^2 = \mathcal{D}_B &= 1 - \mathrm{Tr}\sqrt{\rho_1^{1/2}\rho_2\rho_1^{1/2}} \\
&\simeq 1 - \mathrm{Tr}\sqrt{\rho_1\rho_2} \\
&= 1 - \sqrt{\sum_{k,j}\langle k|\rho_1|j\rangle\langle j|\rho_2|k\rangle} \\
&= 1 - \sqrt{\sum_{k}\langle k|\rho_1\rho_2|k\rangle}.
\end{aligned} \tag{24}
$$

After substituting (19) and (20) into (12) and making use of (13)–(15) and (21) and (24), it becomes

$$
\begin{aligned}
ds^2 &= 1 - \sqrt{\frac{1}{Z}\frac{1}{Z(\rho+d\rho,\phi+d\phi)}\sum_{\alpha,\alpha'} e^{-\beta(\Delta_\alpha+\Delta_{\alpha'})}|\langle\Psi_\alpha(\rho,\phi)|\Psi_{\alpha'}(\rho+d\rho,\phi+d\phi)\rangle|^2} \\
&= 1 - \sqrt{\frac{1}{Z}\frac{1}{Z(\rho+d\rho,\phi+d\phi)}\sum_{\alpha} e^{-2\beta\Delta_\alpha}[1-\frac{1}{8\delta^2}(d\rho^2+\sinh^2\rho d\phi^2)]^2} \\
&= 1 - f(\beta)[1-\frac{1}{8\delta^2}(d\rho^2+\sinh^2\rho d\phi^2)],
\end{aligned} \tag{25}
$$

where

$$
f(\beta) = \sqrt{\frac{1}{Z}\frac{1}{Z(\rho+d\rho,\phi+d\phi)}\sum_{\alpha} e^{-2\beta\Delta_\alpha}}. \tag{26}
$$

If we treat $(\delta\lambda)^2 \equiv d\rho^2 + \sinh^2\rho d\phi^2$ as small perturbations in the parameters $\lambda = (\rho,\phi)$, then $\mathcal{F}(\beta,\lambda_1,\lambda_2) = \mathrm{Tr}\sqrt{\rho_1^{1/2}\rho_2\rho_1^{1/2}}$ by definition is the quantum fidelity of the field [35]. The explicit expression of $f(\beta)$ can be calculated in this way. Noticing that the metric is given by $ds^2 = 1 - \mathcal{F}(\beta,\lambda_1,\lambda_2)$, Eq. (B.5) implies that

$$
\begin{aligned}
\mathcal{F}(\beta,\lambda_1,\lambda_2) \sim e^{-\frac{\beta}{8}(\delta\lambda)^2\chi} &\simeq 1 - \frac{\beta}{8}(d\rho^2+\sinh^2\rho d\phi^2)\chi \\
&= 1 - \frac{\beta}{8\delta^2}(d\rho^2+\sinh^2\rho d\phi^2),
\end{aligned} \tag{27}
$$

where $\chi = 1/\delta^2$ has been used. This implies that

$$
f(\beta) \sim 1 + \frac{1-\beta}{8\delta^2}(d\rho^2+\sinh^2\rho d\phi^2), \tag{28}
$$

after plugging it into (25), the metric of time slice turns out to be[4] (keeping only quadratic terms)

$$
ds^2 \simeq \frac{\beta}{8\delta^2}(d\rho^2+\sinh^2\rho d\phi^2). \tag{29}
$$

This is the spatial part of the AdS metric in global coordinate up to a constant factor, but now it involves temperature through the parameter β, and possibly can be viewed as thermal AdS for low temperature (large β). However, for temperature larger than its critical value the Hawking–Page transition [36] will be induced, and it becomes a BTZ black hole. In this case, we define a set of new coordinates

$$
r = r_+\cosh\rho,\ \hat{t} = \frac{i\sqrt{\beta}\phi}{2\sqrt{2}r_+\delta},\ \theta = \frac{\sqrt{\beta}t}{2\sqrt{2}r_+\delta}, \tag{30}
$$

the full thermal metric (29) after adding g_{tt} (which is a prior given by assumption) can be recast as

$$
ds^2 = -\left(r^2 - r_+^2\right)d\hat{t}^2 + \frac{\beta}{8\delta^2}\frac{dr^2}{r^2-r_+^2} + r^2 d\theta^2. \tag{31}
$$

This is exactly the BTZ metric as expected. With the help of the thermofield double formalism, above procedures allow us knowing little information about the thermal CFT, which is one of the main merits of the proposal.

We would like to make a comment here. The fact that the BTZ metric relates to the thermal AdS merely through a coordinate transformation (30) is based on an observation that BTZ geometry is locally equivalent to AdS_3 [32,33]. However, this does not mean the BTZ black hole is trivially the same as the AdS spacetime—they have different topology [32,33] which can be influenced by temperature. Therefore, there is finite-temperature phase transition between AdS and BTZ black hole which is called Hawking–Page phase transition [36]. Our method in this paper is to construct bulk local excited sates to probe bulk metric. Thus, we can only access to local information. The global topology, however, cannot be obtained in the present approach. To probe global properties, e.g. Hawking–Page phase transition, we need non-local objects such as operator product expansion (OPE) block constructed by Czech et al. [37], which is a new challenge and will be discussed in the future work.

4. Conclusions

In this paper we have studied emergent geometries from CFT at finite temperature in the setup of the surface/state correspondence. We propose a general framework through which thermal geometries emerge from boundary CFTs. Instead of introducing a truncated level to the MERA tensor network, our proposal is realized by applying the thermofield double formalism to the SS-duality, and the thermal correlations are read off by tracing over one of the copies of the CFT. The main advantage of this framework is that the details of the thermal correlations of the CFT are not required. As an explicit example, we computed the information metric for a locally excited mixed state of two dimensional CFT at finite temperature, and showed that the emergent spacetimes are either thermal AdS or BTZ black hole as expected.

In the present framework, the information metric only depends on the behavior of two nearby states. This implies, according to the SS-duality, emergent bulk metric can be obtained by merely knowing the local information. This is one of the advantages of this framework. However, the other side of the coin is that nonlocal information is missing and the global symmetry cannot emerge from local operators. Although it is important, it is a tough difficulty and is out of the scope of the present paper. One possible clue is to resort to the entwinement (long geodesics which is no longer the one in the Ryu–Takayanagi prescription in BTZ case) which extracts non-local information as shown in [38,39], where the authors developed tools for constructing bulk curves in spacetimes beyond pure AdS, such as conical geometry and BTZ black hole, further, entwinement can be described well by kinematic space, and together with entanglement, can be used to reconstruct bulk geometry [37, 40,41].

Another important future problem which is of close correlation is to find which factor determines the emergent spacetime to be thermal AdS or BTZ black hole, or equivalently, the Hawking–Page phase transition [36]. That is to say, how to determine the critical temperature of the transition? In our previous work [22] we have found a cMERA description of the Hawking–Page phase transition in the framework of the truncated MERA. In the present framework, however, its solution obviously depends on the details of the global behavior, implying that the entwinement can be a candidate. Nevertheless, the full picture is far from being achieved

[4] This is correct only when $\beta^3\delta\lambda^3 \ll 1$ as explained in Appendix B.

currently.[5] We hope in the future work we can find its description in this framework.

Acknowledgements

We are very grateful to Bartlomiej Czech and Tadashi Takayanagi for careful reading of the first version of this paper and giving us valuable comments. We also thank Xi Dong, Kanato Goto, Yuting Hu, Yi Ling, Masamichi Miyaji, Xiao-Liang Qi, Edward Witten and Shao-Feng Wu for useful conversations. We are grateful for stimulating discussions to participants of "Holographic duality for condensed matter physics" held in KITPC at CAS and the conference "Strings 2016" held in YMSC, Tsinghua. This work was supported in part by the National Natural Science Foundation of China under Grant No. 11465012, the Natural Science Foundation of Jiangxi Province under Grant No. 20142BAB202007 and the 555 talent project of Jiangxi Province.

Appendix A. Orthogonal relation of the fidelity susceptibility

In our main letter we show that $\langle\Psi_\alpha|l_{-1}l_1+l_1l_{-1}|\Psi_{\alpha'}\rangle$ can be interpreted as fidelity susceptibility between two boundary states associated with two different primary states $|\alpha\rangle$ and $|\alpha'\rangle$. Following [8], we have

$$\langle\Psi_\alpha|l_{-1}l_1+l_1l_{-1}|\Psi_{\alpha'}\rangle=2\langle\Psi_\alpha|l_{-1}l_1|\Psi_{\alpha'}\rangle$$
$$=2(2+e^{2\delta}+e^{-2\delta})\langle\Psi_\alpha|L_{-1}L_1|\Psi_{\alpha'}\rangle\simeq\frac{2}{\delta}\langle\Psi_\alpha|L_0|\Psi_{\alpha'}\rangle, \quad (A.1)$$

where

$$\langle\Psi_\alpha|L_0|\Psi_{\alpha'}\rangle$$
$$=\langle J_\alpha|e^{-i\frac{\pi}{2}(L_0+\tilde{L}_0)}e^{-\delta(L_0+\tilde{L}_0)}L_0e^{-\delta(L_0+\tilde{L}_0)}e^{i\frac{\pi}{2}(L_0+\tilde{L}_0)}|J_{\alpha'}\rangle$$
$$=\langle J_\alpha|e^{-2\delta(L_0+\tilde{L}_0)}L_0|J_{\alpha'}\rangle. \quad (A.2)$$

Using the commutation relation $[L_0,L_{-1}]=L_{-1},[\tilde{L}_0,\tilde{L}_{-1}]=\tilde{L}_{-1}$ repeatedly, we have

$$L_0|J_{\alpha'}\rangle=\sum_k(k+\Delta_{\alpha'})(L_{-1})^k|\alpha'\rangle(\tilde{L}_{-1})^k|\alpha'\rangle, \quad (A.3)$$

and

$$e^{-2\delta(L_0+\tilde{L}_0)}L_0|J_{\alpha'}\rangle$$
$$=\sum_k(k+\Delta_{\alpha'})e^{-4\delta\Delta_{\alpha'}}\left(\frac{1}{e^{2\delta}+1}\right)^{2k}|k\rangle_{\alpha'}|k\rangle_{\alpha'}. \quad (A.4)$$

We therefore get

$$\langle J_\alpha|e^{-2\delta(L_0+\tilde{L}_0)}L_0|J_{\alpha'}\rangle$$
$$=\sum_{k,k'}\langle k|_\alpha\langle k|_\alpha(k'+\Delta_{\alpha'})e^{-4\delta\Delta_{\alpha'}}\left(\frac{1}{e^{2\delta}+1}\right)^{2k'}|k'\rangle_{\alpha'}|k'\rangle_{\alpha'}$$
$$=\sum_{k,k'}\langle k|_\alpha\langle k|_\alpha(k'+\Delta_\alpha)e^{-4\delta\Delta_\alpha}\left(\frac{1}{e^{2\delta}+1}\right)^{2k'}|k'\rangle_\alpha|k'\rangle_\alpha\delta_{\alpha\alpha'}$$
$$=\langle J_\alpha|e^{-2\delta(L_0+\tilde{L}_0)}L_0|J_\alpha\rangle\delta_{\alpha\alpha'}, \quad (A.5)$$

where we have used the orthogonality between the highest weight states $\langle\alpha|\alpha'\rangle=\delta_{\alpha\alpha'}$. In the end, one has

$$\langle\Psi_\alpha|l_{-1}l_1+l_1l_{-1}|\Psi_{\alpha'}\rangle=\langle\Psi_\alpha|l_{-1}l_1+l_1l_{-1}|\Psi_\alpha\rangle\delta_{\alpha\alpha'}=\frac{1}{\delta^2}\delta_{\alpha\alpha'}, \quad (A.6)$$

which is the formula (23) in our main letter.

Appendix B. Simplification of noncommutative density matrix

In this section, we would like to give a simplification of $\mathrm{Tr}\sqrt{\rho_1^{1/2}\rho_2\rho_1^{1/2}}$ (physically this is field fidelity) under a general perturbation ($\delta\lambda$) in the parameter space. Generally speaking, ρ_1 and ρ_2 do not commute. Formally, we can write ρ_1 and ρ_2 in terms of Hamiltonian

$$\rho_1=\frac{e^{-\beta H(\lambda_1)}}{Z(\beta,\lambda_1)},\ \rho_2=\frac{e^{-\beta H(\lambda_2)}}{Z(\beta,\lambda_2)} \quad (B.1)$$

where λ_i $(i=1,2)$ denotes the parameters with $\lambda_2=\lambda_1+\delta\lambda$. The Trotter–Suzuki formula [42] can be used to give an approximation

$$\left\|\rho_1^{1/2}\rho_2\rho_1^{1/2}-\frac{e^{-\beta H(\lambda_1)+H(\lambda_2)}}{Z(\beta,\lambda_1)Z(\beta,\lambda_2)}\right\|$$
$$<\beta^3\Delta_2[H(\lambda_1),H(\lambda_2)]e^{\beta\|H(\lambda_1)\|+\|H(\lambda_2)\|}\sim(\beta\delta\lambda)^3, \quad (B.2)$$

where

$$\Delta_2[H(\lambda_1),H(\lambda_2)]=\frac{1}{12}(\|\,[[H(\lambda_1),H(\lambda_2)],H(\lambda_2)]\,\|$$
$$+\|\,[H(\lambda_1),H(\lambda_2)],H(\lambda_1)]\,\|)\,. \quad (B.3)$$

If $\beta^3\delta\lambda^3\ll 1$, then the fidelity becomes

$$\mathcal{F}(\beta,\lambda_1,\lambda_2)\equiv\mathrm{Tr}\sqrt{\rho_1^{1/2}\rho_2\rho_1^{1/2}}$$
$$\approx\mathrm{Tr}\sqrt{\frac{e^{-\beta H(\lambda_1)+H(\lambda_2)}}{Z(\beta,\lambda_1)Z(\beta,\lambda_2)}}=\mathrm{Tr}\sqrt{\rho_1\rho_2}. \quad (B.4)$$

It was shown in [35] that in this limit, the fidelity has the following behavior

$$\mathcal{F}(\beta,\lambda_1,\lambda_2)\approx e^{-\frac{\beta(\delta\lambda^2)\chi}{8}}, \quad (B.5)$$

where χ is the fidelity susceptibility.

References

[1] J.M. Maldacena, The large-N limit of superconformal field theories and supergravity, Adv. Theor. Math. Phys. 2 (1998) 231, Int. J. Theor. Phys. 38 (1999) 1113, arXiv:hep-th/9711200.
[2] S.S. Gubser, I.R. Klebanov, A.M. Polyakov, Gauge theory correlators from non-critical string theory, Phys. Lett. B 428 (1998) 105, arXiv:hep-th/9802109.
[3] E. Witten, Anti-de Sitter space and holography, Adv. Theor. Math. Phys. 2 (1998) 253, arXiv:hep-th/9802150.
[4] S. Ryu, T. Takayanagi, Holographic derivation of entanglement entropy from the anti-de Sitter space/conformal field theory correspondence, Phys. Rev. Lett. 96 (2006) 181602, arXiv:hep-th/0603001.
[5] S. Ryu, T. Takayanagi, Aspects of holographic entanglement entropy, J. High Energy Phys. 08 (2006) 045, arXiv:hep-th/0605073.
[6] V.E. Hubeny, M. Rangamani, T. Takayanagi, A covariant holographic entanglement entropy proposal, J. High Energy Phys. 07 (2007) 062, arXiv:0705.0016.
[7] M. Miyaji, T. Takayanagi, Surface/state correspondence as a generalized holography, Prog. Theor. Exp. Phys. 73 (2015) 073B03, arXiv:1503.03542.
[8] M. Miyaji, T. Numasawa, N. Shiba, T. Takayanagi, K. Watanabe, Continuous multiscale entanglement renormalization ansatz as holographic surface-state correspondence, Phys. Rev. Lett. 115 (2015) 171602, arXiv:1506.01353.
[9] G. Vidal, Entanglement renormalization, Phys. Rev. Lett. 99 (2007) 220405, arXiv:cond-mat/0512165.
[10] G. Vidal, Class of quantum many-body states that can be efficiently simulated, Phys. Rev. Lett. 101 (2008) 110501, arXiv:cond-mat/0605597.
[11] G. Evenbly, G. Vidal, Algorithms for entanglement renormalization, Phys. Rev. B 79 (2009) 144108, arXiv:0707.1454.
[12] G. Vidal, Entanglement renormalization: an introduction, arXiv:0912.1651.

[5] Since the exact definition of entwinement in the CFT side is still missing, the exact mapping of this quantity between gravity and gauge theory has not been obtained.

[13] R.N.C. Pfeifer, G. Evenbly, G. Vidal, Entanglement renormalization, scale invariance, and quantum criticality, Phys. Rev. A 79 (2009) 040301(R), arXiv:0810.0580.
[14] B. Swingle, Entanglement renormalization and holography, Phys. Rev. D 86 (2012) 065007, arXiv:0905.1317.
[15] G. Evenbly, G. Vidal, Tensor network states and geometry, J. Stat. Phys. 145 (2011) 891, arXiv:1106.1082.
[16] B. Swingle, Constructing holographic spacetimes using entanglement renormalization, arXiv:1209.3304.
[17] X.L. Qi, Exact holographic mapping and emergent space–time geometry, arXiv:1309.6282.
[18] C.H. Lee, X.L. Qi, Exact holographic mapping in free fermion systems, Phys. Rev. B 93 (2016) 035112, arXiv:1503.08592.
[19] J. Haegeman, T.J. Osborne, H. Verschelde, F. Verstraete, Entanglement renormalization for quantum fields in real space, Phys. Rev. Lett. 110 (2013) 100402, arXiv:1102.5524.
[20] M. Nozaki, S. Ryu, T. Takayanagi, Holographic geometry of entanglement renormalization in quantum field theories, J. High Energy Phys. 1210 (2012) 193, arXiv:1208.3469.
[21] N. Bao, C.J. Cao, S.M. Carroll, A. Chatwin-Davies, N. Hunter-Jones, J. Pollack, G.N. Remmen, Consistency conditions for an AdS/MERA correspondence, Phys. Rev. D 91 (2015) 125036, arXiv:1504.06632.
[22] W.-C. Gan, F.-W. Shu, M.-H. Wu, Thermal geometry from CFT at finite temperature, Phys. Lett. B 760 (2016) 796, arXiv:1605.05999.
[23] W. Israel, Thermo-field dynamics of black holes, Phys. Lett. A 57 (1976) 107.
[24] J.M. Maldacena, Eternal black holes in anti-de Sitter, J. High Energy Phys. 0304 (2003) 021, arXiv:hep-th/0106112.
[25] H. Matsueda, M. Ishihara, Y. Hashizume, Tensor network and black hole, Phys. Rev. D 87 (2013) 066002, arXiv:1208.0206.
[26] T. Hartman, J. Maldacena, Time evolution of entanglement entropy from black hole interiors, J. High Energy Phys. 1305 (2013) 014, arXiv:1303.1080.
[27] J. Molina-Vilaplana, Holographic Geometries of one-dimensional gapped quantum systems from Tensor Network States, J. High Energy Phys. 1305 (2013) 024, arXiv:1210.6759.
[28] J. Molina-Vilaplana, J. Prior, Entanglement, tensor networks and black hole horizons, Gen. Relativ. Gravit. 46 (2014) 1823, arXiv:1403.5395.
[29] A. Mollabashi, M. Nozaki, S. Ryu, T. Takayanagi, Holographic geometry of cMERA for quantum quenches and finite temperature, J. High Energy Phys. 1403 (2014) 098, arXiv:1311.6095.
[30] M. Miyaji, S. Ryu, T. Takayanagi, X. Wen, Boundary states as holographic duals of trivial spacetimes, J. High Energy Phys. 05 (2015) 152, arXiv:1412.6226.
[31] M. Fujita, T. Takayanagi, E. Tonni, Aspects of AdS/BCFT, J. High Energy Phys. 1111 (2011) 043, arXiv:1108.5152.
[32] M. Banados, C. Teitelboim, J. Zanelli, Black hole in three-dimensional spacetime, Phys. Rev. Lett. 69 (1992) 1849, arXiv:hep-th/9204099.
[33] M. Banados, M. Henneaux, C. Teitelboim, J. Zanelli, Geometry of the 2+1 black hole, Phys. Rev. D 48 (1993) 1506, arXiv:gr-qc/9302012.
[34] S. Åminneborg, I. Bengtsson, S. Holst, P. Peldán, Making anti-de Sitter black holes, Class. Quantum Gravity 13 (1996) 2707, arXiv:gr-qc/9604005.
[35] H.T. Quan, F.M. Cucchietti, Quantum fidelity and thermal phase transition, Phys. Rev. E 79 (2009) 031101, arXiv:0806.4633.
[36] S.W. Hawking, D.N. Page, Thermodynamics of black holes in anti-de Sitter space, Commun. Math. Phys. 87 (1983) 577.
[37] B. Czech, L. Lamprou, S. McCandlish, B. Mosk, J. Sully, A stereoscopic look into the bulk, J. High Energy Phys. 07 (2016) 129, arXiv:1604.03110.
[38] B. Czech, L. Lamprou, Holographic definition of points and distances, Phys. Rev. D 90 (2014) 106005, arXiv:1409.4473.
[39] V. Balasubramanian, B.D. Chowdhury, B. Czech, J. de Boer, Entwinement and the emergence of spacetime, J. High Energy Phys. 1501 (2015) 048, arXiv:1406.5859.
[40] B. Czech, L. Lamprou, S. McCandlish, J. Sully, Integral geometry and holography, J. High Energy Phys. 10 (2015) 175, arXiv:1505.05515.
[41] B. Czech, L. Lamprou, S. McCandlish, J. Sully, Tensor networks from kinematic space, J. High Energy Phys. 1607 (2016) 100, arXiv:1512.01548.
[42] M. Suzuki, Decomposition formulas of exponential operators and Lie exponentials with some applications to quantum mechanics and statistical physics, J. Math. Phys. 26 (1985) 601.

15

Entanglement entropy and complexity for one-dimensional holographic superconductors

Mahdi Kord Zangeneh [a,b,c,d], Yen Chin Ong [a,d,*], Bin Wang [d,a]

[a] Center for Gravitation and Cosmology, College of Physical Science and Technology, Yangzhou University, Yangzhou 225009, China
[b] Research Institute for Astronomy and Astrophysics of Maragha (RIAAM), Maragha, P.O. Box 55134-441, Iran
[c] Physics Department and Biruni Observatory, Shiraz University, Shiraz 71454, Iran
[d] Center for Astronomy and Astrophysics, Department of Physics and Astronomy, Shanghai Jiao Tong University, Shanghai 200240, China

ARTICLE INFO

Editor: N. Lambert

Keywords:
Holographic complexity
Holographic entanglement entropy
Holographic superconductors
AdS/CFT correspondence

ABSTRACT

Holographic superconductor is an important arena for holography, as it allows concrete calculations to further understand the dictionary between bulk physics and boundary physics. An important quantity of recent interest is the holographic complexity. Conflicting claims had been made in the literature concerning the behavior of holographic complexity during phase transition. We clarify this issue by performing a numerical study on one-dimensional holographic superconductor. Our investigation shows that holographic complexity does not behave in the same way as holographic entanglement entropy. Nevertheless, the universal terms of both quantities are finite and reflect the phase transition at the same critical temperature.

1. Introduction: holographic complexity and phase transition

AdS/CFT correspondence, or holography, has shown a deep connection between gravity in an asymptotically anti-de Sitter (AdS) spacetime ("the bulk") and the quantum field theory that lives on its conformal boundary [1]. In recent years, quantum information has been applied in the context of gravitational physics, notably in the context of black hole information paradox [2]. Two quantities of the boundary field theory, which play important roles in quantum information, are entanglement entropy and (quantum) complexity.

As it turns out, both of these quantities are reflected in the bulk geometry. The entanglement entropy between the degrees of freedom inside a closed region $\mathcal{A}$ with that of its exterior (both of which are on the boundary), is proportional to the area of an extremal surface, $\gamma_{\mathcal{A}}$, that anchors on the boundary of $\mathcal{A}$, i.e., $\partial\mathcal{A} = \partial\gamma_{\mathcal{A}}$, and extends into the bulk[1] (if there are more than one such surfaces, the one with minimal area is chosen). Specifically, the Ryu–Takayanagi formula [4,5] states that the (regularized) holographic entanglement entropy (HEE) is given by, in the units $c = \hbar = k_B = 1$, and G being the Newton's constant,

$$\mathcal{S} = \frac{\text{Area}(\gamma_{\mathcal{A}})}{4G}. \tag{1}$$

The holographic complexity (HC) for this subregion $\mathcal{A}$ is conjectured to be holographically related to the volume enclosed by the aforementioned minimal surface. Specifically [6], up to a constant factor (the factor 8π is merely a convention),

$$\mathcal{C} = \frac{\text{Volume}(\gamma_{\mathcal{A}})}{8\pi\mathcal{R}G}, \tag{2}$$

where $\mathcal{R}$ is the radius of curvature of the background spacetime, e.g., the AdS curvature radius, as in this work. Note that both $\mathcal{S}$ and $\mathcal{C}$ are dimensionless quantities in our choice of units.

Essentially, HEE is related to the content of information encoded in the subsystem (for example, information starts to leak out from a black hole during the Page time [7–9], which is the moment when the HEE of the Hawking radiation starts to decrease). On the other hand, HC has to do with how difficult it is to perform an operation. In the context of Hawking radiation, this is related to the difficulty of decoding and extracting the highly scrambled to decode and extract the highly scrambled information from the Hawking radiation [10–12]. In holography, HC of a field theory can be interpreted as the minimum number of gates to implement a

* Corresponding author.
E-mail addresses: mkzangeneh@shirazu.ac.ir (M. Kord Zangeneh), ongyenchin@gmail.com (Y.C. Ong), wang_b@sjtu.edu.cn (B. Wang).

[1] This simple statement hides a lot of mathematical subtleties [3].

certain unitary operator, to turn a pure reference state into another pure state [13]. For mixed states the interpretation in terms of gates is not as straightforward, we will return to this in the next section.

Due to the ambiguity of choosing the correct length scale $\mathcal{R}$ for different backgrounds, it has been recently proposed that one should use instead the Einstein–Hilbert action in the so-called "Wheeler–DeWitt patch" as the holographic dual of complexity. However, with the right choice of the length scale, the original "complexity=volume" conjecture yields essentially the same result as that of the more recent "complexity=action" conjecture [14,15].

In this work we will focus on holographic complexity as defined in Eq. (2) above, which remains the form that is widely focused on in the holography literature. We will focus on the time-independent subregion holographic complexity, i.e., the minimal surface $\gamma_{\mathcal{A}}$ is entirely outside of the black hole horizon. (See [16] for more discussions on the properties of such time-independent volume in various circumstances.)

Since complexity essentially measures the difficulty of turning a quantum state into another, it is conceivable that a phase transition on the boundary field theory could be reflected in the HC. This possibility is further supported by the recent proposal that the complexity is deeply connected with fidelity susceptibility [17–20], which is known to be able to probe phase transition, even without prior knowledge of the local order parameter [21–24].

Our work was motivated by the discrepancies between existing results in the literature: Momeni et al. [25] claimed that during the phase transition of a one dimensional holographic superconductor, there is a divergent behavior in the HC. (This divergence should, of course, not to be confused with the trivial divergence that could be removed via regularization.) However, Roy and Sarkar found that as far as phase transitions are concerned, HC captures essentially the same information as HEE [26], which had previously been investigated in [27]. (For recent discussion regarding the relations between HEE and HC, see, e.g., [28].) Although the latter does not investigate a 1-dimensional holographic superconductor, but rather a thermodynamics phase transition of a Reissner-Nordström–AdS black hole, it does suggest that the behavior of HC during phase transition should be carefully re-examined.

Furthermore, a divergence in the universal terms of HC during phase transition is rather problematic for the following reason. Quantum complexity can be understood, in the language of circuits, as the minimum number of gates that is required to implement a certain unitary operator, to turn a pure reference state into another pure state [13]. For subregions, such as the one we discussed here, the states are mixed, and so its interpretation is somewhat subtle. Consider the density matrix $\rho_{\mathcal{A}}$ associated with the subregion $\mathcal{A}$. One then prepares $\rho_{\mathcal{A}}$ with a completely positive trace-preserving (CPTP) map acting on the reference state [29]. In doing so, a number of "ancillary" and "erasure" gates, which add and remove additional degrees of freedom, are added to the circuits [30,31]. Effectively, this means that we can interpret the subregion complexity in the following way: one first extend the Hilbert space of $\mathcal{A}$ with new ancillary degrees of freedom, which would purify the mixed state $\rho_{\mathcal{A}}$. The subregion complexity is then the minimum number of (universal) gates required to turn a reference pure state into the required pure state [29].

This point of view would mean that during phase transition, the field theory becomes so complex that one requires an infinite number of gates. So, if correct, the result of [25] would mean that not only does HC respond to phase transition, it also does so *extremely drastically*. An infinite amount of complexity does not appear to be physically plausible.

We therefore return to the model investigated in [25], which is a fully backreacted 1-dimensional holographic superconductor, and perform a numerical analysis to further investigate this issue. Indeed, it was mentioned in [25] that such a numerical analysis is interesting and should be carried out to supplement their analytic analysis.

Our numerical investigation shows *conclusively* that notwithstanding the claim of [25], during the phase transition of a 1-dimensional holographic superconductor, the universal terms of HC remains finite and well-defined, just like HEE. In particular, both HC and HEE show that the superconducting phase intersect with the normal phase at the same critical temperature. We will further explain why our numerical result is inconsistent with [25]. However, in contrast to the claim made in [26] that HC contains the same information as HEE as far as phase transitions are concerned, we found that HC and HEE can behave quite differently. There is, however, no conflict with the results in [26] since the system studied therein is substantially different from ours.

2. One-dimensional holographic superconductor: the set-up

In this section, we first introduce the background geometry of a black hole coupled with a charged complex scalar field, and explain how to take backreaction of the matter field into account to model a holographic superconductor numerically. (Readers who are unfamiliar with holographic superconductors may consult [32] for details.)

The holographic setup of a one-dimensional[2] superconductor involves an asymptotically anti-de Sitter bulk geometry, which is governed by the $(2+1)$-dimensional action [33,34]

$$S=\int \mathrm{d}^3x\sqrt{-g}\left[\frac{1}{2\kappa^2}\left(R+\frac{2}{l^2}\right)-\frac{1}{4}F_{ab}F^{ab}\right.\\ \left.-\left|\nabla\psi-iqA\psi\right|^2-m^2\left|\psi\right|^2\right]. \tag{3}$$

Here R and g are, respectively, the Ricci scalar and the determinant of the metric; $\kappa=\sqrt{8\pi G_3}$ is the $(2+1)$-dimensional gravitational constant with G_3 being the $(2+1)$-dimensional Newton's constant, and l is the asymptotic AdS curvature radius. Also in the action one finds the electromagnetic tensor $F_{ab}=\nabla_{[a}A_{b]}$, where A_b is the usual vector gauge potential. The gauge field is coupled to a charged complex scalar field ψ, with m and q being the mass and the charge of the scalar field, respectively.

In order to study the fully backreacted holographic superconductor, we consider an ansatz of the form

$$\mathrm{d}s^2=-f(z)\,e^{-\chi(z)}\mathrm{d}t^2+\left(z^4f(z)\right)^{-1}\mathrm{d}z^2+(zl)^{-2}\,\mathrm{d}x^2, \tag{4}$$

where $\{t,z,x\}$ are the usual Poincaré-type coordinates in asymptotically AdS spacetime.

We are interested in static, translationally invariant solutions, thus we consider the ansatz for the gauge potential and the charged scalar field to be [35], respectively,

$$A=\phi(z)\,\mathrm{d}t, \qquad \psi=\psi(z). \tag{5}$$

Since the Maxwell equations imply that $\psi(z)$ has a constant phase, it can be considered as a real function without loss of generality. In this setup, the black hole horizon and the charge of scalar field can be fixed as unity by virtue of scaling symmetries [33,34,36].

The dual one-dimensional superconductor lives on the boundary at $z=0$. It can be shown that, with backreaction governed by κ which is treated as a parameter,[3] the field equations read

[2] By one-dimensional, we meant one spatial dimension, i.e., the superconductor lives in a $(1+1)$-dimensional spacetime.

[3] We would like to consider the backreaction of the bulk fields on the background metric. The bulk fields are the scalar ψ and the gauge potential coefficient ϕ. If one

$$0=\psi''+\psi'\left(\frac{f'}{f}+\frac{1}{z}-\frac{\chi'}{2}\right)+\frac{\psi}{z^4 f}\left(\frac{e^{\chi}\phi^2}{f}-m^2\right), \tag{6}$$

$$0=\phi''+\left(\frac{1}{z}+\frac{\chi'}{2}\right)\phi'-\frac{2\phi\psi^2}{z^4 f}, \tag{7}$$

$$0=f'-\kappa^2\left[\frac{2\psi^2}{z^3}\left(m^2+\frac{\phi^2 e^{\chi}}{f}\right)\right.$$
$$\left.+zf\left(2\psi'^2+\frac{\phi'^2 e^{\chi}}{f}\right)\right]+\frac{2}{l^2 z^3}, \tag{8}$$

$$0=\chi'-4\kappa^2 z\left(\frac{\phi^2\psi^2 e^{\chi}}{z^4 f^2}+\psi'^2\right), \tag{9}$$

where prime denotes the derivative with respect to z.

Our aim in this letter is to study numerically the behaviors of holographic entanglement entropy (HEE) and complexity (HC) for a subregion with length L of a fully backreacted one-dimensional superconductor. We will consider both the normal and superconducting phases. We will employ the shooting method to carry out our numerical calculation. In order to do this, we need to know the behaviors of the functions of the above setup at both the black hole horizon and the boundary. At the horizon, we can Taylor expand the field equations to find the expansion coefficients of the following functions

$$f(z)=f_1(1-z)+f_2(1-z)^2+\cdots,$$
$$\psi(z)=\psi_0+\psi_1(1-z)+\psi_2(1-z)^2+\cdots,$$
$$\phi(z)=\phi_1(1-z)+\phi_2(1-z)^2+\cdots,$$
$$\chi(z)=\chi_0+\chi_1(1-z)+\chi_2(1-z)^2+\cdots. \tag{10}$$

Note that in the expansions above, we impose the condition that the metric function f and the gauge potential ϕ should both vanish on the black hole horizon. The latter is applied so that the norm of the gauge potential, $A_\mu A^\mu$, is finite at the horizon.

In order to perform the shooting method, we will find the coefficients ψ_0, ϕ_1 and χ_0 such that the desired values for some parameters on the boundary is attained. In our study, we will focus on the case $m=0$. For this case, the various functions at the boundary are approximately given by

$$\chi\approx\chi_-, \qquad f\approx(zl)^{-2},$$
$$\phi\approx\rho+\mu\ln(z), \qquad \psi\approx\psi_-+\psi_+z^2, \tag{11}$$

where χ_-, ρ, μ, ψ_- and ψ_+ are some constants. According to the holographic dictionary, μ corresponds to the chemical potential of the dual superconductor. The quantity ψ_+ is related to the expectation value of the order parameter $\langle\mathcal{O}_+\rangle$ of the dual superconductor, whereas ψ_- is considered as the source of this order parameter.

To apply the shooting method, we change the value of the coefficient ϕ_1 and set ψ_0 and χ_0 at the horizon so that both ψ_- and χ_- vanish at the boundary. The former is required since we treat the superconducting phase transition as a spontaneous symmetry breaking phenomenon (i.e., the symmetry breaking is entirely due to the low temperature, not induced by a source), while the latter is allowed by the rescaling symmetry

$$e^{\chi}\to a^2e^{\chi}, \qquad \phi\to\phi/a, \qquad t\to at, \tag{12}$$

for some constant a.

The temperature of holographic superconductor is given by the Hawking temperature of the black hole in the bulk [25,34], which can be easily checked by the usual method of Wick-rotating the metric in Eq. (4) to Euclidean signature and imposing regularity at the Euclidean horizon:

$$T=\left.\frac{e^{-\chi/2}f'}{4\pi}\right|_{z=1}. \tag{13}$$

The HEE $\mathcal{S}$ associated to the subregion $\mathcal{A}$ of a dual field theory is proportional to the area of a minimal co-dimension two surface $\gamma_{\mathcal{A}}$ such that their boundaries are the same, i.e., $\partial\gamma_{\mathcal{A}}=\partial\mathcal{A}$ [4,5]. Therefore, for a strip subregion with length L, we have [25]

$$\mathcal{S}=\frac{2\pi}{\kappa^2}\int_{-L/2}^{L/2}\frac{dx}{z^2}\sqrt{\frac{1}{f}\left(\frac{dz}{dx}\right)^2+\frac{z^2}{l^2}}. \tag{14}$$

The minimality condition implies

$$\frac{dz}{dx}=\pm l^{-1}\sqrt{\left(z_*^2-z^2\right)f}, \tag{15}$$

in which the constant z_* satisfies the stationary condition $dz/dx|_{z=z_*}=0$. This can be verified using the Euler–Lagrange variation method. Next, setting $x_\mp(z_*)=0$, we find

$$x_\mp(z)=\mp l\int_z^{z_*}\frac{dz}{\sqrt{\left(z_*^2-z^2\right)f}}. \tag{16}$$

This satisfies, with a UV cutoff ϵ,

$$x_\mp(\epsilon\to 0)=\mp L/2. \tag{17}$$

Notice that $+$ (respectively, $-$) in Eq. (15) corresponds to the region $-L/2<x\leqslant 0$ (respectively, $0\leqslant x<L/2$) while in Eq. (16), it corresponds to $0\leqslant x<L/2$ (respectively, $-L/2<x\leqslant 0$). Using Eq. (15), one can rewrite Eq. (14) as

$$\mathcal{S}=\frac{4\pi}{\kappa^2}\int_{\epsilon\to 0}^{z_*}\frac{z_*dz}{z^2 l\sqrt{\left(z_*^2-z^2\right)f}}. \tag{18}$$

On the other hand, the complexity corresponding to $\mathcal{A}$ is holographically related to the volume in the bulk enclosed by $\gamma_{\mathcal{A}}$, namely [6,25]

$$\mathcal{C}=\frac{2}{l^2\kappa^2}\int_{\epsilon\to 0}^{z_*}\frac{x_+(z)\,dz}{z^3\sqrt{f}}. \tag{19}$$

Notice that, according to the scaling symmetry

$$(t,z,x)\to b^{-1}(t,z,x), \qquad f\to b^2 f, \qquad \phi\to b\phi, \tag{20}$$

the quantities T, L, μ, $\mathcal{S}$ and $\mathcal{C}$, scale as

$$T\to bT, \qquad L\to b^{-1}L, \qquad \mu\to b\mu,$$
$$\mathcal{S}\to\mathcal{S}, \qquad \mathcal{C}\to\mathcal{C}. \tag{21}$$

Therefore, to study the physics, it is useful to employ the dimensionless quantities T/μ, μL, $\mathcal{S}$ and $\mathcal{C}$.

At this point, let us give some comments about the formally diverging terms of HEE and HC before regularization. The diverging term of HEE (Eq. (18)) caused by the pure AdS geometry $f\to(zl)^{-2}$ near the UV cutoff ϵ is $\left(4\pi\ln\epsilon^{-1}\right)/\kappa^2$. Subtracting

re-scales them to $q\psi$ and $q\phi$ in the action, then although the Maxwell and scalar equations remain invariant, the gravitational coupling is re-scaled by $\kappa^2\mapsto\kappa^2/q^2$. Fixing the charge, one can vary κ, which now serves as a backreaction parameter. Note that the limit $\kappa\to 0$ corresponds to the probe limit, i.e., there is no backreaction. Alternatively, if one fixes κ, then $q\to\infty$ limit corresponds to the probe limit [37,38].

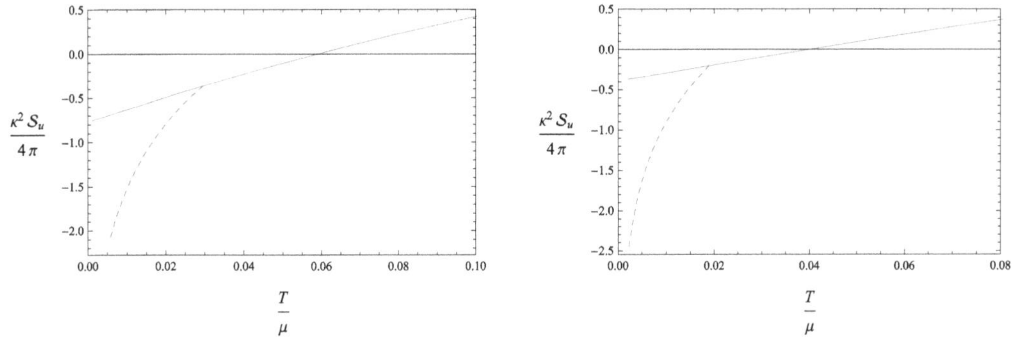

(a) $\kappa^2 = 0.1$ (b) $\kappa^2 = 0.2$

Fig. 1. The behavior of $\kappa^2 \mathcal{S}_u/4\pi$ versus T/μ for $l=1$ and $\mu L/2=1$. Blue (solid) and red (dashed) curves correspond respectively to normal and superconducting phases. For $\kappa^2=0.1$ and 0.2, the critical temperatures per chemical potential are $T_c/\mu=0.0295$ and 0.0189, respectively [34]. (For interpretation of the references to color in this figure legend, the reader is referred to the web version of this article.)

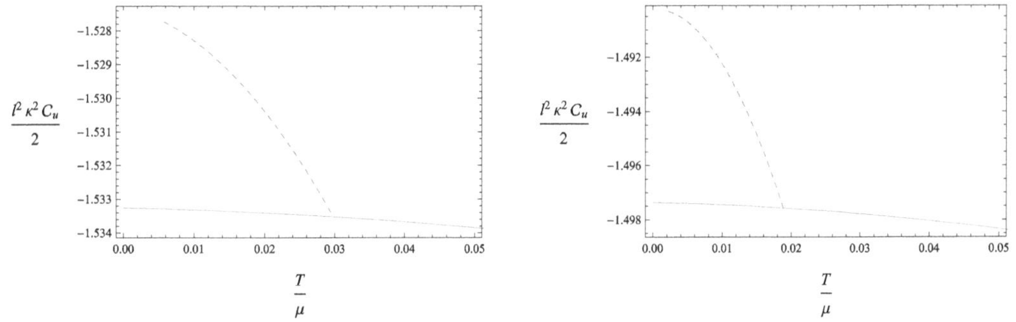

(a) $\kappa^2 = 0.1$ (b) $\kappa^2 = 0.2$

Fig. 2. The behavior of $l^2\kappa^2 \mathcal{C}_u/2$ versus T/μ for $l=1$ and $\mu L/2=1$. Blue (solid) and red (dashed) curves correspond respectively to normal and superconducting phases. For $\kappa^2=0.1$ and 0.2, the critical temperatures per chemical potential are $T_c/\mu=0.0295$ and 0.0189, respectively [34]. (For interpretation of the references to color in this figure legend, the reader is referred to the web version of this article.)

this diverging term from $\mathcal{S}$ in Eq. (18), one can find the universal term of HEE, $\mathcal{S}_u$. As for the HC, the diverging term corresponding to the pure AdS geometry is $2z_*/\left(l^2\kappa^2\epsilon\right)$; it *cannot* be used to subtract off the divergence in every situation. For instance, for normal phase ($\psi=\chi=0$), the diverging term includes $\tanh^{-1}(z_*)/\epsilon$ for the $\kappa\to 0$ case (see Appendix A). Indeed, the diverging term may include different *functions* of z_* under different situations. It is not possible to find a general form for HC diverging term analytically. Fortunately, this is not necessary, since we can overcome this problem numerically. The HC includes a universal term $\mathcal{C}_u$ and a diverging term in the form of $\mathcal{F}(z_*)/\epsilon$. Since the value of universal term should not change for different cutoffs, subtracting HC in Eq. (19) for two different values of cutoff ϵ_1 and ϵ_2, one finds $\left(\epsilon_1^{-1}-\epsilon_2^{-1}\right)\mathcal{F}(z_*)$. Therefore, the value of $\mathcal{F}(z_*)$ in different situations can be found numerically. (The regularization of HC is discussed in great details in [39]. See also [29] for a discussion on the geometry related to the UV divergence in the HC.)

In the rest of this letter, we will study the behavior of (the universal parts of) HEE and HC of a 1-dimensional superconductor numerically. To do this, we will first evaluate z_* numerically using Eq. (17). Then, we will obtain $x_+(z)$ numerically from Eq. (16) and from this, calculate HC from Eq. (19). We can also compute the HEE given by Eq. (18).

3. Numerical results for HEE and HC

In this section, we will study HEE and HC for a strip subregion $\mathcal{A}$ of the 1-dimensional dual system. We first show below the numerical results: the plots of the universal terms of HEE, $\mathcal{S}_u$ (Fig. 1), and HC, $\mathcal{C}_u$ (Fig. 2), against the ratio of temperature to chemical potential, T/μ, for both normal and superconducting phases.

As mentioned in the Introduction, we are motivated by the inconsistency in the literature: on one hand is the claim by Momeni et al. [25] that during the phase transition of a 1-dimensional holographic superconductor, there is a divergent behavior in HC. On the other hand, Roy and Sarkar found that during the phase transition of a Reissner–Nordström–AdS black hole, HC behaves in the same manner as HEE [26]. Granted that the latter does not investigate a 1-dimensional holographic superconductor, it is quite suggestive that there is a conflict between the two results.

Here, our numerical investigation shows *conclusively* that, during phase transition of a 1-dimensional holographic superconductor, the universal terms of holographic complexity is still finite and well-defined. Indeed, as can be seen from Fig. 1, the points where the plots of $\mathcal{S}_u$ for the normal phase (blue solid curves) intersect with that of the superconducting phase (red dashed curves), occur at critical temperatures $T_c/\mu=0.0295$ and $T_c/\mu=0.0189$, for backreaction parameters $\kappa^2=0.1$ and $\kappa^2=0.2$, respectively. This agrees with the results in Fig. 2, which are plots for the universal terms in HC. Varying κ^2 does not change the qualitative behavior of these plots.

In other words, the critical temperature of the phase transition can be read from the plot of $\mathcal{C}_u$, which agrees with the critical temperature read from the plot of $\mathcal{S}_u$. This means that HC does indeed responds to phase transitions, just like the HEE would. It is worth noting that increasing the strength of backreaction makes condensation harder, i.e., it occurs at a lower temperature.

In Fig. 3, we have also plotted the logarithms of $-l^2\mathcal{C}_u$ and $-\mathcal{S}_u$ as functions of T/μ. The aim here is to show that the opening an-

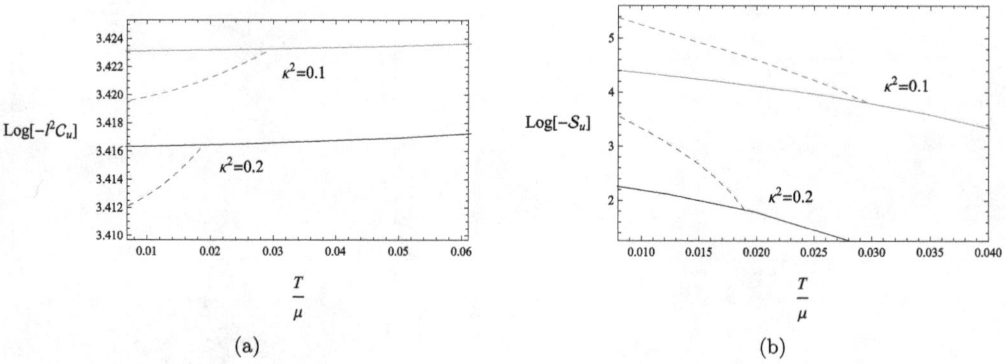

Fig. 3. The logarithms of $-l^2\mathcal{C}_u$ and $-\mathcal{S}_u$ plotted against T/μ. The top set of curves correspond to $\kappa^2 = 0.1$, while the lower set of curves correspond to $\kappa^2 = 0.2$. The curves for $\kappa^2 = 0.2$ have been shifted vertically to fit both set of curves into a single plot.

gles between the normal phase and superconducting phase (for both HEE and HC) increases as we increase the strength of backreaction, κ^2. (In order to fit both set of curves into a single plot, we have shifted the curves for $\kappa^2 = 0.2$ vertically.) Thus, for a fixed value of T/μ, a larger value of κ^2 means that there is a larger difference between the values of HEE and HC of the normal phase compared to those of the superconducting phase.

We now compare our results to Roy and Sarkar [26], and Momeni et al. [25]. Roy and Sarkar found that HC contains the same information as HEE, as far as phase transitions are concerned. To be more precise, they investigated the phase transition of a $(3+1)$-dimensional spherical Reissner–Noström AdS black hole in Section 5 of their work [26], in which they plotted the graphs of renormalized complexity for fixed charge ensemble and fixed opening angle θ_0 (the entangling region being a spherical cap defined by $\theta \leqslant \theta_0$). It turned out that complexity behaves in the same way as entanglement entropy, whose plots are shown in Section 5 of [27]. In our case however, it is clear that Fig. 1 are not similar to Fig. 2: $\mathcal{S}_u$ increases with T/μ, while $\mathcal{C}_u$ decreases with T/μ. So our result shows that HC and HEE need *not* behave in the same manner in the context of phase transitions (in either the normal phase or the superconducting phase). This does not contradict the results in [26] since there are a few differences between our set-up and theirs: our bulk spacetime is $(2+1)$-dimensional, and our subsystem is strip-shaped, whereas in [26] they considered a $(3+1)$-dimensional bulk with spherical horizon, and furthermore their subsystem is circular. Note that HC and HEE behave differently in our work *even during normal phase*, so this suggests that the differences between our results and that of [26] is not simply due to us considering a different kind of phase transition (superconductor instead of a thermodynamical one).

On the other hand, our result, which shows no divergence in the behavior of HC, is clearly inconsistent with Momeni et al. [25]. Going back to their analytic calculation, we found that their analysis was not performed carefully. The expression of HC is, up to a positive dimensionful factor, of the form (see Eq. (28) of their work for the full expression)

$$\mathcal{C}|_{T\to T_c} \sim \left[\frac{1}{\mu-\mu_c}\left(\frac{T}{T_0}-1\right)+\text{const.}\right]^2 + \cdots, \tag{22}$$

which they claimed to be divergent in the limit $\mu \to \mu_c$. Here T_0 is the Hawking temperature of a pure BTZ black hole. However, according to the discussion below Eq. (24) of their paper, one also has $T_0 \to T_c$ near the critical point, so that in the phase transition limit the expression above becomes indeterminate. One therefore cannot conclude whether there is a divergent behavior from this analysis alone, though it potentially could still happen.

Our numerical work shows that this does not happen. Indeed, the location of the black hole horizon is, up to a positive dimensionful factor,[4]

$$z_+ \sim \left[\frac{1}{\mu-\mu_c}\left(\frac{T}{T_0}-1\right)+\text{const.}\right]^{-1}. \tag{23}$$

(See Eq. (27) of [25]; but with their coordinate $r = 1/z$.) This expression will tend to zero if HC is indeed divergent. However, to be well-defined in the context of holography, the horizon should be well inside the bulk and thus z_+ should always be bounded away from 0. This is only possible if the potential divergence is avoided due to an indeterminate form as remarked above.

4. Summary and discussion

In this work, we conducted a numerical analysis of the holographic complexity (HC) and the holographic entanglement entropy (HEE) for a fully-backreacted 1-dimensional holographic superconductor. We showed that both quantities reflect the presence of a phase transition. We found no divergent behavior in the HC during phase transition, contrary to the claim in [25], whose analytic analysis contains a mistake. Despite these mistakes, the analytic work of [25] is nontrivial and important.

Furthermore, our results demonstrated that in the context of a 1-dimensional holographic superconductor, the universal part of the entanglement entropy, $\mathcal{S}_u$, is increasing with T/μ. On the other hand, the universal part of the complexity, $\mathcal{C}_u$, is a decreasing function of T/μ. Therefore, these two quantities behave quite differently for phase transitions that involve a superconducting phase, in contrast to thermodynamical phase transitions that were investigated in [26], in which HC and HEE behave, qualitatively, in the same manner.

Some physical interpretations are useful at this point. During the normal phase, as temperature decreases, order in the system increases due to cooling, and so the effective degrees of freedom also decreases. This is consistent with the decrease in the entanglement entropy, since entanglement entropy *is* a measure of the degrees of freedom in the field theory. Note that in the normal phase, the order parameter $\langle\mathcal{O}_+\rangle$ is zero because ψ is zero. The quantity $\langle\mathcal{O}_+\rangle$ only determines the condensation in the superconducting phase. For superconducting phase, HEE decreases as T/μ decreases because HEE is related to the degrees of freedom in the field theory, and so as Cooper pairs formed, it is expected that

[4] In our setup, as mentioned in the previous section, we have fixed both the horizon and charge of the scalar field to be unity by utilizing scaling symmetries. Thus, Eq. (23) only applies to the work of [25], not the ones carried out in this paper.

HEE would decrease. For the same reason, the HEE for the superconducting phase is lower than that of the normal phase [40].

On the other hand, HC is related to the number of unitary operators that are required to reach some quantum state [13]. A larger (but finite) complexity at lower temperatures is therefore related to the quantum state of the system becoming more complicated towards the critical temperature. The detailed underlying physics remains to be investigated. Regardless, both $\mathcal{C}_u$ and $\mathcal{S}_u$ reflect the presence of superconducting phase transition, *at the same critical temperature*. We also found that, with T/μ fixed, increasing κ^2 means that there is a larger difference between the values of HEE and HC of the normal phase compared to the superconducting phase. See Fig. 3.

These findings indicate that HC and HEE can behave in different ways, in both the normal phase as well as the superconducting phase. Since the results of [26] showed that during a thermodynamical phase transition of a different system, HC and HEE *do* behave in the same manner, it would be interesting to further investigate the sufficient and necessary conditions for HC and HEE to behave in the same manner, as well as the effects of different spacetime dimensions and the geometry of the underlying subsystem on the behaviors of HC and HEE.

Lastly, it would also be interesting to compare the behavior of HC to that of fidelity susceptibility during phase transitions of various systems, so as to further investigate the recently proposed connection between holographic complexity and (reduced) fidelity susceptibility [18,19], or "RFS/HC duality" for short.

Holographic entanglement entropy has proven to be a useful concept, which has allowed us to further understand the quantum information theoretic aspects of gravity and holography. We expect that holographic complexity, being a relatively new and novel concept, has the potential to offer even more nontrivial insights into the deep and subtle connection between gravity and field theory [41].

Acknowledgements

MKZ would like to thank Shanghai Jiao Tong University for the warm hospitality during his visit. He also thanks the Research Council of Shiraz University. The work of MKZ has been supported financially by Research Institute for Astronomy & Astrophysics of Maragha (RIAAM) under research project No. 1/4717-169. YCO and BW acknowledge the support by the National Natural Science Foundation of China (NNSFC). The authors thank Xiao-Mei Kuang for useful discussions.

Appendix A. HC for normal phase without backreaction

In this appendix, we shall discuss the diverging term of the holographic complexity for a strip-shaped subregion. Related discussions concerning a ball-shaped subregion can be found in, e.g., [6,19].

Setting $l = 1$, let us consider the normal phase case ($\psi = \chi = 0$), of which we know the solution,

$$f(z) = z^{-2} - 1 + \kappa^2\mu^2 \ln(z) \quad \text{and} \quad \phi(z) = \mu \ln(z).$$

In the probe limit $\kappa \to 0$, i.e., in the absence of backreaction, one finds:

$$x_+(z) = \int_z^{z_*} \frac{dz}{\sqrt{(z_*^2 - z^2) f}} = \int_z^{z_*} \frac{dz}{\sqrt{(z_*^2 - z^2)(z^{-2} - 1)}} = \coth^{-1}\left(\sqrt{\frac{1 - z^2}{z_*^2 - z^2}}\right). \tag{A.1}$$

Notice that $x_+(z_*) = 0$ as required. By setting the strip length as L, we can find z_* via the boundary condition $x_+(\epsilon \to 0) = L/2$. Finally, using Eq. (A.1), we can calculate HC as

$$\frac{l^2\kappa^2\mathcal{C}}{2} = \int_{\epsilon\to 0}^{z_*} \frac{x_+(z)\,dz}{z^3\sqrt{f}} = \int_{\epsilon\to 0}^{z_*} \frac{\coth^{-1}\left(\sqrt{\frac{1-z^2}{z_*^2-z^2}}\right) dz}{z^3\sqrt{(z^{-2} - 1)}} = -\frac{\pi}{2} + \frac{\tanh^{-1}(z_*)}{\epsilon} + O(\epsilon). \tag{A.2}$$

If one performs the same procedure for pure AdS background, in which $f = z^{-2}$, one will find the diverging term in this case to be z_*/ϵ, which is obviously different from the result above.

This shows that the diverging term of HC for a generic asymptotically AdS geometry is not the same as that of a pure AdS spacetime. In the case of nonvanishing κ, we have a logarithmic term in $f(z)$, and so it is not possible to find the diverging term analytically. The same problem also arises in the superconducting phase. In Sec. 2, we have explained how to overcome this problem numerically.

References

[1] Juan M. Maldacena, The large-N limit of superconformal field theories and supergravity, Adv. Theor. Math. Phys. 2 (1998) 231, arXiv:hep-th/9711200.
[2] Daniel Harlow, Jerusalem lectures on black holes and quantum information, Rev. Mod. Phys. 88 (2016) 15002, arXiv:1409.1231 [hep-th].
[3] Netta Engelhardt, Aron C. Wall, Extremal surface barriers, J. High Energy Phys. 1403 (2014) 068, arXiv:1312.3699 [hep-th].
[4] Shinsei Ryu, Tadashi Takayanagi, Holographic derivation of entanglement entropy from AdS/CFT, Phys. Rev. Lett. 96 (2006) 181602, arXiv:hep-th/0603001.
[5] Shinsei Ryu, Tadashi Takayanagi, Aspects of holographic entanglement entropy, J. High Energy Phys. 0608 (2006) 045, arXiv:hep-th/0605073.
[6] Mohsen Alishahiha, Holographic complexity, Phys. Rev. D 92 (2015) 126009, arXiv:1509.06614 [hep-th].
[7] Don N. Page, Average entropy of a subsystem, Phys. Rev. Lett. 71 (1993) 1291, arXiv:gr-qc/9305007.
[8] Don N. Page, Information in black hole radiation, Phys. Rev. Lett. 71 (1993) 3743, arXiv:hep-th/9306083.
[9] Don N. Page, Time dependence of Hawking radiation entropy, J. Cosmol. Astropart. Phys. 1309 (2013) 028, arXiv:1301.4995 [hep-th].
[10] Daniel Harlow, Patrick Hayden, Quantum computation vs. firewalls, J. High Energy Phys. 06 (2013) 085, arXiv:1301.4504 [hep-th].
[11] Leonard Susskind, Black hole complementarity and the Harlow–Hayden conjecture, arXiv:1301.4505 [hep-th].
[12] Yen Chin Ong, Brett McInnes, Pisin Chen, Cold black holes in the Harlow–Hayden approach to firewalls, Nucl. Phys. B 891 (2015) 627, arXiv:1403.4886 [hep-th].
[13] Leonard Susskind, Computational complexity and black hole horizons, arXiv:1402.5674 [hep-th].
[14] Adam R. Brown, Daniel A. Roberts, Leonard Susskind, Brian Swingle, Ying Zhao, Holographic complexity equals bulk action?, Phys. Rev. Lett. 116 (2016) 191301, arXiv:1509.07876 [hep-th].
[15] Adam R. Brown, Daniel A. Roberts, Leonard Susskind, Brian Swingle, Ying Zhao, Complexity, action, and black holes, Phys. Rev. D 93 (2016) 086006, arXiv:1512.04993 [hep-th].
[16] Omer Ben-Ami, Dean Carmi, On volumes of subregions in holography and complexity, J. High Energy Phys. 11 (2016) 129, arXiv:1609.02514 [hep-th].
[17] Davood Momeni, Mir Faizal, Kairat Myrzakulov, Ratbay Myrzakulov, Fidelity susceptibility as holographic PV-criticality, Phys. Lett. B 765 (2017) 154, arXiv:1604.06909 [hep-th].
[18] Mario Flory, A complexity/fidelity susceptibility g-theorem for $AdS_3/BCFT_2$, arXiv:1702.06386 [hep-th].
[19] Wen-Cong Gan, Fu-Wen Shu, Holographic complexity: a tool to probe the property of reduced fidelity susceptibility, arXiv:1702.07471 [hep-th].
[20] Masamichi Miyaji, Tokiro Numasawa, Noburo Shiba, Tadashi Takayanagi, Kento Watanabe, Gravity dual of quantum information metric, Phys. Rev. Lett. 115 (2015) 261602, arXiv:1507.07555 [hep-th].
[21] Lei Wang, Ye-Hua Liu, Jakub Imriška, Ping Nang Ma, Matthias Troyer, Fidelity susceptibility made simple: a unified quantum Monte Carlo approach, Phys. Rev. X 5 (2015) 031007, arXiv:1502.06969 [cond-mat.stat-mech].
[22] Huan-Qiang Zhou, John Paul Barjaktarevič, Fidelity and quantum phase transitions, arXiv:cond-mat/0701608.

[23] Huan-Qiang Zhou, Renormalization group flows and quantum phase transitions: fidelity versus entanglement, arXiv:0704.2945 [cond-mat.stat-mech].
[24] Shi-Jian Gu, Fidelity approach to quantum phase transitions, Int. J. Mod. Phys. B 24 (2010) 4371, arXiv:0811.3127 [quant-ph].
[25] Davood Momeni, Seyed Ali Hosseini Mansoori, Ratbay Myrzakulov, Holographic complexity in gauge/string superconductors, Phys. Lett. B 756 (2016) 354, arXiv:1601.03011 [hep-th].
[26] Pratim Roy, Tapobrata Sarkar, A note on subregion holographic complexity, arXiv:1701.05489 [hep-th].
[27] Clifford V. Johnson, Large *N* phase transitions, finite volume, and entanglement entropy, J. High Energy Phys. 1403 (2014) 047, arXiv:1306.4955 [hep-th].
[28] Davood Momeni, Mir Faizal, Ratbay Myrzakulov, Holographic cavalieri principle as a universal relation between holographic complexity and holographic entanglement entropy, arXiv:1703.01337 [hep-th].
[29] Dean Carmi, Robert C. Myers, Pratik Rath, Comments on holographic complexity, J. High Energy Phys. 1703 (2017) 118, arXiv:1612.00433 [hep-th].
[30] Dorit Aharonov, Alexei Kitaev, Noam Nisan, Quantum circuits with mixed states, in: Proceedings of the Thirtieth Annual ACM Symposium on Theory of Computing, 1998, pp. 20–30, arXiv:quant-ph/9806029.
[31] John Watrous, Quantum computational complexity, in: R.A. Meyers (Ed.), Encyclopedia of Complexity and Systems Science, 2009, pp. 7174–7201, arXiv: 0804.3401 [quant-ph].
[32] Gary T. Horowitz, Introduction to holographic superconductors, arXiv: 1002.1722 [hep-th].
[33] Jie Ren, One-dimensional holographic superconductor from AdS_3/CFT_2 correspondence, J. High Energy Phys. 1011 (2010) 055, arXiv:1008.3904 [hep-th].
[34] Yunqi Liu, Qiyuan Pan, Bin Wang, Holographic superconductor developed in BTZ black hole background with backreactions, Phys. Lett. B 702 (2011) 94, arXiv:1106.4353 [hep-th].
[35] Sean A. Hartnoll, Christopher P. Herzog, Gary T. Horowitz, Building an AdS/CFT superconductor, Phys. Rev. Lett. 101 (2008) 031601, arXiv:0803.3295 [hep-th].
[36] Sean A. Hartnoll, Christopher P. Herzog, Gary T. Horowitz, Holographic superconductors, J. High Energy Phys. 0812 (2018) 015, arXiv:0810.1563 [hep-th].
[37] Luke Barclay, Ruth Gregory, Sugumi Kanno, Paul Sutcliffe, Gauss-Bonnet holographic superconductors, J. High Energy Phys. 1012 (2010) 029, arXiv: 1009.1991 [hep-th].
[38] Qiyuan Pan, Jiliang Jing, Bin Wang, Songbai Chen, Analytical study on holographic superconductors with backreactions, J. High Energy Phys. 06 (2012) 087, arXiv:1205.3543 [hep-th].
[39] Run-Qiu Yang, Chao Niu, Keun-Young Kim, Surface counterterms and regularized holographic complexity, arXiv:1701.03706 [hep-th].
[40] Xiao-Mei Kuang, Eleftherios Papantonopoulos, Bin Wang, Entanglement entropy as a probe of the proximity effect in holographic superconductors, J. High Energy Phys. 05 (2014) 130, arXiv:1401.5720 [hep-th].
[41] Leonard Susskind, Entanglement is not enough, arXiv:1411.0690 [hep-th].

Entanglement entropy in a non-conformal background

M. Rahimi, M. Ali-Akbari *, M. Lezgi

Department of Physics, Shahid Beheshti University G.C., Evin, Tehran 19839, Iran

ARTICLE INFO

Editor: N. Lambert

ABSTRACT

We use gauge-gravity duality to compute entanglement entropy in a non-conformal background with an energy scale Λ. At zero temperature, we observe that entanglement entropy decreases by raising Λ. However, at finite temperature, we realize that both $\frac{\Lambda}{T}$ and entanglement entropy rise together. Comparing entanglement entropy of the non-conformal theory, $S_{A(N)}$, and of its conformal theory at the UV limit, $S_{A(C)}$, reveals that $S_{A(N)}$ can be larger or smaller than $S_{A(C)}$, depending on the values of Λ and T.

1. Introduction

The AdS/CFT correspondence states that type IIB string theory on the $AdS_5 \times S^5$ background is dual to $\mathcal{N} = 4$ $SU(N_c)$ superconformal gauge theory in a four-dimensional Minkowski space-time living on the boundary of the AdS_5 background [1,2]. This outstanding correspondence is a strong–weak duality which makes it possible to investigate various strongly coupled systems. As a matter of fact, in the large number of colors and large 't Hooft coupling constant limit, the gauge theory is still a quantum theory but strongly coupled. However, the string theory reduces to a classical gravity which is a weakly coupled theory. Therefore, different questions in the strongly coupled gauge theory can be translated into corresponding problems in the classical gravity. This duality has been frequently applied to study various aspects of the strongly coupled systems such as quantum chromodynamics, quark–gluon plasma and condense matter, for instance see [3–5].

Since the AdS/CFT correspondence, or more generally gauge-gravity duality, applies to the non-conformal gauge theories as well as conformal ones, studying various effects of the non-conformal behavior on the physical quantities is always an attractive problem. A new family of solutions of a five-dimensional gravity model, including Einstein gravity coupled to a scalar field with a non-trivial potential, has been recently introduced and studied in [6]. The corresponding four-dimensional strongly coupled gauge theory is not conformal and the theory has conformal fixed points at IR as well as at UV. This means that these solutions are asymptotic to the AdS_5 in the UV and IR limits with different radii. Different properties of the above background such as thermodynamics and relaxation channels have been studied [6].

One interesting physical quantity, on the gauge theory side, is entanglement entropy [8]. In the literature, gauge-gravity duality has been applied to investigate entanglement entropy successfully. For example, entanglement entropy is also helpful to probe a confinement–deconfinement phase transition at zero temperature in confining theories [11]. Then search for transition has also been extended to non-conformal gauge theories at finite temperature [12]. It is shown that no transition takes place at finite temperature. In this paper we study the effect of introducing an energy scale on the entanglement entropy and to check the possibility of a phase transition in such a case.[1]

2. Model

Here we review the non-conformal background introduced in [6]. The background is a solution of five-dimensional gravity theory coupled to a scalar field with a non-trivial potential. The action of the gravity theory is given by

$$S = \frac{2}{G_5^2} \int d^5x \sqrt{-g} \left(\frac{1}{4}\mathcal{R} - \frac{1}{2}(\nabla\phi)^2 - V(\phi) \right), \tag{1}$$

where G_5 is the five-dimensional Newton constant. The particular form of the potential is

* Corresponding author.
E-mail addresses: me_rahimi@sbu.ac.ir (M. Rahimi), m_aliakbari@sbu.ac.ir (M. Ali-Akbari), mahsalezgee@yahoo.com (M. Lezgi).

[1] It is also important to notice that, in [13], it is argued that entanglement entropy is more relevant timescale for the approach to equilibrium than two-point function and Wilson loop.

$$R_{UV}^2 V = -3 - \frac{3\phi^2}{2} - \frac{\phi^4}{3} + \left(\frac{1}{3\phi_M^2} + \frac{1}{2\phi_M^4}\right)\phi^6 - \frac{\phi^8}{12\phi_M^4}. \quad (2)$$

An important point is that the potential has a maximum at $\phi = 0$ and a minimum at $\phi = \phi_M$. It is shown that the resulting solution is asymptotically AdS_5 in the UV($\phi = 0$) limit with radius R_{UV}. Moreover the solution near $\phi = \phi_M$, in the IR limit, approaches AdS_5 as well with a different radius R_{IR}. The relation between the radii of the AdS_5 backgrounds is given by

$$R_{\rm IR} = \frac{1}{1 + \frac{\phi_M^2}{6}} R_{\rm UV}, \quad (3)$$

which clearly indicates that $R_{\rm IR} < R_{\rm UV}$. According to gauge-gravity duality, the number of degrees of freedom in the gauge theory is related to the radius of the background. Thus a smaller number of degrees of freedom lives in the IR limit.

The vacuum solution for arbitrary ϕ_M can be analytically expressed in the form

$$ds^2 = e^{2A(r)}\left(-dt^2 + dx^2\right) + dr^2, \quad (4)$$

where

$$e^{2A} = \frac{\phi_0^2}{\phi^2}\left(1 - \frac{\phi^2}{\phi_M^2}\right)^{\frac{\phi_M^2}{6}+1} e^{-\frac{\phi^2}{6}},$$

$$\phi(r) = \frac{\phi_0 e^{\frac{-r}{R_{uv}}}}{\sqrt{1 + \frac{\phi_0^2}{\phi_M^2} e^{\frac{-2r}{R_{uv}}}}}. \quad (5)$$

ϕ_0 is a constant corresponding to the source Λ of the scalar operator on the gauge theory side. It is also related to an energy scale Λ via $\Lambda = \phi_0 / R_{UV}$. After two successive change of coordinates as follows.

$$u = e^{-r/R_{uv}},$$

$$z(u) = \int_0^u du \frac{R_{UV}}{u} e^{-A}, \quad (6)$$

we finally obtain

$$ds^2 = \frac{R_{\rm eff}(z)^2}{z^2}\left(-dt^2 + dx^2 + dz^2\right), \quad (7)$$

where $R_{\rm eff}(z) = ze^A$. Parameter $R_{\rm eff}(z)$ varies between the radii of the AdS_5 backgrounds in the UV and IR limits corresponding to $R_{\rm UV}$ and $R_{\rm IR}$, respectively.

At finite temperature, the solution in the Eddington–Finkelstein coordinate is

$$ds^2 = e^{2A}\left(-h(\phi)d\tau^2 + dx^2\right) - 2e^{A+B} R_{UV} d\tau d\phi, \quad (8)$$

with $h(\phi)$ vanishing at the horizon, i.e. $h(\phi_h) = 0$. Solving Einstein's equations obtained from (1), the different metric components are given by [6]

$$A(\phi) = -\log\left(\frac{\phi}{\phi_0}\right) + \int_0^\phi d\tilde{\phi}\left(G(\tilde{\phi}) + \frac{1}{\tilde{\phi}}\right), \quad (9a)$$

$$B(\phi) = \log(|G(\phi)|) + \int_0^\phi d\tilde{\phi} \frac{2}{3G(\tilde{\phi})}, \quad (9b)$$

$$h(\phi) = -\frac{e^{2B(\phi)} L^2 \left(4V(\phi) + 3G(\phi)V'(\phi)\right)}{3G'(\phi)}. \quad (9c)$$

The function G must satisfy the following equation

$$\frac{G'(\phi)}{G(\phi) + \frac{4V(\phi)}{3V'(\phi)}} = \frac{d}{d\phi}\log\left[\frac{1}{3G(\phi)} - 2G(\phi) + \frac{G'(\phi)}{2G(\phi)} - \frac{G'(\phi)}{2\left(G(\phi) + \frac{4V(\phi)}{3V'(\phi)}\right)}\right], \quad (10)$$

where, using the Einstein's equations, its behavior near horizon is

$$G(\phi) = -\frac{4V(\phi_H)}{3V'(\phi_H)} + \frac{2}{3}(\phi - \phi_H)\left(\frac{V(\phi_H)V''(\phi_H)}{V'(\phi_H)^2} - 1\right), \quad (11)$$

up to second order in $\phi - \phi_H$. Computing the Hawking temperature of the above solution yields

$$\frac{T}{\Lambda} = -\frac{R_{UV}^2 V(\phi_H)}{3\pi\phi_H} \exp\left\{\int_0^{\phi_H} d\phi\left(G(\phi) + \frac{1}{\phi} + \frac{2}{3G(\phi)}\right)\right\}. \quad (12)$$

Based on numerical results [6], at high temperature ($T \gg \Lambda$), the gauge theory behaves as a conformal theory although the trace of the stress tensor is not zero. In the opposite limit, at low temperature ($T \ll \Lambda$), the gauge theory is conformal as well. Furthermore, it is clearly seen that ϕ_M evaluates the non-conformality of the theory or in the other words the larger ϕ_M the larger deviation from conformality. Finally a significant result is time ordering of relaxation times. In fact the system under consideration can either be firstly isotropised, meaning that all pressures become equal or be firstly equilibrated, meaning that the equation of state becomes applicable. Quasi-normal mode calculation indicates that at high temperature the system first isotropises and subsequently equilibrates. At low temperature the time ordering is reversed. For more details, we refer the reader to the original paper.

Before closing this part, we would like to emphasize that various time scales of relaxation for an out-of-equilibrium system has been firstly studied in [7].

3. Entanglement entropy

Let us consider a quantum system with many degrees of freedom at zero temperature which is described by a pure ground state $|\psi\rangle$. Therefore the density matrix is given by $\rho = |\psi\rangle\langle\psi|$. One can divide the mentioned system into two subsystems A and B. The observer who is restricted to live in the subsystem A does not have access to the degrees of freedom of subsystem B. Thus its density matrix can be found by taking the trace over these degrees of freedom, i.e. $\rho_A = tr_B \rho$. Then entanglement entropy of the subsystem A is defined as $S_A = -tr_A(\rho_A \log \rho_A)$. This quantity states how much information is lost when an observer is restricted to the subsystem A. For a gauge theory in $d > 2$ space-time, dimension the leading divergence of S_A is proportional to the area of the subsystem A. For a two-dimensional conformal gauge theory, where the subsystem A is an interval of length l, the entanglement entropy can be analytically calculated as a universal result $S_l = \frac{c}{3}\log(\frac{l}{a})$ where c and a are central charge and the UV cut-off of the field theory, respectively.

On the holographic side, entanglement entropy calculation has a simple prescription. It is proposed that entanglement entropy S_A can be computed from the following formula

$$S_A = \frac{\text{Area}(\gamma_A)}{4G_5}, \quad (13)$$

where γ_A is a three-dimensional minimal area surface in asymptomatically AdS_5 background whose boundary is given by ∂A

(which is the boundary of the subsystem A). This prescription perfectly produces well-known results, such as entanglement entropy in two-dimensional conformal field theory, and therefore it is reliable to compute the entanglement entropy in the strongly coupled gauge theories. For more details see for example [8–10] and references therein.

4. Numerical results

We start with a general form for the background as

$$ds^2 = -f_1(z)dt^2 + f_2(z)dz^2 + f_3(z)d\vec{x}^2, \tag{14}$$

where z is the radial direction. The background is asymptotically AdS_5 and its boundary is located at $z = 0$. In order to determine entanglement entropy, we have to divide the boundary region into two subsystems A and B. Subsystem B is defined by $-\frac{l}{2} < x_1(\equiv x) < \frac{l}{2}$ and $x_2, x_3 \in (-\infty, +\infty)$ at a given time. Then the minimal area of γ_A, which is proportional to entanglement entropy of subsystem A, is obtained by minimizing the following area

$$S_A = \frac{1}{4G_5}\int d^3x\sqrt{g_{in}}, \tag{15}$$

where g_{in} is induced metric on γ_A. Using (14), it is easy to see that

$$S_A = \frac{V_2}{4G_5}\int_{\frac{-l}{2}}^{\frac{l}{2}} dx\sqrt{f_3^3(z) + f_3^2(z)f_2(z)z'^2}, \tag{16}$$

where V_2 is the area of two-dimensional surface of x_2 and x_3 and $z' = \frac{dz}{dx}$. The above area does not depend on x explicitly and thus the corresponding Hamiltonian is a constant of motion

$$\frac{f_3^2(z)}{\sqrt{f_3(z) + f_2(z)z'^2}} = \text{const} = f_3^{\frac{3}{2}}(z_*), \tag{17}$$

where z_* is the minimal value of z, i.e. $z(x = 0) = z_*$, and $z'(x = 0) = 0$. Hence, from (17), we find

$$z' = \sqrt{\frac{f_3(z)}{f_2(z)}}\sqrt{\frac{f_3^3(z)}{f_3^3(z_*)} - 1}, \tag{18}$$

and then the relation between l and z_* can be easily obtained

$$l = 2\int_0^{z_*}\sqrt{\frac{f_2(z)}{f_3(z)}}\frac{dz}{\sqrt{\frac{f_3^3(z)}{f_3^3(z_*)} - 1}}. \tag{19}$$

Finally, by substituting (18) in (16), we have

$$S_A = \frac{V_2}{2G_5}\int_0^{z_*}\frac{f_3^{\frac{5}{2}}(z)f_2^{\frac{1}{2}}(z)}{\sqrt{f_3^3(z) - f_3^3(z_*)}}dz. \tag{20}$$

As it is obvious the factor behind the integral in (20) is a constant and as a result we only need to compute the integral which is proportional to the entanglement entropy.

At zero temperature, according to (7), it is clear that $f_1(z) = f_2(z) = f_3(z) = \frac{R_{eff}^2(z)}{z^2}$. In Fig. 1, we have plotted the difference of the entanglement entropies for non-conformal and conformal (in the UV limit) field theories, i.e. $\Delta S \equiv G_5\Delta S = S_{A(N)} - S_{A(C)}$, as a function of l. Since we would like that a sizable non-conformality emerge in the field theory, the value of ϕ_M is chosen to be large. As usual, the non-conformal field theory becomes conformal in the UV limit, which is probed by very small l, and therefore ΔS must go to zero in this limit. In other words, on the gravity side, z_* is close to the boundary for very small values of l. Therefore, near boundary region, which is AdS_5 with good accuracy, contributes more to $S_{A(N)}$. By increasing l, z_* reaches deeper in the bulk and thus the deviation of geometry from AdS_5 can be realized further by minimal area or equivalently entanglement entropy. On the field theory side, since in our model the number of degrees of freedom decreases from boundary to the bulk and considering the point that entanglement entropy is proportional to the number of degrees of freedom, for a given value of l the non-conformal entanglement entropy $S_{A(N)}$ is smaller in comparison to its value in the UV limit, $S_{A(C)}$. As a result, ΔS is always negative at zero temperature. For large enough values of l, z_* approaches AdS_5 in the IR limit with smaller radius R_{IR}, meaning that the number of degrees of freedom becomes constant and proportional to R_{IR}. Therefore the difference between the number of degrees of freedom in the UV and IR limit becomes constant and accordingly ΔS goes to a constant value. The above discussion is confirmed by Fig. 1. Furthermore, by increasing energy scale Λ, the entanglement entropy of the non-conformal theory decreases. It is also significant to note that ΔS does not change for large enough values of l.

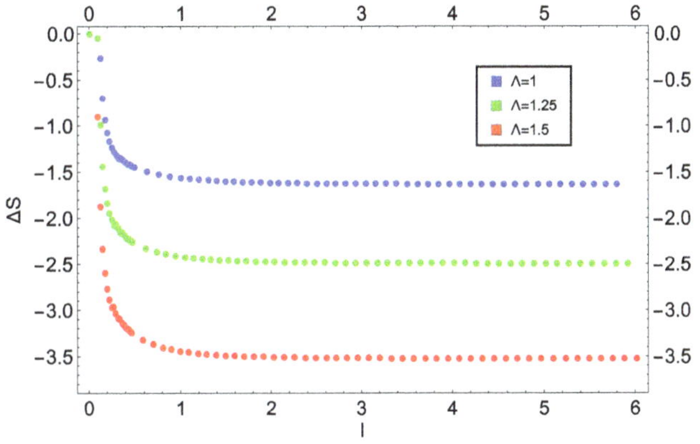

Fig. 1. ΔS for three different energy scales as a function of l with $\phi_M = 100$.

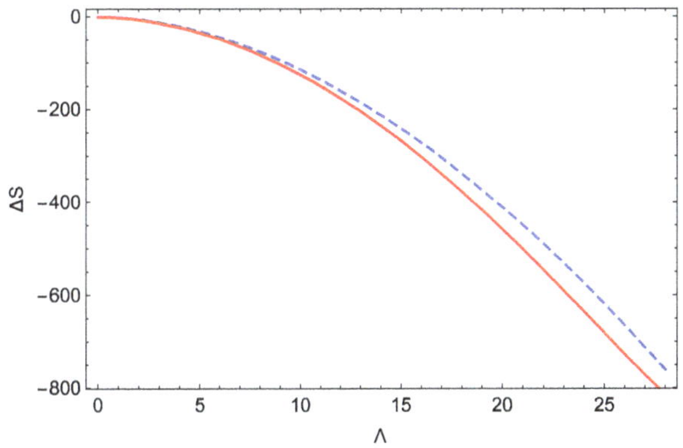

Fig. 2. ΔS in terms of energy scale Λ for $\phi_M = 1$ (blue) and $\phi_M = 100$ (red) and $l = 2.2$. (For interpretation of the references to color in this figure legend, the reader is referred to the web version of this article.)

The mere introduction of an energy scale Λ breaks conformal symmetry in the model and the non-conformality is controlled by ϕ_M. In Fig. 2, for given l, we have plotted ΔS in terms of Λ for two values of ϕ_M. This figure shows that non-conformal entanglement entropy decreases by raising Λ, in agreement with Fig. 1. For small enough values of energy scale, ΔS is almost independent of non-conformality. However, for larger values of Λ there is a decrease in the non-conformal entanglement entropy due to

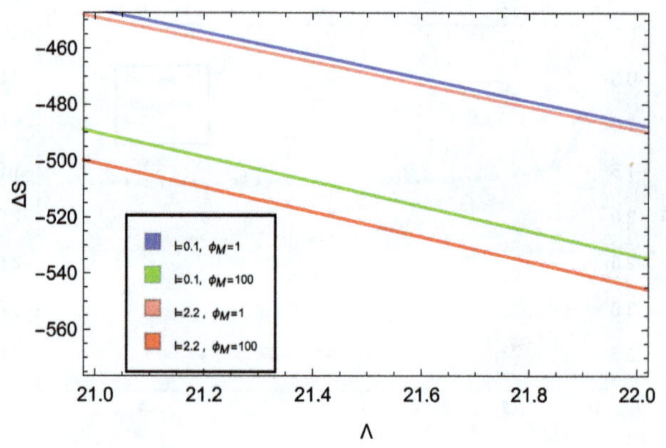

Fig. 3. ΔS in terms of energy scale Λ.

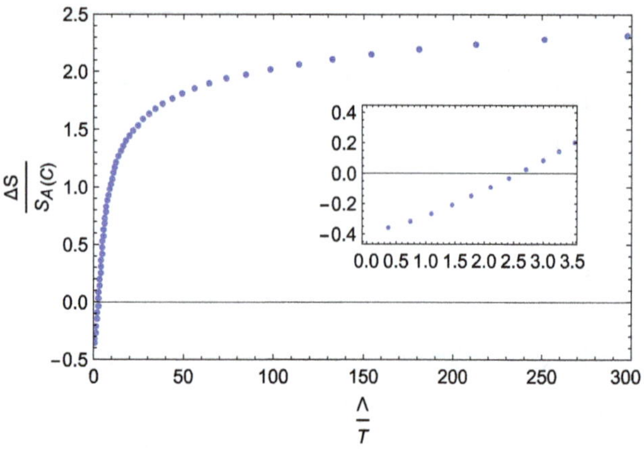

Fig. 4. $\Delta S = S_A(\Lambda = 0.8, \phi_H \geq 0.1) - S_{A(C)}(\phi_H = 0.1)$ as a function of $\frac{\Lambda}{T}$ for $\phi_M = 100$ and $l = 0.21$.

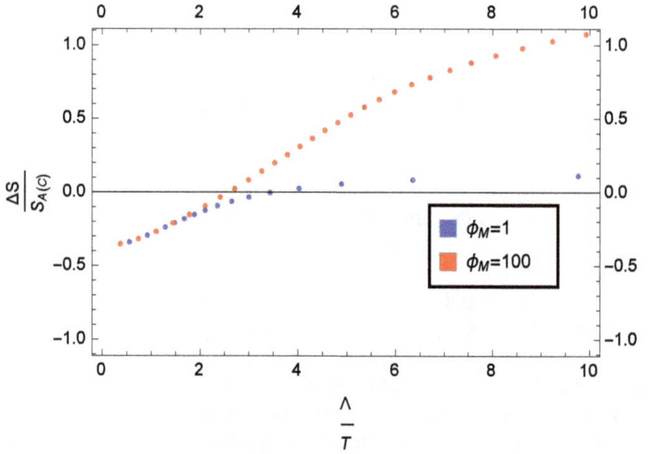

Fig. 5. $\Delta S = S_A(\Lambda = 0.8, \phi_H \geq 0.1) - S_{A(C)}(\phi_H = 0.1)$ as a function of $\frac{\Lambda}{T}$ for $l = 0.21$.

non-conformality. Note that there is a slight difference between the two curves at large Λ. This behavior is confirmed for different values of l by our numerical results. In particular, in Fig. 3, ΔS is plotted in the range of $\Lambda = 21$–22. At fixed Λ, ΔS is smaller for the case with $l = 0.1$, as expected.

At finite temperature, we have

$$f_2(\phi) = \frac{R_{UV}^2}{h(\phi)} e^{2B(\phi)},$$
$$f_3(\phi) = e^{2A(\phi)}. \tag{21}$$

Using (12), (20) and (21), $\frac{\Delta S}{S_{A(C)}}$ is plotted as a function of $\frac{\Lambda}{T}$ for given l. We, first, focus on $\frac{\Lambda}{T} \ll 1$. By carrying out a simple numerical analysis, it turns out that $\frac{\Lambda}{T}$ goes to zero when $\phi_H \to 0$. As a result, for very small values of $\frac{\Lambda}{T}$ the horizon is close to the boundary which makes it difficult to find a reliable numerical solution. But there is an intuitive explanation for this high temperature region. Since in the high temperature limit the non-conformal background approaches AdS_5 with radius R_{UV}, we expect that $\Delta S \to 0$, similar to the zero temperature case. Now let us consider a small enough value for T in such a way that $\Lambda \ll T$. Taking these assumptions, the case at hand and zero temperature case are nearly alike and hence one expects that ΔS increase from negative values to zero by decreasing Λ, as it is clearly seen in Fig. 2.

Apart from the mentioned region in the above paragraph, our numerical results are shown in Fig. 4. By varying $\frac{\Lambda}{T}$ we observe that $\frac{\Delta S}{S_{A(C)}}$ changes sign at a point denoted by $(\frac{\Lambda}{T})_*$. For $\frac{\Lambda}{T} < (\frac{\Lambda}{T})_*$, ΔS is always negative and therefore $S_{A(N)}$ is smaller than $S_{A(C)}$. The discussion for negative values of ΔS is similar to the case with $T = 0$ that we do not repeat it here. At $\frac{\Lambda}{T} = (\frac{\Lambda}{T})_*$, $S_{A(N)} = S_{A(C)}$. Although the number of degrees of freedom is not equal in the non-conformal and conformal field theories, the above equality indicates that $\frac{\Lambda}{T}$ would compensate for the difference in the number of degrees of freedom. For larger values of $\frac{\Lambda}{T}$, ΔS becomes positive and hence $S_{A(N)} > S_{A(C)}$. Note that, on the gravity side, in the low temperature limit, $\frac{\Lambda}{T} \to \infty$, the geometry approaches AdS_5 with radius R_{IR}. It seems that for large enough values of $\frac{\Lambda}{T}$, $S_{A(N)}$ does not vary significantly. We also observe that by increasing Λ, $(\frac{\Lambda}{T})_*$ decreases. Therefore, ΔS is a function of Λ and T and not only $\frac{\Lambda}{T}$.

In Fig. 5 we have plotted the $\frac{\Delta S}{S_{A(c)}}$ in terms of $\frac{\Lambda}{T}$ for two different values of ϕ_M. Since a sizable non-conformality emerges in the field theory for larger values of ϕ_M, ΔS is substantially smaller for $\phi_M = 1$ and large enough values of $\frac{\Lambda}{T}$. When $\frac{\Lambda}{T}$ is small enough, the field theory is almost conformal for both cases and therefore ΔS's, for $\phi_M = 1$ and $\phi_M = 100$, are equal. This figure also indicates that $(\frac{\Lambda}{T})_*^{\phi_M=1} > (\frac{\Lambda}{T})_*^{\phi_M=100}$. Since in our calculation Λ is fixed, one can conclude that $T_*^{\phi_M=1} < T_*^{\phi_M=100}$. In other words, for $\phi_M = 100$, the effect of non-conformality appears at higher temperature which is reasonable.

Another point we would like to make here is the relation between time orderings presented in [6] and our observation of changing sign of ΔS. At high temperatures it is shown that $t_{iso} < t_{eq}$ and in the same limit we see $\Delta S < 0$. In the opposite limit, the situation is reversed, i.e. $t_{iso} > t_{eq}$ and $\Delta S > 0$. Based on these observations, one may speculate that there is a connection between time ordering and the sign of ΔS. This idea needs further investigation which we postpone to a future work.

As a final point, it is instructive to investigate the possibility of a transition due to energy scale Λ. In order to do so, we have to consider two types of surfaces: connected and piecewise smooth. The connected surface has been already obtained in (20) for an arbitrary background (14). The second surface is defined as

$$x = -\frac{l}{2}, \quad \phi = \phi_H, \quad x = \frac{l}{2}, \tag{22}$$

and then, using (15), it is easy to find

$$\hat{S}_A = \frac{V_2}{4G_5}\left(2\int_0^{\phi_H} d\phi\, f_3(\phi)\sqrt{f_2(\phi)} + l\sqrt{f_3^3(\phi_H)}\right). \tag{23}$$

Now let us consider the following expression

$$\bar{\Delta} S = S_A - \hat{S}_A. \tag{24}$$

Computing difference in entanglement entropies in confining theories with gravity duals reveals a deconfinement transition [11]. An extension of above idea to the thermal backgrounds shows that entanglement entropy does not exhibit a phase transition (expect for geometry of the near horizon limit of D6-branes) [12]. In our case, transition does not take place at finite temperature as well as zero temperature. In other words, $\bar{\Delta} S$ is always negative, i.e. $S_A < \hat{S}_A$, and as a result the connected surface is favorable at all times.

Acknowledgements

We would like to thank School of Physics of Institute for research in fundamental sciences (IPM) for the research facilities and environment. The authors would also like to thank H. Ebrahim, M.R. Mohammadi Mozaffar and M.M. sheikh Jabbari for useful comments.

References

[1] J.M. Maldacena, The large N limit of superconformal field theories and supergravity, Adv. Theor. Math. Phys. 2 (1998) 231, arXiv:hep-th/9711200; Int. J. Theor. Phys. 38 (1999) 1113;
S.S. Gubser, I.R. Klebanov, A.M. Polyakov, Gauge theory correlators from non-critical string theory, Phys. Lett. B 428 (1998) 105, arXiv:hep-th/9802109;
E. Witten, Anti-de Sitter space and holography, Adv. Theor. Math. Phys. 2 (1998) 253, arXiv:hep-th/9802150.

[2] M. Ammon, J. Erdmenger, Gauge/Gravity Duality: Foundations and Applications, Cambridge University Press, Cambridge, UK, 2015.

[3] J. Casalderrey-Solana, H. Liu, D. Mateos, K. Rajagopal, U.A. Wiedemann, Gauge/string duality, hot QCD and heavy ion collisions, arXiv:1101.0618 [hep-th].

[4] M. Natsuume, AdS/CFT Duality User Guide, Lect. Notes Phys., vol. 903, 2015, arXiv:1409.3575 [hep-th].

[5] G. Camilo, Expanding plasmas from Anti de Sitter black holes, arXiv:1609.07116 [hep-th];
M. Ali-Akbari, F. Charmchi, A. Davody, H. Ebrahim, L. Shahkarami, Evolution of Wilson loop in time-dependent $N = 4$ super Yang–Mills plasma, Phys. Rev. D 93 (8) (2016) 086005, arXiv:1510.00212 [hep-th];
S. Amiri-Sharifi, M. Ali-Akbari, A. Kishani-Farahani, N. Shafie, Double relaxation via AdS/CFT, Nucl. Phys. B 909 (2016) 778, arXiv:1601.04281 [hep-th];
M. Ali-Akbari, F. Charmchi, A. Davody, H. Ebrahim, L. Shahkarami, Time-dependent meson melting in an external magnetic field, Phys. Rev. D 91 (2015) 106008, arXiv:1503.04439 [hep-th];
M. Ali-Akbari, D. Allahbakhshi, Meson life time in the anisotropic quark–gluon plasma, J. High Energy Phys. 1406 (2014) 115, arXiv:1404.5790 [hep-th];
M. Ali-Akbari, H. Ebrahim, Chiral symmetry breaking: to probe anisotropy and magnetic field in quark–gluon plasma, Phys. Rev. D 89 (6) (2014) 065029, arXiv:1309.4715 [hep-th].

[6] M. Attems, J. Casalderrey-Solana, D. Mateos, I. Papadimitriou, D. Santos-Oliván, C.F. Sopuerta, M. Triana, M. Zilhão, Thermodynamics, transport and relaxation in non-conformal theories, arXiv:1603.01254 [hep-th].

[7] M. Ali-Akbari, F. Charmchi, H. Ebrahim, L. Shahkarami, Various time-scales of relaxation, Phys. Rev. D 94 (4) (2016) 046008, arXiv:1602.07903 [hep-th].

[8] T. Nishioka, S. Ryu, T. Takayanagi, Holographic entanglement entropy: an overview, J. Phys. A 42 (2009) 504008, arXiv:0905.0932 [hep-th].

[9] M. Srednicki, Entropy and area, Phys. Rev. Lett. 71 (1993) 666, arXiv:hep-th/9303048.

[10] D. Harlow, Jerusalem lectures on black holes and quantum information, Rev. Mod. Phys. 88 (2016) 15002, Rev. Mod. Phys. 88 (2016) 15002, arXiv:1409.1231 [hep-th].

[11] I.R. Klebanov, D. Kutasov, A. Murugan, Entanglement as a probe of confinement, Nucl. Phys. B 796 (2008) 274, arXiv:0709.2140 [hep-th].

[12] I. Bah, A. Faraggi, L.A. Pando Zayas, C.A. Terrero-Escalante, Holographic entanglement entropy and phase transitions at finite temperature, Int. J. Mod. Phys. A 24 (2009) 2703, arXiv:0710.5483 [hep-th].

[13] J.F. Pedraza, Evolution of nonlocal observables in an expanding boost-invariant plasma, Phys. Rev. D 90 (4) (2014) 046010, arXiv:1405.1724 [hep-th].

17

Exact microstate counting for dyonic black holes in AdS_4

Francesco Benini [a,b], Kiril Hristov [c], Alberto Zaffaroni [d,e,*]

[a] SISSA, INFN, Sezione di Trieste, via Bonomea 265, 34136 Trieste, Italy
[b] Blackett Laboratory, Imperial College London, London SW7 2AZ, United Kingdom
[c] INRNE, Bulgarian Academy of Sciences, Tsarigradsko Chaussee 72, 1784 Sofia, Bulgaria
[d] Dipartimento di Fisica, Università di Milano-Bicocca, I-20126 Milano, Italy
[e] INFN, sezione di Milano-Bicocca, I-20126 Milano, Italy

ARTICLE INFO

Editor: N. Lambert

ABSTRACT

We present a counting of microstates of a class of dyonic BPS black holes in AdS_4 which precisely reproduces their Bekenstein–Hawking entropy. The counting is performed in the dual boundary description, that provides a non-perturbative definition of quantum gravity, in terms of a twisted and mass-deformed ABJM theory. We evaluate its twisted index and propose an extremization principle to extract the entropy, which reproduces the attractor mechanism in gauged supergravity.

1. Introduction

Supersymmetric black holes in string theory constitute important models to test fundamental questions about quantum gravity in a relatively simple setting. The main question we would like to address here is the origin of the black hole (BH) entropy, which statistically is expected to count the number of degenerate BH configurations. String theory provides a microscopic explanation for the entropy of a class of asymptotically flat black holes [1]. Much less is known about asymptotically AdS ones in four or more dimensions.

In principle AdS/CFT [2] provides a non-perturbative definition of quantum gravity in asymptotically AdS space, as a dual boundary quantum field theory (QFT). The BH microstates appear as particular states in the boundary description. The difficulty with this approach is the need to perform computations in a strongly coupled QFT, but the development of exact non-perturbative techniques makes progress possible. We recently reported [3] on a particular example of magnetically charged BPS black holes in AdS_4 [4] with a known field theory dual—topologically twisted ABJM theory. Using the technique of supersymmetric localization we were able to calculate in an independent way the (regularized) number of ground states of the theory and successfully match it with the leading macroscopic entropy of the black holes.

In this Letter we discuss a function $Z(u_a)$ that encodes the quantum entropies of static dyonic BPS black holes in AdS_4, computed non-perturbatively from the dual QFT description, and show that its leading behavior reproduces the Bekenstein–Hawking entropy [5,6].

In particular we show that, at leading order, the entropy of BPS black holes with magnetic charges $\mathfrak{p}_a$, electric charges $\mathfrak{q}_a$ and asymptotic to $AdS_4 \times S^7$ can be obtained by extremizing the quantity

$$\mathcal{I} = \log Z(u_a) - i \sum\nolimits_a u_a \mathfrak{q}_a \tag{1}$$

with respect to a set of complexified chemical potentials u_a for the global $U(1)$ flavor symmetries of the boundary theory. $Z(u_a)$ is the topologically twisted index [7] of the ABJM theory [8] which explicitly depends on the magnetic charges $\mathfrak{p}_a$ (see [9,10] for other examples). The entropy is given by $S = \mathcal{I}(\hat{u})$ evaluated at the extremum, with a constraint on the charges that S be real positive.

As we will see, the extremization of $\mathcal{I}$ is equivalent to the attractor mechanism for AdS_4 black holes in gauged supergravity. We also argue, generalizing [3], that the extremization of $\mathcal{I}$ selects the exact R-symmetry of the superconformal quantum mechanics dual to the AdS_2 horizon region. We notice strong similarities between our formalism and those based on Sen's entropy functional [11] and the OSV conjecture [12].

2. The black holes

We consider dyonic BPS BHs that can be embedded in M-theory and are asymptotic to $AdS_4 \times S^7$. They are more easily described as solutions in the STU model, a four-dimensional $\mathcal{N}=2$ gauged supergravity with three vector multiplets, which is a consistent trun-

* Corresponding author.
E-mail addresses: fbenini@sissa.it (F. Benini), khristov@inrne.bas.bg (K. Hristov), alberto.zaffaroni@mib.infn.it (A. Zaffaroni).

cation both of M-theory on S^7, and of the 4d maximal $\mathcal{N}=8$ $SO(8)$ gauged supergravity [13]. The model contains four Abelian vector fields (one is the graviphoton) corresponding to the $U(1)^4 \subset SO(8)$ isometries of S^7.

In 4d $\mathcal{N}=2$ supergravities with n_V vector multiplets, one can use the standard machinery of special geometry [14–16]. The Lagrangian $\mathscr{L}$ of the theory is completely specified by the prepotential $\mathcal{F}(X^\Lambda)$, which is a homogeneous holomorphic function of sections X^Λ, and the vector of Fayet–Iliopoulos (FI) terms $\mathcal{G} = (g^\Lambda, g_\Lambda)$. The symplectic index $\Lambda = 0, 1, \ldots, n_V$ runs over the graviphoton and the n_V vectors in vector multiplets. The scalars z^i in vector multiplets, with $i = 1, \ldots, n_V$, parametrize a special Kähler manifold $\mathcal{M}$ and X^Λ are sections of a symplectic Hodge vector bundle on $\mathcal{M}$. The formalism is covariant with respect to symplectic $Sp(2n_V+2)$ transformations. Indicating as (A^Λ, A_Λ) the $2n_V+2$ components of a symplectic vector A, the scalar product is $\langle A, B\rangle = A_\Lambda B^\Lambda - A^\Lambda B_\Lambda$. One defines the covariantly-holomorphic sections

$$\mathcal{V} = e^{\mathcal{K}(z,\bar{z})/2} \begin{pmatrix} X^\Lambda(z) \\ \mathcal{F}_\Lambda(z) \end{pmatrix} \tag{2}$$

on $\mathcal{M}$, where $\mathcal{K}$ is the Kähler potential and $\mathcal{F}_\Lambda \equiv \partial_\Lambda \mathcal{F}$. They satisfy $D_{\bar{i}}\mathcal{V} \equiv \left(\partial_{\bar{i}} - \frac{1}{2}\partial_{\bar{i}}\mathcal{K}\right)\mathcal{V} = 0$. The Kähler potential is then determined by $\langle \mathcal{V}, \overline{\mathcal{V}}\rangle = -i$.

The ansatz for dyonic black holes is of the form

$$ds^2 = -e^{2U(r)}dt^2 + e^{-2U(r)}\left(dr^2 + V(r)^2 ds^2_{\Sigma_\mathfrak{g}}\right) \tag{3}$$

where $\Sigma_\mathfrak{g}$ is a Riemann surface of genus $\mathfrak{g}$, and the scalar fields z^i are assumed to only have radial dependence. We can write the metric on $\Sigma_\mathfrak{g}$ locally as

$$ds^2_{\Sigma_\mathfrak{g}} = d\theta^2 + f_\kappa^2(\theta)\, d\varphi^2\,, \qquad f_\kappa(\theta) = \begin{cases} \sin\theta & \kappa = 1 \\ \theta & \kappa = 0 \\ \sinh\theta & \kappa = -1 \end{cases} \tag{4}$$

where $\kappa = 1$ for S^2, $\kappa = 0$ for T^2, and $\kappa = -1$ for $\Sigma_\mathfrak{g}$ with $\mathfrak{g} > 1$. The scalar curvature is 2κ and the volume is

$$\mathrm{Vol}(\Sigma_\mathfrak{g}) = 2\pi\eta\,, \qquad \eta = \begin{cases} 2|\mathfrak{g}-1| & \text{for } \mathfrak{g} \neq 1 \\ 1 & \text{for } \mathfrak{g} = 1\,. \end{cases} \tag{5}$$

The magnetic and electric charges of the black hole are

$$\int_{\Sigma_\mathfrak{g}} F^\Lambda = \mathrm{Vol}(\Sigma_\mathfrak{g})\, p^\Lambda\,, \qquad \int_{\Sigma_\mathfrak{g}} G_\Lambda = \mathrm{Vol}(\Sigma_\mathfrak{g})\, q_\Lambda\,, \tag{6}$$

where $G_\Lambda = 8\pi G_N\, \delta(\mathscr{L}\mathrm{dvol}_4)/\delta F^\Lambda$ and G_N is the Newton constant. This particular normalization ensures that the BPS equations are independent of $\mathfrak{g}$ (besides a linear constraint). The charges are collected in the vector $\mathcal{Q} = (p^\Lambda, q_\Lambda)$. The vector $\mathcal{G}$ of FI terms controls the gauging and determines the charges of the gravitini under the gauge fields. In a frame with purely electric gauging g_Λ, the lattice of electro-magnetic charges is

$$\eta\, g_\Lambda\, p^\Lambda \in \mathbb{Z}\,, \qquad \frac{\eta}{4G_N g_\Lambda} q_\Lambda \in \mathbb{Z} \tag{7}$$

not summed over Λ. It turns out that the BPS equations fix the more stringent condition

$$\langle \mathcal{G}, \mathcal{Q}\rangle = -\kappa\,, \tag{8}$$

that we call the linear constraint.

It has been noticed in [17] that the BPS equations of gauged supergravity for the near-horizon geometry can be put in the form of "attractor equations". One defines the central charge of the black hole $\mathcal{Z}$ and the superpotential $\mathcal{L}$:

$$\begin{aligned} \mathcal{Z} &= \langle \mathcal{Q}, \mathcal{V}\rangle = e^{\mathcal{K}/2}\left(q_\Lambda X^\Lambda - p^\Lambda \mathcal{F}_\Lambda\right) \\ \mathcal{L} &= \langle \mathcal{G}, \mathcal{V}\rangle = e^{\mathcal{K}/2}\left(g_\Lambda X^\Lambda - g^\Lambda \mathcal{F}_\Lambda\right). \end{aligned} \tag{9}$$

The BPS equations for the near-horizon geometry

$$ds^2_{\mathrm{nh}} = -\frac{r^2}{R_A^2}dt^2 + \frac{R_A^2}{r^2}dr^2 + R_S^2\, ds^2_{\Sigma_\mathfrak{g}} \tag{10}$$

with constant scalar fields z^i imply the following two equations [17]: $\mathcal{Z} - iR_S^2\mathcal{L} = 0$ and $D_j\left(\mathcal{Z} - iR_S^2\mathcal{L}\right) = 0$, where $D_j = \partial_j + \frac{1}{2}\partial_j\mathcal{K}$, besides $\langle \mathcal{G}, \mathcal{Q}\rangle = -\kappa$. These equations can be rewritten as

$$\partial_j \frac{\mathcal{Z}}{\mathcal{L}} = 0\,, \qquad -i\frac{\mathcal{Z}}{\mathcal{L}} = R_S^2\,. \tag{11}$$

In other words, the scalars z^i at the horizon take a value such that the quantity $-i\mathcal{Z}/\mathcal{L}$ has a critical point on $\mathcal{M}$ and then its value is proportional to the Bekenstein–Hawking black hole entropy.

Notice that a condition to have BHs with smooth horizon is that $-i\mathcal{Z}/\mathcal{L}$ be real positive at the critical point. Since the critical-point equations already fix the values of the scalars, this condition becomes a second (non-linear) constraint on the charges. Therefore the domain of allowed electro-magnetic charges has real dimension $2n_V$ (before imposing quantization). There are other inequalities to be satisfied by the charges, for instance to ensure that also R_A^2 be positive.

In the case of very special Kähler geometry, *i.e.* that the prepotential takes the form $\mathcal{F} = d_{ijk}X^iX^jX^k/X^0$ or symplectic transformations thereof, general solutions to the near-horizon BPS equations as well as full BH solutions have been found in [18–20]. That analysis guarantees that all near-horizon solutions can be completed into full BH solutions.

Our focus is on the STU model, which has $n_V = 3$ and prepotential

$$\mathcal{F} = -2i\sqrt{X^0X^1X^2X^3}\,, \tag{12}$$

with purely electric gauging $g_\Lambda \equiv g$, $g^\Lambda = 0$. Then the AdS$_4$ vacuum has radius $L^2 = 1/2g^2$. Note that all dyonic BH solutions with electric charges have complex profiles for the scalars, *i.e.* the axions are turned on.

3. The dual field theory

M-theory on AdS$_4 \times S^7$ has a dual holographic description as a three-dimensional supersymmetric gauge theory, the ABJM theory [8], which provides a non-perturbative definition thereof. In $\mathcal{N}=2$ notation, the ABJM theory is a $U(N)_1 \times U(N)_{-1}$ Chern–Simons theory (the subscripts are the levels) with bi-fundamental chiral multiplets A_i and B_j, $i, j = 1, 2$, transforming in the $(N, \overline{N})$ and $(\overline{N}, N)$ representations of the gauge group, respectively, and with superpotential $W = \varepsilon^{ik}\varepsilon^{jl}\,\mathrm{Tr}\, A_iB_jA_kB_l$. The theory has $\mathcal{N}=8$ superconformal symmetry and $SO(8)$ R-symmetry. The identification between gravitational and QFT parameters is

$$\frac{L^2}{G_N} = \frac{1}{2g^2G_N} = \frac{2\sqrt{2}}{3}N^{3/2}\,. \tag{13}$$

The "topologically twisted index" of an $\mathcal{N}=2$ three-dimensional theory is its supersymmetric Euclidean partition function on $S^2 \times S^1$ with a topological twist on S^2 [7]. Its higher-genus generalization, namely the twisted partition function on $\Sigma_\mathfrak{g} \times S^1$, has been constructed as well [21,22]. They depend on a set of integer magnetic fluxes $\mathfrak{p}_a$ and complex fugacities y_a, along the Cartan generators of the flavor symmetry group.

In the present case, to make the enhanced symmetry more manifest, we introduce an index $a = 1, 2, 3, 4$ that simultaneously runs over the four ABJM chiral fields and the four Abelian symmetries $U(1)^4 \subset SO(8)$. This is done by introducing a basis of four R-symmetries R_a, each acting with charge 2 on one of the chiral fields and zero on the others. Then the magnetic fluxes identify a $U(1)$ subgroup of $SO(8)$ used to twist, and are required by supersymmetry to satisfy $\sum_a \mathfrak{p}_a = 2g - 2$. The complex fugacities $y_a = \exp iu_a$ must satisfy $\prod_a y_a = 1$ ($\sum_a u_a \in 2\pi\mathbb{Z}$) and encode background values for the flavor symmetries. Writing $u_a = \Delta_a + i\beta\sigma_a$ (where β is the length of S^1), we can identify Δ_a with flavor flat connections and σ_a with real masses.

The Hamiltonian definition of the index is [7]

$$Z(u_a, \mathfrak{p}_a) = \mathrm{Tr}\,(-1)^F\, e^{i\sum_{a=1}^{3}\Delta_a J_a}\, e^{-\beta H}\,, \tag{14}$$

where $J_a = \frac{1}{2}(R_a - R_4)$ are the three independent flavor symmetries and H is the twisted Hamiltonian on S^2, explicitly dependent on the magnetic charges $\mathfrak{p}_a$ and the real masses σ_a. Due to the supersymmetry algebra $Q^2 = H - \sum_{a=1}^{3}\sigma_a J_a$, the index $Z(u_a, \mathfrak{p}_a)$ is a meromorphic function of y_a. For simplicity, we will keep the dependence on $\mathfrak{p}_a$ implicit and use the shorthand notation $\sigma J = \sum_{a=1}^{3}\sigma_a J_a$. We stress that, in general, (14) is well-defined only for complex u_a while the index for $\sigma_a = 0$ is defined by analytic continuation. We would like to see how we can extract the BH entropies from Z.

4. Statistical interpretation

The partition function $Z(u)$ describes a supersymmetric ensemble which is canonical with respect to the magnetic charges (*i.e.* all states have the same, fixed, magnetic charges) but grand canonical with respect to the electric charges (*i.e.* it is a sum over all electric charge sectors, with fixed chemical potentials u_a). A similar viewpoint in BH physics is advocated in [23]. We can decompose Z as a sum over sectors with fixed charges $\mathfrak{q}_a$ under $R_a/2$ (then the lattice of charges is such that both $J_a, R_a \in \mathbb{Z}$, up to a possible zero-point shift in the vacuum):

$$Z(u) = \sum_{\mathfrak{q}} e^{i\sum_{a=1}^{3} u_a(\mathfrak{q}_a - \mathfrak{q}_4)}\, Z_{\mathfrak{q}}\,. \tag{15}$$

We would like to identify $S_{\mathfrak{q}} \equiv \mathbb{R}\mathrm{e}\log Z_{\mathfrak{q}}$ with the leading entropy of a BH of fixed electric charges $\mathfrak{q}_a$. We take the real part to remove the effect of a possible overall sign. An important assumption is that $(-1)^F$ in the trace (14) does not cause dangerous cancelations at leading order.

We can Fourier transform the previous expression with respect to the three independent Δ_a to obtain

$$\sum_{\mathfrak{q}_4}{}' Z_{\mathfrak{q}} = \int \frac{d^3\Delta_a}{(2\pi)^3}\, e^{-i\sum_{b=1}^{3}\Delta_b(\mathfrak{q}_b - \mathfrak{q}_4)}\, Z(u)\,, \tag{16}$$

where prime means that the sum is taken at fixed integer $\mathfrak{q}_a - \mathfrak{q}_4$. As we will see, for supergravity BHs both the electric charges $\mathfrak{q}_a$ and $\log Z$ are of order $N^{3/2}$, therefore the previous expression can be evaluated at large N using a saddle point approximation:

$$\sum_{\mathfrak{q}_4}{}' Z_{\mathfrak{q}} = \exp\Big[\log Z(\hat{u}) - i\sum_{a=1}^{3}\hat{u}_a(\mathfrak{q}_a - \mathfrak{q}_4)\Big] \tag{17}$$

at leading order, where $\hat{u}_a$ is a solution for u_a to

$$\frac{\partial}{\partial u_a}\Big[\log Z(u) - i\sum_{b=1}^{3} u_b(\mathfrak{q}_b - \mathfrak{q}_4)\Big] = 0 \tag{18}$$

with $a = 1, 2, 3$. This saddle point in general gives complex values for $\hat{u}_a$. The sum on the LHS of (17) will also be dominated by a specific value of $\mathfrak{q}_4$, corresponding to the electric R-charge of the black hole. For that value:

$$S_{\mathfrak{q}} = \mathbb{R}\mathrm{e}\Big[\log Z(\hat{u}) - i\sum_{a=1}^{3}\hat{u}_a(\mathfrak{q}_a - \mathfrak{q}_4)\Big]\,. \tag{19}$$

We can restore the permutation symmetry between the charges, part of the Weyl group of $SO(8)$, by introducing

$$\mathcal{I}(u) \equiv \log Z(u) - i\sum_{a=1}^{4} u_a\mathfrak{q}_a\,. \tag{20}$$

Eqn. (18) is equivalent to extremization of $\mathcal{I}$ and the entropy is given by $S_{\mathfrak{q}} = \mathbb{R}\mathrm{e}\,\mathcal{I}(\hat{u})$.

This argument does not determine the R-charge of the BH, essentially because the index $Z(u)$ lacks a chemical potential for it. However from the attractor equations (11) it follows that, for given magnetic charges $\mathfrak{p}_a$ and flavor electric charges $\mathfrak{q}_a - \mathfrak{q}_4$, there is at most one value of $\mathfrak{q}_4$ leading to a large smooth BH. Our argument then gives an unambiguous prediction for the leading entropy of that BH.

5. RG flow interpretation

We can extract more information from the index if we interpret the BH as an holographic RG flow. The near-horizon geometry of BPS black holes contains an AdS_2 factor permeated by constant electric flux, where the super-isometry algebra is enhanced to $\mathfrak{su}(1,1|1)$. Thus we can think of the BH solution as a holographic RG flow from the 3d theory on S^2 to an ensemble of $\mathfrak{su}(1,1|1)$-invariant states in a 1d system. The bosonic subalgebra is $\mathfrak{sl}(2,\mathbb{R}) \times \mathfrak{u}(1)_c$ where the second factor is the IR superconformal R-symmetry, which is some linear combination of $U(1)^4 \subset SO(8)$. In the near-horizon canonical ensemble this implies that all BH states have zero $U(1)_c$ charge (by an argument similar to that in [24,25]). We will assume that there are no other contributions outside the horizon.

The asymptotic behavior of electrically charged BH solutions with axions turned on suggests that the dual ABJM theory is also deformed by real masses σ_a. In general, they lift a possible vacuum degeneracy of the Hamiltonian H. The presence of AdS_2 with constant electric flux, though, indicates that there should be a large vacuum degeneracy for a modified Hamiltonian H_{nh} in which the energy of states gets an extra contribution linear in the charge: $H_{\mathrm{nh}}(\sigma) = H(\sigma) - \sigma$. From the supersymmetry algebra $Q^2 = H_{\mathrm{nh}}$ we conclude that $H_{\mathrm{nh}} \geq 0$, and the index gets contribution only from its ground states. We can rewrite the index in (14) as

$$Z(\Delta, \sigma) = \mathrm{Tr}'\, e^{i\pi R_{\mathrm{trial}}(\Delta)}\, e^{-\beta\sigma J}\,, \tag{21}$$

where $\mathrm{Tr}' = \mathrm{Tr}_{H_{\mathrm{nh}}=0}$. We introduced a trial R-current $R_{\mathrm{trial}}(\Delta) \equiv R_0 + \Delta J/\pi$ that parametrizes the mixing of the R-symmetry with the flavor symmetries, with R_0 a reference R-symmetry such that $e^{i\pi R_0} = (-1)^F$.

We want to argue, generalizing [3], that the superconformal R-symmetry R_c of the Hamiltonian H_{nh} can be found by extremizing $Z(\Delta, \sigma)$ for fixed values of σ_a. Let $\hat{\Delta}_a$ be the value such that $R_{\mathrm{trial}}(\hat{\Delta}) = R_c$. One computes $\partial\log Z/\partial\Delta_a\big|_{\hat{\Delta}_a} = i\langle J_a e^{-\beta\sigma J}\rangle/\langle e^{-\beta\sigma J}\rangle$, using that at zero temperature the density matrix is uniformly distributed over the ground states of H_{nh}, and that $R_c = 0$ in those states as argued above. The expression on the right is imaginary, implying that $\hat{\Delta}_a$ are determined by extremizing the index with respect to Δ_a at fixed σ_a:

$$\frac{\partial\,\mathbb{R}\mathrm{e}\log Z(\Delta, \sigma)}{\partial\Delta_a}\bigg|_{\hat{\Delta}} = 0\,. \tag{22}$$

This is the generalization of the $\mathcal{I}$-extremization principle proposed in [3]. Assuming the large N factorization $\langle Je^{-\beta\sigma J}\rangle = \langle J\rangle\langle e^{-\beta\sigma J}\rangle$, we also have

$$\frac{\partial \mathbb{I}\mathrm{m} \log Z(\Delta,\sigma)}{\partial \Delta_a}\Big|_{\hat{\Delta}} = i\langle J_a\rangle \equiv i(\mathfrak{q}_a - \mathfrak{q}_4)\,, \tag{23}$$

where $\langle J_a\rangle$ is the charge of the vacuum density matrix. This determines the relation between the flavor charges $\mathfrak{q}_a - \mathfrak{q}_4$ and σ_a. Since $Z(\Delta,\sigma)$ is a holomorphic function of $u_a = \Delta_a + i\beta\sigma_a$, we can summarize the result in the complex equation

$$\frac{\partial \log Z(u)}{\partial u_a}\Big|_{\hat{u}} = i(\mathfrak{q}_a - \mathfrak{q}_4)\,, \tag{24}$$

which determines both $\hat{\Delta}_a$ and σ_a as functions of $\mathfrak{q}_a$.

From eqn. (21), at the critical point

$$Z(\hat{\Delta},\sigma) = e^{-\beta\sigma\langle J\rangle}\,\mathrm{Tr}'\,1 = e^{-\sum_{a=1}^4 \beta\sigma_a\mathfrak{q}_a}\, e^{S_\mathfrak{q}}\,. \tag{25}$$

The real part of the logarithm of this expression reproduces the result of the statistical argument, namely

$$S_\mathfrak{q} = \mathbb{R}\mathrm{e}\Big[\log Z(\hat{u}) - i\sum\nolimits_{a=1}^4 \hat{u}_a\mathfrak{q}_a\Big]\,. \tag{26}$$

An advantage of this derivation is that we can argue, at least at leading order, that $e^{S_\mathfrak{q}}$ is the number of ground states, without dangerous signs that could cause cancelations.

We can also write the entropy in a slightly different form and make a conjecture for the value of the fourth charge. Since $\hat{u}$ only depends on the differences $\mathfrak{q}_a - \mathfrak{q}_4$ and $\sum_a u_a \in 2\pi\mathbb{Z}$, we can always shift the integer charges $\mathfrak{q}_a$ and write the entropy in the permutationally symmetric and holomorphic form

$$S_\mathfrak{q} = \log Z(\hat{u}) - i\sum\nolimits_{a=1}^4 \hat{u}_a\mathfrak{q}_a = \mathcal{I}(\hat{u})\,, \tag{27}$$

up to $\mathcal{O}(N^0)$ terms which are invisible in the large N limit. The determination of the logarithm is such that $\log Z$ is real for $\sigma_a = 0$ and extended by continuity. The requirement that (27) be real positive fixes the fourth charge. Interestingly, this is precisely the constraint (11) that comes from supergravity.

6. Explicit match for ABJM

The large N expression for the index of ABJM was found in [3, 21] for the case of real u_a, and we can extend it to the complex plane using holomorphy:

$$\log Z = \frac{N^{3/2}}{3}\sqrt{2u_1u_2u_3u_4}\sum\nolimits_{a=1}^4 \frac{\mathfrak{p}_a}{u_a}\,. \tag{28}$$

This is valid for $\sum_a u_a = 2\pi$ and $0 < \mathbb{R}\mathrm{e}\, u_a < 2\pi$. The $\mathcal{I}$-extremization principle (24) is equivalent to the extremization of

$$\mathcal{I}_{\text{QFT}} = \sum\nolimits_{a=1}^4 \left(\frac{N^{3/2}}{3}\sqrt{2u_1u_2u_3u_4}\,\frac{\mathfrak{p}_a}{u_a} - i\mathfrak{q}_a u_a\right)\,. \tag{29}$$

Then the entropy is given by $S_\mathfrak{q} = \mathcal{I}_{\text{QFT}}(\hat{u})$, with the constraint on the charges that $\mathcal{I}_{\text{QFT}}(\hat{u})$ be positive.

In supergravity, the BH entropy is determined by

$$S_{\text{BH}} = \frac{\text{Area}}{4G_{\text{N}}} = -i\frac{\mathcal{Z}}{\mathcal{L}}\frac{2\pi\eta}{4G_{\text{N}}} \equiv \mathcal{I}_{\text{SUGRA}} \tag{30}$$

using (12), and $\mathcal{I}_{\text{SUGRA}}$ should be extremized with respect to X^Λ. We can identify the index $\Lambda = \{0,1,2,3\}$ with $a = \{1,2,3,4\}$, as well as $2\pi X^a/\sum_b X^b = u_a$ since they have the same domain and constraint:

$$\mathcal{I}_{\text{SUGRA}} = \frac{\eta}{4gG_{\text{N}}}\sum\nolimits_{a=1}^4 \left(\sqrt{u_1u_2u_3u_4}\,\frac{p^a}{u_a} - iq_a u_a\right)\,. \tag{31}$$

Identifying the integers in (7) with the charges $\mathfrak{p}_a$, $\mathfrak{q}_a$, respectively, and using (13) we obtain a perfect match $\mathcal{I}_{\text{QFT}} = \mathcal{I}_{\text{SUGRA}}$. The field theory extremization principle corresponds to the supergravity attractor mechanism: they lead to the same entropy and non-linear constraint on the charges.

Acknowledgements

We thank J. de Boer, A. Gnecchi, N. Halmagyi and S. Murthy for instructive clarifications. FB is supported by the MIUR-SIR grant RBSI1471GJ. AZ is supported by the MIUR-FIRB grant RBFR10QS5J.

References

[1] A. Strominger, C. Vafa, Microscopic origin of the Bekenstein–Hawking entropy, Phys. Lett. B 379 (1996) 99–104, http://dx.doi.org/10.1016/0370-2693(96)00345-0.
[2] J.M. Maldacena, The large N limit of superconformal field theories and supergravity, Int. J. Theor. Phys. 38 (1999) 1113–1133, http://dx.doi.org/10.1023/A:1026654312961.
[3] F. Benini, K. Hristov, A. Zaffaroni, Black hole microstates in AdS$_4$ from supersymmetric localization, J. High Energy Phys. 05 (2016) 054, http://dx.doi.org/10.1007/JHEP05(2016)054.
[4] S.L. Cacciatori, D. Klemm, Supersymmetric AdS$_4$ black holes and attractors, J. High Energy Phys. 01 (2010) 085, http://dx.doi.org/10.1007/JHEP01(2010)085.
[5] J.D. Bekenstein, Black holes and entropy, Phys. Rev. D 7 (1973) 2333–2346, http://dx.doi.org/10.1103/PhysRevD.7.2333.
[6] S.W. Hawking, Particle creation by black holes, Commun. Math. Phys. 43 (1975) 199–220, http://dx.doi.org/10.1007/BF02345020.
[7] F. Benini, A. Zaffaroni, A topologically twisted index for three-dimensional supersymmetric theories, J. High Energy Phys. 07 (2015) 127, http://dx.doi.org/10.1007/JHEP07(2015)127.
[8] O. Aharony, O. Bergman, D.L. Jafferis, J. Maldacena, $\mathcal{N}=6$ superconformal Chern–Simons-matter theories, M2-branes and their gravity duals, J. High Energy Phys. 10 (2008) 091, http://dx.doi.org/10.1088/1126-6708/2008/10/091.
[9] S.M. Hosseini, A. Zaffaroni, Large N matrix models for 3d $\mathcal{N}=2$ theories: twisted index, free energy and black holes, J. High Energy Phys. 08 (2016) 064, http://dx.doi.org/10.1007/JHEP08(2016)064.
[10] S.M. Hosseini, N. Mekareeya, Large N topologically twisted index: necklace quivers, dualities, and Sasaki–Einstein spaces, J. High Energy Phys. 08 (2016) 089, http://dx.doi.org/10.1007/JHEP08(2016)089.
[11] A. Sen, Entropy function and AdS$_2$/CFT$_1$ correspondence, J. High Energy Phys. 11 (2008) 075, http://dx.doi.org/10.1088/1126-6708/2008/11/075.
[12] H. Ooguri, A. Strominger, C. Vafa, Black hole attractors and the topological string, Phys. Rev. D 70 (2004) 106007, http://dx.doi.org/10.1103/PhysRevD.70.106007.
[13] M.J. Duff, J.T. Liu, Anti-de Sitter black holes in gauged $\mathcal{N}=8$ supergravity, Nucl. Phys. B 554 (1999) 237–253, http://dx.doi.org/10.1016/S0550-3213(99)00299-0.
[14] B. de Wit, A. Van Proeyen, Potentials and symmetries of general gauged $\mathcal{N}=2$ supergravity: Yang–Mills models, Nucl. Phys. B 245 (1984) 89–117, http://dx.doi.org/10.1016/0550-3213(84)90425-5.
[15] A. Strominger, Special geometry, Commun. Math. Phys. 133 (1990) 163–180, http://dx.doi.org/10.1007/BF02096559.
[16] L. Andrianopoli, et al., $\mathcal{N}=2$ supergravity and $\mathcal{N}=2$ super Yang–Mills theory on general scalar manifolds: symplectic covariance, gaugings and the momentum map, J. Geom. Phys. 23 (1997) 111–189, http://dx.doi.org/10.1016/S0393-0440(97)00002-8.
[17] G. Dall'Agata, A. Gnecchi, Flow equations and attractors for black holes in $\mathcal{N}=2$ $U(1)$ gauged supergravity, J. High Energy Phys. 03 (2011) 037, http://dx.doi.org/10.1007/JHEP03(2011)037.
[18] N. Halmagyi, BPS black hole horizons in $\mathcal{N}=2$ gauged supergravity, J. High Energy Phys. 02 (2014) 051, http://dx.doi.org/10.1007/JHEP02(2014)051.
[19] N. Halmagyi, Static BPS black holes in AdS$_4$ with general dyonic charges, J. High Energy Phys. 03 (2015) 032, http://dx.doi.org/10.1007/JHEP03(2015)032.
[20] S. Katmadas, Static BPS black holes in U(1) gauged supergravity, J. High Energy Phys. 09 (2014) 027, http://dx.doi.org/10.1007/JHEP09(2014)027.
[21] F. Benini, A. Zaffaroni, Supersymmetric partition functions on Riemann surfaces, arXiv:1605.06120.

[22] C. Closset, H. Kim, Comments on twisted indices in 3d supersymmetric gauge theories, J. High Energy Phys. 08 (2016) 059, http://dx.doi.org/10.1007/JHEP08(2016)059.
[23] S.W. Hawking, S.F. Ross, Duality between electric and magnetic black holes, Phys. Rev. D 52 (1995) 5865–5876, http://dx.doi.org/10.1103/PhysRevD.52.5865.
[24] A. Sen, Arithmetic of quantum entropy function, J. High Energy Phys. 08 (2009) 068, http://dx.doi.org/10.1088/1126-6708/2009/08/068.
[25] A. Dabholkar, J. Gomes, S. Murthy, A. Sen, Supersymmetric index from black hole entropy, J. High Energy Phys. 04 (2011) 034, http://dx.doi.org/10.1007/JHEP04(2011)034.

18

Exponentiating Higgs

Marco Matone

Dipartimento di Fisica e Astronomia "G. Galilei", Istituto Nazionale di Fisica Nucleare, Università di Padova, Via Marzolo, 8-35131 Padova, Italy

ARTICLE INFO

Editor: B. Grinstein

ABSTRACT

We consider two related formulations for mass generation in the $U(1)$ Higgs–Kibble model and in the Standard Model (SM). In the first formulation there are no scalar self-interactions and, in the case of the SM, the formulation is related to the normal subgroup of $G = SU(3) \times SU(2) \times U(1)$, generated by $(e^{2\pi i/3}I, -I, e^{\pi i/3}) \in G$, that acts trivially on all the fields of the SM. The key step of our construction is to relax the non-negative definiteness condition for the Higgs field due to the polar decomposition. This solves several stringent problems, that we will shortly review, both at the non-perturbative and perturbative level. We will show that the usual polar decomposition of the complex scalar doublet Φ should be done with $U \in SU(2)/\mathbb{Z}_2 \simeq SO(3)$, where $\mathbb{Z}_2$ is the group generated by $-I$, and with the Higgs field $\phi \in \mathbb{R}$ rather than $\phi \in \mathbb{R}_{\geq 0}$. As a byproduct, the investigation shows how Elitzur theorem may be avoided in the usual formulation of the SM. It follows that the simplest lagrangian density for the Higgs mechanism has the standard kinetic term in addition to the mass term, with the right sign, and to a linear term in ϕ. The other model concerns the scalar theories with normal ordered exponential interactions. The remarkable property of these theories is that for $D > 2$ the purely scalar sector corresponds to a free theory.

1. Introduction

The Higgs mechanism [1–6] is a basic step in the formulation of the Standard Model (SM) [7,8]. This has been confirmed by the spectacular experimental results at LHC [9,10]. Despite this, there are still some open questions. The most important one is that the vev of the Higgs field is evaluated at the classical level. On the other other hand, there are models with non-trivial minima for the classical potential, with no order parameter. The point is that Spontaneous Symmetry Breaking (SSB) is a strictly non-perturbative phenomenon, concerning infinitely many degrees of freedom. As such, even radiative corrections to $\langle\phi\rangle$ should be considered with particular attention. In this respect, one should also recall that, against the evidence coming from the perturbative expansion, there is strong evidence that $\lambda\phi^4$ is a free theory, that is, the renormalized coupling constant vanishes in the limit of large cut-off [11]. This is a particular case of the so-called quantum triviality: all four-dimensional scalar theories are believed to be free. Although the situation is more subtle in the case in which the Higgs is coupled to other particles, the question of triviality of the purely scalar sector of the SM still holds once one considers the perturbative expansion in the gauge coupling constants keeping λ fixed. This is known as the Higgs triviality problem. On the other hand, since the Higgs mass m is 125 GeV, and $\langle\phi\rangle = m/(\sqrt{2\lambda}) \simeq 246$ GeV, it follows that according to the usual formulation of the SM one should have for the scalar self-coupling

$$\lambda \simeq 0.13\,. \tag{1.1}$$

It should be stressed that the check of self-interactions terms in the Higgs legrangian are a priority in the LHC experiments. This is a hot topic, for example in [12] it has been proposed that the Higgs trilinear self-coupling can be tested in a near future via single Higgs production at LHC.

A natural question, which is also of considerable experimental interest, and suggested by the above analysis, is whether it is possible to have a Higgs mechanism from a scalar model where the unique self-interaction is represented by the mass term. Presumably such a possibility has not been considered because of the non-negative definiteness of the Higgs field ϕ and the related assumption that the potential should be a function of $\Phi^\dagger\Phi \in \mathbb{R}_{\geq 0}$, with Φ the complex scalar doublet Φ. We will see that these questions can be solved.

We begin the investigation with the $U(1)$ Higgs–Kibble model, showing that one may generate the mass term with a free potential and without SSB. In this respect, we note that an ingenious formulation of the Higgs phenomenon without SSB has been proposed by Fröhlich, Morchio and Strocchi in [13,14]. Then we will

E-mail address: matone@pd.infn.it.

extend the analysis to the SM, formulating a mass generation without SSB. We will start with the analysis of the so-called polar decomposition of Φ

$$\begin{pmatrix}\phi^+\\ \phi^0\end{pmatrix}=U\begin{pmatrix}0\\ \frac{\phi}{\sqrt{2}}\end{pmatrix}, \qquad \phi\in\mathbb{R}_{\geq 0}, \qquad U\in SU(2). \tag{1.2}$$

Note that this parametrization does not imply any gauge choice and should not be confused with the unitary gauge. Also note that $\phi=\sqrt{2\Phi^\dagger\Phi}$ is gauge invariant.

A key point related to the parametrization (1.2) is that the normal subgroup of $G=SU(3)\times SU(2)\times U(1)$, generated by $(e^{2\pi i/3}I,-I,e^{\pi i/3})\in G$, acts trivially on all the fields of the SM [15,16]. In particular, recalling that the action of the $U(1)$ is $e^{i3Y\gamma}$, and that $Y(\Phi)=1$, one sees that in the case of Φ, on which $SU(3)$ acts trivially, the $U(1)$ transformation with $\gamma=\pi/3$ times $-I\in SU(2)$ is the identity. This suggests considering the representation (1.2) with $U(x)\in SU(2)/\mathbb{Z}_2\simeq SO(3)$, where $\mathbb{Z}_2$ is the group generated by $-I$, and with $\phi\in\mathbb{R}$ rather than $\phi\in\mathbb{R}_{\geq 0}$. This solves the problem of the non-negative definiteness of ϕ of the polar decomposition. It follows that while in the standard polar decomposition $\phi=\sqrt{2\Phi^\dagger\Phi}$, in the case of $\mathbb{R}\ni\phi\neq\sqrt{2\Phi^\dagger\Phi}\in\mathbb{R}_{\geq 0}$, there are potentials[1] $U(\phi)$ not equivalent to $U(\sqrt{2\Phi^\dagger\Phi})$. In particular, one may consider potentials not constrained by the parity condition $U(-\phi)=U(\phi)$ and with $\phi\in\mathbb{R}$.

The outcome is that the simplest model for the Higgs mechanism, free of the above mentioned problems of the standard formulation, is the one whose lagrangian density has the standard kinetic term in addition to the mass term, with the right sign, and to a linear term in ϕ. As a byproduct, we will also show that the investigation provides the way to avoid Elitzur theorem in the usual formulation of the SM.

In the remanent part of the paper we investigate a related model that still considers $\phi\in\mathbb{R}$ and concerns the exponential interactions. Recently, in [17], the exponential interaction has been considered as a master model to derive other scalar theories. Interestingly, exponentiation of the Higgs also arises in the framework of skyrmions [18].

The motivation for studying exponential interactions is that in the four-dimensional case, in agreement with the Higgs triviality problem, such theories turns out to be free. In this sense such models are related to the above proposed model. Nevertheless, the precise correspondence between such free models is still unknown, for this reason, considered the nice properties of the exponential interactions, it makes sense to investigate their possible role.

Let us recall that scalar theories with exponential interactions are non-renormalizable. As emphasized in [19], the difficulties in quantizing some non-renormalizable field theories, concern the non-uniqueness of the solution, rather than its existence. In such a context, let us remind that in [20], using the ultraviolet cut-off γ^{-N}, $\gamma>1$, $N>0$, have been investigated scalar theories with interaction $\lambda:\exp(\alpha\phi):$. It turns out that for $D>2$, for all α, and for $D=2$, with $|\alpha|>\alpha_0$, the Schwinger functions converge to the free Schwinger functions. The essential point in the investigation of [20] is that $\Delta_{F,\Lambda}(0)$, with $\Delta_{F,\Lambda}(x)$ the Feynman propagator with cut-off on the momenta, grows sufficiently fast to kill the fluctuations of ϕ, so that $:\exp(\alpha\phi):=\exp(-\frac{\alpha^2}{2}\Delta_{F,\Lambda}(0))\exp(\alpha\phi)$ vanishes in the limit $\Lambda\to\infty$.

In the present paper, we consider the D-dimensional euclidean scalar theory with potential

$$V=\mu^D\exp(-\alpha\phi(x)). \tag{1.3}$$

It turns out that the functional generator associated to such a potential is the first term of an expansion $W[J]=W_R[J]+\ldots$, where

$$W_R[J]=e^{-Z_R[J]}=\langle 0|:e^{-\int d^Dx V(\phi)}:|0\rangle_J, \tag{1.4}$$

and one may easily check that [17]

$$Z_R[J]=Z_0[J]+\mu^D\int d^Dx e^{-\alpha\int d^Dy J(y)\Delta(y-x)}, \tag{1.5}$$

where $\Delta(x-y)$ is the Feynman propagator. It turns out that $Z_R[J]$ generates the lowest order contributions in α to the N-point point function. In particular

$$-\frac{\delta Z_R[J]}{\delta J(x)}|_{J=0}=\frac{\alpha\mu^D}{m^2}. \tag{1.6}$$

Next, as done in the previous model, we will parameterize the scalar doublet with $U(x)\in SU(2)/\mathbb{Z}_2$ and $\phi\in\mathbb{R}$ and consider the lagrangian density

$$\mathcal{L}_\Phi=(D_\mu\Phi)^\dagger(D^\mu\Phi)-\frac{1}{2}m^2\phi^2+2\nu m^3\sinh\left(\frac{\phi}{\nu}\right). \tag{1.7}$$

We will see that this leads to

$$\langle\phi\rangle=2m+\mathcal{O}(\nu^{-1}). \tag{1.8}$$

2. The $U(1)$ Higgs–Kibble model and Elitzur's theorem

Let us consider the lagrangian density of the $U(1)$ Higgs–Kibble model

$$\mathcal{L}=-\frac{1}{4}F_{\mu\nu}F^{\mu\nu}+\frac{1}{2}(D_\mu\varphi)^\dagger(D^\mu\varphi)-U(|\varphi|), \tag{2.1}$$

where φ is a complex scalar and $D_\mu=\partial_\mu-ieA_\mu$. The fact that the potential depends only on $\rho=|\varphi|$ naturally selects the two independent fields, ρ and θ, where $e^{i\theta}=\varphi/\rho$. So that (2.1) is identical to

$$\mathcal{L}=-\frac{1}{4}F_{\mu\nu}F^{\mu\nu}+\frac{1}{2}e^2\rho^2W_\mu W^\mu+\frac{1}{2}\partial_\mu\rho\partial^\mu\rho-U(\rho), \tag{2.2}$$

where $W_\mu=A_\mu+e^{-1}\partial_\mu\theta$. Note that a gauge transformation corresponds to $\theta\to\theta+\alpha$ and $A_\mu\to A_\mu-e^{-1}\partial_\mu\alpha$, so that, like ρ, even W_μ is gauge invariant. In this way, without performing any gauge choice, one passes from the degrees of freedom φ and A_μ, to ρ and W_μ.

The usual treatment of (2.2) is to consider a semiclassical approximation around the minimum ρ_0 of $U(\rho)$. Set $\chi=\rho-\rho_0$. In such an approximation, considering only the terms quadratic in χ and W_μ, one gets the lagrangian density

$$\tilde{\mathcal{L}}=-\frac{1}{4}F_{\mu\nu}F^{\mu\nu}+\frac{1}{2}e^2\rho_0^2W_\mu W^\mu+\frac{1}{2}\partial_\mu\chi\partial^\mu\chi-\frac{1}{2}U''(\rho_0)\chi^2, \tag{2.3}$$

showing that W_μ and χ have square masses $e^2\rho_0^2$ and $U''(\rho_0)$ respectively.

In the lucid analysis in [21] have been discussed the main problems with such a model. The first point is that in passing from (2.1) to (2.3) one has to fix the condition $\rho\in\mathbb{R}_{\geq 0}$, a difficult task even at the classic level because this should be consistent with the time evolution. The problem is even more difficult in considering the semi-classical approximation because one should keep χ bounded by ρ_0. At the quantum level there is the problem of treating the term $|\varphi|$. In a rigorous QFT formulation φ is a distribution and the

[1] In the following the "potential" $U(\phi)$ denotes the mass term together with the true potential part $V(\phi)$.

modulus of a distribution is a ill-defined quantity. This implies that ρ cannot be considered a quantum field.

An alternative approach is to make the decomposition $\varphi = \varphi_1 + i\varphi_2$, and then considering the semi-classical expansion

$$\varphi_1 = \varphi_0 + \chi_1 \,, \qquad \varphi_2 = \chi_2 \,, \tag{2.4}$$

with χ_1 and χ_2 considered as small fluctuations. The resulting lagrangian density is still (2.3) with $W_\mu = A_\mu + e^{-1}\partial_\mu \chi_2$ and ρ_0 and χ replaced by φ_0 and χ_1 respectively.

The problem with such a formulation is that while perturbation theory leads to $\langle\varphi\rangle \neq 0$, at the non-perturbative level one has, according to Elitzur's theorem [22],

$$\langle\varphi\rangle = 0 \,. \tag{2.5}$$

Another possibility is to map φ to a real field $\varphi_r \in \mathbb{R}$ by a gauge transformation. Nevertheless, it turns out that there is a residual $\mathbb{Z}_2$ gauge symmetry that gives, even in this case, $\langle\varphi_r\rangle = 0$ [21]. However, there is a key point which leads to a well-defined solution. Namely, note that such a $\mathbb{Z}_2$ symmetry is the consequence of the tacitely assumed invariance of the potential under $\varphi \to -\varphi$. On the other hand, one may interpret $e^{i\theta}$ and ρ as independent fields, so that with $\rho \in \mathbb{R}$, and $\theta \in (-\pi, \pi]$. An interesting alternative is to consider $\varphi = e^{i\theta}\rho$ as unique complex scalar field, so that $\rho \in \mathbb{R}$ requires $\theta \in (-\pi/2, \pi/2]$. In the next section we will see that, in the case of the SM, the latter possibility is also suggested by the presence of a normal subgroup of the gauge group leaving the fields invariant.

We can then consider ρ to take real values and choose the potential in (2.1) to be

$$U(\rho) = \frac{1}{2}m^2\rho^2 - m^2\rho_0\rho \,, \tag{2.6}$$

so that the lagrangian density now reads

$$\mathcal{L} = -\frac{1}{4}F_{\mu\nu}F^{\mu\nu} + \frac{1}{2}(D_\mu\varphi)^\dagger(D^\mu\varphi) - \frac{1}{2}m^2\rho^2 + m^2\rho_0\rho \,, \tag{2.7}$$

where $\varphi = e^{i\theta}\rho$, $\rho \in \mathbb{R}$, and ρ_0 is a real constant. The purely scalar sector is now a free theory, so that $\langle\rho\rangle$ coincides with the value of ρ that minimizes (2.6). Therefore, we have

$$\langle\rho\rangle = \rho_0 \,. \tag{2.8}$$

Setting $\eta = \rho - \rho_0$ leads to the lagrangian density with the mass term for the gauge field without any SSB. In particular, since ρ is gauge invariant, the Elitzur theorem is avoided simply because it concerns the vacuum expectation value of gauge non-invariant quantities.

3. Trivial Higgs

Let us now consider the Higgs mechanism in the SM. By (1.2) we have that even in this case ϕ takes non-negative values. This means that $\eta = \phi - v$, $v = \langle\phi\rangle$, is bounded by $-v$, so that the path integral on η should be

$$\int_{\eta \geq -v} D\eta e^{iS} \,. \tag{3.1}$$

Nevertheless, the field η is usually considered as taking all real values. The standard argument to justify $\eta \in \mathbb{R}$ is that one is considering small oscillations around a minimum of the potential. We saw that this is a subtle point for several reasons. In the case of the SM the condition $\eta \geq v$ has effects on all the terms of the lagrangian density of the SM, the kinetic one, the mass and the η^3 terms, and the Yukawa interactions. Furthermore, even in doing perturbation theory, one should replace the Feynman propagator by the one coming from the path integral on field configurations bounded by $-v$.

Since the physical fields in the SM are the ones identified once one considers the polar decomposition, it is clear that one should understand if an why one can choose the Higgs field η to take real values. We now show that one may in fact relax the condition $\eta \geq -v$, a result that leads to the free model. To this end, let us first recall that the reason why in (1.2) one can choose $U \in SU(2)$, that is

$$U = \frac{\sqrt{2}}{\phi}\begin{pmatrix} \bar{\phi}^0 & \phi^+ \\ -\bar{\phi}^+ & \phi^0 \end{pmatrix} , \tag{3.2}$$

is because the first column of U in the polar decomposition is completely arbitrary. Such an arbitrariness implies that the action on Φ of a $U(1)$ transformation can be always represented by a matrix with the same determinant of U. In other words,

$$e^{i\beta}\begin{pmatrix} a & b \\ c & d \end{pmatrix}\begin{pmatrix} 0 \\ 1 \end{pmatrix} = \begin{pmatrix} e^{-i\beta}a & e^{i\beta}b \\ e^{-i\beta}c & e^{i\beta}d \end{pmatrix}\begin{pmatrix} 0 \\ 1 \end{pmatrix} .$$

It follows that any $U(1)$ transformation of Φ is equivalent to a map from $U \in SU(2)$ to $SU(2)$. In turn, this implies that the $SU(2) \times U(1)$ and $SU(2)$ orbits of Φ are the same. In other words, any $SU(2) \times U(1)$ gauge transformation of Φ corresponds to a map of $U \in SU(2)$ to $U' \in SU(2)$, that is the gauge transformations act on U only. However, one should note that the identity transformation $\Phi \to \Phi$ is obtained in two different ways, by the simultaneous action of the $U(1)$ and the $SU(2)$ identities and by acting with $-1 \in U(1)$ and $-I \in SU(2)$. This is related to the fact that the order 6 normal subgroup N of $SU(3) \times SU(2) \times U(1)$, generated by

$$(e^{2\pi i/3}I, -I, e^{\pi i/3}) \in SU(3) \times SU(2) \times U(1) \,, \tag{3.3}$$

acts trivially on all the fields of the SM (recall that $U(1) = e^{i3Y\gamma}$ and $Y(\Phi) = 1$). Therefore, the non-trivial part of the gauge group of the SM is [15,16]

$$SU(3) \times SU(2) \times U(1)/N \,. \tag{3.4}$$

This leads to consider $U(x)$ and $\phi \in \mathbb{R}$ as fully independent degrees of freedom, with ϕ that, as in the polar decomposition, is gauge invariant. The point is to use the following parametrization

$$\Phi(x) = U(x)\begin{pmatrix} 0 \\ \dfrac{\phi(x)}{\sqrt{2}} \end{pmatrix} , \qquad \phi(x) \in \mathbb{R} \,,$$
$$U(x) \in SU(2)/\mathbb{Z}_2 \simeq SO(3) \,, \tag{3.5}$$

so that now $\phi = \pm\sqrt{2(|\phi^\dagger|^2 + |\phi^0|^2)}$. This suggests a possible role of $\mathbb{Z}_2$ monopoles [23–26]. Note that the analogous representation for a complex number $z = x + iy$ is

$$z = \chi e^{i\theta} \,, \qquad \chi \in \mathbb{R} \,, \qquad \theta \in (-\pi/2, \pi/2] \,, \tag{3.6}$$

that should be compared with the polar decomposition $z = \rho e^{i\alpha}$, $\rho \geq 0$. The principal part of $\arg z = \alpha$, denoted $\operatorname{Arg} z \in (-\pi, \pi]$, corresponds to $\arctan(y/x)$ for $x \geq 0$. In the case $x < 0$ one has $\operatorname{Arg} z = \arctan(y/x) + \pi$ if $y \geq 0$ and $\operatorname{Arg} z = \arctan(y/x) - \pi$ if $y < 0$. Comparison with (3.6) shows that the natural choice is to set $\theta = \arctan(y/x) + 2k\pi$, $k \in \mathbb{Z}$, that is

$$z = \chi e^{i[\arctan(y/x) + 2k\pi]} \,, \qquad \chi \in \mathbb{R} \,. \tag{3.7}$$

Note that

$$\chi = \rho e^{i[\operatorname{Arg} z - \arctan(y/x) + 2k\pi]} = \pm\rho \,. \tag{3.8}$$

As a result, Φ factorizes in a unitary field $U(x)$ times the gauge invariant field $\phi(x) \neq \sqrt{2\Phi^\dagger\Phi}$, which now takes values in the full real axis, so that it can be considered a quantum field. In the following we show that this leads to a simple gauge invariant lagrangian providing a non-trivial vev $\langle\phi\rangle$.

In the usual formulation the aspects related to the non-negative definiteness arise in two contexts. The first one concerns the choice of the range of ϕ discussed above. The other one is the tacit assumption that the potential should be a function of $\Phi^\dagger\Phi$. As done in the previous section we relax such a condition by considering the lagrangian density

$$\mathcal{L}_\Phi = (D_\mu\Phi)^\dagger(D^\mu\Phi) - U(\phi) , \tag{3.9}$$

without the constraint $U(-\phi) = U(\phi)$ that would be implied if one chooses $U(\sqrt{2\Phi^\dagger\Phi})$.

As in the case of the $U(1)$ model, the above analysis indicates that there is a natural candidate which is free of the problems associated to the formulation of the Higgs mechanism. Let us choose $U(\phi) = \frac{1}{2}m^2\phi^2 - 2m^3\phi$, so that the lagrangian density of the purely scalar sector is

$$\mathcal{L}_\phi = \frac{1}{2}\partial_\mu\phi\partial^\mu\phi - \frac{1}{2}m^2\phi^2 + 2m^3\phi , \tag{3.10}$$

$\phi \in \mathbb{R}$. A nice consequence is that the contributions to $\langle\phi\rangle$ from the purely scalar sector can be evaluated exactly. Namely, since the theory is the free one, it follows that the vev $\langle\phi\rangle$, evaluated with respect to (3.10), coincides with the value of ϕ that minimizes $\phi^2 - 4m\phi$, that is

$$\langle\phi\rangle = 2m . \tag{3.11}$$

Note that this choice is in agrement with the experimental data $\langle\phi\rangle \approx 2m$, with the difference $2m - \langle\phi\rangle$ which may fit the corrections, that we discuss below, due to the contributions to $\langle\phi\rangle$ coming from the other fields in the SM. Setting $\eta = \phi - 2m \in \mathbb{R}$ the lagrangian density of the purely scalar part becomes

$$\mathcal{L}_\eta = \frac{1}{2}\partial_\mu\eta\partial^\mu\eta - \frac{1}{2}m^2\eta^2 . \tag{3.12}$$

The exact value of $\langle\phi\rangle$ can be evaluated by first considering the path integration on ϕ taking into account that the full contributions to the quadratic and linear terms in ϕ include fermions, gauge bosons and the remanent bosonic fields in Φ. Denoting by $F_1\phi$ and $F_2\phi^2/2$ the contributions of such fields, the complete lagrangian density for ϕ has the form

$$\frac{1}{2}\partial_\mu\phi\partial^\mu\phi + \frac{1}{2}(F_2 - m^2)\phi^2 + (F_1 + 2m^3)\phi , \tag{3.13}$$

giving the classical equation of motion

$$(\partial_\mu\partial^\mu + m^2 - F_2(x))\phi(x) = F_1(x) + 2m^3 . \tag{3.14}$$

It follows that the vacuum expectation value, $\bar\phi$, of ϕ, obtained by integrating only over ϕ is

$$\bar\phi(x) = \int d^4y G(x,y)(F_1(y) + 2m^3) , \tag{3.15}$$

where $G(x, y)$ is the Green function for the operator $(\partial_\mu\partial^\mu + m^2 - F_2(x))$. Note that neglecting F_1 and F_2, $G(x, y)$ reduces to $-\Delta(y - x)$, where $\Delta(y - x)$ is the minkowskian Feynman propagator. Using $\int d^4y\Delta(y - x) = -1/m^2$, one may check that in this case $\bar\phi(x)$ reproduces (3.11). The exact value of $\langle\phi\rangle$ is then given by

$$\langle\phi(x)\rangle = \int d^4y\langle G(x,y)(F_1(y) + 2m^3)\rangle_\chi , \tag{3.16}$$

where the subscript χ denotes the path integration over the remanent fields.

A byproduct of the previous analysis is that the parametrization with ϕ taking all real values can be extended also to the usual formulation of the SM. In particular, even if $\langle\phi\rangle \neq 0$, there is no contradiction with the Elitzur theorem because ϕ is now a genuine quantum field and gauge invariant, so that there is no SSB. This arises only at the perturbative level by introducing the gauge fixing term. Therefore, the Higgs lagrangian density of the SM can be expressed in the form

$$\mathcal{L}_\Phi = (D_\mu\Phi)^\dagger(D^\mu\Phi) + \frac{1}{2}\mu^2\phi^2 - \frac{\lambda}{4}\phi^4 , \tag{3.17}$$

with $U \in SU(2)/\mathbb{Z}_2$ and $\phi \in \mathbb{R}$.

4. Exponential interactions

In the following we investigate, in the euclidean space, a model that considers again $\phi \in \mathbb{R}$ and concerns the exponential interactions. The motivation for such an analysis is that such theories are free for $D > 2$ [20], so that they are related to the above proposed model.

Let us shortly review the investigation in [17]. The notation follows the one in [27]. Define

$$\langle f(x_1, \dots, x_n)\rangle_{x_j \dots x_k} \equiv \int d^Dx_j \dots d^Dx_k f(x_1, \dots, x_n) , \tag{4.1}$$

and denote by $\langle f(x_1, \dots, x_n)\rangle$ integration of f over $x_1, \dots, x_n$. Let

$$\Delta(x - y) = \int \frac{d^Dp}{(2\pi)^D}\frac{e^{ip(x-y)}}{p^2 + m^2} , \tag{4.2}$$

be the Feynman propagator and set

$$Z_0[J] = -\frac{1}{2}\langle J(x)\Delta(x - y)J(y)\rangle . \tag{4.3}$$

To compute $W[J]$ we use Schwinger's method

$$W[J] = Ne^{-\langle V(\frac{\delta}{\delta J})\rangle}e^{-Z_0[J]} . \tag{4.4}$$

The first step in [17] has been the observation that exponential interactions can be obtained by acting on $\exp(-Z_0[J])$ with power series in the operator $\langle\exp(-\alpha\delta_J)\rangle$ whose action corresponds to a translation of J. Consider the potential investigated in [17] with the opposite sign of α

$$V(\phi) = \mu^De^{-\alpha\phi} . \tag{4.5}$$

The corresponding generating functional (we drop the constant N) is

$$\begin{aligned} W[J] &= \exp\Big[-\mu^D\langle\exp(-\alpha\frac{\delta}{\delta J})\rangle\Big]\exp(-Z_0[J]) \\ &= \sum_{k=0}^{\infty}\frac{(-\mu^D)^k}{k!}\langle\exp(-\alpha\frac{\delta}{\delta J})\rangle^k\exp(-Z_0[J]) . \end{aligned} \tag{4.6}$$

By [17]

$$\begin{aligned} \exp(-\alpha\frac{\delta}{\delta J(x)})\exp(-Z_0[J]) &= \exp(-Z_0[J - \alpha_x])\exp(-\alpha\frac{\delta}{\delta J(x)}) \\ &= \exp(-Z_0[J - \alpha_x]) , \end{aligned} \tag{4.7}$$

where

$$\begin{aligned} Z_0[J - \alpha_x] &= -\frac{1}{2}\int d^Dy\int d^Dz(J(y) - \alpha\delta(x - y))\Delta(y - z)(J(z) \\ &\quad - \alpha\delta(x - z)) \\ &= Z_0[J] - \frac{\alpha^2}{2}\Delta(0) + \alpha\int d^DyJ(y)\Delta(y - x) , \end{aligned} \tag{4.8}$$

one gets

$$W[J] = \exp(-Z_0[J]) \sum_{k=0}^{\infty} \Big[\frac{(-\mu^D)^k}{k!} \exp\big(\frac{k\alpha^2}{2}\Delta(0)\big)$$

$$\int d^D z_1 \ldots \int d^D z_k \exp\Big(-\alpha \int d^D z J(z) \sum_{j=1}^{k} \Delta(z - z_j)$$

$$+ \alpha^2 \sum_{j>l}^{k} \Delta(z_j - z_l)\Big)\Big] . \qquad (4.9)$$

Let us show that the Feynman propagators appearing in this expression are related to normal ordering. Let us focus on $\exp\big(\frac{k\alpha^2}{2}\Delta(0)\big)$ and $\exp(\alpha^2 \sum_{j>l}^{k} \Delta(z_j - z_l))$. In this respect, note that (4.6) corresponds to the expansion of $\exp\big(-\int d^D x V(\phi)\big)$ in the time-ordered vev, that is

$$W[J] = \langle 0|T e^{-\mu^D \int d^D x \exp(-\alpha\phi(x))}|0\rangle_J$$

$$= \sum_{k=0}^{\infty} \frac{(-\mu^D)^k}{k!} \int d^D x_1 \ldots$$

$$\int d^D x_k \langle 0|T e^{-\alpha\phi(x_1)} \ldots e^{-\alpha\phi(x_k)}|0\rangle_J , \qquad (4.10)$$

where the vacua are the ones of the free scalar theory coupled to the external source J. Comparison with (4.9) fixes the expression of $\langle 0|T e^{-\alpha\phi(x_1)} \ldots e^{-\alpha\phi(x_k)}|0\rangle_J$. The fact that the normal ordering problem is the cause of some of the infinities arising in perturbation theory, suggests considering

$$W_R[J] = \langle 0| : e^{-\mu^D \int d^D x \exp(-\alpha\phi(x))} : |0\rangle_J$$

$$= \sum_{k=0}^{\infty} \frac{(-\mu^D)^k}{k!} \int d^D x_1 \ldots$$

$$\int d^D x_k \langle 0| : e^{-\alpha\phi(x_1)} \ldots e^{-\alpha\phi(x_k)} : |0\rangle_J . \qquad (4.11)$$

Note that $: e^{-\alpha\phi(x)} := e^{-\frac{\alpha^2}{2}\Delta(0)} e^{-\alpha\phi(x)}$, and

$$T : e^{-\alpha\phi(x_1)} : \ldots : e^{-\alpha\phi(x_k)} :$$

$$= e^{\alpha^2 \sum_{j>l}^{k} \Delta(x_j - x_l)} : e^{-\alpha\phi(x_1)} \ldots e^{-\alpha\phi(x_k)} : . \qquad (4.12)$$

Therefore,

$$: e^{-\alpha\phi(x_1)} \ldots e^{-\alpha\phi(x_k)} :$$

$$= e^{-\alpha^2\left(\frac{k}{2}\Delta(0) + \sum_{j>l}^{k} \Delta(x_j - x_l)\right)} T e^{-\alpha\phi(x_1)} \ldots e^{-\alpha\phi(x_k)} . \qquad (4.13)$$

It follows that the expansion on the right hand side of (4.11) exponentiates. Actually, (4.9), (4.10), (4.11) and (4.13) yield

$$W_R[J] = \exp(-Z_R[J]) , \qquad (4.14)$$

where

$$Z_R[J] = Z_0[J] + \mu^D \int d^D x e^{-\alpha \int d^D y J(y)\Delta(y-x)} . \qquad (4.15)$$

Interestingly, removing the term $\exp(\alpha^2 \sum_{j>l}^{k} \Delta(x_j - x_l))$, coming from the normal ordering in (4.12), is equivalent to remove a term $\langle \exp(-\alpha \frac{\delta}{\delta J})\rangle$ in (4.6). To show this, recall that for any suitable function F, if A and B are operators, then $A^{-1}F(B)A = F(A^{-1}BA)$. Therefore,

$$W[J] = \exp\Big[-\mu^D \langle \exp(-\alpha \frac{\delta}{\delta J})\rangle\Big] \exp(-Z_0[J])$$

$$= \exp(-Z_0[J]) \exp\Big[-\mu^D \exp(Z_0[J]) \langle \exp(-\alpha \frac{\delta}{\delta J})\rangle$$

$$\exp(-Z_0[J])\Big]$$

$$= \exp(-Z_0[J]) \exp\Big[-\mu_0^D \langle \exp\big(-\alpha \langle J(y)\Delta(x-y)\rangle_y\big)\rangle_x$$

$$\langle \exp(-\alpha \frac{\delta}{\delta J})\rangle\Big] , \qquad (4.16)$$

where in the last equality we used (4.7) and (4.8), and

$$\mu_0^D = \mu^D \exp\Big(\frac{\alpha^2}{2}\Delta(0)\Big) . \qquad (4.17)$$

Eq. (4.16) differs from $W_R[J]$ by the term $\langle \exp(-\alpha \frac{\delta}{\delta J})\rangle$ in the last member, and by the relabeling of μ_0. The latter is equivalent to consider the normal ordering of $\exp(-\alpha\phi)$. Therefore,

$$W[J, : e^{-\alpha\phi} :] = W_R[J] + \ldots , \qquad (4.18)$$

where the dots denote the terms in (4.9) coming from the expansion

$$\sum_{n=1}^{\infty} \frac{\alpha^{2n}}{n!} \Big(\sum_{j>l}^{k} \Delta(z_j - z_l)\Big)^n . \qquad (4.19)$$

Consider the field

$$\phi_{\text{cl}}(x) := -\frac{\delta Z_R[J]}{\delta J(x)} , \qquad (4.20)$$

and note that by (4.15)

$$\phi_{\text{cl}}(x) = \langle J(y)\Delta(x-y)\rangle_y + \alpha\mu^D \langle \Delta(y-x)$$

$$\exp\big(-\alpha\langle J(z)\Delta(y-z)\rangle_z\big)\rangle_y , \qquad (4.21)$$

that satisfies the equation of motion

$$(-\partial_\mu\partial_\mu + m^2)\phi_{cl}(x) = J(x) + \alpha\mu^D \exp(-\alpha\langle J(y)\Delta(x-y)\rangle_y) . \qquad (4.22)$$

By (4.21) it follows that $\Gamma_R[\phi_{\text{cl}}] = Z_R[J] - \langle J(x)\phi_{\text{cl}}(x)\rangle_x$, reads

$$\Gamma_R[\phi_{\text{cl}}] = Z_R[J] - \langle J(x)\Delta(x-y)J(y)\rangle_{xy} - \alpha\mu^D \langle J(x)\Delta(x-y)$$

$$\exp(-\alpha\langle J(z)\Delta(z-y)\rangle_z)\rangle_{xy} \qquad (4.23)$$

Furthermore, at the first order in α

$$\langle 0|\phi(x)|0\rangle = -\frac{\delta Z_R[J]}{\delta J(x)}\Big|_{J=0} = \frac{\alpha\mu^D}{m^2} , \qquad (4.24)$$

where we used $\langle \Delta(x-y)\rangle_y = 1/m^2$. It follows that the higher derivatives of $Z_R[J]$, evaluated at $J = 0$, correspond, to the lowest order contribution in the α expansion, to the connected Green functions associated to

$$\eta(x) = \phi(x) - \frac{\alpha\mu^D}{m^2} , \qquad (4.25)$$

that is

$$-\frac{\delta^N Z_R[J]}{\delta J(x_1)\ldots\delta J(x_N)}\Big|_{J=0} = \langle 0|T\eta(x_1)\ldots\eta(x_N)|0\rangle_c , \qquad (4.26)$$

and by (4.15), for $N > 1$,

$$\langle 0|T\eta(x_1)\dots\eta(x_N)|0\rangle_c$$
$$= \delta_{N2}\Delta(x_1 - x_2) + \alpha^N\mu^D \int d^D y\Delta(y - x_1)\cdots\Delta(y - x_N) . \quad (4.27)$$

Note that higher order contributions in α come from the expansion (4.19).

5. $\sinh(\phi/\nu)$

The above model can be extended to more general interactions, such as

$$V(\phi) = \sum_{k=1}^{n} \mu_k^D \exp(\alpha_k\phi) . \quad (5.1)$$

In order to find the explicit expression of $W_R[J]$ in the case of the potential (5.1), one first notes that the exact generating functional

$$W[J] = \exp(-Z[J])$$
$$= \Big[\prod_{k=1}^{n} \exp[-\mu_k^D\langle\exp(\alpha_k\frac{\delta}{\delta J})\rangle]\Big]\exp(-Z_0[J]) , \quad (5.2)$$

and then uses (4.16) iteratively. The first step is

$$\exp\Big[-\mu_n^D\langle\exp(\alpha_n\frac{\delta}{\delta J})\rangle\Big]\exp(-Z_0[J])$$
$$= \exp(-Z_0[J])\exp\Big[-\mu_{n0}^D\langle\exp\big(\alpha_n\langle J(y)\Delta(x-y)\rangle_y\big)\rangle_x$$
$$\langle\exp(\alpha_n\frac{\delta}{\delta J})\rangle\Big] . \quad (5.3)$$

Repeating this for the remaining $n-1$ terms in (5.2), makes it clear that

$$W_R[J] = \exp(-Z_R[J]) = \langle 0| : e^{-\int d^D x\sum_{k=1}^n \mu_k^D \exp(\alpha_k\phi(x))} : |0\rangle_J , \quad (5.4)$$

is obtained from $W[J]$ by removing, from the final expression, the term $\langle\exp\big(\sum_{k=1}^n \alpha_k\frac{\delta}{\delta J}\big)\rangle$ on the right hand side, and by canceling the $\exp\big(\sum_{k=1}^n \alpha_k^2\Delta(0)\big)$ term. Such a cancelation is equivalent to relabel each μ_{k0} by μ_k. It follows that

$$Z_R[J] = Z_0[J] + \int d^D x\sum_{k=1}^{n}\mu_k^D e^{\alpha_k\int d^D y J(y)\Delta(y-x)} . \quad (5.5)$$

We note that taking the normal ordering of $\exp(-\int d^D x V(\phi))$ may lead to well-defined $Z_R[J]$, even in cases when $V(\phi)$ is unbounded below. A particularly interesting case is the four-dimensional potential

$$V(\phi) = -2\nu m^3\sinh\left(\frac{\phi}{\nu}\right) . \quad (5.6)$$

By (5.5), we have

$$Z_R[J] = Z_0[J] - 2\nu m^3\int d^4x\sinh\left(\frac{\phi_c(x)}{\nu}\right) , \quad (5.7)$$

where

$$\phi_c(x) = \int d^4 y J(y)\Delta(y - x) , \quad (5.8)$$

that satisfies the free classical equation of motion in the presence of the external source J, is a key quantity in the dual representation of $W[J]$ recently introduced in [28]. Repeating the analysis leading to (4.24), at the zero order in ν^{-1}, (5.7) yields

$$\langle\phi\rangle = 2m , \quad (5.9)$$

so that, at the same order,

$$2^{-1/4}G_F^{-1/2} = 2m , \quad (5.10)$$

in agreement with the LHC data. Note that

$$\lim_{\nu\to\infty} V(\phi) = -2m^3\phi , \quad (5.11)$$

so that, in this limit, (5.9) corresponds to the value of ϕ that minimizes $m^2\phi^2/2 + V(\phi)$.

Making the expansion in powers of ν^{-1}, one sees that the lowest order contribution to the $(2N+1)$-point functions is generated by $Z_R[J]$, so that, at this order

$$\langle 0|T\eta(x_1)\dots\eta(x_{2N+1})|0\rangle$$
$$= 2^{2N+1}\nu^{-2N}m^3\int d^4y\Delta(y - x_1)\cdots\Delta(y - x_{2N+1}) , \quad (5.12)$$

where $\eta(x) = \phi(x) - \langle 0|\phi(x)|0\rangle$. In the case of the $2N$-point functions, $Z_R[J]$ contributes to the lowest-order of the two-point function only, so that it gives the free propagator.

6. Conclusions and perspectives

We proposed two models for mass generation for fermions and gauge bosons without SSB. One is based on a scalar sector without self-interactions and the other with exponential interactions. In both cases there is no need to start with an imaginary mass term, so that even the initial lagrangian density is physically meaningful.

The first model has the mass term and a linear term in ϕ. The model is also suggested by the strong evidence that $\lambda\phi^4$ is a free theory, that is, the renormalized coupling constant vanishes in the large cut-off limit. This is a particular case of the mentioned Higgs triviality problem. As a byproduct, we saw how Elitzur theorem may be avoided in the usual formulation of the SM.

We also investigated the exponential interactions. We then focused on the potential (5.6). In the limit $\nu\to\infty$ this explicitly corresponds to the free theory. At the next order, the theory is described by Z_R, that, besides the propagator of the free theory, generates only the lowest order contributions to the $(2N+1)$-point functions, $N > 1$. As such, for large ν, the model is well-described by $Z_R[J]$, so that $W_R[J]$ can be seen as describing an effective theory. Interestingly, exponentiation of the Higgs has been also considered in the framework of skyrmions [18].

An intriguing feature of the investigation is that the parametrization of the Higgs may be related to string theory. The reason is that such a parametrization is related to the normal subgroup of the SM that acts trivially on the fields, which in turn, is related to Calabi–Yau manifolds [16].

Another aspect of the formulation that should be investigated concerns the induced electroweak symmetry breaking model in [29–31]. In particular, the potential in (3.10) is reminiscent of the effective Higgs potential obtained by integrating out the heavy mass eigenstate.

Other investigations suggested by the proposed model concern the possible connection with $\mathbb{Z}_2$ monopoles, see for example [23–26].

We note that the absence of the cubic and quartic Higgs self-interactions can be tested experimentally. In particular, it is reasonable that, in a near future, LHC will give some evidence about the possible absence of the η^3 term in the lagrangian density. This can be checked in the production of two Higgs, with a virtual Higgs decaying in two real Higgs. The process to be investigated at LHC is of course $p + p \to H + H$. Their absence would be a fundamental check of the present model for the Higgs mechanism. We also note that possible precision tests may be suggested by the present model.

Let us conclude by mentioning that it would be interesting to investigate whether the free model we proposed could be related to an effective theory in which the Higgs field is a fermionic condensate. Of course this is suggested by the BCS theory of superconductivity that greatly motivated the original papers on the Higgs mechanism. In fact there is a strong analogy with superconductivity, whose lower energy states can be described by the analogue of the Higgs field. In this respect, it is interesting to note that the energy gap above the Fermi sphere is described by an exponential function that provides an example of energy hierarchy scale (see the excellent book by Strocchi [32] for an account on the BCS theory). A key feature of the superconductivity is that the condensation energy, that is the difference between the ground state energy in the superconducting state and the conducting state, is of order 10^{-7}–10^{-8} eV per electron, much less than the other energy scales of the metal, which are of order of 1–10 eV. In [33] it has been used a mechanism reminiscent of the BCS theory to propose a non-perturbative mechanism explaining the problem of gauge hierarchies. The analogy between the BCS theory and Higgs mechanism in the SM has been also stressed in the recent review by Peskin [34].

Acknowledgements

It is a pleasure to thank Franco Strocchi for key comments on the Higgs model and the anonymous referee for bringing Ref. [29] to my attention. I also thank Antonio Bassetto, Giulio Bonelli, Tommaso Dorigo, Giuseppe Degrassi, Kurt Lechner, Pieralberto Marchetti, Paride Paradisi, Paolo Pasti, Javi Serra, Dima Sorokin, Mario Tonin, Roberto Volpato and Andrea Wulzer for interesting discussions.

References

[1] F. Englert, R. Brout, Phys. Rev. Lett. 13 (1964) 321.
[2] P.W. Higgs, Phys. Lett. 12 (1964) 132.
[3] P.W. Higgs, Phys. Rev. Lett. 13 (1964) 508.
[4] G.S. Guralnik, C.R. Hagen, T.W.B. Kibble, Phys. Rev. Lett. 13 (1964) 585.
[5] P.W. Higgs, Phys. Rev. 145 (1966) 1156.
[6] T.W.B. Kibble, Phys. Rev. 155 (1967) 1554.
[7] S. Weinberg, Phys. Rev. Lett. 19 (1967) 1264.
[8] A. Salam, Conf. Proc. C 680519 (1968) 367.
[9] G. Aad, et al., ATLAS Collaboration, Phys. Lett. B 716 (2012) 1.
[10] S. Chatrchyan, et al., CMS Collaboration, Phys. Lett. B 716 (2012) 30.
[11] R. Fernandez, J. Frohlich, A.D. Sokal, Texts and Monographs in Physics, Springer, Berlin, Germany, 1992, 444 pp.
[12] G. Degrassi, P.P. Giardino, F. Maltoni, D. Pagani, Probing the Higgs self coupling via single Higgs production at the LHC, J. High Energy Phys. 1612 (2016) 080, http://dx.doi.org/10.1007/JHEP12(2016)080, arXiv:1607.04251 [hep-ph].
[13] J. Fröhlich, G. Morchio, F. Strocchi, Higgs phenomenon without a symmetry breaking order parameter, Phys. Lett. B 97 (1980) 249, http://dx.doi.org/10.1016/0370-2693(80)90594-8.
[14] J. Fröhlich, G. Morchio, F. Strocchi, Higgs phenomenon without symmetry breaking order parameter, Nucl. Phys. B 190 (1981) 553, http://dx.doi.org/10.1016/0550-3213(81)90448-X.
[15] J. Hucks, Global structure of the standard model, anomalies, and charge quantization, Phys. Rev. D 43 (1991) 2709, http://dx.doi.org/10.1103/PhysRevD.43.2709.
[16] J.C. Baez, Calabi–Yau manifolds and the standard model, arXiv:hep-th/0511086.
[17] M. Matone, Quantum field perturbation theory revisited, Phys. Rev. D 93 (2016) 065021, http://dx.doi.org/10.1103/PhysRevD.93.065021, arXiv:1506.00987 [hep-th].
[18] M. Atiyah, Symmetry Breaking and Solitons, Talk at the Higgs Symposium, 2013.
[19] K. Pohlmeyer, Nonrenormalizable quantum field theories, in: Proceedings, Renormalization Theory, Erice 1975, Springer, Dordrecht, 1976, pp. 461–482.
[20] S. Albeverio, G. Gallavotti, R. Hoegh-Krohn, Some results for the exponential interaction in two or more dimensions, Commun. Math. Phys. 70 (1979) 187.
[21] F. Strocchi, Symmetry breaking, Lect. Notes Phys. 732 (2008) 1, http://dx.doi.org/10.1007/978-3-540-73593-9.
[22] S. Elitzur, Impossibility of spontaneously breaking local symmetries, Phys. Rev. D 12 (1975) 3978, http://dx.doi.org/10.1103/PhysRevD.12.3978.
[23] E. Tomboulis, SU(2) versus SU(2)/Z_2 lattice gauge theory and the crossover from weak to strong coupling, Phys. Lett. B 108 (1982) 209, http://dx.doi.org/10.1016/0370-2693(82)91177-7.
[24] K. Nomura, A. Kitazawa, SU(2)/Z_2 symmetry of the BKT transition and twisted boundary condition, J. Phys. A 31 (1988) 7341, http://dx.doi.org/10.1088/0305-4470/31/36/008, arXiv:cond-mat/9711294.
[25] R.V. Gavai, M. Mathur, Z_2 monopoles, vortices and the universality of the SU(2) deconfinement transition, Phys. Lett. B 458 (1999) 331, http://dx.doi.org/10.1016/S0370-2693(99)00625-5, arXiv:hep-lat/9905030.
[26] D. Tong, Line operators in the standard model, arXiv:1705.01853 [hep-th].
[27] P. Ramond, Field theory. A modern primer, Front. Phys. 51 (1981) 1.
[28] M. Matone, Dual representation for the generating functional of the Feynman path-integral, Nucl. Phys. B 910 (2016) 309, http://dx.doi.org/10.1016/j.nuclphysb.2016.07.003, arXiv:1511.07408 [hep-th].
[29] J. Galloway, M.A. Luty, Y. Tsai, Y. Zhao, Induced electroweak symmetry breaking and supersymmetric naturalness, Phys. Rev. D 89 (7) (2014) 075003, http://dx.doi.org/10.1103/PhysRevD.89.075003, arXiv:1306.6354 [hep-ph].
[30] S. Chang, J. Galloway, M. Luty, E. Salvioni, Y. Tsai, J. High Energy Phys. 1503 (2015) 017, http://dx.doi.org/10.1007/JHEP03(2015)017, arXiv:1411.6023 [hep-ph].
[31] R. Contino, D. Greco, R. Mahbubani, R. Rattazzi, R. Torre, arXiv:1702.00797 [hep-ph].
[32] F. Strocchi, Elements of Quantum Mechanics of Infinite Systems, SISSA Lecture Ser., vol. 3, World Scientific, Singapore, 1985, 179 pp.
[33] G. Morchio, F. Strocchi, A natural non-perturbative mechanism for gauge hierarchies, Phys. Lett. B 104 (1981) 277.
[34] M.E. Peskin, On the trail of the Higgs boson, Ann. Phys. 528 (1–2) (2016) 20, http://dx.doi.org/10.1002/andp.201500225, arXiv:1506.08185 [hep-ph].

Axion mass bound in very special relativity

R. Bufalo [a,*], S. Upadhyay [b]

[a] Departamento de Física, Universidade Federal de Lavras, Caixa Postal 3037, 37200-000 Lavras, MG, Brazil
[b] Centre for Theoretical Studies, Indian Institute of Technology, Kharagpur, Kharagpur 721302, WB, India

ARTICLE INFO

Editor: A. Ringwald

Keywords:
Very special relativity
Axion electrodynamics
Axion mass
Classical solutions

ABSTRACT

In this paper we propose a very special relativity (VSR)-inspired description of the axion electrodynamics. This proposal is based upon the construction of a proper study of the SIM(2)–VSR gauge-symmetry. It is shown that the VSR nonlocal effects give a health departure from the usual axion field theory. The axionic classical dynamics is analyzed in full detail, first by a discussion of its solution in the presence of an external magnetic field. Next, we compute photon–axion transition in VSR scenario by means of Primakoff interaction, showing the change of a linearly polarized light to a circular one. Afterwards, duality symmetry is discussed in the VSR framework.

1. Introduction

The axion is a hypothetical light and weakly interacting elementary particle postulated by Peccei–Quinn in 1977 as a solution to the strong CP problem in quantum chromodynamics (QCD) associated with a new $U(1)$ symmetry [1–3]. Although it had been initially thought that the *invisible* axion solves the strong CP problem without being amenable to verification by experiments, we have witnessed 40 years of intensive research on axion physics, based on either astrophysical observations or pure laboratory based experiments [4–6]. So far, unfortunately, none was able to yield a positive signature for the axion or an axion-like particle.

Besides being originally proposed as a solution to the strong CP problem, axion-like particles play an important part in explaining unanswered questions of cosmology [7]. Moreover, due to its weakness of their interactions with a sufficiently small mass, axions go as one of the prominent candidates to account for the dark matter in the Universe [8,9].

Notice, however, that non-trivial QCD vacuum effects (e.g., instantons) spoil the Peccei–Quinn symmetry explicitly and provide a small mass for the axion. Hence, the axion is viewed actually as a pseudo-Nambu–Goldstone boson [2,3], with a non-vanishing, but parametrically small mass. On the other hand, instead of considering the traditional Peccei–Quinn mechanism, we will approach the axion dynamics from an alternative point of view, where Lorentz violating effects are responsible to engender massive effects. In this sense, we shall focus in exploring features of VSR [10,11] in this paper.

The cornerstone from the VSR proposal is that the laws of physics are not invariant under the whole Poincaré group but rather under subgroups of the Poincaré group preserving the basic elements of special relativity, but at the same time enhancing the Lorentz algebra by modifying the dynamics of particles. In particular, within this proposal, it is useful in the realization of VSR the use of representations of the full Lorentz group but supplemented by a Lorentz-violating factor, such that the symmetry of the Lagrangian is then reduced to one of the VSR subgroups of the Lorentz group. These effects can then be encoded in the form Lorentz-violating terms in the Lagrangian that are necessarily nonlocal.

As an example, one can observe that a VSR-covariant Dirac equation has the form

$$\left(i\gamma^{\mu}\tilde{\partial}_{\mu} - M\right)\Psi(x) = 0, \tag{1}$$

where the wiggle operator is defined such as $\tilde{\partial}_{\mu} = \partial_{\mu} + \frac{1}{2}\frac{m^2}{n.\partial}n_{\mu}$, with the chosen preferred null direction $n_{\mu} = (1, 0, 0, 1)$ so that it transforms multiplicatively under a VSR transformation. So, by squaring the VSR-covariant Dirac equation we find

$$\left(\partial^{\mu}\partial_{\mu} + \mathscr{M}^2\right)\Psi(x) = 0, \quad \mathscr{M}^2 = M^2 + m^2. \tag{2}$$

* Corresponding author.
E-mail addresses: rodrigo.bufalo@dfi.ufla.br (R. Bufalo), sudhakerupadhyay@gmail.com, sudhaker@iitkgp.ac.in (S. Upadhyay).

We thus immediately realize that conservation laws and the usual relativistic dispersion relation are preserved; moreover, an interesting observable consequence of VSR is to provide a novel mechanism for introducing neutrino masses without the need for new particles [11]. Moreover, the VSR parameter m sets the scale for the VSR effects.

Let us now explain how axion dynamics can be defined in order to encompass Lorentz violating effects. Due to its sensitive tests, photons are always good candidates as test particles in order to probe a physical system [12]. In this sense, we can explore the fact that axions can be converted to photons and vice versa in the presence of magnetic fields [13–17] in order to detect modifications in the axion dynamics, more precisely to probe prominent VSR effects in the theory's dynamics in a significant and novel manner.

As it concerns our interest, VSR-effects have been discussed in the context of electromagnetic theories: Abelian and non-Abelian Maxwell theories [18,19], Chern–Simons theory [20–23], Born-Infeld electrodynamics [24] and higher-spin gauge fields [25–27]. So, in this paper, we shall consider the conversion of photons into axions in the presence of a background magnetic field, in the sense of Primakoff effect, where the VSR will play a part in the photon and axion sectors and will be responsible to engender massive nonlocal effects. It should be stressed that there are ongoing efforts in order to establish axion effects due to an electromagnetic probing [28–31].

In this paper, we will examine the Axion electrodynamics in a VSR setting. We start Sec. 2 by establishing the VSR-axion electromagnetic dynamics main aspects and reviewing the SIM(2)–VSR gauge invariance, which allow us to determine the VSR-modified Abelian field-strength to be used in our analysis. Moreover, we do first compute the solution for the axion field θ in the presence of an external magnetic field in terms of a plane wave solution. In Sec. 3, we compute explicitly the VSR photon–axion transition rate in a Primakoff framework, showing the change of a linearly polarized light to a circular one. Afterwards, in Sec. 4 duality symmetry for the VSR axion electrodynamics is established. In Sec. 5 we summarize the results, and present our final remarks.

2. VSR axion mass

We define the Lagrangian for the axion electrodynamics without source term in VSR as given by

$$\mathcal{L}_{\text{axion}} = -\frac{1}{4}\tilde{F}_{\mu\nu}\tilde{F}^{\mu\nu} + \frac{\kappa}{4}\theta\tilde{F}_{\mu\nu}\tilde{G}^{\mu\nu} + \frac{1}{2}\tilde{\partial}_\mu\theta\tilde{\partial}^\mu\theta, \tag{3}$$

where κ is the dimensionful parameter characterizing the strength of the axion–photon coupling, $\theta(x)$ is a pseudo-scalar field known as the axion-like field, wiggle derivative is defined as before by $\tilde{\partial}_\mu = \partial_\mu + \frac{1}{2}\frac{m^2}{n\cdot\partial}n_\mu$, $\tilde{F}^{\mu\nu}$ and $\tilde{G}^{\mu\nu} = \frac{1}{2}\varepsilon^{\mu\nu\rho\sigma}\tilde{F}_{\rho\sigma}$ are the field-strength and the dual field-strength, respectively. The axion electrodynamics in Lorentz invariant case admits a new internal (gauge) symmetry of the axion-electromagnetic field Lagrangian due to duality transformation [36], between the axion field and the gauge potential, which in turn leads to a conserved current. We discuss this point from a VSR perspective in later section 4.

Notice the absence of a potential for the axion field. The $\theta\tilde{F}_{\mu\nu}\tilde{G}^{\mu\nu}$ term is responsible to provide a solution the strong CP problem, known as Peccei–Quinn solution. It is also known as the effective potential for the axion field, and it is related to the axion mass $m_a^2 = \left\langle\frac{\partial^2 V_{\text{eff}}}{\partial\theta^2}\right\rangle$, generated due to the spontaneous breaking of the $U(1)_{PQ}$ symmetry. As discussed before, we replace this mechanism by VSR nonlocal point-of-view defined in (3), which encompass Lorentz violating effects and are responsible to engender mass for the axion field in such a way that the axion mass has nothing to do with axion–photon coupling κ.

Besides, in order to write-down an expression for $F_{\mu\nu}$ we make use of the usual definition of the raw field-strength $[D_\mu, D_\nu]\phi = -iF_{\mu\nu}\phi$. This is ensured by the construction of a gauge invariant quantity, where the covariant derivative is given by [19]

$$D_\mu\phi = \partial_\mu\phi - iA_\mu\phi + \frac{i}{2}m^2 n_\mu\left(\frac{1}{(n\cdot\partial)^2}(n\cdot A)\right)\phi,$$

which satisfies the transformation law $\delta(D_\mu\phi) = i\Lambda(D_\mu\phi)$, where $\delta A_\mu = \partial_\mu\Lambda$. On the other hand, the raw field-strength $F_{\mu\nu}$ does not coincide with the wiggle operator

$$\tilde{F}_{\mu\nu} = \tilde{\partial}_\mu A_\nu - \tilde{\partial}_\nu A_\mu$$

However, we can realize that the difference between the raw and wiggle field-strength must be gauge invariant as well. So that the wiggle in terms of the usual derivative can be written as [19]

$$\tilde{F}_{\mu\nu} = \partial_\mu A_\nu + \frac{m^2}{2}n_\mu\left(\frac{1}{(n\cdot\partial)^2}\partial_\nu(n\cdot A)\right) - \mu\leftrightarrow\nu, \tag{4}$$

which is gauge invariant and it will be used to describe massive gauge fields.

Lagrangian (3) will now be extensively explored in order to establish some features concerning axion physics, basically it describes how axions can be converted into photons, and vice versa. This basic process, known as Primakoff process, arising from the electromagnetic anomaly and expressed in the effective interaction with coupling constant, underpins many constraints on axions.

First, we will determine solutions for the axion field equation in the presence of an external magnetic field, that can work as a source axion produced in laboratory due to the conversion of photons into axions, which might be seen as an inverse Primakoff process [17]. In the next section, Sec. 3, we will discuss axion–photon interaction via direct Primakoff process, in which we observe the variation of the polarization state of a light wave interacting with the axion field in the presence of an external magnetic field [16].

The sourceless dynamical field equations can be obtained from (3), and for the electromagnetic and axion fields they read

$$\tilde{\partial}_\mu\tilde{F}^{\mu\nu} - \kappa\tilde{G}^{\mu\nu}\tilde{\partial}_\mu\theta = 0, \tag{5}$$

$$\left(\Box + m^2\right)\theta = \frac{\kappa}{4}\tilde{F}_{\mu\nu}\tilde{G}^{\mu\nu}, \tag{6}$$

where the differential identity $\tilde{\Box} = \Box + m^2$ and the Bianchi identity $\tilde{\partial}_\mu\tilde{G}^{\mu\nu} = 0$ have been used. Please notice the presence of massive excitations in (6) that are engendered by VSR effects. If we now make use of the definitions for the electric and magnetic fields $E^i = \tilde{F}^{i0}$ and $B^i = \frac{1}{2}\varepsilon^{ijk}\tilde{F}_{jk}$, respectively, we have

$$\tilde{\nabla}\cdot\mathbf{E} - \kappa\mathbf{B}\cdot\tilde{\nabla}\theta = 0, \tag{7}$$

$$\tilde{\nabla}\times\mathbf{B} - \tilde{\partial}_0\mathbf{E} + \kappa\mathbf{B}\tilde{\partial}_0\theta - \kappa\left(\mathbf{E}\times\tilde{\nabla}\theta\right) = 0, \tag{8}$$

$$\left(\Box + m^2\right)\theta = -\kappa\mathbf{B}\cdot\mathbf{E}. \tag{9}$$

In the VSR setting, $\tilde{\partial}_\mu\tilde{G}^{\mu\nu} = 0$, the complementary electromagnetic field equations read

$$\tilde{\nabla}\cdot\mathbf{B} = 0, \tag{10}$$

$$\tilde{\nabla}\times\mathbf{E} + \tilde{\partial}_0\mathbf{B} = 0. \tag{11}$$

In order to establish the framework of observing axions produced due to a electromagnetic wave we consider a strong uniform background magnetic field $\mathbf{B}_0$, orthogonal to the wave propagation. This can be achieved by means of

$$\tilde{\mathscr{F}}_{\mu\nu} = F^{\text{ext}}_{\mu\nu} + \tilde{F}_{\mu\nu} \tag{12}$$

where $F^{\text{ext}}_{\mu\nu}$ represents the external magnetic field. In this context the field equations Eqs. (7)–(9) are written as

$$\left(\Box + m^2\right)\mathbf{E} - \kappa \mathbf{B}_0 \tilde{\partial}_0^2 \theta = 0, \tag{13}$$

$$\left(\Box + m^2\right)\theta = -\kappa \mathbf{B}_0 \cdot \mathbf{E}. \tag{14}$$

It is important to emphasize that here, the axion mass m is entirely due to VSR effects, and has nothing to do with axion–photon coupling κ. Furthermore, in this setting, Eqs. (13) and (14), both photon and axion fields have the same mass, displaying screened profiles.

A simple setup to determine the solution for the axion field is to take it propagating along the $\hat{x}$ direction, $\phi(x,t)$. Moreover, we can decompose the electric field $\mathbf{E}$ into components perpendicular $E_\perp$ and parallel $E_\parallel$ to the external field $\mathbf{B}_0$, respectively. Within this framework, we get the following coupled field equations

$$\left(\Box + m^2\right)E_\perp = 0, \tag{15}$$

$$\left(\Box + m^2\right)E_\parallel + \kappa B_0 \tilde{\partial}_0^2 \theta = 0 \tag{16}$$

$$\left(\Box + m^2\right)\theta = -\kappa E_\parallel B_0. \tag{17}$$

Notice that the perpendicular component $E_\perp$ does not couple to the axion field.

By simplicity, we can also consider that the background magnetic field $\mathbf{B}_0$ is limited to a region $0 \le x \le L$, while is vanishing outside this region. In this case, we easily see that we can represent the axion field as free plane waves in the noninteracting regions [17].

Now in the interacting region we can make use of the plane wave decomposition

$$E_\parallel = E_0 e^{-i(\omega t - kx)}, \quad \theta = \theta_0 e^{-i(\omega t - kx)}. \tag{18}$$

We then get

$$\left(-\omega^2 + k^2 + m^2\right)E_0 - \kappa \tilde{\omega}^2 B_0 \theta_0 = 0 \tag{19}$$

$$\left(-\omega^2 + k^2 + m^2\right)\theta_0 + \kappa E_0 B_0 = 0, \tag{20}$$

where we have defined $\tilde{\omega}^2 = \omega^2 - \frac{\omega m^2}{n \cdot k} + \frac{m^4}{4(n \cdot k)^2}$. A solution to the axion field θ in this case reads

$$\theta_0(\omega, k) = \kappa \frac{E_0 B_0}{\omega^2 - \omega_m^2}, \tag{21}$$

where $\omega_m = \pm\sqrt{k_m^2 + m^2}$. It is important to notice that this solution exists provided ω and ω_m satisfy the condition

$$\left(-\omega^2 + \omega_m^2\right)\left(\omega^2 - \omega_m^2\right) - \kappa^2 B_0^2 \tilde{\omega}^2 = 0 \tag{22}$$

In particular, if we realize that $n \cdot k = \omega$ we can write

$$\omega^6 - a\omega^4 + b\omega^2 + c = 0 \tag{23}$$

where we have identified $a = 2\omega_m^2 - \kappa^2 B_0^2$, $b = \omega_m^4 - \kappa^2 B_0^2 m^2$ and $c = \frac{\kappa^2}{4} B_0^2 m^4$.

From the three dispersion relation solutions for (23), two of them have an imaginary part, showing a damped behavior of the plane wave in the given region in both solutions. The real and complex dispersion relations read

$$\omega^2 = \frac{a}{3} - \frac{2^{\frac{1}{3}}}{3\Xi^{\frac{1}{3}}}\left(3b - a^2\right) + \frac{\Xi^{\frac{1}{3}}}{32^{\frac{1}{3}}}, \tag{24}$$

$$\omega_\pm^2 = \frac{a}{3} - \frac{1 \pm i\sqrt{3}}{62^{\frac{1}{3}}}\Xi^{\frac{1}{3}} + \frac{1 \pm i\sqrt{3}}{32^{\frac{2}{3}}\Xi^{\frac{1}{3}}}\left(3b - a^2\right) \tag{25}$$

where

$$\Xi = 2a^3 - 9ab - 27c + 3\sqrt{3}\sqrt{27c^2 + 18abc - 4a^3 c + 4b^3 - a^2 b^2}.$$

So general solutions for the axion field, satisfying (17), in the free and interacting regions with the background magnetic field are

$$\begin{cases} \theta_I(t,x) = c_1 e^{-i(\omega_m t + kx)}, \\ \theta_{II}(t,x) = c_2 e^{-i(\omega_m t + kx)} + c_3 e^{-i(\omega_m t - kx)} + \kappa \frac{E_0 B_0}{\omega^2 - \omega_m^2} e^{-i(\omega t - kx)}, \\ \theta_{III}(t,x) = c_4 e^{-i(\omega_m t - kx)}. \end{cases} \tag{26}$$

All the amplitudes in this solution can be uniquely obtained by imposing the continuity conditions at $x = 0$ and $x = L$,[1]

$$\begin{cases} c_1 = \frac{1}{2}\frac{\kappa E_0 B_0}{k_m(k + k_m)}\left(e^{i(k + k_m)L} - 1\right), \\ c_2 = \frac{1}{2}\frac{\kappa E_0 B_0}{k_m(k + k_m)} e^{i(k + k_m)L}, \\ c_3 = -\frac{1}{2}\frac{\kappa E_0 B_0}{k_m(k - k_m)}, \\ c_4 = \frac{1}{2}\frac{\kappa E_0 B_0}{k_m(k - k_m)}\left(e^{i(k - k_m)L} - 1\right). \end{cases} \tag{27}$$

The axion field solutions (26) represent waves outgoing from the interaction region.

A possible use of the solutions (26) is to get an estimate of the axion flux density attainable from an artificial source [17]. This can be achieved by solving the axion and "electric field" coupled equations iteratively, Eqs. (19) and (20), starting from the decoupled equations and considering the first-order contribution in κ onto the axion field. This results into

$$J \simeq \frac{E_e \kappa^2 B_0^2 \omega_m}{m^4}, \tag{28}$$

where $E_e = E_0^2/2$ is the irradiance of a linearly polarized electromagnetic wave (Poynting vector).

In turn, Eq. (28) can be applied to estimate possible production by electromagnetic fields of general axion-like particles, and to collect general information on axion parameters, e.g. their masses and coupling constants. However, it should be noticed that within VSR framework both photon and axion masses have the same origin and are strictly the same, i.e. due to Lorentz violating effects in VSR, see Eqs. (15)–(17).

In particular, it should be emphasized that bounds on photon mass $m_\gamma \le 1.8 \times 10^{-14}$ eV [32] are much more stronger than those in axion mass $m_a \le 1 \times 10^{-2}$ eV [33]. Since, both the masses have the same signature, the stringent bound on photon mass can also be imposed onto the axion's as well.

3. VSR photon–axion transition

We turn our attention to the analysis of photon production due to an axion source. In the presence of a magnetic field, the Primakoff interaction between axions and photons allows for the vacuum to become birefringent and dichroic [16,34]. These effects cause the polarization plane of linearly polarized light to be rotated as it propagates.

[1] Notice that in these expressions we have made use of the notation in terms of the wave numbers k and k_m instead of frequencies ω and ω_m.

In order to investigate the phenomenon, we proceed to compute the photon–axion conversion rate in the VSR scenario. For this matter, we return to the equations of motion (5) and (6), but written now in terms of the vector potential $\mathbf{A}$ instead of the electric field $\mathbf{E}$. In our analysis we consider the radiation gauge, $A_0 = 0$ and $\nabla \cdot \mathbf{A} = 0$. Thus, keeping only linear terms in $\mathbf{A}$ and θ, the classical field equations are written as

$$\left(\Box + m^2\right)\mathbf{A} + \kappa \mathbf{B}\tilde{\partial}_0\theta = 0 \tag{29}$$

$$\left(\Box + m^2\right)\theta - \kappa \mathbf{B} \cdot \tilde{\partial}_0\mathbf{A} = 0 \tag{30}$$

Moreover in the small perturbations $\omega \approx k$ regime (i.e. a WKB limit where we assume that the amplitude varies slowly), we find the linearized system of equations

$$\left(-2\left(\omega + i\partial_x\right) + m^2\right) A_\perp = 0 \tag{31}$$

$$\left(-2\left(\omega + i\partial_x\right) + m^2\right) A_\parallel - i\kappa\tilde{\omega}B_0\theta = 0 \tag{32}$$

$$\left(-2\left(\omega + i\partial_x\right) + m^2\right) \theta + i\kappa\tilde{\omega}B_0 A_\parallel = 0 \tag{33}$$

where we have once again decomposed the potential $\mathbf{A}$ into components perpendicular $A_\perp$ and parallel $A_\parallel$ to the external field. If we introduce a vector

$$\Psi = \begin{pmatrix} \theta_0(x) \\ A_\parallel(x) \\ A_\perp(x) \end{pmatrix} e^{-i\omega x} \tag{34}$$

and identify $\tilde{\omega} = \omega - \frac{m^2}{2\omega}$, we can rewrite the above equations, Eqs. (31)–(33), in a more suitable Schrödinger-like form

$$i\frac{d}{dx}\Psi = \begin{pmatrix} \frac{m^2}{2\omega} & -i\frac{\kappa\tilde{\omega}B_0}{2\omega} & 0 \\ i\frac{\kappa\tilde{\omega}B_0}{2\omega} & \frac{m^2}{2\omega} & 0 \\ 0 & 0 & \frac{m^2}{2\omega} \end{pmatrix} \Psi = \mathbf{M}\Psi \tag{35}$$

where we have defined $\mathbf{M}$ as the mixing matrix.

In order to highlight the VSR effects, let us consider the photon–axion conversion by considering that the axion can only convert in the parallel component $A_\parallel$. Hence,

$$\begin{aligned} i\frac{d}{dx}\begin{pmatrix} \theta_0(x) \\ A_\parallel(x) \end{pmatrix} &= \begin{pmatrix} \frac{m^2}{2\omega} & -i\frac{\kappa\tilde{\omega}B_0}{2\omega} \\ i\frac{\kappa\tilde{\omega}B_0}{2\omega} & \frac{m^2}{2\omega} \end{pmatrix} \begin{pmatrix} \theta_0(x) \\ A_\parallel(x) \end{pmatrix} \\ &= \begin{pmatrix} \Delta_a & -i\Delta_m \\ i\Delta_m & \Delta_\parallel \end{pmatrix} \Phi \end{aligned} \tag{36}$$

Usually the diagonal term involving the vector potential $A_\parallel$ is related to its effective mass due to the Euler–Heisenberg effective Lagrangian, plasma effect (since, in general, the photon does not propagate in vacuum) and Cotton–Mouton effect, i.e. the birefringence of gases and liquids in presence of a magnetic field, so that $\Delta_\parallel = \Delta_{VSR} + \Delta_{EH} + \Delta_{\text{plasma}} + \Delta_{CM}$ [34], that is, it receives further contribution than the VSR one $\Delta_{VSR} = \frac{m^2}{2\omega}$. However, we can see that a photon effective mass is naturally encompassed in the VSR framework, as well as the axion. On the other hand, this might be seen as its bare mass, being corrected by further effects as mentioned.

Now, to compute the photon–axion conversion probability we must first diagonalize the above mixing matrix, whose eigenvalues read

$$\chi_\pm = \frac{\left(\Delta_a + \Delta_\parallel\right) \pm \sqrt{\left(\Delta_a - \Delta_\parallel\right)^2 + 4\Delta_m^2}}{2} \tag{37}$$

However, in the bare case, i.e. taking into account solely VSR effects, we have that

$$\Delta_a = \Delta_\parallel = \frac{m^2}{2\omega} \equiv \Delta. \tag{38}$$

This gives us the following simple relation

$$\chi_\pm = \Delta \pm \Delta_m \tag{39}$$

in which we can assume that the above matrix is diagonalized through an orthonormal transformation $\tilde{\Phi} = \mathbf{O}\Phi$, or even $\mathbf{O}^\dagger\mathbf{M}_\parallel\mathbf{O} = \mathbf{M}_D$,

$$\mathbf{O} = \begin{pmatrix} \cos\varphi & \sin\varphi \\ -\sin\varphi & \cos\varphi \end{pmatrix} \tag{40}$$

where φ is the mixing angle. We thus obtain the following solution for the axion θ_0 and photon $\tilde{A}_\parallel$ fields

$$\begin{aligned} \theta_0(x) &= \left(\cos^2\varphi e^{-i\chi_+ x} + \sin^2\varphi e^{-i\chi_- x}\right)\theta_0(0) \\ &\quad + \cos\varphi\sin\varphi\left(e^{-i\chi_+ x} - e^{-i\chi_- x}\right)A_\parallel(0) \end{aligned} \tag{41}$$

$$\begin{aligned} A_\parallel(x) &= \cos\varphi\sin\varphi\left(e^{-i\chi_+ x} - e^{-i\chi_- x}\right)\theta_0(0) \\ &\quad + \left(\sin^2\varphi e^{-i\chi_+ x} + \cos^2\varphi e^{-i\chi_- x}\right)A_\parallel(0) \end{aligned} \tag{42}$$

With the solutions in hands, we can easily compute the probability of oscillation of a photon after make a distance x starting from the initial state, in which we consider the initial state as $\theta_0(0) = 0$ and $A_\parallel(0) = 1$. Hence, the photon–axion conversion probability can be evaluated as

$$\begin{aligned} P(\gamma \to a) &= \left|\langle A_\parallel(0) | \theta_0(x)\rangle\right|^2 \\ &= \sin^2(2\varphi)\sin^2\left(\frac{(\chi_+ - \chi_-)x}{2}\right) \end{aligned} \tag{43}$$

We can characterize the transition by introducing the oscillation length $\ell_{\text{osc}} = 2\pi/\Delta_{\text{osc}}$, where the oscillation wavenumber reads

$$\Delta_{\text{osc}} = \chi_+ - \chi_- \tag{44}$$

Some remarks are in place. Now, in general, if we had $\Delta_a \neq \Delta_\parallel$, we would have

$$\Delta_{\text{osc}} = \sqrt{\left(\Delta_a - \Delta_\parallel\right)^2 + 4\Delta_m^2} = \frac{2\Delta_m}{\sin(2\varphi)} \tag{45}$$

Notice that a complete transition between a photon and an axion is only possible when the mixing is maximal, i.e. when $\varphi = \pi/4$. However, in our case, notice that due to VSR effects we have the equality $\Delta_a = \Delta_\parallel$, which implies

$$\sin(2\varphi) = \frac{2\Delta_m}{\sqrt{\left(\Delta_a - \Delta_\parallel\right)^2 + 4\Delta_m^2}} = 1 \tag{46}$$

This shows that in the VSR framework we naturally have the strong mixing regime: $\Delta_{\text{osc}} = 2\Delta_m$. Hence, the transition rate (43) is written in its final form

$$P(\gamma \to a) = \sin^2(\Delta_m x) \sim (\Delta_m x)^2 \tag{47}$$

Since only one of the photon components can mix with the axion, in our case $A_\parallel$, so the photon–axion conversion can affect polarization of the photon. This can be analyzed by means of the Stokes parameters [35]

$$I(x) = A_\parallel(x)A_\parallel^*(x) + A_\perp(x)A_\perp^*(x)$$
$$Q(x) = A_\parallel(x)A_\parallel^*(x) - A_\perp(x)A_\perp^*(x)$$
$$U(x) = A_\parallel(x)A_\perp^*(x) + A_\perp(x)A_\parallel^*(x)$$
$$V(x) = i\left(A_\parallel(x)A_\perp^*(x) - A_\perp(x)A_\parallel^*(x)\right)$$

The degree of polarization can be readily defined in terms of such parameters, the circular polarization reads

$$\Pi_C = \frac{|V(x)|}{I(x)} \tag{48}$$

while the linear polarization is

$$\Pi_L = \frac{\sqrt{Q^2(x) + U^2(x)}}{I(x)} \tag{49}$$

In the VSR case, we have that the photon circular polarization is

$$\Pi_C = \left| \frac{U_0(\Delta_m/2)}{I_0 - \frac{1}{4}(I_0 + Q_0)(1 - \cos(2\Delta_m x))} \times \left[\frac{\sin(\chi_+ x)}{\chi_+} - \frac{\sin(\chi_- x)}{\chi_-}\right]\right| \tag{50}$$

where I_0, Q_0, U_0 represent the initial Stokes parameters, we took $V_0 = 0$ in the above relation.

We then explicitly see that in the VSR framework the change of linearly polarized light to be rotated in the Eq. (50) is due to the Primakoff interaction.

4. Duality symmetry and conserved current for VSR axion

In order to conclude our discussion, we shall now present an analysis of the duality symmetry of the axion electrodynamics [36] but now in the VSR framework. It is well known that the sourceless dynamical field equation for the electromagnetic field and their complementary field equations without axion field is invariant under $SO(2)$ rotation by an angle ζ,

$$\begin{pmatrix} \tilde{F}'^{\mu\nu} \\ \tilde{G}'^{\mu\nu} \end{pmatrix} = \begin{pmatrix} \cos\zeta & \sin\zeta \\ -\sin\zeta & \cos\zeta \end{pmatrix} \begin{pmatrix} \tilde{F}^{\mu\nu} \\ \tilde{G}^{\mu\nu} \end{pmatrix}. \tag{51}$$

Now, to analyze the duality symmetry in the VSR modified axion electrodynamics Eq. (3), we apply the $SO(2)$ transformation with $\zeta = \pi/2$ in Eq. (5)

$$\tilde{\partial}_\mu \tilde{G}^{\mu\nu} - \kappa \tilde{F}^{\mu\nu}\tilde{\partial}_\mu\theta = 0. \tag{52}$$

This leads to the new following set of field equations

$$\tilde{\nabla}\cdot\mathbf{B} + \kappa\mathbf{E}\cdot\tilde{\nabla}\theta = 0,$$
$$\tilde{\nabla}\times\mathbf{E} + \kappa(\mathbf{B}\times\tilde{\nabla}\theta) + \tilde{\partial}_0\mathbf{B} + \kappa\mathbf{E}\tilde{\partial}_0\theta = 0. \tag{53}$$

We can obtain the gauge field equations of VSR axion electrodynamics theory by defining the electric and magnetic fields in terms of a new gauge potential $\hat{A}_\mu$ as

$$\mathbf{B} + \kappa\theta\mathbf{E} \equiv \hat{\mathbf{B}} = \tilde{\nabla}\times\hat{\mathbf{A}},$$
$$\mathbf{E} - \kappa\theta\mathbf{B} \equiv \hat{\mathbf{E}} = -\tilde{\partial}_0\hat{\mathbf{A}} - \tilde{\nabla}\hat{A}0. \tag{54}$$

Notice the use of the wiggle derivatives in the above definition. In order to study the conserved current of axion electrodynamics in VSR, we consider that the vector field has the following configuration

$$\mathbf{A} = A_x\hat{i} + B_x y\hat{k}, \tag{55}$$

where A_x depends on time only, while B_x is a constant and equals to the magnitude of the magnetic field, $\tilde{\nabla}\times\mathbf{A} = B_x\hat{i}$. In this scenario, the action describing the axion electrodynamics for θ and the field A_x in VSR is[2]

$$S = \frac{1}{2}\int d^4x\big[\partial_\mu\theta\partial^\mu\theta - m^2\theta^2 + \partial_\mu A_x\partial^\mu A_x - m^2A_x^2 + \kappa B_x(\theta\dot{A}_x - A_x\dot{\theta})\big]. \tag{56}$$

The equations of motion for A_x and θ are given, respectively, by

$$(\Box + m^2)A_x + kB_x\partial_0\theta = 0,$$
$$(\Box + m^2)\theta - \kappa B_x\partial_0 A_x = 0. \tag{57}$$

Here, we observe that the mass terms appear naturally due to VSR effects, without the need of a potential term. The action (56) is invariant under the following gauge symmetry:

$$\delta\theta = A_x\eta,$$
$$\delta A_x = -\theta\eta, \tag{58}$$

where η is an infinitesimal (dimensionless) constant parameter. It is important to emphasize that in the usual framework the transformations (58) are a symmetry of the action (56) only when one of the two conditions is satisfied: either i) photon has a bare mass, equal to the axion mass, or ii) photon and axion fields are massless [36]. Notice that the first condition is naturally satisfied in the VSR framework, showing hence that VSR axion electrodynamics has duality symmetry by construction.

At last, utilizing the Noether' theorem, we are able to calculate conserved charge and current. These are

$$J^0 = \theta\partial_0 A_x - A_x\partial_0\theta + \frac{\kappa}{2}B_x(A_x^2 + \theta^2),$$
$$J^i = \theta(\partial^i A_x) - (\partial^i\theta)A_x. \tag{59}$$

From the above expressions, the conservation of current $\partial_\mu J^\mu = 0$ is evident.

5. Conclusion

In this paper, we have studied a VSR inspired modification of the axion electrodynamics. The analysis consisted in first formulation a SIM(2)–VSR axion electrodynamics, with the expectation that the nonlocal (Lorentz violating) effects would contribute in a novel way showing a distinct departure from the usual theory. Due to the results obtained, a natural extension of the present analysis would be a study concerning QCD, more precisely the strong CP problem in the VSR setting, where the Lorentz violating effects might play an interesting part in the Peccei–Quinn mechanism.

We started with a brief review on the VSR formalism for the Abelian gauge sector so that we have a proper formulation of the VSR axion electrodynamics. In order to extract physical features of the model, we have chosen to exploit Primakoff interaction, i.e. the photon–axion transition. First, we have considered the inverse Primakoff process, the production of axions due to a photons source. In this case, we have fully established the axion field solution in the presence of an external magnetic field.

Next, we have considered the production of photons due to axions source, more precisely we computed the photon–axion conversion probability. In particular, we have shown that in the VSR framework we naturally have the strong mixing regime, i.e. the maximum production of photons due to axions. Besides, we have

[2] Notice, however, that the mass term here is due to VSR effects, i.e. $\tilde{\partial}_\mu\phi\tilde{\partial}^\mu\phi = \partial_\mu\phi\partial^\mu\phi - m^2\phi^2$.

computed the photon circular polarization by means of Stokes parameters, showing in the Primakoff process in a VSR framework the change of linearly polarized light to a circular one.

At last, we have discussed the duality symmetry in the VSR setting. It is remarkable to notice that due the fact that both photon and axion acquire the same mass m due to VSR effects, VSR axion electrodynamics is by construction invariant by duality symmetry.

Acknowledgement

R.B. gratefully acknowledges CNPq for partial support, Project No. 304241/2016-4.

References

[1] R.D. Peccei, H.R. Quinn, CP conservation in the presence of instantons, Phys. Rev. Lett. 38 (1977) 1440.
[2] S. Weinberg, A new light boson?, Phys. Rev. Lett. 40 (1978) 223.
[3] F. Wilczek, Problem of strong p and t invariance in the presence of instantons, Phys. Rev. Lett. 40 (1978) 279.
[4] M.S. Turner, Windows on the axion, Phys. Rep. 197 (1990) 67.
[5] M. Kuster, G. Raffelt, B. Beltran, Axions: Theory, Cosmology, and Experimental Searches, Lect. Notes Phys., vol. 741, Springer-Verlag, Berlin, Heidelberg, 2008.
[6] J. Jaeckel, A. Ringwald, The low-energy frontier of particle physics, Annu. Rev. Nucl. Part. Sci. 60 (2010) 405, arXiv:1002.0329 [hep-ph].
[7] D.J.E. Marsh, Axion cosmology, Phys. Rep. 643 (2016) 1, arXiv:1510.07633 [astro-ph.CO].
[8] A. Ringwald, Exploring the role of axions and other WISPs in the dark universe, Phys. Dark Universe 1 (2012) 116, arXiv:1210.5081 [hep-ph].
[9] G. Ballesteros, J. Redondo, A. Ringwald, C. Tamarit, Unifying inflation with the axion, dark matter, baryogenesis and the seesaw mechanism, Phys. Rev. Lett. 118 (7) (2017) 071802, arXiv:1608.05414 [hep-ph].
[10] A.G. Cohen, S.L. Glashow, Very special relativity, Phys. Rev. Lett. 97 (2006) 021601, arXiv:hep-ph/0601236.
[11] A.G. Cohen, S.L. Glashow, A Lorentz-violating origin of neutrino mass?, arXiv: hep-ph/0605036.
[12] S.L. Adler, Photon splitting and photon dispersion in a strong magnetic field, Ann. Phys. 67 (1971) 599.
[13] P. Sikivie, Experimental tests of the invisible axion, Phys. Rev. Lett. 51 (1983) 1415; Phys. Rev. Lett. 52 (1984) 695 (Erratum).
[14] D.B. Kaplan, Opening the axion window, Nucl. Phys. B 260 (1985) 215.
[15] M. Srednicki, Axion couplings to matter. 1. CP conserving parts, Nucl. Phys. B 260 (1985) 689.
[16] L. Maiani, R. Petronzio, E. Zavattini, Effects of nearly massless, spin zero particles on light propagation in a magnetic field, Phys. Lett. B 175 (1986) 359.
[17] M. Gasperini, Axion production by electromagnetic fields, Phys. Rev. Lett. 59 (1987) 396.
[18] S. Cheon, C. Lee, S.J. Lee, SIM(2)-invariant modifications of electrodynamic theory, Phys. Lett. B 679 (2009) 73, arXiv:0904.2065 [hep-th].
[19] J. Alfaro, V.O. Rivelles, Non Abelian fields in very special relativity, Phys. Rev. D 88 (2013) 085023, arXiv:1305.1577 [hep-th].
[20] J. Vohánka, M. Faizal, Super-Yang–Mills theory in SIM(1) superspace, Phys. Rev. D 91 (4) (2015) 045015, arXiv:1409.6334 [hep-th].
[21] J. Vohánka, M. Faizal, Chern–Simons theory in SIM(1) superspace, Eur. Phys. J. C 75 (12) (2015) 592, arXiv:1503.04761 [hep-th].
[22] A.C. Nayak, R.K. Verma, P. Jain, Effect of VSR invariant Chern–Simon Lagrangian on photon polarization, J. Cosmol. Astropart. Phys. 1507 (2015) 07, 031, arXiv:1504.04921 [hep-ph].
[23] R. Bufalo, SIM(1)–VSR Maxwell–Chern–Simons electrodynamics, Phys. Lett. B 757 (2016) 216, arXiv:1604.00213 [hep-th].
[24] R. Bufalo, Born–Infeld electrodynamics in very special relativity, Phys. Lett. B 746 (2015) 251, arXiv:1505.02483 [hep-th].
[25] S. Upadhyay, Reducible gauge theories in very special relativity, Eur. Phys. J. C 75 (2015) 593, arXiv:1511.01063 [hep-th].
[26] S. Upadhyay, P.K. Panigrahi, Quantum gauge freedom in very special relativity, Nucl. Phys. B 915 (2017) 168, arXiv:1608.03947 [hep-th].
[27] S. Upadhyay, M.B. Shah, P.A. Ganai, Lorentz violating p-form gauge theories in superspace, Eur. Phys. J. C 77 (3) (2017) 157, arXiv:1702.05755 [hep-th].
[28] P. Sikivie, N. Sullivan, D.B. Tanner, Proposal for axion dark matter detection using an LC circuit, Phys. Rev. Lett. 112 (13) (2014) 131301, arXiv:1310.8545 [hep-ph].
[29] B.T. McAllister, S.R. Parker, M.E. Tobar, Axion dark matter coupling to resonant photons via magnetic field, Phys. Rev. Lett. 116 (16) (2016) 161804, arXiv: 1512.05547 [hep-ph]; Phys. Rev. Lett. 117 (15) (2016) 159901 (Erratum).
[30] S. Villalba-Chávez, T. Podszus, C. Müller, Polarization-operator approach to optical signatures of axion-like particles in strong laser pulses, Phys. Lett. B 769 (2017) 233, arXiv:1612.07952 [hep-ph].
[31] V. Anastassopoulos, et al., CAST Collaboration, New CAST limit on the axion–photon interaction, Nat. Phys. 13 (2017) 584, arXiv:1705.02290 [hep-ex].
[32] L. Bonetti, J. Ellis, N.E. Mavromatos, A.S. Sakharov, E.K.G. Sarkisyan-Grinbaum, A.D.A.M. Spallicci, Photon mass limits from fast radio bursts, Phys. Lett. B 757 (2016) 548, arXiv:1602.09135 [astro-ph.HE].
[33] Georg G. Raffelt, Astrophysical axion bounds, in: Axions: Theory, Cosmology, and Experimental Searches, Springer, Berlin, Heidelberg, 2008, pp. 51–71.
[34] C. Deffayet, D. Harari, J.P. Uzan, M. Zaldarriaga, Dimming of supernovae by photon pseudoscalar conversion and the intergalactic plasma, Phys. Rev. D 66 (2002) 043517, arXiv:hep-ph/0112118.
[35] William H. McMaster, Matrix representation of polarization, Rev. Mod. Phys. 33 (1961) 8.
[36] L. Visinelli, Axion-electromagnetic waves, Mod. Phys. Lett. A 28 (35) (2013) 1350162, arXiv:1401.0709 [physics.class-ph].

20

Glueball–baryon interactions in holographic QCD

Si-Wen Li

Department of Modern Physics, University of Science and Technology of China, Hefei 230026, Anhui, China

ARTICLE INFO

Editor: M. Cvetič

ABSTRACT

Studying the Witten–Sakai–Sugimoto model with type IIA string theory, we find the glueball–baryon interaction is predicted in this model. The glueball is identified as the 11D gravitational waves or graviton described by the M5-brane supergravity solution. Employing the relation of M-theory and type IIA string theory, glueball is also 10D gravitational perturbations which are the excited modes by close strings in the bulk of this model. On the other hand, baryon is identified as a D4-brane wrapped on S^4 which is named as baryon vertex, so the glueball–baryon interaction is nothing but the close string/baryon vertex interaction in this model. Since the baryon vertex could be equivalently treated as the instanton configurations on the flavor brane, we identify the glueball–baryon interaction as "graviton–instanton" interaction in order to describe it quantitatively by the quantum mechanical system for the collective modes of baryons. So the effective Hamiltonian can be obtained by considering the gravitational perturbations in the flavor brane action. With this Hamiltonian, the amplitudes and the selection rules of the glueball–baryon interaction can be analytically calculated in the strong coupling limit. We show our calculations explicitly in two characteristic situations which are "scalar and tensor glueball interacting with baryons". Although there is a long way to go, our work provides a holographic way to understand the interactions of baryons in hadronic physics and nuclear physics by the underlying string theory.

1. Introduction

The underlying fundamental theory QCD for nuclear physics and particle physics has achieved great successes. However, nuclear physics remains one of the most difficult and intriguing branches of high energy physics because physicists are still unable to analytically predict the behavior of nuclei or even a single proton. The key problem is that the behavior in the strong-coupling regime of QCD is less clear theoretically. Fortunately, gauge/gravity (gauge/string) duality (see, e.g., [1–5] for a review) has become a revolutionary and powerful tool for studying the strongly coupled quantum field theory. Particularly, the Witten–Sakai–Sugimoto (WSS) model [6–8], as one of the most famous models, has been proposed to holographically study the non-perturbative QCD for a long time [9–19]. Therefore, in this paper, we are going to extend the previous works to study the interactions in holographic QCD.

The holographic glueball–meson interaction has been studied in [20–23] by naturally considering the gravitational waves or graviton in the bulk of this model. Since the gravitational waves or graviton signals the glueball states holographically and mesons are excited by the open string on the flavor branes, the close/open string (on the flavor brane) interaction is definitely interpreted as glueball–meson interaction. And the effective action could be derived by taking account of the gravitational perturbation in the flavor brane action.

On the other hand, in the WSS model, baryon could be identified as a D4′-brane[1] wrapped on S^4, which is named as "baryon vertex" [24,25]. The D4′-brane has to attach the ends of N_c fundamental strings since the S^4 is supported by N_c units of a R–R flux in the supergravity (SUGRA) solution. Such a D4′-brane is realized as a small instanton configuration in the world-volume theory of the flavor branes in this model. Basically, the baryon states could be obtained by quantizing the baryon vertex. In the strong coupling limit (i.e. the t' Hooft coupling constant $\lambda \gg 1$), the two-flavor case (i.e. $N_f = 2$) has been studied in [9] and it turns out that baryons can be described by the $SU(2)$ Belavin–Polyakov–Schwarz–Tyupkin (BPST) instanton solution with a $U(1)$ potential in the world-volume theory of the flavor branes. And employing the soliton picture, baryon states could be obtained by a

E-mail address: cloudk@mail.ustc.edu.cn.

[1] In order to distinguish from N_c D4-branes who are responsible for the background geometry, we denote the baryon vertex as "D4′-brane" in this paper.

holographic quantum mechanical system for collective modes, see also Appendix B.

Accordingly, there must be close string/D4′-brane interaction if the baryon vertex is taken into account, which could be interpreted as the glueball–baryon interaction in this model. Thus we will explore whether or not it is able to describe this interaction by the quantum mechanical system (B.7). So the main contents of this paper are: First, we find that there must be the glueball–baryon interaction in this holographic model. Second, we use the holographic quantum mechanical system in [9] (or Appendix B) to describe the glueball–baryon interaction quantitatively in the $\lambda \gg 1$ limit. Since the analytical instanton configuration with generic numbers of the flavors is not known, only the two-flavor case (i.e. $N_f = 2$) [9] is considered in this paper. The outline of this paper is very simple. In Section 2, we discuss the glueball–baryon interaction in this model and how to describe it quantitatively in the strong coupling limit. Section 3 is the Summary and discussion. Since there are many papers and lectures about the WSS model (such as [6,14]), we will not review this model systematically. Only the relevant parts of this model are collected in Appendix A and Appendix B on which our discussions and calculations are based. Appendix C shows some details of the calculation in our manuscript.

2. The equivalent description of the glueball–baryon interactions

In this section, we will explore that how the "close string interacting with baryon vertex" can be interpreted as the "glueball interacting with baryon" and how to describe it by the holographic quantum mechanical system in [9] (or in Appendix B). First of all, let us take a look at the most general aspects about the interaction of the graviton (close string) in this model.

As a gravity theory, it is very natural to consider the gravitational waves (or graviton) in the bulk of this D4/D8 system (WSS model). According to [20–23], such a gravitational perturbation signals the glueball states and can definitely interact with the open string on the flavor branes. Thus such close/open string interactions have been holographically interpreted as "glueball–meson interactions" or "glueball decays to the mesons" because mesons are excited by the open strings on the flavor branes in this model. Interestingly, once the baryon vertex is taken into account, the interaction between baryon vertex and graviton (or gravitational waves) must occur since the graviton is excited by the close string in the bulk. Hence there must be close string/baryon vertex interaction which could be interpreted as glueball–baryon interaction in this model. It provides a holographic way for understanding and can be treated as a parallel mechanism to "glueball–meson interaction" proposed in [20–23].

Basically, the "glueball–baryon interaction" in this model is nothing but the close/open string (D-brane, baryon vertex) interaction. However, it is not easy to quantitatively describe the close/open string or close string/D-brane interaction by the underlying string theory in a generic spacetime, in order to describe glueball interacting with baryons. Fortunately, according to [6,9,10, 12–14,24,25], baryon vertex is equivalently described by the instanton configuration in world-volume of the D8/$\overline{\text{D8}}$-branes with the BPST solution (B.2). Therefore, the "glueball–baryon interaction" could be identified as the "gravitons (or gravitational waves) interacting with instantons" in the world-volume theory of the flavor branes.

With this idea, let us consider a gravitational perturbation in the bulk geometry since the glueball states are signaled by the graviton in this model, i.e. replace the metric as,

$$g_{MN} \to g_{MN}^{(0)} + h_{MN}, \tag{2.1}$$

where $g_{MN}^{(0)}$ is the background metric (A.1) and h_{MN} is a perturbative tensor which satisfies $h_{MN} \ll g_{MN}^{(0)}$. Then, we consider the interaction (coupling) between graviton (glueball) and instantons in the world-volume of the flavor branes. The dynamic in the world-volume theory of the flavor branes is described by the Yang–Mills action (A.5) plus the Chern–Simons (CS) action (A.8). Since the CS action (A.8) is independent of the metric, it remains to be (A.8) even if (2.1) is imposed. However the Yang–Mills action (A.5) contains additional terms which depend on h_{MN} as,

$$\begin{aligned}
S_{YM} &= S_{YM}^{(0)} + S_{YM}^{(1)} + \mathcal{O}\left(h_{MN}^2\right), \\
S_{YM}^{(0)} &= -\frac{1}{4}(2\pi\alpha')^2 T_8 \int\limits_{\text{D8}/\overline{\text{D8}}} d^4x dU d\Omega_4 e^{-\Phi^{(0)}} \sqrt{-\det g_{ab}^{(0)}} \\
&\quad \times \operatorname{Tr}\left[g^{(0)ac} g^{(0)bd} \mathcal{F}_{ab}\mathcal{F}_{cd}\right], \\
S_{YM}^{(1)} &= \frac{1}{4}(2\pi\alpha')^2 T_8 \int\limits_{\text{D8}/\overline{\text{D8}}} d^4x dU d\Omega_4 e^{-\Phi^{(0)}} (1-\delta\Phi) \sqrt{-\det g_{ab}^{(0)}} \\
&\quad \times \operatorname{Tr}\left[\left(h^{ac} g^{(0)bd} + h^{bd} g^{(0)ac}\right)\mathcal{F}_{ab}\mathcal{F}_{cd}\right]
\end{aligned} \tag{2.2}$$

Notice that only the linear perturbation of h_{MN} is considered in (2.2). So if h_{MN} is the mode of gravitational waves propagating in the bulk, it must depend on time which means $h_{MN} = h_{MN}(t)$. Furthermore, because the baryon states are given by the quantum mechanical system (Appendix B), so once we evaluate the potential term (B.7) in the moduli space by using (B.6) with (2.2), it implies there must be an additionally time-dependent term $H(t)$ to the Hamiltonian (B.7). Therefore, the transition amplitude can be calculated by the standard technique of time-dependent perturbation in quantum mechanics in order to describe the glueball–baryon interaction quantitatively. So let us evaluate the perturbed Hamiltonian $H(t)$ and the transition amplitude explicitly by taking account of two characteristic situations in the following subsections.

2.1. Interactions with scalar glueball

In this section, let us consider the "scalar glueball interacting with baryons". In order to evaluate the perturbed Hamiltonian, we need to write the explicit formulas of the gravitational perturbation first. In this model, the scalar glueball can be described by the gravitational polarization [21–23]. So let us consider the gravitational polarization in 11D SUGRA because the WSS model is based on type IIA SUGRA which can be reduced from M5-brane solution of 11D SUGRA for M-theory (see [26,27] for a complete review). For scalar glueball, the gravitational polarization in 11D SUGRA takes the following forms,

$$\begin{aligned}
H_{44} &= -\frac{r^2}{L^2} F(r) H_E(r) G_E(x), \\
H_{\mu\nu} &= \frac{r^2}{L^2} H_E(r)\left[\frac{1}{4}\eta_{\mu\nu} - \left(\frac{1}{4} + \frac{3r_{KK}^6}{5r^6 - 2r_{KK}^6}\right)\frac{\partial_\mu\partial_\nu}{M_E^2}\right] G_E(x), \\
H_{55} &= \frac{r^2}{4L^2} H_E(r) G_E(x), \\
H_{rr} &= -\frac{L^2}{r^2}\frac{1}{F(r)}\frac{3r_{KK}^6}{5r^6 - 2r_{KK}^6} H_E(r) G_E(x), \\
H_{r\mu} &= \frac{90 r^7 r_{KK}^6}{M_E^2 L^2 \left(5r^6 - 2r_{KK}^6\right)^2} H_E(r)\partial_\mu G_E(x),
\end{aligned} \tag{2.3}$$

We use G_{AB}, H_{AB} to represent the 11D metric and gravitational polarizations in oder to distinguish 10D metric g_{MN} and perturbation h_{MN}. The 11 coordinates correspond to $\left\{x^\mu, x^4, x^5, r, \Omega_4\right\}$ and

$x = \{x^\mu\}$ in our convention. The function $F(r)$ and the relation between 11D (r, r_{KK}, L) and 10D variables (U, z, U_{KK}, R) are given as,

$$F(r) = 1 - \frac{r_{KK}^6}{r^6},\ U = \frac{r^2}{2L},\ 1 + \frac{z^2}{U_{KK}^2} = \frac{r^6}{r_{KK}^6} = \frac{U^3}{U_{KK}^3},\ L = 2R. \tag{2.4}$$

Since the near-horizon solution of M5-branes in 11D is $AdS_7 \times S^4$, the 11D metric satisfies the equations of motion from the following action with the integration on S^4,

$$S_{11D} = \frac{1}{2\kappa_{11}^2}\left(\frac{L}{2}\right)^4 V_4 \int d^7x\sqrt{-\det G}\left(\mathcal{R}_{11D} + \frac{30}{L^2}\right). \tag{2.5}$$

Imposing the near-horizon solution of M5-branes with (2.3) to (2.5), we obtain the eigenvalue equation for $H_E(r)$ as,

$$\frac{1}{r^3}\frac{d}{dr}\left[r\left(r^6 - r_{KK}^6\right)\frac{d}{dr}H_E(r)\right] + \left[\frac{432r^2r_{KK}^{12}}{\left(5r^6 - 2r_{KK}^6\right)^2} + L^4M_E^2\right]H_E(r) = 0, \tag{2.6}$$

and the kinetic action of the function $G_E(x)$ which is,

$$S_{G_E(x)} = C_E\int d^4xdx^4dx^5\frac{1}{2}\left[\left(\partial_\mu G_E\right)^2 + M_E^2G_E^2\right], \tag{2.7}$$

with

$$C_E = \int_{r_{KK}}^{\infty} dr\frac{r^3}{L^3}\frac{5}{8}H_E^2(r). \tag{2.8}$$

Obviously, (2.7) shows why (2.3) signals scalar glueball field. Then we have to translate 11D gravitational polarization (2.3) into 10D WSS model in order to evaluate the Hamiltonian for collective modes. Employing the dimensional reduction as [26,27], the components of 10D h_{MN} are collected by subtracting $g_{MN}^{(0)}$. As a result, they are,

$$\begin{aligned} h_{\mu\nu} &= \left(\frac{U}{R}\right)^{3/2}\left[\frac{R}{2U}H_{55}\eta_{\mu\nu} + \frac{R}{U}H_{\mu\nu}\right], \\ h_{44} &= \left(\frac{U}{R}\right)^{1/2}\left[H_{44} + \frac{1}{2}f(U)H_{55}\right], \\ h_{zz} &= \frac{4R^{3/2}U_{KK}}{9U^{5/2}}\left(\frac{R}{2U}H_{55} + \frac{U_{KK}z^2}{RU^2}H_{rr}\right), \\ h_{z\mu} &= \frac{2U_{KK}z}{3U^2}H_{r\mu},\ h_{\Omega\Omega} = \frac{R^{5/2}}{2U^{1/2}}H_{55}, \end{aligned} \tag{2.9}$$

with the dilaton,

$$e^{4\Phi/3} = \frac{U}{R}\left(1 + \frac{R}{U}H_{55}\right). \tag{2.10}$$

For the reader convenience, we give the explicit form of the equation (2.6) in the z coordinate, which is,

$$0 = H_E''(z) + \frac{U_{KK}^2 + 3z^2}{z\left(U_{KK}^2 + z^2\right)}H_E'(z) + \frac{432U_{KK}^{13/3}\left(U_{KK}^2 + z^2\right)^{1/3} + 4R^3M_E^2\left(3U_{KK}^2 + 5z^2\right)^2}{9U_{KK}^{1/3}\left(U_{KK}^2 + z^2\right)^{4/3}\left(3U_{KK}^2 + 5z^2\right)^2}H_E(z). \tag{2.11}$$

While (2.11) is difficult to solve, we have to search for a solution for H_E in order to evaluate the perturbed Hamiltonian for collective modes. Sine only the $\mathcal{O}(\lambda^0)$ of the Hamiltonian (B.5), (B.7) is the concern in our paper, we need to solve (2.11) up to $\mathcal{O}(\lambda^{-1})$. Rescale (2.11) as (B.1), we obtain the following equation[2] (derivatives are w.r.t. $\mathbf{z}$),

$$H_E''(\mathbf{z}) + \left(\frac{1}{\mathbf{z}} + \frac{2\mathbf{z}}{\lambda}\right)H_E'(\mathbf{z}) + \frac{16 + 3M_E^2}{3\lambda}H_E(\mathbf{z}) + \mathcal{O}\left(\lambda^{-2}\right) = 0. \tag{2.12}$$

In order to compare our calculations with [9], we have employed the unit of $M_{KK} = U_{KK} = 1$ so that $R^3 = 9/4$. The (2.12) is easily to solve in terms of hypergeometric function and Meijer G function which is,

$$\begin{aligned} H_E(\mathbf{z}) = {} & C_1\text{Hypergeometric}_1\text{F}_1\left[\frac{4}{3} + \frac{M_E^2}{4}, 1, -\frac{\mathbf{z}^2}{\lambda}\right] \\ & + C_2\text{MeijerG}\left[\{\#\}, -\frac{1}{3} - \frac{M_E^4}{4}, \{0, 0, \{\#\}\}, \frac{\mathbf{z}^2}{\lambda}\right], \end{aligned} \tag{2.13}$$

where C_1, C_2 are two integration constants. Since the background is the bubble solution of D4-branes, the metric in our model must be regular everywhere. Accordingly, we have to set $C_2 = 0$ because Meijer G function diverges at $U = U_{KK} = 1$. On the other hand, C_1 has to consistently satisfy $C_1 \ll 1$, because H_E appearing in (2.9) should also be the perturbation to the background metric $g_{MN}^{(0)}$. Therefore the solution of H_E is only valid up to $\mathcal{O}(\lambda^{-1})$ according to (2.12). Hence, we have the solution of H_E in the large λ expansion,

$$H_E(\mathbf{z}) \simeq C_1 - C_1\frac{\left(16 + 3M_E^2\right)\mathbf{z}^2}{12\lambda} + \mathcal{O}\left(\lambda^{-2}\right). \tag{2.14}$$

Next, we will evaluate the perturbed Hamiltonian with (2.9) additional to (B.7). Using (B.6) and (2.2),

$$S_{YM}^{(0)} + S_{YM}^{(1)} + S_{CS} = -\int dt\left[U^{(0)}(X^\alpha) - H\left(t, X^\alpha\right)\right]. \tag{2.15}$$

$U^{(0)}(X^\alpha)$ is obtained by evaluating $S_{YM}^{(0)} + S_{CS}$ which is the exact forms of the potential in (B.7). Therefore we need to evaluate $S_{YM}^{(1)}$ in order to obtain the perturbed Hamiltonian $H\left(t, X^\alpha\right)$ in (2.15) and the procedures are as follows,

1. We decompose the $U(2)$ gauge field in $S_{YM}^{(1)}$ (2.2) as (A.9) and use (B.2) to represent the instanton (baryon) in the worldvolume of D8/$\overline{\text{D8}}$-branes.
2. Insert (2.9) into $S_{YM}^{(1)}$ (2.2), rescale the obtained formula of $S_{YM}^{(1)}$ by imposing (B.1) and then expand the result up to $\mathcal{O}(\lambda^{-1})$.
3. Finally, we use (2.15) to evaluate the perturbed Hamiltonian for the collective modes up to $\mathcal{O}(\lambda^{-1})$.

While the above procedures are quite straightforward, the calculation is very messy. So let us give the resultant formula here.[3] The perturbed Hamiltonian can be written as,

$$H_{Scalar}\left(t, X^\alpha\right) = \mathcal{A}\kappa\int d^3\mathbf{x}d\mathbf{z}d\Omega_4\mathcal{K}\left(t, \mathbf{x}, \mathbf{z}, X^\alpha\right), \tag{2.16}$$

[2] "$\mathbf{z}$" is the rescaled coordinate defined as in (B.1).

[3] κ is given in (B.8) and more details of the calculation for (2.17) are given in the Appendix C.

where the function $\mathcal{K}(t, \mathbf{x}, \mathbf{z}, X^{\alpha})$ is given in (C.4) in the unit of $U_{KK} = M_{KK} = 1$. Hence the Hamiltonian is calculated as,

$$\begin{aligned}
&H_{Scalar}(t, X^{\alpha}) \\
&= \mathcal{A}\kappa C_1 \cos(\omega t) \left\{ \frac{9}{2}\pi^2 \left(2 - \frac{5\omega^2}{M_E^2}\right) + \right. \\
&+ \left[\left(\frac{40k^2 - 18M_E^4 + 420\omega^2}{16M_E^2} + \frac{48 + 45\omega^2}{16} \right) \right. \\
&\times \left(2Z^2 + \rho^2\right)\pi^2 \\
&+ \left. \frac{9M_E^2 - 90\omega^2 - 45k^2}{1280M_E^2 a^2\pi^2\rho^2} + 9k^2\rho^2\pi^2 \left(\frac{5\omega^2}{8M_E^2} - \frac{1}{4}\right) \right] \lambda^{-1} \\
&\left. + \mathcal{O}\left(\lambda^{-2}\right) \right\}.
\end{aligned} \tag{2.17}$$

$\mathcal{A}$ is a constant independent on λ.[4] k and ω is the 3-momentum and the frequency of the glueball field $G_E(t, \mathbf{x})$.[5] Notice that (2.17) is suitable to be a perturbation since C_1 has to satisfy $C_1 \ll 1$. Hence, with the quantum mechanical system of baryons (B.7), the average transition amplitude $\mathcal{M}$ and the probability of transition $\mathcal{P}$ can be evaluated by the standard technique in the quantum mechanics with a time-dependent perturbation, which is,

$$\begin{aligned}
\mathcal{P}_{i \to f} &= \left| \int_0^t \langle H(t', X^{\alpha}) e^{-iE_{if}t'} \rangle dt' \right|^2, \\
&= \left| \int_0^t e^{i(E_{if} - \omega)t'} \mathcal{M}(i \to j) dt' \right|,
\end{aligned} \tag{2.18}$$

where $E_{ij} = E\left(l', n'_\rho, n'_z\right) - E\left(l, n_\rho, n_z\right)$ is defined by (B.9). For simplification, let us consider the case of small k limit i.e. $k \to 0$ which means the glueball field, as an external field, is homogeneous. In this limit, it implies $\omega \simeq M_E$ since the "classical glueball field $G_E(t, \mathbf{x})$" means the onshell condition $\omega^2 - k^2 = M_E^2$ has to be satisfied. Thus in small k limit, we can simplify (2.17) as,

$$\begin{aligned}
&H_{Scalar}(t, X^{\alpha}) \\
&= \mathcal{A}\kappa C_1 \cos(\omega t) \left\{ -\frac{27}{2}\pi^2 \right. \\
&+ \left[-\frac{81}{1280a^2\pi^2\rho^2} + \frac{27M_E^2 + 468}{16}\pi^2 \left(2Z^2 + \rho^2\right) \right] \lambda^{-1} \\
&\left. + \mathcal{O}\left(\lambda^{-2}\right) \right\}.
\end{aligned} \tag{2.19}$$

By analyzing the eigenfunctions (B.9) of the Hamiltonian (B.7), we find the following selection rules,

$$\begin{cases} \tilde{l}' = \tilde{l}\ (l' = l) \\ n'_z = n_z \end{cases} \text{or} \begin{cases} \tilde{l}' = \tilde{l}\ (l' = l) \\ n'_z = n_z \pm 2 \\ n'_\rho = n_\rho \end{cases}. \tag{2.20}$$

Working out (2.18), it is easy to find another constraint of the transition which is $\omega = E_{ij}$. Interestingly, our holographical quantum mechanical system is very similar as the atomic spectrum of hydrogen. The "holographic baryon interacting with glueball" behaves similarly as the "electron interacting with photon" in the hydrogen atomic. Both of them can be described by the quantum mechanics which means the baryon (electron) is described by the quantum mechanics while the glueball (photon), as a classically external field, is described by the classical gravity theory (classical electrodynamics) respectively.

Furthermore, let us examine whether or not the transition procedures, in this quantum mechanical system with the constraints and selection rule discussed above, are really possible to occur. We consider the low energy (small momentum) limit as the most simple case which is $k \to 0$, so that $\omega \simeq M_E$. Since M_E represents the mass spectrum of the scalar glueball in (2.11), it reads with the WKB approximation [20] ($\bar{\beta} = 2\pi$ in the unit of $M_{KK} = U_{KK} = 1$),

$$M_E(j) \simeq \frac{8.12}{\bar{\beta}} \sqrt{j\left(j + \frac{5}{2}\right)}. \tag{2.21}$$

Consequently, we find the following transitions,

$$\begin{aligned}
&\frac{E(l = 1, 3; n_\rho = 3; n_z = 0, 1, 2, 3) - E(l = 1, 3; n_\rho = 0; n_z = 0, 1, 2, 3)}{M_E(j = 1)} \\
&\simeq 1.013,
\end{aligned} \tag{2.22}$$

and

$$\begin{aligned}
&\frac{E(l = 1, 3; n_\rho = 5; n_z = 0, 1, 2, 3) - E(l = 1, 3; n_\rho = 0; n_z = 0, 1, 2, 3)}{M_E(j = 2)} \\
&\simeq 1.053,
\end{aligned} \tag{2.23}$$

are possible to occur according to the above selection rules and constraint. Notice that the WSS model is a low-energy effective theory for baryons or mesons, so it may not be very consistent to consider the high energy states of baryons in this model. Using (2.17) and (2.18), the transition amplitude corresponding to (2.22) (2.23) can be calculated, respectively, as,

$$\begin{aligned}
&\mathcal{M}\left(n_\rho = 3 \to n'_\rho = 0\right)\Big|_{l = l', n_z = n'_z} \\
&= \frac{\left(27M_E^2 - 270\omega^2 - 135k^2\right)\mathcal{A}\kappa C_1}{44800M_E^2 a^2\pi^2 m_y^4 \omega_\rho^4 \lambda} \\
&\simeq -\frac{243\mathcal{A}\kappa C_1}{44800a^2\pi^2 m_y^4 \omega_\rho^4 \lambda} + \mathcal{O}\left(k^2\right), \\
&\mathcal{M}\left(n_\rho = 5 \to n'_\rho = 0\right)\Big|_{l = l', n_z = n'_z} \\
&= \frac{\left(9M_E^2 - 90\omega^2 - 45k^2\right)\mathcal{A}\kappa C_1}{53760M_E^2 a^2\pi^2 m_y^4 \omega_\rho^4 \lambda} \\
&\simeq -\frac{81\mathcal{A}\kappa C_1}{63760a^2\pi^2 m_y^4 \omega_\rho^4 \lambda} + \mathcal{O}\left(k^2\right).
\end{aligned} \tag{2.24}$$

2.2. Interactions with tensor glueball

Let us consider another special example for the interaction with tensor glueball. In the bulk, the 11D gravitational polarization of the tensor glueball could be simply chosen as [23],

$$H_{11} = -H_{22} = -\frac{r^2}{L^2} H_T(r) G_T(x), \tag{2.25}$$

where the equation of motion for the radial function H_T is,

---

[4] In the unit of $M_{KK} = U_{KK} = 1$, $\mathcal{A}$ should be $\mathcal{A} = \frac{243}{512\pi^2 l_s^2}$.

[5] Since our theory is symmetrically rotated in the 3d x^i-space, we assume the momentum k of the glueball field $G(t, \mathbf{x})$ is along x^3 direction.

$$\frac{1}{r^3}\frac{d}{dr}\left[r\left(r^6-r_{KK}^6\right)\frac{d}{dr}H_T(r)\right]+L^4M_T^2H_T(r)=0. \tag{2.26}$$

While (2.25) has to be reduced into 10D metric, it satisfies the traceless condition,

$$h_{11}=-h_{22}. \tag{2.27}$$

Inserting (2.27) into (2.2), we can immediately find that the perturbed Hamiltonian of the collective coordinates is vanished. However, the perturbed Hamiltonian from the tensor glueball should be $H_{Tensor}(t,X^\alpha)\sim\mathcal{O}\left(H_{AB}^2\right)$. Since the gravitational polarization (2.25) is solved by the linear gravity perturbation, it would be inconsistent to consider the contribution from $\mathcal{O}\left(H_{AB}^2\right)$ to the Hamiltonian of the collective coordinates.

3. Summary and discussion

In this paper, we consider the linearly gravitational perturbation in the bulk of the Witten–Sakai–Sugimoto model. According to [20–23], such gravitational perturbations signal the glueball states. On the other hand, baryon can be identified as wrapped D-brane which is named as the "baryon vertex" as [24,25]. So in the viewpoints of the string theory, there must be the glueball–baryon interaction if the baryon vertex is taken into account. Therefore the glueball–baryon interaction is nothing but the close string/D-brane (baryon vertex) interaction in this model. Since baryons can be treated as instanton configurations in the world-volume of the flavor branes, we identify the glueball–baryon interaction as "graviton–instanton" interaction as an equivalent description. With the BPST instanton configuration, we find the perturbed Hamiltonian for the collective modes of the baryons could be evaluated quantitatively in the strong coupling limit. Hence the amplitude and the selection rules of the glueball–baryon interaction can be accordingly calculated. In order to quantitatively clarify our idea, we show our methods in two characteristic situations which are "scalar and tensor glueball interacting with baryons". Particularly, the perturbed Hamiltonian of "tensor glueball–baryon" interaction is vanished in the linearly gravitational perturbation which means it should be non-linear interaction in the gravity side.

Our work should be an application of strongly Maldacena's conjecture since we have considered the quantum effect (graviton) in the gravity side. So our conclusions may also be suitable with finite N_c and λ. Moreover, if combining [20–23] with our work, it shows the complete glueball–meson–baryon interaction. Although these holographic approaches are a little different from traditional theories, they show us an analytical way to study the strongly coupled interactions in hadronic physics and nuclear physics by the string theory.

Acknowledgements

I would like to thank Dr. Chao Wu and Prof. Qun Wang for helpful discussions. I was supported partially by the Major State Basic Research Development Program in China under the Grant No. 2015CB856902 and the National Natural Science Foundation of China under the Grant No. 11125524 for this work.

Appendix A. The geometry of the Witten–Sakai–Sugimoto model

In the WSS model, there are N_c coincident D4-branes representing "colors" of QCD, wrapped on a supersymmetry breaking compact circle. The background geometry produced by these D4-branes is described by 10-dimensional type IIA supergravity in the near horizon limit. The metric reads [6],

$$ds^2=\left(\frac{U}{R}\right)^{3/2}\left[\eta_{\mu\nu}dx^\mu dx^\nu+f(U)\left(dx^4\right)^2\right]+\left(\frac{R}{U}\right)^{3/2}\left[\frac{dU^2}{f(U)}+U^2d\Omega_4^2\right], \tag{A.1}$$

which is the bubble geometry of the D4-brane solution. And the dilaton, Romand–Romand 4-form field, the function $f(U)$ are given as,

$$e^\phi\equiv e^{\Phi-\Phi_0}=g_s\left(\frac{U}{R}\right)^{3/4},\quad F_4=dC_3=\frac{2\pi N_c}{V_4}\epsilon_4,$$
$$f(U)=1-\frac{U_{KK}}{U^3}, \tag{A.2}$$

where $x^\mu,\mu=0,1,2,3$ and x^4 are the directions which the D4-branes are extended along. U is the coordinate of the holographic radius and U_{KK} is the coordinate radius of the bottom of the bubble. The relation between R and the string coupling g_s with string length l_s is given as $R^3=\pi g_sN_cl_s^3$. Respectively, $d\Omega_4^2$, ϵ_4 and $V_4=8\pi^2/3$ are the line element, the volume form and the volume of an S^4 with unit radius. We have used x^4 to denote the periodic direction where the D4-branes are wrapped on as $x^4\sim x^4+\delta x^4$ with $\delta x^4=\frac{4\pi}{3}R^{3/2}/U_{KK}^{1/2}$. Accordingly, the Kaluza–Klein mass can be defined as $M_{KK}=2\pi/\delta x^4=\frac{3}{2}U_{KK}^{1/2}/R^{3/2}$. Hence the parameters R, U_{KK}, g_s can be expressed in terms of QCD variables g_{YM}, M_{KK}, l_s as,

$$R^3=\pi g_sN_cl_s^3,\quad U_{KK}=\frac{2}{9}g_{YM}^2N_cM_{KK}l_s^2,\quad g_s=\frac{1}{2\pi}\frac{g_{YM}^2}{M_{KK}l_s}. \tag{A.3}$$

Additionally, the "flavors" of QCD could be introduced into this model by embedding a stack of N_f D8 and anti-D8 branes (D8/$\overline{\text{D8}}$-branes) as probes into the background (A.1). The dynamic of the flavor branes is described by the following action,

$$S_{\mathrm{D8}/\overline{\mathrm{D8}}}=S_{\mathrm{DBI}}+S_{WZ}, \tag{A.4}$$

The first term in (A.4) is the Dirac–Born–Infield (DBI) action and the second term is the Wess–Zumino (WZ) action. The DBI action of D8/$\overline{\text{D8}}$-branes in this model can be expanded in small field strength. Keeping only $\mathcal{O}(\mathcal{F}^2)$, we get the Yang–Mills action for the dual field theory on the flavor branes, which is,[6]

$$S_{\mathrm{DBI}}\simeq S_{YM}+\mathcal{O}(\mathcal{F}^4),$$
$$S_{YM}=-\frac{1}{4}(2\pi\alpha')^2T_8\int_{\mathrm{D8}/\overline{\mathrm{D8}}}d^4xdUd\Omega_4e^{-\Phi}\sqrt{-\det g_{ab}}\times\mathrm{Tr}\left[g^{ac}g^{bd}\mathcal{F}_{ab}\mathcal{F}_{cd}\right]. \tag{A.5}$$

On the other hand, since only C_3 in non-vanished (A.2), the relevant term in WZ action is,

$$S_{WZ}=\frac{1}{3!}\mu_8(2\pi\alpha')^3\int_{D8}C_3\wedge\mathrm{Tr}\mathcal{F}^3=\frac{1}{3!}\mu_8(2\pi\alpha')^3\int_{D8}dC_3\omega_5(\mathcal{A}), \tag{A.6}$$

where $\omega_5(\mathcal{A})$ is Chern–Simons 5-form given as,

$$\omega_5(\mathcal{A})=\mathrm{Tr}\left(\mathcal{A}\mathcal{F}^2-\frac{i}{2}\mathcal{A}^3\mathcal{F}-\frac{1}{10}\mathcal{A}^5\right). \tag{A.7}$$

[6] $\mathcal{F}$ is the dimensionlessful gauge field strength which is defined as $\mathcal{F}=2\pi\alpha'F$.

Since we are going to discuss the two-flavor case i.e. $N_f = 2$, the explicit form of (A.6) after integrating out dC_3 can be written as,

$$S_{WZ} = \frac{N_c}{24\pi^2}\epsilon_{mnpq}\int d^4xdz\Big[\frac{3}{8}\hat{A}_0\mathrm{Tr}(F_{mn}F_{pq}) - \frac{3}{2}\hat{A}_m\mathrm{Tr}(\partial_0 A_n F_{pq}) + \frac{3}{4}\hat{F}_{mn}\mathrm{Tr}(A_0F_{pq}) + \frac{1}{16}\hat{A}_0\hat{F}_{mn}\hat{F}_{pq} - \frac{1}{4}\hat{A}_m\hat{F}_{0n}\hat{F}_{pq} + (\text{total derivatives})\Big] \equiv S_{CS}, \tag{A.8}$$

where the $U(2)$ gauge field $\mathcal{A}$ has been decomposed into its $U(1)$ and $SU(2)$ part as,[7]

$$\mathcal{A} = A^i\frac{\tau^i}{2} + \frac{1}{\sqrt{2N_f}}\hat{A}\times \mathbf{1}_{N_f\times N_f}. \tag{A.9}$$

Notice that (A.8) is expressed in the z coordinate with transformation $U^3 = U_{KK}^3 + U_{KK}z^2$ and the index is defined as $m, n, p, q = 1, 2, 3, z$ in the above equation. τ^is are the Pauli matrices. Hence we have used the Yang–Mills action (A.5) plus Chern–Simons action (A.8) to govern the low energy dynamics on the flavored D8/$\overline{\text{D8}}$-branes in this paper.

Appendix B. Baryon as instanton in the Witten–Sakai–Sugimoto model

In the WSS model, baryon has been provided by a D4′-brane wrapped on S^4, which is named as "baryon vertex". In the world-volume theory of the flavor branes, the coordinates x^M and the $U(2)$ gauge field $\mathcal{A}_M$ need to be rescaled as [9] in order to obtain the variables independent of λ,

$$\begin{aligned} &x^m = \lambda^{-1/2}\mathbf{x}^m,\ x^0 = \mathbf{x}^0,\\ &\mathcal{A}_0(t,x) = \boldsymbol{\mathcal{A}}_0(t,x),\ \mathcal{A}_m(t,x) = \lambda^{1/2}\boldsymbol{\mathcal{A}}_m(t,x),\\ &\mathcal{F}_{0m}(t,x) = \lambda^{1/2}\boldsymbol{\mathcal{F}}_{0m}(t,x),\ \mathcal{F}_{mn}(t,x) = \lambda\boldsymbol{\mathcal{F}}_{mn}(t,x), \end{aligned} \tag{B.1}$$

in the expansion of λ^{-1}. Then by solving the equations of motion the resultantly non-vanished components of the gauge field take the following forms,

$$\begin{aligned} &\hat{\mathbf{A}}_0 = \frac{1}{8\pi^2 a}\frac{1}{\xi^2}\left[1 - \frac{\rho^4}{(\rho^2+\xi^2)^2}\right],\\ &\mathbf{F}_{ij} = Q(\xi,\rho)\epsilon_{ijk}\tau^k,\\ &\mathbf{F}_{zi} = Q(\xi,\rho)\delta_{ij}\tau^j,\\ &Q(\xi,\rho) = \frac{2\rho^2}{(\xi^2+\rho^2)^2},\ a = \frac{1}{216\pi^3}, \end{aligned} \tag{B.2}$$

where

$$\xi^2 = \left(\vec{\mathbf{x}} - \vec{X}\right)^2 + (\mathbf{z} - Z)^2. \tag{B.3}$$

In (B.2), we have used same convention as [9], so that $\vec{\mathbf{x}} = \{\mathbf{x}^i\}$, $i = 1, 2, 3$ represents the 3-spatial coordinates where the baryons or instantons live and ρ represents its size. According to [9] (see [28] for a complete review), the baryon spectrum could be obtained by a quantum mechanical system for the collective coordinates in a moduli space of one instanton. Since we are working in the strong coupling limit (i.e. $\lambda \gg 1$), the contribution of $\mathcal{O}(\lambda^{-1})$ could be neglected. Accordingly the moduli space takes the following topology,

$$\mathcal{M} = \mathbb{R}^4 \times \mathbb{R}^4/\mathbb{Z}_2. \tag{B.4}$$

The first $\mathbb{R}^4$ corresponds to the position of the instanton which is parameterized by the collective coordinates $\left(\vec{X}, Z\right)$ and $\mathbb{R}^4/\mathbb{Z}_2$ is parameterized by the size ρ and the $SU(2)$ orientation of the instanton. $\mathbb{R}^4/\mathbb{Z}_2$ can be parametrized by y_I, $I = 1, 2, 3, 4$ and the size of the instanton corresponds to the radial coordinate i.e. $\rho = \sqrt{y_1^2 + \dots y_4^2}$. The $SU(2)$ orientation is parameterized by $a_I = \frac{y_I}{\rho}$ with the normalized constraint $\sum_{I=1}^4 a_I^2 = 1$.[8] It has been turned out that the Lagrangian of the collective coordinates in such a moduli space is given as,

$$L = \frac{m_X}{2}g_{\alpha\beta}\dot{X}^\alpha\dot{X}^\beta - U(X^\alpha) + \mathcal{O}\left(\lambda^{-1}\right). \tag{B.5}$$

The first term in (B.5) is the line element of the moduli space which corresponds the kinetic term in the Lagrangian while the second term corresponds the potential of this quantum mechanical system. Notice that we have used $X^\alpha = \left(\vec{X}, Z, y_I\right)$, and $m_X = 8\pi^2 a N_c$. The potential term $U(X^\alpha)$ in (B.5) could be calculated by employing the soliton picture as [9,11,28–31], which takes the following form,

$$S_{\mathrm{D8}/\overline{\mathrm{D8}}}^{\text{onshell}} \simeq S_{YM+CS}^{\text{onshell}} = -\int dt U(X^\alpha). \tag{B.6}$$

After quantization, the Hamiltonian corresponding to (B.5) for the collective coordinates is given as,

$$\begin{aligned} &H = M_0 + H_y + H_Z + \mathcal{O}\left(\lambda^{-1}\right),\\ &H_y = -\frac{1}{2m_y}\sum_{I=1}^4\frac{\partial^2}{\partial y_I^2} + \frac{1}{2}m_y\omega_y^2\rho^2 + \frac{\mathcal{Q}}{\rho^2},\\ &H_Z = -\frac{1}{2m_Z}\frac{\partial^2}{\partial Z^2} + \frac{1}{2}m_Z\omega_Z^2 Z^2, \end{aligned} \tag{B.7}$$

where,[9]

$$M_0 = 8\pi^2\kappa,\ \omega_Z^2 = \frac{2}{3},\ \omega_\rho^2 = \frac{1}{6},\ \mathcal{Q} = \frac{N_c}{40\pi^2 a},\ \kappa = \frac{\lambda N_c}{216\pi^3}. \tag{B.8}$$

The eigenfunctions and eigenvalues of (B.7) can be easily evaluated by solving its Schrodinger equation, respectively they are,[10]

$$\begin{aligned} &\psi(y_I) = R(\rho)T^{(l)}(a_I),\\ &R(\rho) = e^{-\frac{m_y\omega_\rho}{2}\rho^2}\rho^{\tilde{l}}\mathit{Hypergeometric}_1F_1\left(-n_\rho, \tilde{l}+2; m_y\omega_\rho\rho^2\right),\\ &E(l, n_\rho, n_z) = \omega_\rho\left(\tilde{l} + 2n_\rho + 2\right) = \sqrt{\frac{(l+1)^2}{6} + \frac{2}{15}N_c^2} + \frac{2(n_\rho + n_z) + 2}{\sqrt{6}}. \end{aligned} \tag{B.9}$$

Notice that $T^{(l)}(a_I)$ is the function of the spherical part which satisfies $\nabla_{S^3}^2 T^{(l)} = -l(l+2)T^{(l)}$ since H_y could be rewritten with the

[7] We have used "ˆ" to represent the Abelian part of the gauge field while the non-Abelian part is expressed without a "ˆ".

[8] Such a parameterization is also used in [11,29,31,30].

[9] Eqs. (B.7)–(B.10) are expressed in the unit of $M_{KK} = U_{KK} = 1$.

[10] The relation of l and $\tilde{l}$ is $\tilde{l} = -1 + \sqrt{(l+1)^2 + 2m_y\mathcal{Q}}$ and the quantum number of the angle momentum can be represented by either l or $\tilde{l}$.

radial coordinate ρ,

$$H_y = -\frac{1}{2m_y}\left[\frac{1}{\rho^3}\partial_\rho(\rho^3\partial_\rho) + \frac{1}{\rho^2}\left(\nabla^2_{S^3} - 2m_y\mathcal{Q}\right)\right] + \frac{1}{2}m_y\omega_\rho^2\rho^2. \tag{B.10}$$

And we have used the quantum numbers $n_z, n_\rho, \tilde{l}$ to denote the eigenvectors of the combined quantum system $H_y + H_Z$ as $|n_z, n_\rho, \tilde{l}\rangle$ in this paper.

Appendix C. Some calculations about the perturbed Hamiltonian

The perturbed Hamiltonian (2.17) can be computed by (2.2) (2.15) or (B.6), equivalently,

$$\begin{aligned}
&-H_{Scalar}\left(t, X^\alpha\right) \\
&= -\frac{1}{4}\left(2\pi\alpha'\right)^2 T_8 \int_{D8/\overline{D8}} d^4xdzd\Omega_4 e^{-\Phi}\sqrt{-\det g_{ab}} \\
&\quad\times \mathrm{Tr}\left[g^{ac}g^{bd}\mathcal{F}_{ab}\mathcal{F}_{cd}\right] \\
&\quad - \left\{-\frac{1}{4}(2\pi\alpha')^2 T_8 \int_{D8/\overline{D8}} d^4xdzd\Omega_4 e^{-\Phi^{(0)}}\sqrt{-\det g^{(0)}_{ab}}\right. \\
&\quad\left.\times \mathrm{Tr}\left[g^{(0)ac}g^{(0)bd}\mathcal{F}_{ab}\mathcal{F}_{cd}\right]\right\}.
\end{aligned} \tag{C.1}$$

With the linear perturbation of gravity, we have,

$$g^{ab} = g^{(0)ab} - h^{ab},\ h^{ab} = g^{(0)ac}g^{(0)bd}h_{cd}. \tag{C.2}$$

Therefore all the functions in (C.1) have been given in (2.3), (2.14), (A.1), (B.2). Rescale the formulas in (C.1) as (B.1), we can obtained the following result by direct computation,

$$H_{Scalar}\left(t, X^\alpha\right) = \frac{1}{4}\left(2\pi\alpha'\right)^2 T_8 \int_{D8/\overline{D8}} d^3\mathbf{x}dzd\Omega_4\mathcal{K}\left(t, \mathbf{x}, \mathbf{z}, X^\alpha\right), \tag{C.3}$$

where

$$\begin{aligned}
&\mathcal{K}\left(t, \mathbf{x}, \mathbf{z}, X^\alpha\right) \\
&= C_1\mathrm{Tr}\left\{\frac{9Q^2\delta_{ij}\tau^i\tau^j}{8M_E^2}\left(2M_E^2 - 5\omega^2\right)G_E(t, \mathbf{x})\right. \\
&\quad + \frac{45k\omega Q\left(\hat{\mathbf{F}}_{02}\tau^1 - \hat{\mathbf{F}}_{01}\tau^2 + \hat{\mathbf{F}}_{0z}\tau^3\right)}{8M_E^2}G_E(x)\lambda^{-1/2} \\
&\quad + \left[-\frac{9}{64M_E^2}\left(\left(5\hat{\mathbf{F}}^2_{0z} - 5\hat{\mathbf{F}}^2_{03}\right)k^2 + \left(7\hat{\mathbf{F}}^2_{0z} - 3\hat{\mathbf{F}}^2_{03}\right)M_E^2\right.\right. \\
&\quad + \left(5\hat{\mathbf{F}}^2_{03} + 5\hat{\mathbf{F}}^2_{0z}\right)\omega^2 \\
&\quad + \left.\left(\hat{\mathbf{F}}^2_{01} + \hat{\mathbf{F}}^2_{02}\right)\left(5k^2 - 3M_E^2 + 5\omega^2\right)\right)G_E(x) \\
&\quad - \frac{15i\omega\mathbf{z}Q\hat{\mathbf{F}}_{0i}\tau^i}{M_E^2}F_E(t, \mathbf{x}) \\
&\quad + \frac{3\mathbf{z}^2Q^2}{32M_E^2}\left(40k^2\left(\left(\tau^1\right)^2 + \left(\tau^2\right)^2 - \left(\tau^3\right)^2\right)\right. \\
&\quad - \left(\left(\tau^1\right)^2 + \left(\tau^2\right)^2 + \left(\tau^3\right)^2\right) \\
&\quad \times \left.\left(6M_E^4 - 140\omega^2 - M_E^2\left(16 + 15\omega^2\right)\right)\right)G_E(t, \mathbf{x})\Big]\lambda^{-1} \\
&\quad + \mathcal{O}\left(\lambda^{-3/2}\right)\Big\}.
\end{aligned} \tag{C.4}$$

Notice that (C.4) is written in the unit of $U_{KK} = M_{KK} = 1$, so that $R^3 = 9/4$. The explicit formula of the glueball field $G_E(t, \mathbf{x})$ is needed in order to work out (2.17). The most simple way is to solve its classical equation of motion from the action (2.7). So we choose the real solution for $G_E(t, \mathbf{x})$ since it also appears in the perturbed metric (2.9) of the bulk geometry, therefore we have

$$\begin{aligned}
G_E(t, \mathbf{x}) &= \frac{e^{-ik_\mu x^\mu} + e^{ik_\mu x^\mu}}{2} = \cos\left(kx^3 - \omega t\right) \\
&= \cos\left(\frac{k\mathbf{x}^3}{\lambda^{1/2}} - \omega t\right),
\end{aligned} \tag{C.5}$$

so that the derivatives of $G_E(t, \mathbf{x})$ in (2.3) can be calculated as,

$$\partial_\mu G_E(t, \mathbf{x}) = \frac{ik_\mu\left(e^{ik_\mu x^\mu} - e^{-ik_\mu x^\mu}\right)}{2} \equiv ik_\mu F_E(t, \mathbf{x}). \tag{C.6}$$

Notice that since the system is rotationally symmetric in x^i-space, we have assumed that the momentum k in (C.5)–(C.6) has only one component along x^3 direction. In order to further simplify (C.4), we calculate the following integrals appearing in (C.3),

$$\begin{aligned}
&\textbf{I)} \int_{-\infty}^{+\infty} d^3\mathbf{x}d\mathbf{z}Q\left(\mathbf{x}, \mathbf{z}\right)^2 G_E(t, \mathbf{x}) \\
&= \frac{1}{3}\frac{k^2\pi^2\rho^2}{\lambda}\mathrm{BesselK}\left[2, \frac{k\rho}{\lambda^{1/2}}\right]\cos(\omega t) \equiv \mathcal{I}_1\cos(\omega t) \\
&\simeq \left[\frac{2}{3}\pi^2 - \frac{1}{6}\frac{\pi^2\rho^2k^2}{\lambda} + \mathcal{O}\left(\lambda^{-2}\right)\right]\cos(\omega t),
\end{aligned} \tag{C.7}$$

$$\begin{aligned}
&\textbf{II)} \int_{-\infty}^{+\infty} d^3\mathbf{x}d\mathbf{z}\mathbf{z}^2 Q\left(\mathbf{x}, \mathbf{z}\right)^2 G_E(t, \mathbf{x}) \\
&= \frac{1}{9}\pi^2\rho^2\left\{\frac{k^2}{\lambda}\left(3Z^2 + \rho^2\right)\mathrm{BesselK}\left[2, \frac{k\rho}{\lambda^{1/2}}\right]\right. \\
&\quad + \left.2\mathrm{MeijerG}\left[\left\{-\frac{1}{2}, \#\right\}; \{0, 1\}, \left\{\frac{1}{2}\right\}; \frac{k^2\rho^2}{4\lambda}\right]\right\}\cos(\omega t) \\
&\equiv \mathcal{I}_2\cos(\omega t) \\
&\simeq \left[\frac{1}{3}\pi^2\left(2Z^2 + \rho^2\right) + \frac{1}{36}\pi^2\rho^2\left(-6Z^2 - 5\rho^2 + 6\gamma\rho^2\right.\right. \\
&\quad + 6\rho^2\log\frac{k}{\lambda^{1/2}} + 3\rho^2\log\frac{\rho^2}{4} \\
&\quad - \left.3\rho^2\mathrm{PolyGamma}\left[0, \frac{3}{2}\right] + 3\rho^2\mathrm{PolyGamma}\left[0, \frac{5}{2}\right]\right)\frac{k^2}{\lambda} \\
&\quad + \left.\mathcal{O}\left(\lambda^{-2}\right)\right]\cos(\omega t),
\end{aligned} \tag{C.8}$$

$$\begin{aligned}
&\textbf{III)} \int_{-\infty}^{+\infty} d^3\mathbf{x}d\mathbf{z}\hat{\mathbf{F}}^2_{03}G_E(t, \mathbf{x}) \\
&= \frac{1}{11520a^2\pi^2\rho^2}\left\{4\mathrm{MeijerG}\left[\left\{-\frac{5}{2}, \#\right\}; \{0, 1\}, \left\{\frac{1}{2}\right\}; \frac{k^2\rho^2}{4\lambda}\right]\right. \\
&\quad + 18\mathrm{MeijerG}\left[\left\{-\frac{3}{2}, \#\right\}; \{0, 2\}, \left\{\frac{1}{2}\right\}; \frac{k^2\rho^2}{4\lambda}\right]
\end{aligned}$$

$$+21\text{MeijerG}\left[\left\{\left\{-\frac{1}{2},\#\right\};\{0,3\},\left\{\frac{1}{2}\right\};\frac{k^2\rho^2}{4\lambda}\right]\right\}\cos(\omega t)$$
$$\equiv \mathcal{I}_3\cos(\omega t)$$
$$\simeq\left[\frac{1}{80a^2\pi^2\rho^2}+\frac{1}{8960a^2\pi^2}\left(-101+70\gamma+70\log\frac{k}{\lambda^{1/2}}+35\log\frac{\rho^2}{4}-35\text{PolyGamma}\left[0,\frac{3}{2}\right]+35\text{PolyGamma}\left[0,\frac{9}{2}\right]\right)\frac{k^2}{\lambda}+\mathcal{O}\left(\lambda^{-2}\right)\right]\cos(\omega t), \quad (C.9)$$

IV) $\int_{-\infty}^{+\infty} d^3\mathbf{x}dz\hat{\mathbf{F}}^2_{01,02,0z}G_E(t,\mathbf{x})$

$$=\frac{1}{11520a^2\pi^2\rho^2}\left\{13\frac{k^3\rho^3}{\lambda^{3/2}}\text{BesselK}\left[3,\frac{k\rho}{\lambda^{1/2}}\right]+16\text{MeijerG}\left[\left\{\left\{-\frac{3}{2},\#\right\};\{0,1\},\left\{\frac{1}{2}\right\};\frac{k^2\rho^2}{4\lambda}\right]+56\text{MeijerG}\left[\left\{\left\{-\frac{1}{2},\#\right\};\{0,2\},\left\{\frac{1}{2}\right\};\frac{k^2\rho^2}{4\lambda}\right]\right\}\cos(\omega t)$$
$$\equiv \mathcal{I}_4\cos(\omega t)$$
$$\simeq\left\{\frac{1}{80a^2\pi^2\rho^2}+\frac{1}{11520a^2\pi^2\rho^2}\left(-49+30\gamma+30\log\frac{k}{\lambda^{1/2}}+15\log\frac{\rho^2}{4}-15\text{PolyGamma}\left[0,\frac{3}{2}\right]+15\text{PolyGamma}\left[0,\frac{7}{2}\right]\right)\frac{k^2}{\lambda}+\mathcal{O}\left(\lambda^{-2}\right)\right\}\cos(\omega t), \quad (C.10)$$

where γ is the Euler–Gamma constant. Using (C.4), (C.7)–(C.10) and $\mathrm{Tr}\{\tau^i\}=0$, we can obtain,

$$H_{Scalar}(t,X^\alpha)=\mathcal{A}\kappa C_1\cos(\omega t)\left\{\frac{27}{4M_E^2}\left(2M_E^2-5\omega^2\right)\mathcal{I}_1+\left[\frac{120k^2-9\left(6M_E^4-140\omega^2-M_E^2\left(16+15\omega^2\right)\right)}{16M_E^2}\mathcal{I}_2+\frac{9\left(3M_E^2-5\omega^2+5k^2\right)}{32M_E^2}\mathcal{I}_3-\frac{9\left(M_E^2+15\omega^2+15k^2\right)}{32M_E^2}\mathcal{I}_4\right]\lambda^{-1}+\mathcal{O}\left(\lambda^{-2}\right)\right\}. \quad (C.11)$$

Inserting the expansion of large λ of $\mathcal{I}_{1,2,3,4}$, then (2.17) can be obtained.

References

[1] O. Aharony, S.S. Gubser, J. Maldacena, H. Ooguri, Y. Oz, Large N field theories, string theory and gravity, Phys. Rep. 323 (2000) 183, arXiv:hep-th/9905111.

[2] J.M. Maldacena, The large-N limit of superconformal field theories and supergravity, Int. J. Theor. Phys. 38 (1999) 1113, Adv. Theor. Math. Phys. 2 (1998) 231, arXiv:hep-th/9711200.

[3] E. Witten, Anti-de Sitter space and holography, Adv. Theor. Math. Phys. 2 (1998) 253, arXiv:hep-th/9802150.

[4] J.L. Petersen, Introduction to the Maldacena conjecture on AdS/CFT, Int. J. Mod. Phys. A 14 (1999) 3597, arXiv:hep-th/9902131.

[5] J. Casalderrey-Solana, H. Liu, D. Mateos, K. Rajagopal, U.A. Wiedemann, Gauge/string duality, hot QCD and heavy ion collisions, arXiv:1101.0618 [hep-th].

[6] T. Sakai, S. Sugimoto, Low energy hadron physics in holographic QCD, Prog. Theor. Phys. 113 (2005) 843, arXiv:hep-th/0412141.

[7] T. Sakai, S. Sugimoto, More on a holographic dual of QCD, Prog. Theor. Phys. 114 (2005) 1083, arXiv:hep-th/0507073.

[8] T. Imoto, T. Sakai, S. Sugimoto, Mesons as open strings in a holographic dual of QCD, Prog. Theor. Phys. 124 (2010) 263, arXiv:1005.0655 [hep-th].

[9] Hiroyuki Hata, Tadakatsu Sakai, Shigeki Sugimoto, Shinichiro Yamato, Baryons from instantons in holographic QCD, Prog. Theor. Phys. 117 (2007) 1157, arXiv: hep-th/0701280.

[10] K. Hashimoto, T. Sakai, S. Sugimoto, Nuclear force from string theory, Prog. Theor. Phys. 122 (2009) 427, arXiv:0901.4449 [hep-th], [hep-th]].

[11] K. Hashimoto, N. Iizuka, Y. Piljin, A matrix model for baryons and nuclear forces, J. High Energy Phys. 1010 (2010) 003, arXiv:1003.4988 [hep-th].

[12] Oren Bergman, Gilad Lifschytz, Matthew Lippert, Holographic nuclear physics, J. High Energy Phys. 0711 (2007) 056, arXiv:0708.0326.

[13] Si-wen Li, Andreas Schmitt, Qun Wang, From holography towards real-world nuclear matter, Phys. Rev. D 92 (2015) 026006, arXiv:1505.04886.

[14] Anton Rebhan, The Witten–Sakai–Sugimoto model: a brief review and some recent results, arXiv:1410.8858.

[15] Francesco Bigazzi, Aldo L. Cotrone, Holographic QCD with dynamical flavors, J. High Energy Phys. 01 (2015) 104, arXiv:1410.2443.

[16] Si-wen Li, Tuo Jia, Dynamically flavored description of holographic QCD in the presence of a magnetic field, arXiv:1604.07197.

[17] Florian Preis, Anton Rebhan, Andreas Schmitt, Inverse magnetic catalysis in dense holographic matter, J. High Energy Phys. 1103 (2011) 033, arXiv:1012.4785.

[18] Moshe Rozali, Hsien-Hang Shieh, Mark Van Raamsdonk, Jackson Wu, Cold nuclear matter in holographic QCD, J. High Energy Phys. 0801 (2008) 053, arXiv: 0708.1322.

[19] Kazuo Ghoroku, Kouki Kubo, Motoi Tachibana, Tomoki Taminato, Fumihiko Toyoda, Holographic cold nuclear matter as dilute instanton gas, Phys. Rev. D 87 (6) (2013) 066006, arXiv:1211.2499.

[20] Neil R. Constable, Robert C. Myers, Spin-two glueballs, positive energy theorems and the AdS/CFT correspondence, J. High Energy Phys. 9910 (1999) 037, arXiv:hep-th/9908175.

[21] Richard C. Brower, Samir D. Mathur, Chung-I. Tan, Glueball spectrum for QCD from AdS supergravity duality, Nucl. Phys. B 587 (2000) 249–276, arXiv:hep-th/0003115.

[22] Koji Hashimoto, Chung-I Tan, Seiji Terashima, Glueball decay in holographic QCD, Phys. Rev. D 77 (2008) 086001, arXiv:0709.2208.

[23] Frederic Brünner, Denis Parganlija, Anton Rebhan, Glueball decay rates in the Witten–Sakai–Sugimoto model, Phys. Rev. D 91 (2015) 106002, arXiv:1501.07906.

[24] Edward Witten, Baryons and branes in anti-de Sitter space, J. High Energy Phys. 9807 (1998) 006, arXiv:hep-th/9805112.

[25] David J. Gross, Hirosi Ooguri, Aspects of large N gauge theory dynamics as seen by string theory, Phys. Rev. D 58 (1998) 106002, arXiv:hep-th/9805129.

[26] Edward Witten, Anti-de Sitter space, thermal phase transition, and confinement in gauge theories, Adv. Theor. Math. Phys. 2 (1998) 505–532, arXiv:hep-th/9803131.

[27] Edward Witten, String theory dynamics in various dimensions, Nucl. Phys. B 443 (1995) 85–126, arXiv:hep-th/9503124.

[28] David Tong, TASI lectures on solitons: instantons, monopoles, vortices and kinks, arXiv:hep-th/0509216, 2005.

[29] Koji Hashimoto, Norihiro Iizuka, Three-body nuclear forces from a matrix model, J. High Energy Phys. 1011 (2010) 058, arXiv:1005.4412.

[30] Si-wen Li, Tuo Jia, Matrix model and holographic baryons in the D0–D4 background, Phys. Rev. D 92 (2015) 046007, arXiv:1506.00068.

[31] Si-wen Li, Tuo Jia, Three-body force for baryons from the D0–D4/D8 matrix model, Phys. Rev. D 93 (2016) 065051, arXiv:1602.02259.

Hadronic structure functions in the $e^+e^- \to \bar{\Lambda}\Lambda$ reaction

Göran Fäldt *, Andrzej Kupsc

Division of Nuclear Physics, Department of Physics and Astronomy, Uppsala University, Box 516, 75120 Uppsala, Sweden

A R T I C L E   I N F O

Editor: J.-P. Blaizot

Keywords:
Hadron production in e^-e^+ interactions
Hadronic decays

A B S T R A C T

Cross-section distributions are calculated for the reaction $e^+e^- \to J/\psi \to \bar{\Lambda}(\to \bar{p}\pi^+)\Lambda(\to p\pi^-)$, and related annihilation reactions mediated by vector mesons. The hyperon-decay distributions depend on a number of structure functions that are bilinear in the, possibly complex, psionic form factors G_M^ψ and G_E^ψ of the Lambda hyperon. The relative size and relative phase of these form factors can be uniquely determined from the unpolarized joint-decay distributions of the Lambda and anti-Lambda hyperons. Also the decay-asymmetry parameters of Lambda and anti-Lambda hyperons can be determined.

1. Introduction

Two hadronic form factors, commonly called $G_M(s)$ and $G_E(s)$, are needed for the description of the annihilation process $e^-e^+ \to \Lambda\bar{\Lambda}$, Fig. 1a, and by varying the c.m. energy $\sqrt{s}$, their numerical values can in principle be determined for all s values above $\Lambda\bar{\Lambda}$ threshold. For the general case of annihilation via an intermediate photon, the joint $\Lambda(\to p\pi^-)\bar{\Lambda}(\to \bar{p}\pi^+)$ decay distributions were calculated and analyzed in Ref. [1], using methods developed in [2,3]. Recently, a first attempt to calculate the hyperon form factors $G_M(s)$ and $G_E(s)$ in the time-like region was reported in Ref. [4].

Previously, the interesting special case of annihilation through an intermediate J/ψ or $\psi(2S)$, Fig. 1b, has been investigated in several theoretical [5,6] and experimental papers [7–9]. This process has also been used for determination of the anti-Lambda decay-asymmetry parameter and for CP symmetry tests in the hyperon system. A precise knowledge of the Lambda decay-asymmetry parameter is needed for studies of spin polarization in Ω^-, Ξ^-, and Λ_c^+ decays.

Presently, a collected data sample of 1.31×10^9 J/ψ events [10] by the BESIII detector [11] permits high-precision studies of spin correlations.

In the experimental work referred to above, the joint-hyperon-decay distributions considered are not the most general ones possible, but seem to be curtailed. Incomplete distribution functions do not permit a reliable determination of the form factors and we

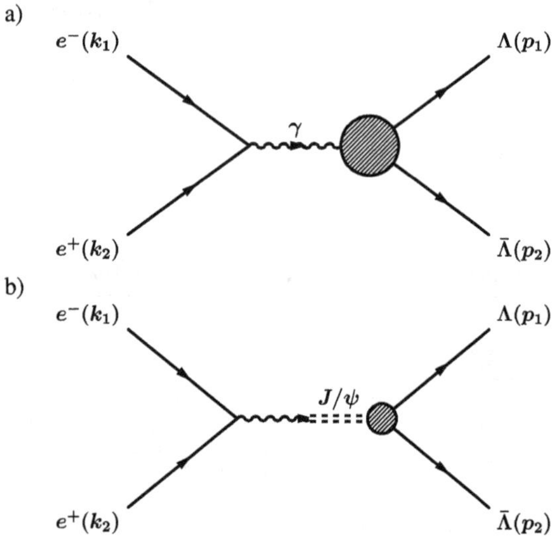

Fig. 1. Graph describing the reaction $e^+e^- \to \bar{\Lambda}\Lambda$; a) general case, and b) mediated by the J/ψ resonance.

therefore suggest to fit the experimental data to the general distribution described in [1], and further elaborated below.

Since the photon and the J/ψ are both vector particles, their corresponding annihilation processes will be similar. In fact, by a simple substitution, the cross-section distributions in Ref. [1], valid in the photon case, are transformed into distributions valid in the J/ψ case, but expressed in the corresponding psionic form factors G_M^ψ and G_E^ψ.

* Corresponding author.
E-mail addresses: goran.faldt@physics.uu.se (G. Fäldt), andrzej.kupsc@physics.uu.se (A. Kupsc).

In order to specify events and compare measured data with theoretical predictions, we need distribution functions expressed in some specific coordinate system. For this purpose we employ the coordinate system introduced in [1]. Many investigations employ different coordinate systems for the Lambda and anti-Lambda decays, a custom which in our opinion can lead to confusion.

Our calculation is performed in two steps. After some preliminaries we turn to the inclusive process of lepton annihilation into polarized hyperons. The results obtained are the starting point for the calculation of exclusive annihilation, i.e. the distribution for the hyperon-decay products. Our method of calculation consists in multiplying the hyperon-production distribution with the hyperon-decay distributions, averaging over intermediate hyperon-spin directions. The method is referred to as folding.

2. Basic necessities

Resolving the hyperon vertex in Fig. 1a uncovers a number of contributions. The one of interest to us is described by the diagram of Fig. 1b, whereby the photon interaction with the hyperons is mediated by the J/ψ vector meson, and the coupling of the initial-state leptons to the J/ψ related to the decay $J/\psi \to e^+e^-$.

For a J/ψ decay through an intermediate photon, tensor couplings can be ignored. Thus, the effective coupling of the J/ψ to the leptons is the same as that for the photon, provided we replace the electric charge e_{em} by a coupling strength e_ψ,

$$\Gamma^e_\mu(k_1,k_2) = -ie_\psi\gamma_\mu, \tag{2.1}$$

with e_ψ determined by the $J/\psi \to e^+e^-$ decay (see Appendix A).

At the J/ψ-hyperon vertex two form factors are possible and they are both considered. We follow Ref. [1] in writing the hyperon vertex as

$$\Gamma^\Lambda_\mu(p_1,p_2) = -ie_g\Big[G^\psi_M\gamma_\mu - \frac{2M}{Q^2}(G^\psi_M - G^\psi_E)Q_\mu\Big], \tag{2.2}$$

with $P = p_1 + p_2$, and $Q = p_1 - p_2$, and M the Lambda mass. The argument of the form factors equals $s = P^2$. The coupling strength e_g in Eq. (2.2) is determined by the hadronic-decay rate for $J/\psi \to \Lambda\bar{\Lambda}$ (see Appendix A).

In Ref. [1] polarizations and cross-section distributions were expressed in terms of structure functions, themselves functions of the form factors G^ψ_M and G^ψ_E. Here, we shall introduce combinations of form factors called D, α, and $\Delta\Phi$, which are employed by the experimental groups [7–9] as well.

The strength of form factors is measured by $D(s)$,

$$D(s) = s\left|G^\psi_M\right|^2 + 4M^2\left|G^\psi_E\right|^2, \tag{2.3}$$

a factor that multiplies all cross-section distributions. The ratio of form factors is measured by α,

$$\alpha = \frac{s\left|G^\psi_M\right|^2 - 4M^2\left|G^\psi_E\right|^2}{s\left|G^\psi_M\right|^2 + 4M^2\left|G^\psi_E\right|^2}, \tag{2.4}$$

with α satisfying $-1 \le \alpha \le 1$. The relative phase of form factors is measured by $\Delta\Phi$,

$$\frac{G^\psi_E}{G^\psi_M} = e^{i\Delta\Phi}\left|\frac{G^\psi_E}{G^\psi_M}\right|. \tag{2.5}$$

The diagram of Fig. 1 represents a J/ψ exchange of momentum P. J/ψ being a vector meson, its propagator takes the form

$$\frac{g_{\mu\nu} - P_\mu P_\nu/m^2_\psi}{s - m^2_\psi + im_\psi\Gamma(\psi)}, \tag{2.6}$$

where m_ψ is the J/ψ mass, and $\Gamma(\psi)$ the full width of the J/ψ. However, since the J/ψ couples to conserved lepton and hyperon currents, the contribution from the $P_\mu P_\nu$ term vanishes. In conclusion, the matrix element for e^+e^- annihilation through a photon will be structurally identical to that for annihilation through a J/ψ provided we make the replacement

$$\frac{e_\psi e_g}{s - m^2_\psi + im^2_\psi\Gamma(\psi)} \to \frac{e^2_{em}}{s}, \tag{2.7}$$

where e_{em} is the electric charge.

3. Cross section for $e^+e^- \to \Lambda(s_1)\bar{\Lambda}(s_2)$

Our first task is to calculate the cross-section distribution for e^+e^- annihilation into polarized hyperons. From the squared matrix element $|\mathcal{M}|^2$ for this process we remove a factor $\mathcal{K}_\psi$, to get

$$d\sigma = \frac{1}{2s}\mathcal{K}_\psi\,|\mathcal{M}_{red}|^2\ \mathrm{dLips}(k_1+k_2;p_1,p_2), \tag{3.8}$$

with dLips the phase-space factor, with $s = P^2$, and with

$$\mathcal{K}_\psi = \frac{e^2_\psi e^2_g}{(s-m^2_\psi)^2 + m^2_\psi\Gamma^2(m_\psi)}. \tag{3.9}$$

The square of the reduced matrix element can be factorized as

$$\left|\mathcal{M}_{red}(e^+e^- \to \Lambda(s_1)\overline{\Lambda}(s_2))\right|^2 = L\cdot K(s_1,s_2), \tag{3.10}$$

with $L(k_1,k_2)$ and $K(p_1,p_2;s_1,s_2)$ lepton and hadron tensors, and s_1 and s_2 hyperon spin four-vectors.

Lepton tensor including averages over lepton spins;

$$\begin{aligned} L_{\nu\mu}(k_1,k_2) &= \tfrac{1}{4}\mathrm{Tr}\big[\gamma_\nu \not{k}_1\gamma_\mu \not{k}_2\big] \\ &= k_{1\nu}k_{2\mu} + k_{2\nu}k_{1\mu} - \tfrac{1}{2}sg_{\nu\mu}. \end{aligned} \tag{3.11}$$

Hadron tensor for polarized hyperons;

$$\begin{aligned} K_{\nu\mu}(s_1,s_2) = \mathrm{Tr}\Big[&\overline{\Gamma}^\Lambda_\nu(\not{p}_1 + M)\tfrac{1}{2}(1+\gamma_5\not{s}_1) \\ &\times \Gamma^\Lambda_\mu(\not{p}_2 - M)\tfrac{1}{2}(1+\gamma_5\not{s}_2)\Big]/e^2_g, \end{aligned} \tag{3.12}$$

with p_1 and s_1 momentum and spin for the Lambda hyperon and p_2 and s_2 correspondingly for the anti-Lambda hyperon. The trace itself is symmetric in the two hyperon variables.

The spin four-vector $s(\mathbf{p},\mathbf{n})$ of a hyperon of mass M, three-momentum $\mathbf{p}$, and spin direction $\mathbf{n}$ in its rest system, is

$$s(\mathbf{p},\mathbf{n}) = \frac{n_\parallel}{M}(|\mathbf{p}|;E\hat{\mathbf{p}}) + (0;\mathbf{n}_\perp). \tag{3.13}$$

Here, longitudinal and transverse designations refer to the $\hat{\mathbf{p}}$ direction; $n_\parallel = \mathbf{n}\cdot\hat{\mathbf{p}}$ and $\mathbf{n}_\perp = \mathbf{n} - \hat{\mathbf{p}}(\mathbf{n}\cdot\hat{\mathbf{p}})$ are parallel and transverse components of the spin vector $\mathbf{n}$. Also, observe that the four-vectors p and s are orthogonal, i.e. $p\cdot s(p) = 0$.

For the evaluation of the matrix element we turn to the global c.m. system where kinematics simplifies. Here, three-momenta $\mathbf{p}$ and $\mathbf{k}$ are defined such that

$$\mathbf{p}_1 = -\mathbf{p}_2 = \mathbf{p}, \tag{3.14}$$

$$\mathbf{k}_1 = -\mathbf{k}_2 = \mathbf{k}, \tag{3.15}$$

and scattering angle by,

$$\cos\theta = \hat{\mathbf{p}}\cdot\hat{\mathbf{k}}. \tag{3.16}$$

The phase-space factor becomes

$$\text{dLips}(k_1+k_2;p_1,p_2)=\frac{p}{32\pi^2 k}\,d\Omega, \tag{3.17}$$

with $p=|\mathbf{p}|$ and $k=|\mathbf{k}|$.

The matrix element in Eq. (3.10) can be written as a sum of four terms that depend on the hyperon spin directions in their respective rest systems, $\mathbf{n}_1$ and $\mathbf{n}_2$,

$$\left|\mathcal{M}_{red}(e^+e^-\to\Lambda(s_1)\overline{\Lambda}(s_2))\right|^2$$
$$=sD(s)\left[H^{00}(0,0)+H^{05}(\mathbf{n}_1,0)+H^{50}(0,\mathbf{n}_2)+H^{55}(\mathbf{n}_1,\mathbf{n}_2)\right]. \tag{3.18}$$

The polarization distributions H^{ab} are each expressed in terms of structure functions that depend on the scattering angle θ, the ratio function $\alpha(s)$, and the phase function $\Delta\Phi(s)$. There are six such structure functions,

$$\mathcal{R}=1+\alpha\cos^2\theta, \tag{3.19}$$

$$\mathcal{S}=\sqrt{1-\alpha^2}\sin\theta\cos\theta\sin(\Delta\Phi), \tag{3.20}$$

$$\mathcal{T}_1=\alpha+\cos^2\theta, \tag{3.21}$$

$$\mathcal{T}_2=-\alpha\sin^2\theta, \tag{3.22}$$

$$\mathcal{T}_3=1+\alpha, \tag{3.23}$$

$$\mathcal{T}_4=\sqrt{1-\alpha^2}\cos\theta\cos(\Delta\Phi). \tag{3.24}$$

The definitions and notations are slightly different from those of Ref. [1]. In particular, a factor $sD(s)$ has been pulled out from the structure functions, and placed in front of the sum of the polarization distributions of Eq. (3.18).

The polarization distributions H^{ab} are,

$$H^{00}=\mathcal{R} \tag{3.25}$$

$$H^{05}=\mathcal{S}\left[\frac{1}{\sin\theta}(\hat{\mathbf{p}}\times\hat{\mathbf{k}})\cdot\mathbf{n}_1\right] \tag{3.26}$$

$$H^{50}=\mathcal{S}\left[\frac{1}{\sin\theta}(\hat{\mathbf{p}}\times\hat{\mathbf{k}})\cdot\mathbf{n}_2\right] \tag{3.27}$$

$$H^{55}=\Big\{\mathcal{T}_1\mathbf{n}_1\cdot\hat{\mathbf{p}}\mathbf{n}_2\cdot\hat{\mathbf{p}}+\mathcal{T}_2\mathbf{n}_{1\perp}\cdot\mathbf{n}_{2\perp}$$
$$+\mathcal{T}_3\mathbf{n}_{1\perp}\cdot\hat{\mathbf{k}}\mathbf{n}_{2\perp}\cdot\hat{\mathbf{k}}$$
$$+\mathcal{T}_4\left(\mathbf{n}_1\cdot\hat{\mathbf{p}}\mathbf{n}_{2\perp}\cdot\hat{\mathbf{k}}+\mathbf{n}_2\cdot\hat{\mathbf{p}}\mathbf{n}_{1\perp}\cdot\hat{\mathbf{k}}\right)\Big\} \tag{3.28}$$

Transverse components $\mathbf{n}_{1\perp}$ and $\mathbf{n}_{2\perp}$ are orthogonal to the Lambda hyperon momentum $\mathbf{p}$ in the global c.m. system. Also, transverse $\mathbf{n}_\perp$ and longitudinal $n_\parallel=\hat{\mathbf{p}}\cdot\mathbf{n}$ polarization components enter differently, since they transform differently under Lorentz transformations.

All polarization observables, single and double, can be directly read off Eqs. (3.25)–(3.28), and there are no other possibilities. The set of scalar products involving $\mathbf{n}_1$ and $\mathbf{n}_2$ is complete. As an example, the Lambda-hyperon polarization is obtained from Eq. (3.26) which shows that the polarization is directed along the normal to the scattering plane, $\hat{\mathbf{p}}\times\hat{\mathbf{k}}$, and that the value of the polarization is

$$P_\Lambda(\theta)=\frac{\mathcal{S}}{\mathcal{R}}=\frac{\sqrt{1-\alpha^2}\cos\theta\sin\theta}{1+\alpha\cos^2\theta}\sin(\Delta\Phi) \tag{3.29}$$

From Eq. (3.27) we conclude that the polarization of the anti-Lambda is exactly the same, but then one should remember that $\mathbf{p}$ is the momentum of the Lambda hyperon but $-\mathbf{p}$ that of the anti-Lambda.

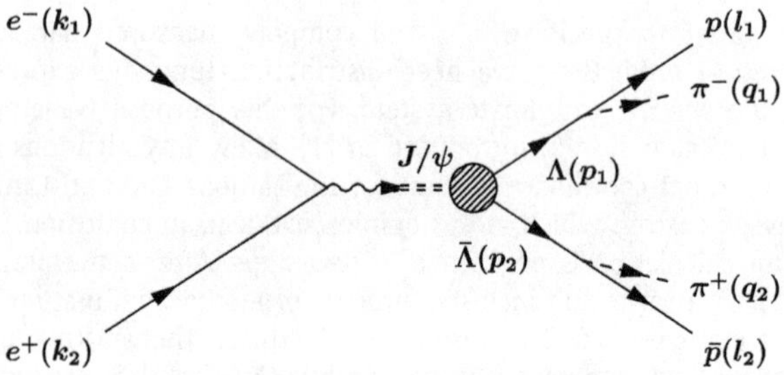

Fig. 2. Graph describing the reaction $e^+e^-\to\Lambda(\to p\pi^-)\bar{\Lambda}(\to\bar{p}\pi^+)$.

4. Folding of distributions

Our next task is to calculate the cross-section distribution for e^+e^- annihilation into hyperon pairs, followed by the hyperon decays into nucleon–pion pairs. This reaction is described by the connected diagram of Fig. 2.

Again, we extract a prefactor, $\mathcal{K}=\mathcal{K}_\psi\mathcal{K}_1\mathcal{K}_2$, from the squared matrix element, writing

$$|\mathcal{M}|^2=\mathcal{K}|\mathcal{M}_{red}|^2. \tag{4.30}$$

The prefactor originates, as before with the propagator denominators. Due to the smallness of the hyperon widths each of the hyperon propagators can, after squaring, be approximated as,

$$\mathcal{K}_i=\frac{1}{(p_i^2-M^2)^2+M^2\Gamma^2(M)}=\frac{2\pi}{2M\Gamma(M)}\delta(p_i^2-M^2). \tag{4.31}$$

Effectively, this approximation puts the hyperons on their mass shells.

Hyperon-decay distributions are obtained by a folding calculation, whereby hyperon-production and -decay distributions are multiplied together and averaged over the intermediate hyperon-spin directions. It was proved in Ref. [2] that the folding prescription gives the same result as the evaluation of the connected-Feynman-diagram expression. Hence, summing over final hadron spins,

$$|\mathcal{M}|^2=\sum_{\pm s_1,\pm s_2}\Big\langle\left|\mathcal{M}(e^+e^-\to\Lambda(s_1)\bar{\Lambda}(s_2))\right|^2$$
$$\times\left|\mathcal{M}(\Lambda(s_1)\to p\pi^-)\right|^2\left|\mathcal{M}(\bar{\Lambda}(s_2)\to\bar{p}\pi^+)\right|^2\Big\rangle_{\mathbf{n}_1\mathbf{n}_2}. \tag{4.32}$$

Production and decay distributions are,

$$\left|\mathcal{M}(e^+e^-\to\Lambda(s_1)\bar{\Lambda}(s_2))\right|^2=L\cdot K(s_1,s_2), \tag{4.33}$$

$$\left|\mathcal{M}(\Lambda(s_1)\to p\pi^-)\right|^2=R_\Lambda\left[1-\alpha_1 l_1\cdot s_1/l_\Lambda\right], \tag{4.34}$$

$$\left|\mathcal{M}(\bar{\Lambda}(s_2)\to\bar{p}\pi^+)\right|^2=R_\Lambda\left[1-\alpha_2 l_2\cdot s_2/l_\Lambda\right], \tag{4.35}$$

with l_Λ the decay momentum in the Lambda rest system. R_Λ is determined by the Lambda decay rate.

The notation in Eq. (4.34) is the following; s_1 denotes the Lambda four-spin vector, l_1 the four-momentum of the decay proton, and α_1 the decay-asymmetry parameter. Similarly for the anti-Lambda hyperon parameters of Eq. (4.35).

We evaluate the hyperon-decay distributions in the hyperon-rest systems, where

$$\left|\mathcal{M}(\Lambda(s_1)\to p\pi^-)\right|^2=R_\Lambda\left[1+\alpha_1\hat{\mathbf{l}}_1\cdot\mathbf{n}_1\right], \tag{4.36}$$

$$\left|\mathcal{M}(\bar{\Lambda}(s_2)\to\bar{p}\pi^+)\right|^2=R_\Lambda\left[1+\alpha_2\hat{\mathbf{l}}_2\cdot\mathbf{n}_2\right], \tag{4.37}$$

where $\hat{\mathbf{l}}_1=\mathbf{l}_1/l_\Lambda$ is the unit vector in the direction of the proton momentum in the Lambda-rest system, and correspondingly for the anti-Lambda case.

Angular averages in Eq. (4.32) are calculated according to the prescription

$$\big\langle(\mathbf{n}\cdot\mathbf{l})\mathbf{n}\big\rangle_{\mathbf{n}} = \mathbf{l}. \tag{4.38}$$

The folding of the production distributions, Eqs. (3.25)–(3.28), with the decay distributions, Eqs. (4.36, 4.37), yields

$$|\mathcal{M}_{red}|^2 = sD(s)R_\Lambda^2\left[G^{00}+G^{05}+G^{50}+G^{55}\right], \tag{4.39}$$

with the G^{ab} functions defined as

$$G^{00} = \mathcal{R}, \tag{4.40}$$

$$G^{05} = \alpha_1\mathcal{S}\left[\frac{1}{\sin\theta}(\hat{\mathbf{p}}\times\hat{\mathbf{k}})\cdot\hat{\mathbf{l}}_1\right], \tag{4.41}$$

$$G^{50} = \alpha_2\mathcal{S}\left[\frac{1}{\sin\theta}(\hat{\mathbf{p}}\times\hat{\mathbf{k}})\cdot\hat{\mathbf{l}}_2\right], \tag{4.42}$$

$$\begin{aligned} G^{55} = \alpha_1\alpha_2\Big\{&\mathcal{T}_1\hat{\mathbf{l}}_1\cdot\hat{\mathbf{p}}\hat{\mathbf{l}}_2\cdot\hat{\mathbf{p}}+\mathcal{T}_2\hat{\mathbf{l}}_{1\perp}\cdot\hat{\mathbf{l}}_{2\perp}\\ &+\mathcal{T}_3\hat{\mathbf{l}}_{1\perp}\cdot\hat{\mathbf{k}}\hat{\mathbf{l}}_{2\perp}\cdot\hat{\mathbf{k}}\\ &+\mathcal{T}_4\Big(\hat{\mathbf{l}}_1\cdot\hat{\mathbf{p}}\hat{\mathbf{l}}_{2\perp}\cdot\hat{\mathbf{k}}+\hat{\mathbf{l}}_2\cdot\hat{\mathbf{p}}\hat{\mathbf{l}}_{1\perp}\cdot\hat{\mathbf{k}}\Big)\Big\}. \end{aligned} \tag{4.43}$$

Thus, we conclude the connection between joint-hadron production and joint-hadron decay distributions simply to be,

$$G^{ab}(\hat{\mathbf{l}}_1;\hat{\mathbf{l}}_2) = H^{ab}(\mathbf{n}_1\to\alpha_1\hat{\mathbf{l}}_1;\mathbf{n}_2\to\alpha_2\hat{\mathbf{l}}_2). \tag{4.44}$$

We repeat the notation; $\mathbf{p}$ and $\mathbf{k}$ are momenta of Lambda and electron in the global c.m. system; $\mathbf{l}_1$ and $\mathbf{l}_2$ are momenta of proton and anti-proton in Lambda and anti-Lambda rest systems; orthogonal means orthogonal to $\mathbf{p}$; and structure functions $\mathcal{R}$, $\mathcal{S}$, and $\mathcal{T}$ are functions of θ, α, and $\Delta\Phi$. The angular functions multiplying the structure functions form a set of seven mutually orthogonal functions, when integrated over the proton and anti-proton decay angles.

5. Cross section for $e^+e^-\to\Lambda(\to p\pi^-)\bar{\Lambda}(\to\bar{p}\pi^+)$

Our last task is to find the properly normalized cross-section distribution. We start from the general expression,

$$\mathrm{d}\sigma = \frac{1}{2s}\mathcal{K}|\mathcal{M}_{red}|^2\,\mathrm{dLips}(k_1+k_2;q_1,l_1,q_2,l_2), \tag{5.45}$$

with dLips the phase-space density for four final-state particles. The prefactor $\mathcal{K}$ contains on the mass shell delta functions for the two hyperons. This effectively separates the phase space into production and decay parts. Repeating the manipulations of Ref. [2] we get

$$\begin{aligned} \mathrm{d}\sigma = &\frac{1}{64\pi^2}\frac{p}{k}\frac{\alpha_g\alpha_\psi}{(s-m_\psi^2)^2+m_\psi^2\Gamma^2(\psi)}\frac{\Gamma_\Lambda\Gamma_{\bar{\Lambda}}}{\Gamma^2(M)}\cdot\\ &\cdot\left(D(s)\sum_{a,b}G^{ab}\right)\mathrm{d}\Omega_\Lambda\mathrm{d}\Omega_1\mathrm{d}\Omega_2, \end{aligned} \tag{5.46}$$

with k and p the initial- and final-state momenta; Ω_Λ the hyperon scattering angle in the global c.m. system; Ω_1 and Ω_2 decay angles measured in the rest systems of Λ and $\bar{\Lambda}$; Γ_Λ and $\Gamma_{\bar{\Lambda}}$ channel widths; and $\Gamma(M)$ and $\Gamma(\psi)$ total widths.

Integration over the angles Ω_1 makes the contributions from the functions G^{05} and G^{55} disappear [2], and correspondingly for the angles Ω_2. Integration over both angular variables results in the cross-section distribution for the reaction $e^+e^-\to J/\psi\to\Lambda\bar{\Lambda}$.

Suppose we integrate over the angles Ω_2. Then, the predicted hyperon-decay distribution becomes proportional to the sum

$$G^{00}+G^{05} = \mathcal{R}\left(1+\alpha_1\mathbf{P}_\Lambda\cdot\hat{\mathbf{l}}_1\right), \tag{5.47}$$

$$P_\Lambda = \frac{\mathcal{S}}{\mathcal{R}}, \tag{5.48}$$

with the polarization P_Λ as in Eq. (3.29), and the polarization vector $\mathbf{P}_\Lambda$ directed along the normal to the scattering plane

6. Differential distributions

We first define our coordinate system. The scattering plane with the vectors $\mathbf{p}$ and $\mathbf{k}$ make up the xz-plane, with the y-axis along the normal to the scattering plane. We choose a right-handed coordinate system with basis vectors

$$\mathbf{e}_z = \hat{\mathbf{p}}, \tag{6.49}$$

$$\mathbf{e}_y = \frac{1}{\sin\theta}(\hat{\mathbf{p}}\times\hat{\mathbf{k}}), \tag{6.50}$$

$$\mathbf{e}_x = \frac{1}{\sin\theta}(\hat{\mathbf{p}}\times\hat{\mathbf{k}})\times\hat{\mathbf{p}}. \tag{6.51}$$

Expressed in terms of them the initial-state momentum

$$\hat{\mathbf{k}} = \sin\theta\,\mathbf{e}_x+\cos\theta\,\mathbf{e}_z. \tag{6.52}$$

This coordinate system is used for fixing the directional angles of the decay proton in the Lambda rest system, and the decay anti-proton in the anti-Lambda rest system. The spherical angles for the proton are θ_1 and ϕ_1, and the components of the unit vector in direction of the decay-proton momentum are,

$$\hat{\mathbf{l}}_1 = (\cos\phi_1\sin\theta_1,\sin\phi_1\sin\theta_1,\cos\theta_1), \tag{6.53}$$

so that

$$\hat{\mathbf{l}}_{1\perp} = (\cos\phi_1\sin\theta_1,\sin\phi_1\sin\theta_1,0). \tag{6.54}$$

The momentum of the decay proton is by definition $\mathbf{l}_1 = l_\Lambda\hat{\mathbf{l}}_1$. This same coordinate system is used for defining the directional angles of the decay anti-proton in the anti-Lambda rest system, with spherical angles θ_2 and ϕ_2.

Now, we have all ingredients needed for the calculation of the G functions of Eqs. (4.40)–(4.43), the functions that in the end determine the cross-section distributions.

An event of the reaction $e^+e^-\to\Lambda(\to p\pi^-)\bar{\Lambda}(\to\bar{p}\pi^+)$ is specified by the five dimensional vector $\boldsymbol{\xi} = (\theta,\Omega_1,\Omega_2)$, and the differential-cross-section distribution as summarized by Eq. (4.39) reads,

$$\mathrm{d}\sigma \propto \mathcal{W}(\boldsymbol{\xi})\,\mathrm{d}\cos\theta\,\mathrm{d}\Omega_1\mathrm{d}\Omega_2.$$

At the moment, we are not interested in the absolute normalization of the differential distribution. The differential-distribution function $\mathcal{W}(\boldsymbol{\xi})$ is obtained from Eqs. (4.40)–(4.43) and can be expressed as,

$$\begin{aligned} \mathcal{W}(\boldsymbol{\xi}) = &\mathcal{F}_0(\boldsymbol{\xi})+\alpha\mathcal{F}_5(\boldsymbol{\xi})\\ &+\alpha_1\alpha_2\left(\mathcal{F}_1(\boldsymbol{\xi})+\sqrt{1-\alpha^2}\cos(\Delta\Phi)\mathcal{F}_2(\boldsymbol{\xi})+\alpha\mathcal{F}_6(\boldsymbol{\xi})\right)\\ &+\sqrt{1-\alpha^2}\sin(\Delta\Phi)\left(\alpha_1\mathcal{F}_3(\boldsymbol{\xi})+\alpha_2\mathcal{F}_4(\boldsymbol{\xi})\right), \end{aligned} \tag{6.55}$$

using a set of seven angular functions $\mathcal{F}_k(\boldsymbol{\xi})$ defined as:

$$\mathcal{F}_0(\boldsymbol{\xi}) = 1$$
$$\mathcal{F}_1(\boldsymbol{\xi}) = \sin^2\theta \sin\theta_1 \sin\theta_2 \cos\phi_1 \cos\phi_2 + \cos^2\theta \cos\theta_1 \cos\theta_2$$
$$\mathcal{F}_2(\boldsymbol{\xi}) = \sin\theta \cos\theta\, (\sin\theta_1 \cos\theta_2 \cos\phi_1 + \cos\theta_1 \sin\theta_2 \cos\phi_2)$$
$$\mathcal{F}_3(\boldsymbol{\xi}) = \sin\theta \cos\theta \sin\theta_1 \sin\phi_1$$
$$\mathcal{F}_4(\boldsymbol{\xi}) = \sin\theta \cos\theta \sin\theta_2 \sin\phi_2$$
$$\mathcal{F}_5(\boldsymbol{\xi}) = \cos^2\theta$$
$$\mathcal{F}_6(\boldsymbol{\xi}) = \cos\theta_1 \cos\theta_2 - \sin^2\theta \sin\theta_1 \sin\theta_2 \sin\phi_1 \sin\phi_2. \tag{6.56}$$

The differential distribution of Eq. (6.55) involves two parameters related to the $e^+e^- \to \Lambda\bar{\Lambda}$ process that can be determined by data: the ratio of form factors α, and the relative phase of form factors $\Delta\Phi$. In addition, the distribution function $\mathcal{W}(\boldsymbol{\xi})$ can be used to extract separately Λ and $\bar{\Lambda}$ decay-asymmetry parameters: α_1 and α_2, and hence allowing a direct test of CP conservation in the hyperon decays.

The term proportional to $\sin(\Delta\Phi)$ in Eq. (6.55) originates with Eqs. (4.41) and (4.42), and can be rewritten as,

$$\mathcal{S}(\theta)\, (\alpha_1 \sin\theta_1 \sin\phi_1 + \alpha_2 \sin\theta_2 \sin\phi_2)\,,$$

with the structure function $\mathcal{S}$ defined by Eq. (3.20). The relation between the structure functions and the polarization $P_\Lambda(\theta)$ was discussed in Sect. 3, where it was shown that the polarization, $P_\Lambda(\theta)$ of Eq. (3.29), and the polarization vector, $\mathbf{e}_y$, are the same for Lambda and anti-Lambda hyperons. This information tells us that Λ is polarized along the normal to the production plane, and that the polarization vanishes at $\theta = 0°$, $90°$ and $180°$. The maximum value of the polarization is for $\cos\theta = \pm 1/(2+\alpha)$, and $|P_\Lambda(\theta)| < \frac{2}{3}\sin(\Delta\Phi)$.

It should be stressed that the simplified distributions used in previous analyses, such as Ref. [9], assume the hyperons to be unpolarized and therefore terms containing $P_\Lambda(\theta)$ are missing. In fact, such decay distributions, only permit the determination of two parameters: the ratio of form factors α, and the product of hyperon-asymmetry parameters $\alpha_1\alpha_2$.

In our opinion, the formulas presented in this letter should be employed for the exclusive analysis of the new BESIII data [10]. Due to huge and clean data samples: (440675 ± 670) $J/\psi \to \Lambda\bar{\Lambda}$ and (31119 ± 187) $\psi(3686) \to \Lambda\bar{\Lambda}$, precision values for the decay-hadronic-form factors could be extracted as well as precision values for Λ and $\bar{\Lambda}$ decay-asymmetry parameters. The formulas presented could easily be generalized to neutron decays of the Λ and to production of other $J = 1/2$ hyperons with analogous decay modes.

Acknowledgements

We are grateful to Tord Johansson who provided the inspiration for this work.

Appendix A

The coupling of the initial-state leptons to the J/ψ vector meson is determined by the decay $J/\psi \to e^+e^-$. Assuming the decay to go via an intermediate photon, Fig. 1b, we can safely ignore any tensor coupling. The vector coupling of the J/ψ to leptons is therefore the same as for the photon, if replacing the electric charge e_{em} by a coupling strength e_ψ. From the decay $J/\psi \to e^+e^-$ one derives

$$\alpha_\psi = e_\psi^2/4\pi = 3\Gamma(J/\psi \to e^+e^-)/m_\psi. \tag{A.57}$$

In a similar fashion we relate the strength e_g of J/ψ coupling to the hyperons to the decay $J/\psi \to \Lambda\bar{\Lambda}$. In analogy with Eq. (A.57) we get

$$\alpha_g = e_g^2/4\pi = 3\left((1 + 2M^2/m_\psi^2)\sqrt{1 - 4M^2/m_\psi^2}\right)^{-1} \times \Gamma(J/\psi \to \Lambda\bar{\Lambda})/m_\psi. \tag{A.58}$$

When the Λ mass M is replaced by the lepton mass $m_l = 0$ we recover Eq. (A.57).

References

[1] G. Fäldt, Eur. Phys. J. A 52 (2016) 141.
[2] G. Fäldt, Eur. Phys. J. A 51 (2015) 74.
[3] H. Czyż, A. Grzelińska, J.H. Kühn, Phys. Rev. D 75 (2007) 074026.
[4] J. Haidenbauer, U.-G. Meißner, Phys. Lett. B 761 (2016) 456.
[5] Hong Chen, Rong-Gang Ping, Phys. Rev. D 76 (2007) 036005.
[6] Bin Zhong, Guangrui Liao, Acta Phys. Pol. 46 (2015) 2459.
[7] D. Pallin, et al., Nucl. Phys. B 292 (1987) 653.
[8] M.H. Tixier, et al., Phys. Lett. B 212 (1988) 523.
[9] M. Ablikim, et al., Phys. Rev. D 81 (2010) 012003.
[10] M. Ablikim, et al., BESIII Collaboration, Phys. Rev. D 95 (2017) 052003.
[11] M. Ablikim, et al., BESIII Collaboration, Nucl. Instrum. Methods A 614 (2010) 345.

Holographic corrections to the Veneziano amplitude

Adi Armoni, Edwin Ireson *

Department of Physics, Swansea University, Swansea SA28PP, United Kingdom

ARTICLE INFO

Editor: N. Lambert

ABSTRACT

We propose a holographic computation of the $2 \to 2$ meson scattering in a curved string background, dual to a QCD-like theory. We recover the Veneziano amplitude and compute a perturbative correction due to the background curvature. The result implies a small deviation from a linear trajectory, which is a requirement of the UV regime of QCD.

1. Introduction

The Veneziano amplitude, aimed at describing meson scattering, marks the birth of string theory [1]. While the Veneziano amplitude exhibits several attractive properties, such as the duality between the s-channel and the t-channel, and a phenomenological success in the Regge regime, it also suffers from an undesired exponential behaviour in the high energy regime.

The properties of the Veneziano amplitude are closely related to the phenomenon of confinement. Makeenko and Olesen [2] showed that the amplitude can be reproduced from large-N QCD, by representing the amplitude as a sum over Wilson loops. A crucial unrealistic assumption, is that all Wilson loops (even small loops) admit an area law.

In a recent attempt to derive the Veneziano amplitude from large-N QCD [3], the sum over Wilson loops of [2] was written by using holography [4] as a sum over string worldsheets. The sum includes all Wilson loops that pass via the positions of the mesons. The Veneziano amplitude was obtained under the assumption that Wilson loops are saturated by flat space configurations that sit on the IR cut-off (hence leading to an area law). The flat space approximation can be achieved by bringing the IR cut-off close to the UV cut-off (the space boundary).

Thus one can attribute the failure of the Veneziano amplitude in the high energy regime to the flat space approximation, which is identical to the assumption that all Wilson loops admit an area law. It is therefore desired to accommodate small Wilson loops. In the dual string theory it means that one has to calculate string amplitudes in curved space. Even if that could be done, an exact dual of large-N QCD is not known, hence it is not clear which non-linear sigma model should be used.

In this paper we propose a method to improve the scattering amplitude, by incorporating curvature effects from the dual geometry. In particular we study the effect of worldsheet fluctuations in the vicinity of the IR cut-off. As the worldsheet fluctuates it probes part of the UV geometry. To this end we include in the string sigma model an interaction term between the flat 4d coordinates and extra dimensions and calculate a two-loop perturbative correction to the propagator. While we carry out the calculation by using Witten's model of the dual of Yang–Mills theory [5], we believe that the sigma model we use characterizes generic confining holographic models. The result of the calculation is a correction to the Veneziano amplitude in the form of a deviation from a linear Regge trajectory $\alpha(s) \sim s^{(1-\rho^3 \log^2 s)}$, with $\rho^3 \log^2 s \ll 1$.

Our approach is distinct from earlier studies of Pomeron scattering using the AdS/CFT correspondence [6]. The novelty, apart from considering meson scattering (open string amplitude), is that we include a perturbative correction to the X^μ propagator.

Our conclusion about the non-linearity of the Regge trajectory at small angular momenta is similar to that of previous studies [7, 8]. It is consistent with empirical data [8].

2. Setting up the duality

The starting point of this framework is the aforementioned identity involving a sum over all sizes of Wilson loops. To compute this four-point function we make use of the Worldline formalism, namely the following equality for the fermion determinant of $SU(N_c)$ QCD with N_f flavours, in terms of expectation values of super-Wilson loops of length T:

$$\mathcal{Z} = \int DA \exp(-S_{\text{YM}}) \left(\det\left(i\not{D}\right)\right)^{N_f},$$

* Corresponding author.
E-mail address: 746616@swansea.ac.uk (E. Ireson).

$$(\det(i\not{D}))^{N_f} = \exp\left(-\frac{N_f}{2}\mathrm{Tr}\int_0^\infty \frac{dT}{T}\langle \mathcal{W}_T[A]\rangle\right), \tag{1}$$

$$\langle \mathcal{W}_T[A]\rangle = \int DxD\psi e^{-\frac{1}{2}\int_0^T d\tau(\dot{x}^2+\psi\cdot\dot{\psi})} e^{i\int_0^T d\tau(\dot{x}\cdot A-\frac{1}{2}\psi\cdot F\cdot\psi)}.$$

This exponential can then be expanded in powers of N_f/N_c, so that in the 't Hooft large N_c limit only the linear term in Wilson loops needs to be considered. We will use this approximation (the so-called "quenched approximation") for the following computation of a generic meson 4-point function [3]:

$$\left\langle \prod_{i=1}^4 q\bar{q}(x_i)\right\rangle = \frac{1}{\mathcal{Z}}\prod_{i=1}^4 \frac{\delta}{\delta J(x_i)}\int DA\exp(-S_{\mathrm{YM}}) \times \left(-\frac{N_f}{2}\mathrm{Tr}\int_0^\infty \frac{dT}{T}\langle \mathcal{W}_T[A,J]\rangle\right)|_{J=0} \tag{2}$$

with $\langle \mathcal{W}_T[A,J]\rangle$ the worldline action of a theory coupled to a meson source $S_J = \int d^4x J\bar{q}q$ [9]. Taking functional derivatives in terms of J creates delta functions constraining the various loops at hand to pass through the points where the operators are inserted. We therefore arrive at the following formal expression for the scattering amplitude

$$\left\langle \prod_{i=1}^4 q\bar{q}(x_i)\right\rangle = \frac{1}{\mathcal{Z}}\int DA\exp(-S_{\mathrm{YM}}) \times \left(-\frac{N_f}{2}\mathrm{Tr}\int_0^\infty \frac{dT}{T}\langle \mathcal{W}_T[A]\rangle|_{x_1,x_2,x_3,x_4}\right), \tag{3}$$

where the sum over all Wilson loops has been imposed to include only those loops passing through the 4 points where the meson operators have been inserted.

The gauge/gravity duality implies that, for the field theory on the boundary, a generic Wilson loop's expectation value is related to the expectation value of a string worldsheet hanging from the loop on the boundary down into the bulk. More precisely, a single Wilson loop expectation value is obtained as the saddle of a sum over string worldsheets, as explained in section 3 of [11]. It was then proposed that the contribution of "quenched flavours" to the gauge theory partition function is dual to a sum over all string worldsheet with a topology of a disk that terminate on the boundary of the AdS space [10]

$$\int DA\exp(-S_{\mathrm{YM}}) \times \left(-\frac{1}{2}\mathrm{Tr}\int_0^\infty \frac{dT}{T}\langle \mathcal{W}_T[A]\rangle\right) = \int [DX][Dg]\exp\left(-\frac{1}{2\pi\alpha'}\int d^2\sigma\, G_{MN}\partial_\alpha X^M\partial_\beta X^N g^{\alpha\beta}\right). \tag{4}$$

This identification is possible because we have fixed a flat worldsheet gauge $g^{\alpha\beta} = \delta^{\alpha\beta}$, we are not interested in higher genus contributions. Then, schematically, the above equality holds because both path integration measures on either side sum up area-behaved integrands over all possible shapes and sizes of their boundaries. We are therefore able to write the following expression for the amplitude [10]

$$\mathcal{A}(k_{i=1\ldots 4}) = \oint \prod_{i=1}^4 d\sigma_i \int [DX] W e^{ik_i^\mu X_\mu(\sigma_i)} \times \exp\left(-\frac{1}{2\pi\alpha'}\int d^2\sigma\, G_{MN}\partial_\alpha X^M\partial^\alpha X^N\right), \tag{5}$$

where $We^{ik_i^\mu X_\mu(\sigma_i)}$ are meson vertex operators in the dual string theory, inserted at different points on the worldsheet, each associated to a momentum k_i. W encodes details of spin ($W=1$ for the tachyon and $W = \epsilon\partial_\sigma X$ for a vector for the bosonic string), which only really affects the intercept α_0 of the Regge trajectory, $\alpha(s) = \alpha_0 + \alpha' s$. As we shall see the calculated perturbative correction is not valid near $s=0$. In addition, there may be additional subleading phenomena contributing to a shift of the intercept, hence we can make no statements thereupon. The expectation values of the vertex operators are taken with respect to the Polyakov-type non-linear sigma model action over the string worldsheet into a curved space. Choosing the correct background is crucial in this matter. Since we have done away with flavour degrees of freedom via the worldline formalism, we only need to pick a space dual to pure Yang–Mills. For our purposes, we take Witten's background of back-reacted $D4$-branes compactified on a thermal circle [5], whose dual, while not pure Yang–Mills, possesses enough similarities (confinement and a mass gap) that we can hope to make generic arguments thereupon. The chief property of this space is its metric G, inducing the following space–time line element where X^μ ($\mu = 0,1,2,3$) are boundary space–time coordinates, τ the compact direction, U the AdS direction and the remaining 4 coordinates parametrise a sphere:

$$ds^2 = g(U)\left(dX^2 + d\tau^2 f(U)\right) + \frac{1}{g(U)}\left(\frac{dU^2}{f(U)} + U^2 d\Omega_4^2\right)$$
$$\text{where } g(U) = \left(\frac{U}{R}\right)^{3/2},\quad f(U) = 1 - \frac{U_{\mathrm{KK}}^3}{U^3}. \tag{6}$$

It admits a metric singularity at the point $U = U_{KK}$, where the space has a horizon. As is usual, string worldsheets hanging from large loops (of size comparable to the horizon position) will accumulate on the horizon, providing an area-law scaling of their expectation value, rather than a perimeter-law, this is the usual tell-tale sign that confinement has taken place. Then, provided the strings do not have to go very far from the boundary to the horizon, the classical saddle of this action is a string whose geometry is mostly flat, spread out on the horizon itself. By pushing U_{KK} close to infinity, this condition is broadly satisfied, turning the classical saddle of the path integral into a mostly-flat worldsheet. This was the claim proposed in a previous work [3], from which the 4-point function we wish to compute fairly naturally reproduces the Veneziano amplitude as the behaviour associated to flat open-string scattering.

These final steps relied on many broad assumptions, namely that we ignore effects coming from the compact directions (justified through the hierarchy of scales at hand), from the fermionic degrees of freedom (justified by spacetime supersymmetry breaking, and by the insertion of purely bosonic operators), to impose that almost all worldsheets in the sum are heavily accumulating on the horizon, discarding those that do not and ignoring the contribution of the edges of those that do (justified by the near-infinite size of U_{KK}), and to assume no string quantum corrections (both genus/ghost and α'). Our goal is to relax the latter, to allow string tension corrections to the computation at hand, which in physical terms corresponds to letting the string fluctuate around its classical position and experience the U direction curvature and (it will be shown) the compact direction τ.

2.1. Setting up an expansion

From the Polyakov action using the metric shown in Eq. (6), we wish to create a perturbation series for values of the AdS coordinate U close to the horizon position, U_{KK}, which for previously explained reasons should be thought of as a large length scale. The first step is to provide an adequate parametrisation of the space, in which the coordinate singularity at the horizon is removed. Typically a set of Kruskal-like coordinates are desirable, as has been done in similar computations in the past [12], but unlike that particular example, they are non-trivial to compute in the case at hand, it is unclear whether a good new radial coordinate can be analytically computed over the entire AdS region. We will therefore assume that the string fluctuations are of small amplitude and only experience a small range of the target space curvature: in practice this means we can perform an expansion of the offending metric elements for values of U parametrically close to U_{KK} and compute near-horizon Kruskal coordinates, rather than try to globally define them.

At this cost, the operation can be done and results in the following coordinate definitions: introducing $\lambda = \frac{U_{KK}}{R}$ as an arbitrary constant, we define new coordinates by

$$\frac{U}{U_{KK}} = 1 + \frac{u^2}{U_{KK}^2}, \quad \Upsilon = u e^{i\sqrt{\frac{9\lambda^3}{4U_{KK}^2}}\tau} e^{\frac{u^2}{4U_{KK}^2}}, \tag{7}$$

all power series in u/U_{KK} can now be rewritten as series in $|\Upsilon/U_{KK}|$, as the relation above is easily invertible. This produces a string action in a very appropriate form for interaction expansions:

$$\mathcal{L} = \lambda^{3/2}\partial_\alpha X^\mu \partial^\alpha X_\mu + \frac{4}{3\lambda^{3/2}}\partial_\alpha \Upsilon^\dagger \partial^\alpha \Upsilon + \frac{6}{TU_{KK}^2}\Upsilon^\dagger \Upsilon$$
$$+ \frac{3\lambda^{3/2}}{2U_{KK}^2}\Upsilon^\dagger \Upsilon \partial_\alpha X^\mu \partial^\alpha X_\mu + \ldots \tag{8}$$

The last term is the direct result of this process, an interaction term between Υ and X^μ, but we have also gained a mass term for this AdS coordinate: it has come from a Jacobian factor multiplying the invariant path integration measure. Note that it is parametrically small as it depends both on the string tension T and U_{KK}, also that this coordinate redefinition characterises not only vertical motion of the worldsheet but also motion in the τ direction. We keep ignoring effects from the additional compact coordinates, seen as an unfortunate artefact of the string background. We also ignore the effects of supersymmetry: certainly, the string theory has explicit broken space–time supersymmetry, but one does not expect to see it explicitly from the worldsheet, and so we should include fermions in the action. However, we note that, firstly, we are inserting purely bosonic operators in the path integral, so we need to focus on interactions featuring those bosonic fields, secondly, that such terms connecting bosonic and fermionic fields in a generic non-linear sigma model ($\Gamma_{\mu\nu\rho}\bar{\psi}^\mu \partial\!\!\!/ X^\nu \psi^\rho$) vanish in our case, through specific properties of the Christoffel symbol in this background. For this reason the coupling to RR background fields is also expected to be subleading as they couple primarily to the worldsheet fermions.

The issue of the dilaton coupling to the worldsheet Ricci tensor (Fradkin–Tseytlin coupling) is more subtle. In our setup we consider the kinematical regime (Regge regime) where the sum over string worldsheets is dominated by large area and flat-space worldsheets, calculated in the vicinity of $U = U_{KK}$. For such worldsheets we may take the approximation $\int d^2\sigma\, \Phi R \approx \int d^2\sigma\, \Phi(U_{KK}) R$, where the contribution is similar to the flat space case and the "area term" $\int \partial X \partial X$ is expected to be the dominant contribution to the action.

We can now proceed to perform perturbation theory around the $U = U_{KK}$ "vacuum" of the effective field theory described by this Lagrangian.

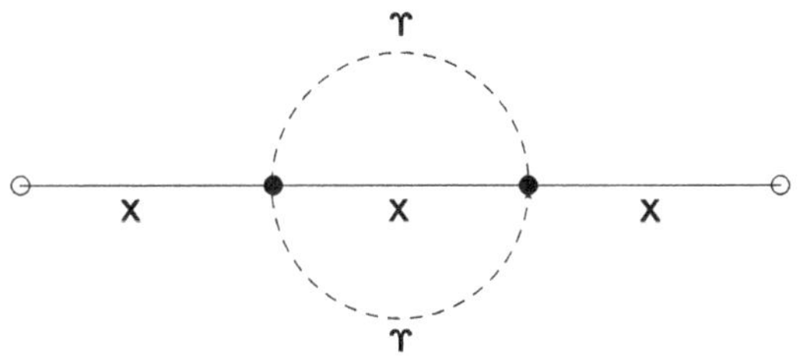

Fig. 1. The first diagram leading to qualitative corrections to the Veneziano amplitude.

3. Evaluation of loop integrals

This action now having a well-defined vacuum to expand around, we set out to compute the quantity we defined previously in Eq. (5). From general considerations on the tower-like structure of string excited states we argue that considering vertex operators of the form $e^{ik_i \cdot X(\sigma_i)}$ should be informative enough (again noting that we are not interested in the intercept). The purely classical approximation immediately reproduces the Beta function behaviour of the amplitude. By adding these operators in the action as a current $\mathcal{J}(\sigma) = \sum_{i=1}^{4} k_i \cdot X(\sigma)\delta(\sigma - \sigma_i)$, this essentially maps the computation onto the calculation of the partition function around a non-zero current (suitably normalised by that without any sourcing). As a result we are computing "vacuum" diagrams with no in/out states, but not vacuum bubbles–rather, all those whose legs are stopped at the positions where the current is being introduced. These effectively 1-leg vertices are dimensionful, thus the more legs a diagram has, the higher its order is in our series expansion, the dimensions being soaked up by powers of $1/U_{KK}$. Thus we will focus on worldsheet 2-point corrections.

Now, given the form of the interaction at hand, and the fact we are dealing with a 2D QFT, we will frequently have to deal with logarithmically divergent integrals, starting with the propagators, and thus have to work to remove regulator poles but also logarithms of vanishing or asymptotic quantities. For this purpose we will use analytic regularisation of the propagators, along with a convenient version of the $\overline{MS}$ scheme. Once this is done, our framework has the particularity of making a large class of diagrams, that we can write with this new interaction vertex, trivial in a sense. The lowest order diagrams in our expansion, the 2-point 1-loop graphs with no momentum transfer, factorise into a number of "vacuum bubble"-like integrals times an overall propagator, our regulation scheme makes such diagrams finite. This corresponds to, equivalently, a finite shift of the wavefunction normalisation, or a finite shift of the string tension T. As the bare value of T is tunable, we should consider that at the end of the procedure, once all the finite shifts in tension at all orders have been applied, the new, effective string tension is the physical QCD one. We will hereafter always refer to the effective string tension when writing T or α'.

Next, we are brought to study a two-loop correction to the two-point amplitude (curiously lower order in U_{KK} than lower-loop, higher-point amplitudes by dimensional analysis) taking the form of a "sunset" diagram (Fig. 1), corresponding to the following integral

$$I = \sum_{i<j} k_i \cdot k_j \frac{81(\pi\alpha')^3\lambda^{3/2}}{64U_{KK}^4}$$
$$\times \int \frac{d^2p\, d^2q_1\, d^2q_2}{(2\pi)^6} \frac{e^{i(p\cdot(\sigma_i - \sigma_j))}(p \cdot (q_1 + q_2))^2}{p^4(p - q_1 - q_2)^2(q_1^2 + m^2)(q_2^2 + m^2)} \tag{9}$$

This integral is not trivial – especially given that only two out of the three internal lines are massive. Some work has been done on the analytics of similar diagrams [13], but not using a regulation scheme adapted to logarithmic divergences as is our case. However, at this point we use the fact that the mass in question is parametrically small: it is of the same order as the coupling constant of our interaction term. Since the massless diagram is well-defined, we should therefore think of it as truly being the lead source of qualitative corrections to the Veneziano regime, the mass inducing higher-order corrections. At this cost, the diagram is then well-suited to treatment in our set-up, and we obtain a finite value for it:

$$I = -\frac{1}{8\pi^3}\frac{27\lambda^{3/2}}{64T^3U_{KK}^4}\sum_{i<j} k_i \cdot k_j \left(\log^3\left((\sigma_i - \sigma_j)^2\right)\right). \quad (10)$$

The computation closes as we insert this result in the overall expectation value and averaging over all possible insertion positions described above in Eq. (5). We encapsulate the expansion parameter and the numerical factors by defining, $\rho^3 = \frac{1}{4\pi^2}\frac{27\lambda^{3/2}}{64T^3U_{KK}^4}$, and we will write scalar products of space–time momenta $k_i \cdot k_j$ as Mandelstam variables s, t: $s = -k_1 \cdot k_2$, $t = -k_1 \cdot k_3$. Systematically these variables come with a factor of α', which we will absorb in the definition of these variables to make them dimensionless.

Then the amplitude becomes

$$\mathcal{A}(s,t) = \int_0^1 dz z^{-s-1}(1-z)^{-t-1} \times \left(1 + \rho^3\left(s\log^3(z) + t\log^3(1-z)\right)\right), \quad (11)$$

which, one can quickly recognize, has a leading term corresponding precisely to the Veneziano Beta-like functional form, with extra logarithmic corrections. The corrective integrals can be computed explicitly and give third derivatives of the Beta function,

$$\mathcal{A}(s,t) = \left(1 - \rho^3\left(s\frac{\partial^3}{\partial s^3} + t\frac{\partial^3}{\partial t^3}\right)\right) B(-s,-t). \quad (12)$$

This expression is actually composed of many terms of varying relevance due to properties of Beta and associated functions.

4. Consequences on the Regge trajectory

To recover the incidence of these corrections on the Regge behaviour, we will express the amplitude in an approximate form, in the so-called Regge regime.

We then take the limit $|s| \gg |t|$, $|t|$ fixed. We expect to have $B(-s,-t) \propto (-s)^t$ (ignoring the purely t-dependent part of the amplitude, which contains the pole at zero in t). This form is still valid in the elastic collision regime, as long as one moves a little away from the positive real s-axis, as is routinely done in such matters. This ensures that the influence of the poles and the zeroes of the function do not disrupt this approximate form too much. The nature of the Beta function means that this limit is quite lax, $|s|$ need not be large in magnitude for it to be adequate, simply comparatively larger than $|t|$. Remarking that every s derivative lowers the order of the asymptotic behaviour of the function, the main contributions come from the t derivatives. Generically either of these derivatives involve the Polygamma functions $\{\Psi_{i\in\mathbb{N}}\}$, all of which vanish as powers of their argument asymptotically, save for the first one, Ψ_0, which diverges logarithmically. These terms will dominate the functional form of the correction in the limit we have chosen. With this prescription we find an approximate form for the amplitude

$$\mathcal{A}(s,t) = \exp\left(t(\log(-s) - \rho^3\log^3(-s))\right) \quad (13)$$

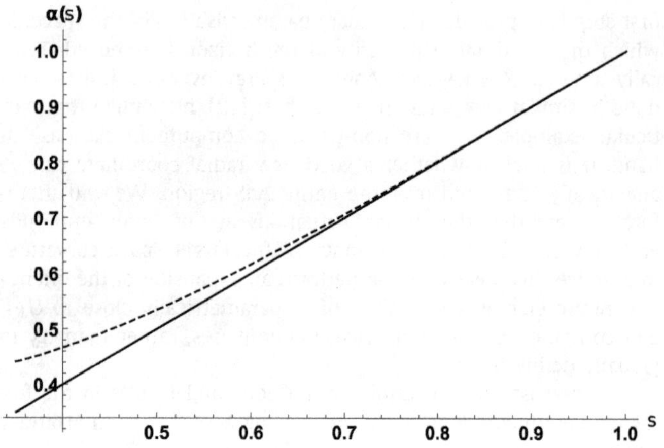

Fig. 2. Comparing the new Regge trajectory (dashed) to linear behaviour (solid), for $\rho^3 = 0.2$.

and correspondingly a Regge trajectory of the form

$$\alpha(-s) \sim (-s)^{1-\rho^3\log^2(-s)} \quad (14)$$

where $\alpha(s) = s$ would be the leading order, purely classical string result. This is a satisfactory result, in that the Regge behaviour for values of $|s| < 1$ starts to deviate away from the linear behaviour, in qualitatively similar ways as in previous efforts, namely, when plotted on the $(s, \alpha(s))$ plane, the curve bends towards the $\alpha(s)$ axis, as shown in Fig. 2.

It is divergent at the origin, but at this point our approximations no longer hold, as is the case generically in similar studies. This does not result in any predictions for the spectrum of the theory.

5. Conclusions and outlook

We have considered a computation of $2 \to 2$ meson scattering using holography, recovering the Veneziano amplitude and computing a correction due to interactions between the spacetime coordinates and bulk coordinates U and τ, whose form we expect to be generic, this can be argued from properties of confining spacetimes.

Our main result is (13), valid in the Regge regime where the tree level Veneziano amplitude is a good approximation. The correction we found affects the low energy regime (low s) of the Regge trajectory, as depicted in Fig. 2. Such a correction is required in QCD, since perfectly linear Regge trajectories mean a linear confining potential even at short distances. This is in conflict with asymptotic freedom.

It is encouraging to note that our results compare well with other studies and empirical data. For example, in ref. [8] the author finds a similar qualitative correction to the Regge trajectory in the low s regime and claims that the empirical data of heavy meson spectra is better fitted by such a non-linear trajectory. Moreover, in ref. [7] the authors calculate a correction to the meson spectrum due to the curvature of the background. They also find a qualitative non-linear behaviour similar to Fig. 2 and fit it to the ρ-meson spectrum. Both cases plot M^2 as a function of J, the mirror of our graph, and get a bending towards J.

Encouraged by our results, we hope to apply our approach to other models and consider other processes such as meson-baryon scattering.

Acknowledgements

We thank Timothy Hollowood, Zohar Komargodski and Carlos Núñez for useful discussions and comments.

References

[1] G. Veneziano, Nuovo Cimento A 57 (1968) 190–197.
[2] Y. Makeenko, P. Olesen, Phys. Rev. D 80 (2009) 026002, arXiv:0903.4114.
[3] A. Armoni, Phys. Lett. B 756 (2016) 328–331, arXiv:1509.03077.
[4] J.M. Maldacena, Int. J. Theor. Phys. 38 (1999) 1113–1133, arXiv:hep-th/9711200, Adv. Theor. Math. Phys. 2 (1998) 231.
[5] E. Witten, Adv. Theor. Math. Phys. 2 (1998) 505–532, arXiv:hep-th/9803131.
[6] J. Polchinski, M.J. Strassler, Phys. Rev. Lett. 88 (2002) 031601, arXiv:hep-th/0109174.
[7] T. Imoto, T. Sakai, S. Sugimoto, Prog. Theor. Phys. 124 (2010) 263–284, arXiv:1005.0655.
[8] J. Sonnenschein, arXiv:1602.00704.
[9] A. Armoni, O. Mintakevich, Nucl. Phys. B 852 (2011) 61.
[10] A. Armoni, Phys. Rev. D 78 (2008) 065017.
[11] N. Drukker, D.J. Gross, H. Ooguri, Phys. Rev. D 60 (1999) 125006, arXiv:hep-th/9904191.
[12] J. Greensite, P. Olesen, J. High Energy Phys. 04 (1999) 001, arXiv:hep-th/9901057.
[13] V.A. Smirnov, Springer Tracts Mod. Phys. 250 (2012) 1–296.

Integration of trace anomaly in 6D

Fabricio M. Ferreira [a,b], Ilya L. Shapiro [a,c,d,*]

[a] Departamento de Física, ICE, Universidade Federal de Juiz de Fora, Campus Universitário – Juiz de Fora, 36036-330, MG, Brazil
[b] Instituto Federal de Educação, Ciência e Tecnologia do Sudeste de Minas Gerais, IF Sudeste MG – Juiz de Fora, 36080-001, MG, Brazil
[c] Tomsk State Pedagogical University, Tomsk, 634041, Russia
[d] National Research Tomsk State University, Tomsk, 634050, Russia

ARTICLE INFO

Editor: M. Cvetič

Keywords:
Conformal anomaly
Effective action
Conformal operators
Topological terms

ABSTRACT

The trace anomaly in six-dimensional space is given by the local terms which have six derivatives of the metric. We find the effective action which is responsible for the anomaly. The result is presented in non-local covariant form and also in the local covariant form with two auxiliary scalar fields.

1. Introduction

Within the modern approach the effective action (EA) is a central object in quantum theory of fields. In particular, evaluation of vacuum EA is one of the main targets in the semiclassical approach to quantum gravity.

Usually complete derivation of EA is impossible, and the main purpose is to develop an approximation which is controllable in the sense we can see what is the physical situation when the given approximation can be applied. The most important examples are the anomaly induced EA, which are derived by integrating trace anomalies and therefore hold full information about the vacuum quantum effects in UV. As far as the original theory possess local conformal symmetry, the notion of UV can be usually extended to most of the physically relevant domains. In some physical situations one can regard masses as small perturbations, extending the area of application of conformal anomaly. Last but not least is that the anomaly induced EA can be derived in a closed, compact and useful form. For this reason the conformal anomaly and the anomaly-induced actions are the main instruments used in cosmology and black hole physics to deal with the quantum effects of vacuum.

In the dimension $D = 2$ the integration of anomaly yields the Polyakov action [1], which proved important for the development of string theory. Since the $D = 2$ result is exact, it also plays the role of a reference for other dimensions, where one has to use approximations. In the $D = 4$ case the analog of Polyakov EA has been obtained by Riegert [2], Fradkin and Tseytlin [3], just three years after the two-dimensional analog. This result proved to be an extremely useful tool for numerous applications [4,5]. The anomaly-induced EA in $D = 4$ is essentially more complicated than the Polyakov action, mainly because in $D = 4$ there is an "integration constant", an arbitrary conformal functional S_c, which can not be determined from anomaly. The problems with Ward identities of the RFT action [2,3] which were discussed in a number of works starting from [6,7] are related to the fact that this conformal functional remains unknown. At the same time, since S_c is not related to the UV sector of the theory, in all known applications this uncertainty proves to be irrelevant (see the discussion in [5]).

Along with the conformal functional, in $D = 4$ there is also one more complication. Compared to $D = 2$, the anomaly includes, along with the topological term, also conformal and surface terms. Furthermore, the coefficients of the most relevant topological and conformal terms possess a mysterious universality of signs, which do not depend on the Grassmann parity, but only on the number of derivatives in the action of quantum fields. There is a very interesting and fruitful statement that this universality of signs and related property of the renormalization group flows holds beyond

* Corresponding author.
E-mail addresses: fabricio.ferreira@ifsudestemg.edu.br (F.M. Ferreira), shapiro@fisica.ufjf.br (I.L. Shapiro).

the one-loop level. This idea opened recently a new area of research which is known as c- and a-theorems (see, e.g., [8–10]).

The c- and a-theorems open the window to the rich mathematical applications concerning the general properties of renormalization group flows. In the part of applications to cosmology and astrophysics, the main purpose is to prove that (and understand why) the sign pattern in the anomaly-induced action is preserved beyond one-loop order, including at the non-perturbative level. In this respect it would be certainly interesting to expand our experience with the sign universality to the qualitatively new theories, and higher dimensions represent a unique possibility in this respect. It is known that the general structure of anomaly does not change when the dimensions $D \geqslant 6$ are considered [11–13]. Our previous experience shows that the consistent analysis of the problem can be achieved only by means of the anomaly-induced EA [11,14]. However, an explicit expression for $D > 4$ was not known until now. One has to note that the integration of anomaly in $D = 6$ always attracted a great deal of attention, but until now there were only particular, albeit very interesting, results [15–17] (see further references therein and [18] for the possible use in $M5$-branes) which do not enable one to obtain the anomaly-induced EA in a closed form.

In this Letter we report on a complete solution of the problem. For the sake of brevity in this first Letter we do not discuss technical and less relevant issues, and only concentrate on the most important part of the results. In particular, an explicit presentation of the local terms in EA is postponed for the more complete paper which is now under preparation.

The work is organized as follows. In Sect. 2 we briefly describe the scheme of integration which can be applied in any dimension D. As one can see there, the three necessary elements are conformal operator (analog of Paneitz operator in $D = 4$), modified topological invariant and its conformal transformation and, finally, the integration of surface terms. The part which requires the most significant efforts is the search of modified topological invariant with the simplest conformal property, and we have this problem solved for $D = 6$. The relevant building blocks of such an EA in $D = 6$ are presented in Sect. 3. Finally, in Sect. 4 we draw our conclusions and describe some of the possible applications.

2. On the general scheme of integrating anomaly

Let us briefly summarize the general scheme of integrating anomaly, as it is described in the review paper [5] for $D = 4$. The changes which are required in higher even dimensions are not relevant for this scheme, regardless of the growth of technical difficulties.

The vacuum part of the trace anomaly in even dimension $D \geqslant 4$ can be always written as [11–13]

$$T = \langle T^{\mu}_{\mu} \rangle = c_r\, W^r_D + a\, E_D + \Xi_D\,, \qquad (1)$$

with the sum over r. Here W^r_D are conformal invariant terms (typically constructed from Weyl tensor). In $D = 2$ there is no conformal term, and in $D = 4$ there is only one, the square of the Weyl tensor. In $D = 6$ there are three such terms, as it was discussed in detail in [19] (see also [20] and further references therein). It will be useful to have the explicit for of these terms in $D = 6$, namely

$$W^1_6 = C_{\mu\nu\rho\sigma}\, C^{\mu\alpha\beta\nu}\, C_{\alpha\cdot\cdot\beta}^{\ \ \rho\sigma}\,,$$
$$W^2_6 = C_{\mu\nu\rho\sigma}\, C^{\rho\sigma\alpha\beta}\, C_{\alpha\beta\cdot\cdot}^{\ \ \ \ \mu\nu}\,,$$
$$W^3_6 = C_{\mu\rho\sigma\lambda}\Big(\delta^{\mu}_{\nu} + 4R^{\mu}_{\nu} - \frac{6}{5}R\delta^{\mu}_{\nu}\Big)C^{\nu\rho\sigma\lambda} + \nabla_{\mu}J^{\mu}\,, \qquad (2)$$

where

$$J_{\mu} = \big(4R_{\mu}^{\cdot\lambda\rho\sigma}\nabla^{\nu} + 3R^{\nu\lambda\rho\sigma}\nabla_{\mu}\big)R_{\nu\lambda\rho\sigma} + \Big(\frac{1}{2}R\nabla_{\mu} - R^{\nu}_{\mu}\nabla_{\nu}\Big)R + R^{\nu\lambda}\big(\nabla_{\nu}R_{\lambda\mu} - 5\nabla_{\mu}R_{\nu\lambda}\big). \qquad (3)$$

Furthermore, Ξ_D in Eq. (1) is a linear combination of the surface terms, $\Xi_D = \sum \gamma_k \chi_k$ in the corresponding dimension. The list of the relevant terms χ_k in $D = 6$ will be given below in Eq. (14). Finally, E_D is the integrand of the topological term,

$$E_D = \varepsilon^{\rho_1\cdots\rho_D}\, \varepsilon^{\sigma_1\ldots\sigma_D}\, R_{\rho_1\sigma_1\rho_2\sigma_2}\cdots R_{\rho_{D-1}\sigma_{D-1}\rho_D\sigma_D}\,. \qquad (4)$$

The classification (1) is a simple consequence of that the anomaly comes from the one-loop divergences and the last satisfy conformal Noether identity. It is easy to see that the terms which satisfy this identity should belong to the three categories described above.

The numerical coefficients a, c and γ_k depend on the number of massless conformal fields of different spins. These quantities have no real concern to us, because we will describe a general solution valid for any values of a, c and γ_k.

Our purpose is to find the anomaly-induced EA Γ_{ind}, such that

$$-\frac{2}{\sqrt{-g}}\, g_{\mu\nu}\, \frac{\delta\Gamma_{ind}}{\delta g_{\mu\nu}} = T\,. \qquad (5)$$

As we already mentioned, the integration of anomaly requires a modified topological invariant

$$\tilde{E}_D = E_D + \sum_k \alpha_k \chi_k\,, \qquad (6)$$

where the values of α_k are chosen to provide the special conformal property of the new topological term. Namely, we require that under the local conformal transformation

$$g_{\mu\nu} = \bar{g}_{\mu\nu}\, e^{2\sigma(x)} \qquad (7)$$

there should be

$$\sqrt{-g}\tilde{E}_D = \sqrt{-\bar{g}}\big(\bar{\tilde{E}}_D + \kappa\bar{\Delta}_D\sigma\big)\,, \qquad (8)$$

where κ is a constant and $\Delta_D = \Box^{D/2} + \ldots$ is the conformal operator acting on a conformally inert scalar. For example, in $D = 4$, the formulas have the well-known form, with Δ_4 being the Paneitz operator [21,22], $\kappa = 4$, and the surface term in (6) is $\alpha_k\chi_k = -(2/3)\Box R$. Some comment is in order. Of course, in $D = 4$ the $\Box R$ is the unique possible surface term, so this part is simple. However, the coefficient $-2/3$ is a little bit mysterious, because it can be established only by a direct calculation. The details can be found in [23], where one can observe that the conformal transformation of both E_4 and $\Box R$ is quite complicated. Nevertheless, the particular combination with the mystic $-2/3$ cancels all terms of second, third and fourth orders in σ and the remaining linear term involves the conformal operator. Indeed, we expect this symmetry in the general even D case, that means

$$\sqrt{-g}\Delta_D\varphi = \sqrt{-\bar{g}}\bar{\Delta}_D\bar{\varphi} \qquad (9)$$

with $\varphi = \bar{\varphi}$ and all other quantities with bar are constructed with the fiducial metric $\bar{g}_{\mu\nu}$.

In order to integrate the anomaly one needs the last element. Namely, there should be a set of *local* metric-dependent Lagrangians $\mathcal{L}_i$, providing that with some coefficients c_{ik} there is an identity

$$-\frac{2}{\sqrt{-g}}\, g_{\mu\nu}\, \frac{\delta}{\delta g_{\mu\nu}} \sum_i c_{ik} \int_x \mathcal{L}_i = \chi_k\,, \qquad (10)$$

where $\int_x \equiv \int d^Dx\sqrt{-g}$, for each of the surface-term components in (1). If the set $\mathcal{L}_i$ is found, the problem of solving (5) is reduced to integrating the first two terms in (1). And it is easy to see that this problem is easily solved by the use of identity (8). In order to verify this statement, let us follow [2] and introduce the conformal Green function $G(x,x')$ of the operator Δ_D, which satisfies

$$\sqrt{-g}\,\Delta_D^x\, G(x,x') = \delta^D(x,x')\,, \quad G = \bar{G}\,. \tag{11}$$

The complete solution for the anomaly-induced EA can be written down in the form

$$\Gamma_{ind} = S_c + \iint\limits_{x\,y} \Big\{\frac{1}{4}c_r W_D^r + \frac{a}{8}\tilde{E}_D(x)\Big\} G(x,y)\,\tilde{E}_D(y) + \sum_k (\gamma_k - \alpha_k)\sum_i c_{ik}\int\limits_x \mathcal{L}_i\,. \tag{12}$$

Here $S_c = S_c[g_{\mu\nu}]$ is an undefined conformal functional, which represents a boundary condition of the variational equation (5). As we have stressed in the Introduction, this term is present in all even dimensions except $D=2$, that means that the anomaly-induced effective action has a known non-uniqueness. Furthermore, the modification of the coefficients γ_k of the anomaly (1) occurs because part of the surface terms were absorbed into $\tilde{E}_D$.

Let us make an observation about the non-local structure of the terms related to conformal invariants (2). It is well-known that the Weyl-square term in the anomaly-induced action in $D=4$ corresponds to the non-local $\log(\Box/\mu^2)$ insertion between the two Weyl tensors (see, e.g., the recent discussion of the physical meaning of this in [24]). The form of the non-local structure in (12) indicates that the non-local logarithmic insertions take place in all three conformal terms in (2), with the coefficients proportional to the corresponding beta functions. At the same time, the detailed distribution of non-local logarithmic insertion can be established only in a direct calculation (see, e.g., [25] for an example of the similar form factors in $D=4$).

Writing the non-local part of the expression (12) in the symmetric form, one can always present the EA as the local covariant expression constructed with the use of two auxiliary fields ψ and φ, as it was suggested in [26,27]

$$\bar{\Gamma} = S_c + \sum_k(\gamma_k-\alpha_k)\sum_i c_{ik}\int\limits_x \mathcal{L}_i + \frac{1}{2}\int\limits_x \Big\{\varphi\Delta_D\varphi - \psi\Delta_D\psi + \sqrt{-a}\,\varphi\,\tilde{E}_D + \frac{1}{\sqrt{-a}}(\psi-\varphi)\,c_r W_D^r(x)\Big\}. \tag{13}$$

In these formulas we assume that $a<0$, as in the $D=4$ case. In case of $a>0$ the expression can be trivially modified by changing the sign $\tilde{E}_D \to -\tilde{E}_D$. The last observation is that one can also write the action in terms of modified auxiliary fields [27,28] or in the simplest non-covariant form in terms of σ and $\bar{g}_{\mu\nu}$ [2]. Since the transition to these forms is not too different compared to the $D=4$ case, we will not consider these issues here.

The expressions (12) and (13) are explicit formal representations of the EA which corresponds to the general dimension-independent structure of trace anomaly established by Deser and Schwimmer in [13]. It is remarkable that the vacuum EA in an arbitrary even dimension can be written in such a simple and general form.

All in all, it is clear that the integration of anomaly needs Eq. (8) at the first place and also Eq. (10) to deal with the local part of induced action. In the next section we present the result for Eq. (8) in $D=6$.

3. Conformal formulas in $D=6$

The candidate terms to the total derivatives in (1) can be reduced to the form [29]

$$\chi_1 = \Box^2 R\,, \quad \chi_{2;3;4} = \Box\big(R^2_{\mu\nu\alpha\beta};\ R^2_{\mu\nu};\ R^2\big)$$
$$\chi_{5;6;7;8} = \nabla_\mu\nabla_\nu\big(R^\mu{}_{\lambda\alpha\beta}R^{\nu\lambda\alpha\beta};\ R_{\alpha\beta}R^{\mu\alpha\nu\beta};\ R^\mu_\alpha R^{\nu\alpha};\ RR^{\mu\nu}\big)\,. \tag{14}$$

After a very long and in fact complicated calculations, we arrived at the following coefficients which guarantee the equations (6) and (8) for $D=6$,

$$\alpha_1 = \frac{3}{5}\,, \quad \alpha_2 = -\frac{9}{10} - \frac{5}{4}\xi_1 + \frac{3}{8}\xi_2\,, \quad \alpha_3 = \xi_1\,,$$
$$\alpha_4 = 0\,, \quad \alpha_5 = \frac{84}{5} + 3\xi_1 + \frac{11}{2}\xi_2\,, \quad \alpha_6 = -\frac{36}{5} - 2\xi_1 - 5\xi_2\,,$$
$$\alpha_7 = -\frac{18}{5} - \xi_1 - \frac{7}{2}\xi_2\,, \quad \alpha_8 = \xi_2\,. \tag{15}$$

Here ξ_1 and ξ_2 are free parameters which remain undetermined by the condition (8). Assuming (15), all the non-linear in σ terms in (8) cancel, and the remaining linear term corresponds to $\kappa = -6$ and the conformal operator

$$\Delta_6 = \Box^3 + 4R^{\mu\nu}\nabla_\mu\nabla_\nu\Box - R\Box^2 + 4(\nabla^\alpha R^{\mu\nu})\nabla_\alpha\nabla_\mu\nabla_\nu + V^{\mu\nu}\nabla_\mu\nabla_\nu + N^\lambda\nabla_\lambda\,, \tag{16}$$

where

$$V^{\mu\nu} = \Big(1+\frac{1}{6}\xi_2\Big)R^2 g^{\mu\nu} - \Big(\frac{29}{5} - \frac{1}{6}\xi_1 + \frac{17}{12}\xi_2\Big)R^2_{\rho\sigma}\,g^{\mu\nu} + \Big(\frac{16}{5} - \frac{1}{3}\xi_1 + \frac{7}{6}\xi_2\Big)R^2_{\rho\sigma\alpha\beta}\,g^{\mu\nu} - \frac{3}{5}(\Box R)\,g^{\mu\nu} + \Big(\frac{64}{5} + \frac{4}{3}\xi_1 + 2\xi_2\Big) \times \Big(R_{\alpha\beta}R^{\mu\alpha\nu\beta} - R^\mu{}_{\alpha\beta\gamma}R^{\nu\alpha\beta\gamma}\Big) + \Big(\frac{78}{5} + \xi_1 + \frac{3}{2}\xi_2\Big)\Big(R^{\mu\alpha}R^\nu_\alpha - \frac{1}{3}RR^{\mu\nu}\Big)$$

and

$$N^\lambda = \frac{2}{5}(\nabla^\lambda\Box R) + \frac{8}{3}(\xi_1-\xi_2)R_{\alpha\beta\rho\sigma}(\nabla^\rho R^{\alpha\beta\sigma\lambda}) + \Big(\frac{64}{5} + \frac{4}{3}\xi_1 + 2\xi_2\Big)R^{\rho\sigma\alpha\lambda}(\nabla_\rho R_{\sigma\alpha}) + \Big(\frac{14}{5} - \frac{1}{3}\xi_1 - \frac{1}{2}\xi_2\Big)R_{\rho\sigma}(\nabla^\rho R^{\sigma\lambda}) + \Big(\frac{6}{5} + \frac{5}{3}\xi_1 - \frac{5}{6}\xi_2\Big)R_{\rho\sigma}(\nabla^\lambda R^{\rho\sigma}) + \Big(\frac{13}{5} + \frac{1}{6}\xi_1 + \frac{1}{4}\xi_2\Big)R^{\rho\lambda}(\nabla_\rho R) - \Big(\frac{3}{5} + \frac{1}{6}\xi_1 - \frac{1}{12}\xi_2\Big)R(\nabla^\lambda R)\,. \tag{17}$$

Let us note that in the literature one can find a general theory for constructing conformal operators (see, e.g., [30,16,17,15]). If we exchange the parameters according to

$$\xi_1 = -\frac{3}{200}(256 + 35\zeta_1 + 60\zeta_2) \quad \text{and}$$
$$\xi_2 = -\frac{3}{100}(128 + 5\zeta_1 - 20\zeta_2)\,, \tag{18}$$

the expression (16) coincides with the one obtained in [17].

The main relation (6) was not derived before, probably due to the complexity of calculations required to get the coefficients (15). We could achieve it by combining hand-made work and the softwares Cadabra [31] and Mathematica [32]. The essential details will be published elsewhere [33], together with the solution for the local terms producing surface terms (10) in the anomaly.

Compared to the main calculation, it is much easier (but still consuming certain time and effort) to check that the operator Δ_6 satisfies the conformal invariance (9) and is self-adjoint, $\int_x \varphi \Delta_6 \chi = \int_x \chi \Delta_6 \varphi$. It is interesting that both conditions do not pose any restriction on the value of arbitrary parameters ξ_1 and ξ_2. We shall discuss the physical consequences of this ambiguity in the last section.

4. Conclusions and discussions

The equations (15) and (16) form the full set of the building blocks for the non-local part of anomaly-induced action (13) in $D = 6$. Together with the previously known examples in $D = 2, 4$ this enables us to draw some general conclusions and discuss the similarities and differences between the new result and the previously known ones. One of the common points is that the anomaly-induced expression is an exact EA for the homogeneous and isotropic metric, where the conformal functional S_c is irrelevant. Assuming that the space-time has six dimensions, and that there are massless conformal fields in the far IR, we arrive at the exact solution for anomaly-induced action in this particular class of metrics.

Qualitatively, the structure of (12) and (13) is the same in all even dimensions, but the complexity of the solution increases with dimension. On the transition from two to four dimensions the main complications were the integration constant S_c and the presence of the two different (conformal and topological) terms in (1) which produce non-local terms in the anomaly-induced action [11]. One of the consequences is that the integrated anomaly can be consistently written in local covariant form only by means of two auxiliary fields [26,27,5], while in $D = 2$ one such field is sufficient. As we have seen in Sect. 2 the number of auxiliary fields remains the same in higher dimensions. At the same time the solution (15), (16) includes a qualitatively new arbitrary parameters ξ_1 and ξ_2. Nothing of this sort takes place in $D = 2, 4$. An interesting possibility is that the ambiguity can be fixed by imposing the consistency conditions [6,34,35], but it is not certain, of course. Another question concerns the possible physical effects of arbitrary parameters $\xi_{1,2}$.

Since the conformal anomaly is the same for any $\xi_{1,2}$, one can simply ignore the ambiguity by fixing some particular value for this parameter. The difference between distinct values can be always absorbed into the conformal functional S_c. The situation is technically similar to the one with the ψ-dependent part of (13), which can be also absorbed into conformal part. However, in the case of ψ-terms this would be a wrong idea. For instance, without the second auxiliary field one can not classify vacuum states in the vicinity of the spherically symmetric black holes [36]. There is no such a problem for the gravitational waves, but maybe only because all known calculations were done for the isotropic cosmological backgrounds [37–40]. Concerning the role of $\xi_{1,2}$, the question is whether it affects the relevant solutions, and this question will remain open until such solutions are explored for he action (13).

The last observation concerns the possible applications of the EA (12) and (13). One can imagine that the explicit form of effective vacuum action for the conformal fields can be useful for verifying the calculations related to holography and AdS/CFT correspondence. Another application is related to the dimensional reduction to $D = 4$, expected to produce a four-dimensional action different from the one coming from integrating anomaly directly in $D = 4$. Due to the universality of the result, the calculation of such a reduced action and the study of its physically relevant solutions may be eventually useful in designing the experimental and/or observational tests for the existence of extra dimensions.

Acknowledgements

I.Sh. is grateful to CNPq, FAPEMIG and ICTP for partial support.

References

[1] A.M. Polyakov, Phys. Lett. B 103 (1981) 207.
[2] R.J. Riegert, Phys. Lett. B 134 (1980) 56.
[3] E.S. Fradkin, A.A. Tseytlin, Phys. Lett. B 134 (1980) 187.
[4] M.J. Duff, Class. Quantum Gravity 11 (1994) 1387, arXiv:hep-th/9308075.
[5] I.L. Shapiro, Class. Quantum Gravity 25 (2008) 103001, arXiv:0801.0216.
[6] L. Bonora, P. Pasti, M. Bregola, Class. Quantum Gravity 3 (1986) 635.
[7] H. Osborn, A.C. Petkou, Ann. Phys. 231 (1994) 311, arXiv:hep-th/9307010; J. Erdmenger, H. Osborn, Nucl. Phys. B 483 (1997) 431, arXiv:hep-th/9605009.
[8] Z. Komargodski, A. Schwimmer, J. High Energy Phys. 1112 (2011) 099, arXiv:1107.3987.
[9] M.A. Luty, J. Polchinski, R. Rattazzi, J. High Energy Phys. 1301 (2013) 152, arXiv:1204.5221.
[10] Z. Komargodski, J. High Energy Phys. 1207 (2012) 069, arXiv:1112.4538.
[11] S. Deser, M.J. Duff, C.J. Isham, Nucl. Phys. B 111 (1976) 45.
[12] M.J. Duff, Nucl. Phys. B 125 (1977) 334.
[13] S. Deser, A. Schwimmer, Phys. Lett. B 309 (1993) 279, arXiv:hep-th/9302047.
[14] S. Deser, Phys. Lett. B 479 (2000) 315, arXiv:hep-th/9911129.
[15] H. Osborn, A. Stergiou, J. High Energy Phys. 1504 (2015) 157, arXiv:1501.01308.
[16] T. Arakelyan, D.R. Karakhanyan, R.P. Manvelyan, R.L. Mkrtchyan, Phys. Lett. B 353 (1995) 52.
[17] K. Hamada, Prog. Theor. Phys. 105 (2001) 673, arXiv:hep-th/0012053.
[18] T. Maxfield, S. Sethi, J. High Energy Phys. 1206 (2012) 075, arXiv:1204.2002.
[19] F. Bastianelli, S. Frolov, A.A. Tseytlin, J. High Energy Phys. 0002 (2000) 013, arXiv:hep-th/0001041.
[20] R.R. Metsaev, J. Phys. A 44 (2011) 175402, arXiv:1012.2079; H. Lü, Yi Pang, C.N. Pope, Phys. Rev. D 84 (2011) 064001, arXiv:1106.4657.
[21] S. Paneitz, MIT preprint, 1983; S. Paneitz, SIGMA 4 (2008) 036, arXiv:0803.4331.
[22] E.S. Fradkin, A.A. Tseytlin, Phys. Lett. B 110 (1982) 117; E.S. Fradkin, A.A. Tseytlin, Nucl. Phys. B 203 (1982) 157.
[23] D.F. Carneiro, E.A. Freiras, B. Gonçalves, A.G. de Lima, I.L. Shapiro, Gravit. Cosmol. 40 (2004) 305, arXiv:gr-qc/0412113.
[24] G. Cusin, F.O. Salles, I.L. Shapiro, Phys. Rev. D 93 (2016) 044039, arXiv:1503.08059.
[25] A.O. Barvinsky, Yu.V. Gusev, G.A. Vilkovisky, V.V. Zhytnikov, Nucl. Phys. B 439 (1995) 561–582, http://dx.doi.org/10.1016/0550-3213(94)00585-3, arXiv:hep-th/9404187.
[26] I.L. Shapiro, A.G. Jacksenaev, Phys. Lett. B 324 (1994) 284.
[27] P.O. Mazur, E. Mottola, Phys. Rev. D 64 (2001) 104022, arXiv:hep-th/0106151.
[28] S. Mauro, I.L. Shapiro, Phys. Lett. B 746 (2015) 372, arXiv:1412.5002.
[29] F.M. Ferreira, I.L. Shapiro, P.M. Teixeira, Eur. Phys. J. Plus 131 (2016) 164, arXiv:1507.03620.
[30] T.P. Branson, Commun. Partial Differ. Equ. 1 (1982) 393; T.P. Branson, Math. Scand. 57 (1985) 293.
[31] K. Peeters, Comput. Phys. Commun. 176 (2007) 550, arXiv:cs/0608005.
[32] Wolfram Research, Mathematica, Version 9.0, Champaign, IL, 2012.
[33] F.M. Ferreira, I.L. Shapiro, Paper in preparation.
[34] F. Bastianelli, G. Cuoghi, L. Nocetti, Class. Quantum Gravity 18 (2001) 793, arXiv:hep-th/0007222.
[35] B. Grinstein, A. Stergiou, D. Stone, J. High Energy Phys. 1311 (2013) 195, arXiv:1308.1096.
[36] R. Balbinot, A. Fabbri, I.L. Shapiro, Phys. Rev. Lett. 83 (1999) 1494, arXiv:hep-th/9904074; R. Balbinot, A. Fabbri, I.L. Shapiro, Nucl. Phys. B 559 (1999) 301, arXiv:hep-th/9904162; P.R. Anderson, E. Mottola, R. Vaulin, Phys. Rev. D 76 (2007) 124028, arXiv:0707.3751 [gr-qc].
[37] A.A. Starobinsky, JETP Lett. 30 (1979) 682, Pis'ma Zh. Eksp. Teor. Fiz. 30 (1979) 719;

A.A. Starobinsky, JETP Lett. 34 (1981) 438, Pis'ma Zh. Eksp. Teor. Fiz. 34 (1981) 460;
A.A. Starobinsky, Sov. Astron. Lett. 9 (1983) 302.
[38] J.C. Fabris, A.M. Pelinson, I.L. Shapiro, Nucl. Phys. B 597 (2001) 539;
J.C. Fabris, A.M. Pelinson, I.L. Shapiro, Nucl. Phys. B 602 (644) (2001) (Erratum), arXiv:hep-th/0009197.
[39] S.W. Hawking, T. Hertog, H.S. Reall, Phys. Rev. D 63 (2001) 083504, arXiv:hep-th/0010232.
[40] J.C. Fabris, A.M. Pelinson, F. de O. Salles, I.L. Shapiro, J. Cosmol. Astropart. Phys. 02 (2012) 019, arXiv:1112.5202.

24

Linking structure and dynamics in (p, pn) reactions with Borromean nuclei: The $^{11}Li(p, pn)^{10}Li$ case

M. Gómez-Ramos [a,*], J. Casal [a,b], A.M. Moro [a]

[a] Departamento de Física Atómica, Molecular y Nuclear, Facultad de Física, Universidad de Sevilla, Apartado 1065, E-41080 Sevilla, Spain
[b] European Centre for Theoretical Studies in Nuclear Physics and Related Areas (ECT*) and Fondazione Bruno Kessler, Villa Tambosi, Strada delle Tabarelle 286, I-38123 Villazzano (TN), Italy

ARTICLE INFO

Editor: J.-P. Blaizot

Keywords:
^{10,11}Li
Transfer to continuum
Overlaps
Three-body

ABSTRACT

One-neutron removal (p, pn) reactions induced by two-neutron Borromean nuclei are studied within a Transfer-to-the-Continuum (TC) reaction framework, which incorporates the three-body character of the incident nucleus. The relative energy distribution of the residual unbound two-body subsystem, which is assumed to retain information on the structure of the original three-body projectile, is computed by evaluating the transition amplitude for different neutron-core final states in the continuum. These transition amplitudes depend on the overlaps between the original three-body ground-state wave function and the two-body continuum states populated in the reaction, thus ensuring a consistent description of the incident and final nuclei. By comparing different ^{11}Li three-body models, it is found that the $^{11}Li(p, pn)^{10}Li$ relative energy spectrum is very sensitive to the position of the $p_{1/2}$ and $s_{1/2}$ states in ^{10}Li and to the partial wave content of these configurations within the ^{11}Li ground-state wave function. The possible presence of a low-lying $d_{5/2}$ resonance is discussed. The coupling of the single particle configurations with the non-zero spin of the ^{9}Li core, which produces a spin-spin splitting of the states, is also studied. Among the considered models, the best agreement with the available data is obtained with a ^{11}Li model that incorporates the actual spin of the core and contains $\sim$31% of $p_{1/2}$-wave content in the n-^{9}Li subsystem, in accord with our previous findings for the $^{11}Li(p, d)^{10}Li$ transfer reaction, and a near-threshold virtual state.

1. Introduction

Two-neutron Borromean nuclei are unique nuclear systems lying at the edge of the neutron drip-line. These are short-lived, weakly bound nuclei, with typically no bound excited states, and whose binary subsystems are unbound. Although some of them, such as ^{6}He and ^{11}Li, had already been identified long ago as products of reactions with stable beams [1,2], it was not until the late eighties that their unusual properties (such as their large size) were realized thanks to the pioneering experiments performed by Tanihata and collaborators [3] using secondary beams of these species and the subsequent theoretical works initiated by Hansen and Jonson [4]. The picture emerging from these studies revealed a very exotic structure, consisting of a relatively compact core surrounded by two loosely bound nucleons forming a dilute halo.

* Corresponding author.
E-mail address: mgomez40@us.es (M. Gómez-Ramos).

Later works have revealed that this fragile structure arises from a delicate interplay of different effects, such as the pairing interaction between the halo neutrons or the coupling of the motion of these nucleons with tensor and collective excitations of the core (e.g. [5–7]). A quantitative account of all these effects is a challenging theoretical problem and, quite often, different models lead to different (sometimes contradictory) predictions of the structure properties, such as energies and spin-parity assignments.

Experimentally, a successful technique to probe the properties of the neutron-core system is by means of (p, pn) reactions at intermediate energies (above $\sim$100 MeV/nucleon), in which the radioactive beam collides with a proton target, removing one neutron, and leaving an (unbound) residual nucleus, which will eventually decay into a neutron and a core [8,9]. Typically, these experiments measure the relative energy spectrum of this neutron-core system, whose prominent structures are associated with virtual states or resonances. Moreover, if the core is left in an excited state, gamma rays will be also emitted [10]. Angular momentum and spin assignment of these structures is often done by comparing these spectra with the profiles expected in the hypotheti-

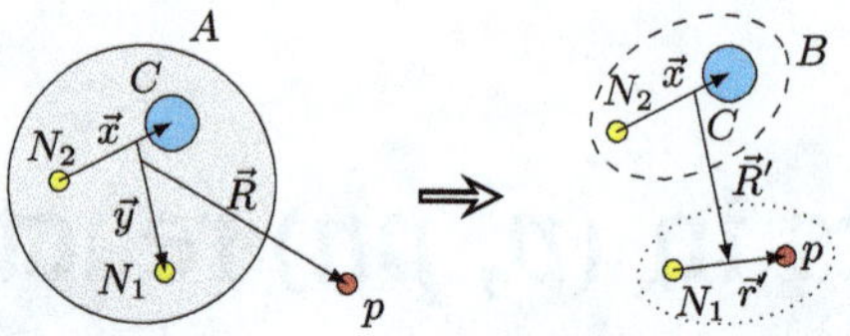

Fig. 1. Diagram for a (p, pN) reaction induced by a three-body projectile in inverse kinematics.

cal neutron-core two-body scattering (e.g. Breit–Wigner functions). This procedure is hampered by a number of limitations. For instance, it ignores completely the effect of the reaction dynamics on the spectra and, therefore, there is no a priori information on the absolute magnitude of the cross sections or, in other words, between the reaction observables and the underlying structure against which the data are confronted. Moreover, due to energy resolution, resonances will appear smeared out or even unresolved.

It is our purpose in this work to propose a new theoretical framework for the analysis of (p, pn) reactions induced by Borromean nuclei in which the structure model of the incident three-body nucleus is incorporated into a reaction formalism, thereby enabling the computation of reaction observables to be directly compared with the reaction data. In particular, we will use a three-body model, which has been very successful in the understanding of the properties of Borromean nuclei (e.g. [11]). For the reaction dynamics, we employ the Transfer-to-the-Continuum (TC) framework, which is formally similar to a CCBA (coupled-channel Born approximation) [12,13] approach populating unbound states of the p-n system. This method has already been applied to (p, pn) reactions with two-body projectiles [14]. We apply this formalism to describe the reaction ^{11}Li$(p, pn)^{10}$Li at 280 MeV/A, which was measured a few years ago at GSI [8].

The paper is organized as follows. In Sec. 2, we present the reaction formalism, which is an extended version of that given in Ref. [14] and will allow us to explain how the three-body structure enters into the calculation. In Sec. 3, the formalism is applied to the ^{11}Li$(p, pn)^{10}$Li reaction, focusing on the relative-energy distribution of the decaying ^{10}Li subsystem. Different structure models are considered and their impact on the reaction observables is discussed. Finally, in Sec. 4 we summarize the main results of the work and outline possible applications and extensions.

2. Reaction formalism

A (p, pN) reaction induced by a three-body projectile comprising an inert core (C) plus two valence neutrons $(N1, N2)$ takes the form,

$$\underbrace{(C + N_1 + N_2)}_{A} + p \rightarrow \underbrace{(C + N_2)}_{B} + N_1 + p, \tag{1}$$

which is schematically depicted in Fig. 1. If the nucleus B does not form bound states (e.g., the composite A is a Borromean system) the products of its decay after one neutron removal will provide spectroscopic information on the original projectile wave function. As in Ref. [15], we describe the process using a *participant/spectator* approximation, assuming that the reaction occurs due to the interaction of the incident proton with a single neutron (N_1) of A, whereas the subsystem $B = N_2 + C$ remains unperturbed. The prior-form transition amplitude of such a process can be formally reduced to an effective few-body problem, leading to

$$\mathcal{T}_{if} = \sqrt{2}\langle \Psi_f^{(-)}(\vec{x}, \vec{R}', \vec{r}')|V_{pN_1} + U_{pB} - U_{pA}|\Phi_A(\vec{x}, \vec{y})\chi_{pA}^{(+)}(\vec{R})\rangle, \tag{2}$$

where Φ_A represents the ground-state wave function of the initial three-body composite, $\chi_{pA}^{(+)}$ is the distorted wave generated by the auxiliary potential U_{pA}, and $\Psi_f^{(-)}$ is the exact four-body wave function for the outgoing pN_1-B system. The $\pm$ superscript refers to the usual ingoing or outgoing boundary conditions. Notice the explicit factor $\sqrt{2}$ arising from the two identical neutrons in the three-body projectile. The origin of this factor is further discussed in Ref. [16].

To reduce Eq. (2) to a tractable form, in the TC method we approximate the exact wave function $\Psi_f^{(-)}$ by the factorized expression,

$$\Psi_f^{(-)}(\vec{x}, \vec{R}', \vec{r}') \approx \varphi_{\vec{q},\sigma_2,\zeta}^{(-)}(\vec{x})\Upsilon_f^{(-)}(\vec{R}', \vec{r}') \tag{3}$$

where $\varphi_{\vec{q},\sigma_2,\zeta}^{(-)}$ is a two-body continuum wave function with wave number $\vec{q}$ and definite spin projections of the binary subsystem B, and $\Upsilon_f^{(-)}$ is a three-body wave function describing the relative motion of the p-N_1-B system in the exit channel. As in Ref. [14], we expand this function in pN_1 continuum states using a binning procedure [17],

$$\Upsilon_f^{(-)}(\vec{R}', \vec{r}') \simeq \sum_{n\mathcal{J}\Pi} \phi_{n\mathcal{J}\Pi}(k_n, \vec{r}')\chi_{n\mathcal{J}\Pi}^{(-)}(K_n, \vec{R}'). \tag{4}$$

Here, $\phi_{n\mathcal{J}\Pi}$ are a set of N discretized pN_1 bins with angular momentum $\mathcal{J}^{\Pi}$,

$$\phi_{n\mathcal{J}\Pi}(k_n, \vec{r}') = \sqrt{\frac{2}{\pi N}} \int_{k_{n-1}}^{k_n} \phi_{\mathcal{J}\Pi}^{(+)}(k, \vec{r}')dk, \tag{5}$$

which are obtained from the scattering eigenstates $\phi_{\mathcal{J}\Pi}^{(+)}$ of the potential V_{pN_1}, and $\chi_{n\mathcal{J}\Pi}^{(-)}$ are distorted waves for the pN_1-B relative motion. Note that the subscript $f = \{n, \mathcal{J}, \Pi\}$ in $\Upsilon_f^{(-)}$ retains the information on the definite final state.

The function $\varphi_{\vec{q},\sigma_2,\zeta}^{(-)}$ in Eq. (3) is the time-reversed of $\varphi_{\vec{q},\sigma_2,\zeta}^{(+)}$, which can be written as (cf. [18], p. 135)

$$\varphi_{\vec{q},\sigma_2,\zeta}^{(+)}(\vec{x}) = \frac{4\pi}{qx} \sum_{LJJ_TM_T} i^L Y_{LM}^*(\widehat{q})\langle LMs_2\sigma_2|JM_J\rangle \\ \times \langle JM_JI\zeta|J_TM_T\rangle f_{LJ}^{J_T}(qx)\left[\mathcal{Y}_{Ls_2J}(\widehat{x}) \otimes \kappa_I\right]_{J_TM_T}, \tag{6}$$

where, for each component, the orbital angular momentum $\vec{L}$ and the spin $\vec{s}_2$ of the neutron N_2 couple to $\vec{J}$, and $\vec{J}_T$ results from coupling $\vec{J}$ with the spin $\vec{I}$ of the core. Note that, in our schematic notation, $\vec{x}$ contains also the internal coordinates of C. The radial functions $f_{LJ}^{J_T}$ are obtained by direct integration of the two-body Schrodinger equation for the $N_2 + C$ system subject to standard scattering boundary conditions.

For the wave function of the projectile nucleus, $\Phi_A^{j\mu}(\vec{x}, \vec{y})$, we use a three-body expansion in hyperspherical harmonics [19–21] using a pseudostate basis for the radial part called analytical transformed harmonic oscillator basis [22–25].

Assuming that the potentials U_{pB} and U_{pA} appearing in the transition amplitude (2) do not change the internal state of B, which is consistent with our participant/spectator approximation, one can perform the integral in the internal coordinates $\vec{x}$, giving rise to the overlap functions [15]

$$\psi_{LJJ_TM_T}(q, \vec{y}) = \int \frac{f_{LJ}^{J_T}(qx)}{x}\left[\mathcal{Y}_{Ls_2J}(\widehat{x}) \otimes \kappa_I\right]_{J_TM_T}^* \Phi_A^{j\mu}(\vec{x}, \vec{y})d\vec{x}, \tag{7}$$

Table 1
Features of the ^{10}Li structure for the different potentials employed in this work. The second column shows the scattering length of the $s_{1/2}$ virtual state, the third column gives the energy of the $p_{1/2}$ resonance, and the fourth column corresponds to the position of the $d_{5/2}$ state. Note that for the model including the spin of the core, P1I, the $s_{1/2}, p_{1/2}$ configurations split into $1^-, 2^-, 1^+, 2^+$. The splitting for the $d_{5/2}$ component is not considered, as this resonance is disregarded in the P1I potential. The next three columns show the partial wave content of ^{10}Li configuration within the ^{11}Li ground state, while the last two columns show its matter and charge radii.

	a (fm)	$E_r[p_{1/2}]$ MeV	$E_r[d_{5/2}]$ (MeV)	$\%s_{1/2}$	$\%p_{1/2}$	$\%d_{5/2}$	r_{mat} (fm)	r_{ch} (fm)
P3	−29.8	0.50	4.3	64	30	3	3.6	2.48
P4	−16.2	0.23	4.3	27	67	3	3.3	2.43
P5	−29.8	0.50	1.5	39	35	23	3.2	2.42
P1I	– −37.9	0.37 0.61	–	67	31	1	3.2	2.41

which contain all the relevant structure information. When used in Eq. (2) one gets

$$\mathcal{T}_{if} = \sqrt{2}\frac{4\pi}{q} \sum_{LJJ_T M_T} (-i)^L Y_{LM}(\widehat{q}) \langle LMs_2\sigma_2 | JM_J \rangle \times \langle JM_J I\zeta | J_T M_T \rangle \mathcal{T}_{if}^{LJJ_T M_T}, \tag{8}$$

depending on a set of auxiliary CCBA-like amplitudes,

$$\mathcal{T}_{if}^{LJJ_T M_T} \equiv \langle \Upsilon_f^{(-)} | V_{pN_1} + U_{pB} - U_{pA} | \psi_{LJJ_T M_T} \chi_{pA}^{(+)} \rangle. \tag{9}$$

These amplitudes enable a consistent description of the process, in which the three-body projectile and the binary sub-system incorporate the same core-nucleon interaction.

From the transition amplitude, and after integrating over the angles $\widehat{q}$ of the relative wave vector $\vec{q}$, the double differential cross section for a given final discretized bin $f = \{n, \mathcal{J}, \Pi\}$ as a function of the C-N_2 relative energy and the scattering angle of B with respect to the incident direction can be written as

$$\frac{d\sigma^2_{n\mathcal{J}\Pi}}{d\Omega_B d\varepsilon_x} = \frac{32\pi^2}{q^2}\rho(\varepsilon_x)\frac{1}{2(2j+1)}\frac{\mu_i\mu_f}{(2\pi\hbar^2)^2}\frac{k_n}{k_i} \times \sum_{LJJ_T}\sum_{M_T\sigma_d}\left|\mathcal{T}_{if}^{LJJ_T M_T}\right|^2, \tag{10}$$

where $\rho(\varepsilon_x) = \mu_x q/[(2\pi)^3\hbar^2]$ is the density of B states as a function of the C-N_2 excitation energy ε_x, $\mu_{i,f}$ the projectile-target reduced mass in the initial and final partitions, and σ_d represents the spin projection of the pN_1 system. Although the non-relativistic expression for the cross section is shown in Eq. (10) for simplicity, relativistic kinematics must be taken into account due to the high beam energy in typical (p, pn) reactions [14]. Note that q is related to the C-N_2 relative energy as $q = \sqrt{2\mu_x\varepsilon}/\hbar$, with μ_x its reduced mass. From Eq. (10), the total differential cross section is obtained as an incoherent sum of the contributions to all pN_1 bins [14].

3. Application to ^{11}Li$(p, pn)^{10}$Li

In this work, the formalism described in the preceding section is applied to the ^{11}Li$(p, pn)^{10}$Li reaction, which has been measured at GSI at 280 MeV/A, using inverse kinematics [8]. For the V_{pn} potential, the Reid93 interaction [26] is chosen. The p-^{11}Li, p-^{10}Li and n-^{10}Li interactions are obtained as in Refs. [14, 15], folding an effective NN interaction with the ground-state density of the composite nucleus. For this purpose, the Paris–Hamburg g-matrix parametrization of the NN interaction [27,28] is employed, while the ground-state densities are computed from Hartree–Fock calculations with the Sk20 effective interaction, using the code OXBASH [29]. This folding is performed making use of the code LEA [30]. Note that, due to the unbound nature of ^{10}Li, the ground-state density of ^{9}Li is used to generate the N-^{10}Li potentials.

As mentioned in the previous section, the p-n continuum is discretized and truncated to a maximum angular momentum. The discretization is performed using a binning procedure [17] with a step of $\Delta E = 15$ MeV, although it was found that the studied observables are rather insensitive to the discretization used. In order to reduce the size of the calculations, we restrict the p-n angular momentum to $\mathcal{J}_{max} = 2$ in the exit channel and, as in Ref. [14], we ignore the couplings between different $\mathcal{J}^\Pi$ states. Test calculations for specific n-^{9}Li relative energies showed that these approximations do not modify significantly the shape of the energy distributions, although they underestimate the magnitude of the total cross section by about 10%. It must be remarked that no rescaling factors need to be applied to our calculations, since absolute cross sections can be obtained from the formalism. This allows us to assess the relative importance of different structure configurations to the cross section.

In the following, we explore the effects of the structure of ^{11}Li on the n-^{9}Li relative energy spectrum after one neutron removal, performing the calculations in the low-energy range where the bulk of the cross section is concentrated. As in [15], different potentials are used to generate the $\psi_{LJJ_T M_T}$ overlaps, leading to different structure properties of the ^{10}Li continuum. In all calculations, we adopt the ^{11}Li ground-state energy of −0.37 MeV [31]. We analyze the effect of virtual and resonant states on the computed spectra, studying in particular the effect of their splitting when the actual spin of the ^{9}Li is included in the calculations.

3.1. Results ignoring the ^{9}Li spin

The spinless-core approximation has been widely used to describe the structure of ^{11}Li [15,32,33]. The analysis of experiments involving ^{10,11}Li usually assumes $I_{^9\mathrm{Li}} = 0$ [8,9,34,35], which simplifies the interpretation of the data. In this picture, the ground state of ^{10}Li is a $s_{1/2}$ virtual state, followed by a low-energy $p_{1/2}$ resonance and, possibly, a $d_{5/2}$ state whose position and width is still unclear [34,35]. In this section we present three different models for ^{11}Li which assume a spinless ^{9}Li and allow us to study the influence of the structure properties on the reaction observables. Here, the n-^{9}Li interaction is modeled with central and spin-orbit Woods-Saxon terms adjusted to produce the ^{10}Li virtual state and resonances at different positions, thus changing the partial wave content in the ^{11}Li(0^+) ground-state wave function. Some properties for ^{10}Li and ^{11}Li resulting from these potentials are shown in Table 1. More details about these structure calculations can be found in Ref. [15].

The results of our calculation for models P3, P4 and P5 are presented in the top, middle and bottom panels of Fig. 2, respectively. On the left side, we show the separate $s_{1/2}$ (dotted), $p_{1/2}$ (dashed) and, in the case of P5, $d_{5/2}$ (dot-dashed) contributions, together

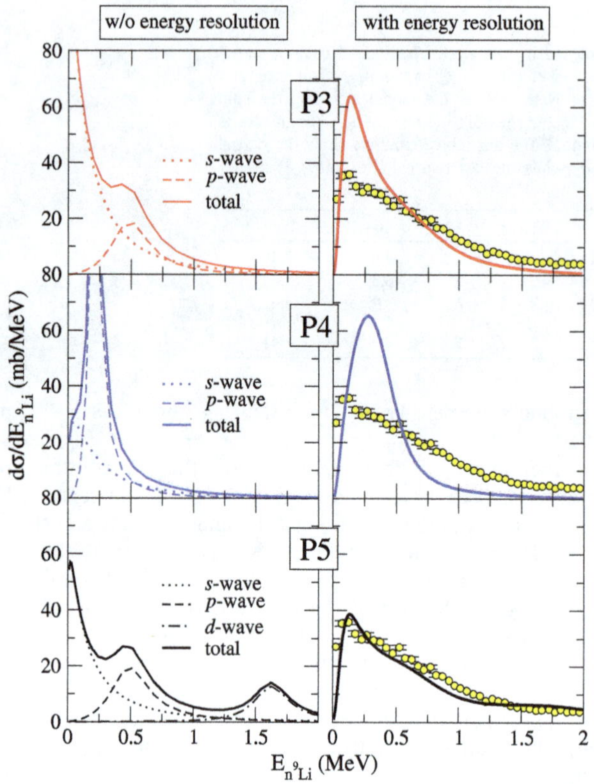

Fig. 2. Relative n-^{9}Li energy spectrum for ^{11}Li$(p, pn)^{10}$Li at 280 MeV/A. Calculations are presented for potentials P3, P4 and P5 with red, blue and black lines, respectively. In the left panels, the contributions for the s and p waves (and d waves for P5) are shown along with their sum. In the right panels, the total cross section is shown after folding with the experimental resolution, along with experimental data from Ref. [8]. (For interpretation of the references to color in this figure legend, the reader is referred to the web version of this article.)

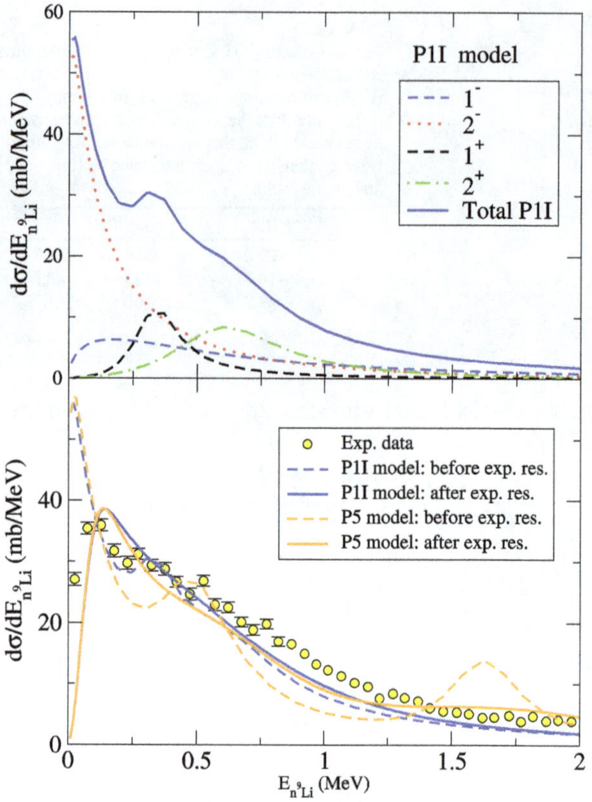

Fig. 3. Effect of the spin-spin splitting in the relative n-^{9}Li energy spectrum. In the top panel, the different contributions for the s $(1^-, 2^-)$ and p $(1^+, 2^+)$ waves within the model P1I are presented. The total cross section is given by the solid blue line. In the bottom panel, the results for models P1I and P5 are shown before and after folding with the experimental resolution. For P5, this includes also the $d_{5/2}$ states. The experimental data are from Ref. [8]. (For interpretation of the references to color in this figure legend, the reader is referred to the web version of this article.)

with their sum (solid). In models P3 and P4, the d-wave content in the ^{11}Li ground state is very small, and the $d_{5/2}$ resonance appears at high energies (see Table 1), thus making this contribution negligible. Note that the height of the peak of the $s_{1/2}$ contribution is related not only to the scattering length, but also to its relative weight in the ^{11}Li ground state. On the right side of Fig. 2, the total cross section is shown for the three models after the convolution with the experimental resolution and compared with the data from [8]. It can be seen that, for P3, the main contribution to the cross section comes from $s_{1/2}$ states, while for P4 it comes from $p_{1/2}$ states. We see that both calculations fail to reproduce the shape of the experimental data, which is heavily influenced by the experimental resolution. In general, both P3 and P4 calculations seem to give too much cross section at low energies while too little at higher energies. The model P5, on the contrary, gives a better agreement with the experimental data by just lowering the position of the $d_{5/2}$ resonance. A similar effect was found in a previous work describing the knockout reaction on carbon [34]. Notice that models P3 and P5 provide the same ^{10}Li states for the $s_{1/2}$ and $p_{1/2}$ configurations, but the partial wave content of ^{11}Li is strongly affected by the presence of the $d_{5/2}$ ^{10}Li resonance at low energies.

However, the d-wave content given by P5, 23%, is rather large compared to the most recent experimental study [9], where it amounts to ~10%. Moreover, the d resonance has been recently identified at higher excitation energies in a (d, p) experiment [35]. This, together with the oversimplification of the ^{10,11}Li models neglecting the spin of the ^{9}Li core, may indicate that the good agreement found for P5 is biased by the low energy resolution of the data.

3.2. Results including the ^{9}Li spin

By considering explicitly the spin of the ^{9}Li core, $I = 3/2^-$, the n-^{9}Li single-particle configurations $s_{1/2}$ and $p_{1/2}$ split in $1^-, 2^-$ and $1^+, 2^+$ states, respectively. Since the s- and p-wave contributions dominate the low-energy relative energy spectra presented in the previous section, these doublets can affect the shape of the distributions. Previous theoretical studies including this effect have reported the existence of one or two s virtual states, and two p resonances [36–38], although no direct experimental evidence of this splitting has been reported [39]. In this section, we perform the same calculation as before, but using a structure model which couples the spin of ^{9}Li to the angular momenta of the two neutrons. We have chosen the potential P1I from Ref. [15], which we found to give a good description of ^{11}Li$(p, d)^{10}$Li data [40]. Some structure features obtained with this model are also shown in Table 1.

In Fig. 3, results for P1I are presented along with those for P5. In the top panel, the contributions from the different two-body configurations, namely, 1^- (thin solid), 2^- (dotted), 1^+ (dashed) and 2^+ (dot-dashed) are presented, together with their sum (thick solid). In this model, the 2^- ground state of ^{10}Li is characterized by a scattering length of $a_s = -37.9$ fm, and the 1^- states correspond to non-resonant continuum. The two p resonances are obtained at 0.37 and 0.61 MeV, providing a doublet that could not be resolved experimentally. As shown in Table 1, the three-body ground-state wave function probabilities in the n-^{9}Li subsystem are given by 31% of $p_{1/2}$ components, 67% of $s_{1/2}$ components and a negligible $d_{5/2}$ contribution. The weights of the individual $1^-, 2^-, 1^+, 2^+$ configurations are 27%, 40%, 12% and 19%, respectively. Taking these values into consideration, the effective scattering length of the 2^- ground state is reduced to $a_{\text{eff}} = -29.3$ fm, and the two p resonances have their centroid at 0.52 MeV. In the bottom panel, the total cross section is presented before and after convoluting with the experimental resolution and compared with the experimental data. Blue lines correspond to model P1I including the spin-spin splitting, while orange lines correspond to the spinless-core model P5. We find the calculation using P1I to provide an even better agreement with the data, when compared to P5. This agreement seems to stem from two sources: first, the $p_{1/2}$ resonance is split, leading to a broader distribution that accommodates the high-energy tail shown by experimental data. Second, the splitting of the $s_{1/2}$ virtual state at low energies enables a reduction of the cross section in this region. Note that, in the present work, these doublets are obtained by introducing a core-spin dependence in the n-^{9}Li potential through a spin-spin term. This allows us to describe the splitting schematically, although the actual mechanism might involve more complex correlations, such as pairing [41], tensor correlations [5] or coupling to excited states of the core [36].

In contrast to P5, the model P1I does not require a $d_{5/2}$ resonance at 1.5 MeV to reproduce the experimental data. The introduction of such a resonance at higher energies (as in models P3, P4 in the preceding section), would have little effect on the partial wave content of ^{11}Li, thus preserving the agreement with the data. However, the coupling of the single-particle configuration $d_{5/2}$ to the $3/2^-$ of the core leads to $1^-, 2^-, 3^-$ and 4^- states for which no information exists, thus complicating the theoretical description of the possible resonances and the interpretation of the data. Clearly, this situation calls for more elaborate theoretical studies and experiments with better energy resolution.

3.3. *Factorization of the cross section*

The analysis of the experimental data presented in Ref. [8] uses a factorization approximation for the cross section. The relative n-^{9}Li energy spectrum is fitted with two distributions corresponding to the $s_{1/2}$ and $p_{1/2}$ states, with their relative weights as parameters. In addition to ignoring the partial wave content of the initial ^{11}Li projectile, which modulates the relevance of the different components, this implies the assumption that the reaction dynamics introduces only a global scaling factor over some structure form factors. Under such considerations, the differential cross section for a given $(LJ)J_T$ configuration can be schematically written as

$$\frac{d\sigma^{LJJ_T}}{d\varepsilon_x} \simeq \mathcal{C}^{LJJ_T} K(\varepsilon_x)\eta^{LJJ_T}(\varepsilon_x), \tag{11}$$

where η^{LJJ_T} represents the structure form factor, K is a kinematic function which contains the density of states and all relevant constants, and $\mathcal{C}^{LJJ_T}$ is a global scaling factor, which contains the effect of the reaction mechanism. In our approach, η^{LJJ_T} are given by the square of the structure overlaps as a function of the energy. This allows us to compare the shape of the cross sections in Eq. (11) with the results from our full TC calculations and, if possible, extract the $\mathcal{C}^{LJJ_T}$ factors. This is shown in Fig. 4 for models P3 and P1I. The rescaled overlaps are rather similar to the full TC calculations, with only small deviations in the shape. This deviation is more significant for the s-wave contributions, which can be associated with the extended halo wave function of ^{11}Li playing a

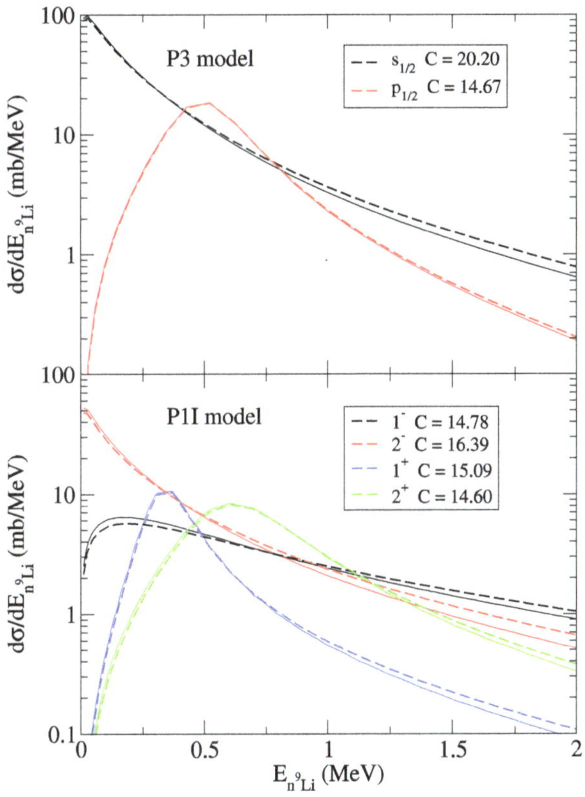

Fig. 4. Comparison between the TC calculations (solid lines) and the rescaled structure overlaps (dashed lines) for models P3 (top panel; $s_{1/2}$ in black lines, $p_{1/2}$ in red lines) and P1I (bottom panel; $1^-, 2^-, 1^+, 2^+$ in black, red, blue and green lines respectively) in logarithmic scale. The factor $\mathcal{C}^{LJJ_T}$ for each contribution is given in the legend. Calculations are shown without convoluting with the experimental resolution. (For interpretation of the references to color in this figure legend, the reader is referred to the web version of this article.)

role for the reaction dynamics, but it is still a minor effect. The resulting $\mathcal{C}^{LJJ_T}$ factors are similar in both models for the p-wave components and differ significantly for the s-wave. For models P4 and P5, the s- and p-wave scaling factors are found to be almost identical to those for P3, while the d-wave scaling in the model P5 is similar to that for p-waves. However, even though at first order the reaction dynamics introduces only J_T-dependent factors, a proper reaction formalism is required to obtain them unambiguously. This, together with the role played by the weights of the different configurations in the ground state of the projectile, indicates that only a consistent description of both the structure and dynamics can provide a reliable interpretation of the data.

4. Summary and conclusions

We have presented a new method to study (p, pn) reactions induced by three-body Borromean nuclei. The formalism is a natural extension of the Transfer-to-the-Continuum method [14], recently proposed and applied to two-body projectiles, to the case of three-body projectiles. The model assumes a participant/spectator picture, in which the proton target knocks out one of the halo neutrons (the participant), while leaving unperturbed the remaining neutron-core subsystem (the spectator). A key feature of the model is the use of structure overlaps obtained from a three-body model of the ground-state wave-function of the Borromean nucleus and the two-body scattering states of the neutron-core residual system. These overlaps are used in a CCBA-like prior-form transition amplitude, thus providing a connection between the structure model and the reaction observables without the need of introducing arbitrary scaling factors. In particular, the formalism provides double differential cross sections as a function of the scattering angle of the residual two-body system and the relative energy between its constituents.

The model has been applied to the $^{11}Li(p,pn)^{10}Li$ reaction at 280 MeV/A, comparing with available data for the neutron-^{9}Li system relative-energy distribution. Several structure models of ^{11}Li have been compared, differing on the position of the assumed $s_{1/2}$ virtual states and $p_{1/2}$ resonances, and on the inclusion or not of the ^{9}Li spin which, in turn, give rise to different relative weights for these partial waves in the ground state of ^{11}. The calculated reaction observable is found to be very sensitive to these structure properties. Among the considered models, the best agreement with the data is obtained using a ^{11}Li model that incorporates the actual spin of the core and contains ~31% of $p_{1/2}$-wave content in the n-^{9}Li subsystem and a near-threshold virtual state with an effective scattering length of about -29 fm. The agreement stems from the splitting of the $s_{1/2}$ virtual state and the $p_{1/2}$ resonance. This splitting was obtained thanks to a spin-spin interaction in this work, although its actual origin may arise from more complex correlations. Interestingly, this model was found to provide also a good description of the recent $^{11}Li(p,d)^{10}Li$ transfer data measured at TRIUMF.

We have discussed also the possible presence of a low-lying d-wave resonance in ^{10}Li. An overall good agreement with the data can be obtained in the model ignoring the ^{9}Li spin by forcing a $d_{5/2}$ resonance to appear at E_r =1.5 MeV, which also reduces the s-wave content in ^{11}Li. However, in view of other experimental evidences, the agreement might be merely accidental. In fact, such a resonance is not required in the model including the spin of ^{9}Li to achieve a good description of the data. Due to the smearing effect produced by the energy resolution of the experiment, it is clear that further data, more sensitive to higher excitation energies and with better energy resolution, will certainly help in extracting robust conclusions on the $d_{5/2}$ states.

The formalism presented could be applied to study (p,pn) or $(p,2p)$ reactions induced by other three-body nuclei. Calculations of this kind for ^{14}Be, including also the effect of core excitations, are in progress and will be presented elsewhere.

Acknowledgements

This work has received funding from the Spanish Ministerio de Economía y Competitividad under Project No. FIS2014-53448-C2-1-P and by the European Union Horizon 2020 research and innovation program under Grant Agreement No. 654002. M.G.-R. acknowledges support from the Spanish Ministerio de Educación, Cultura y Deporte, Research Grant No. FPU13/04109.

References

[1] T. Bjerge, K.J. Brostrøm, Nature 138 (1936) 400, http://dx.doi.org/10.1038/138400b0.
[2] A.M. Poskanzer, S.W. Cosper, E.K. Hyde, J. Cerny, Phys. Rev. Lett. 17 (1966) 1271, http://dx.doi.org/10.1103/PhysRevLett.17.1271.
[3] I. Tanihata, H. Hamagaki, O. Hashimoto, S. Nagamiya, Y. Shida, N. Yoshikawa, O. Yamakawa, K. Sugimoto, T. Kobayashi, D. Greiner, N. Takahashi, Y. Nojiri, Phys. Lett. B 160 (1985) 380, http://dx.doi.org/10.1016/0370-2693(85)90005-X.
[4] P.G. Hansen, B. Jonson, Europhys. Lett. 4 (1987) 409, http://dx.doi.org/10.1209/0295-5075/4/4/005.
[5] T. Myo, K. Katō, H. Toki, K. Ikeda, Phys. Rev. C 76 (2007) 024305, http://dx.doi.org/10.1103/PhysRevC.76.024305.
[6] F. Barranco, P.F. Bortignon, R.A. Broglia, G. Colò, E. Vigezzi, Eur. Phys. J. A 11 (2001) 385, http://dx.doi.org/10.1007/s100500170050.
[7] K. Ikeda, T. Myo, K. Kato, H. Toki, Di-neutron clustering and deuteron-like tensor correlation in nuclear structure focusing on ^{11}Li, in: Clusters in Nuclei, vol. 1, Springer, Berlin, Heidelberg, 2010, pp. 165–221.
[8] Y. Aksyutina, et al., Phys. Lett. B 666 (2008) 430, http://dx.doi.org/10.1016/j.physletb.2008.07.093.
[9] Y. Aksyutina, et al., Phys. Rev. C 87 (2013) 064316, http://dx.doi.org/10.1103/PhysRevC.87.064316.
[10] Y. Kondo, et al., Phys. Lett. B 690 (2010) 245, http://dx.doi.org/10.1016/j.physletb.2010.05.031.
[11] M. Zhukov, B. Danilin, D. Fedorov, J. Bang, I. Thompson, J. Vaagen, Phys. Rep. 231 (1993) 151, http://dx.doi.org/10.1016/0370-1573(93)90141-Y.
[12] R.J. Ascuitto, N.K. Glendenning, Phys. Rev. 181 (1969) 1396, http://dx.doi.org/10.1103/PhysRev.181.1396.
[13] R.J. Ascuitto, C.H. King, L.J. McVay, B. Sørensen, Nucl. Phys. A 226 (1974) 454, http://dx.doi.org/10.1016/0375-9474(74)90495-3.
[14] A.M. Moro, Phys. Rev. C 92 (2015) 044605, http://dx.doi.org/10.1103/PhysRevC.92.044605.
[15] J. Casal, M. Gómez-Ramos, A. Moro, Phys. Lett. B 767 (2017) 307, http://dx.doi.org/10.1016/j.physletb.2017.02.017.
[16] N.K. Glendenning, Distorted-wave born approximation, in: Direct Nuclear Reactions, Academic Press, 1983, pp. 45–60, Chapter 5.
[17] N. Austern, Y. Iseri, M. Kamimura, M. Kawai, G. Rawitscher, M. Yahiro, Phys. Rep. 154 (1987) 125, http://dx.doi.org/10.1016/0370-1573(87)90094-9.
[18] G. Satchler, Direct Nuclear Reactions, Int. Ser. Monogr. Phys., Clarendon Press, 1983.
[19] P. Descouvemont, C. Daniel, D. Baye, Phys. Rev. C 67 (2003) 044309, http://dx.doi.org/10.1103/PhysRevC.67.044309.
[20] I. Thompson, F. Nunes, B. Danilin, Comput. Phys. Commun. 161 (2004) 87, http://dx.doi.org/10.1016/j.cpc.2004.03.007.
[21] M. Rodríguez-Gallardo, J.M. Arias, J. Gómez-Camacho, A.M. Moro, I.J. Thompson, J.A. Tostevin, Phys. Rev. C 72 (2005) 024007, http://dx.doi.org/10.1103/PhysRevC.72.024007.
[22] J. Casal, M. Rodríguez-Gallardo, J.M. Arias, Phys. Rev. C 88 (2013) 014327, http://dx.doi.org/10.1103/PhysRevC.88.014327.
[23] J. Casal, M. Rodríguez-Gallardo, J.M. Arias, I.J. Thompson, Phys. Rev. C 90 (2014) 044304, http://dx.doi.org/10.1103/PhysRevC.90.044304.
[24] J. Casal, M. Rodríguez-Gallardo, J.M. Arias, Phys. Rev. C 92 (2015) 054611, http://dx.doi.org/10.1103/PhysRevC.92.054611.
[25] J. Casal, E. Garrido, R. de Diego, J.M. Arias, M. Rodríguez-Gallardo, Phys. Rev. C 94 (2016) 054622, http://dx.doi.org/10.1103/PhysRevC.94.054622.
[26] V.G.J. Stoks, R.A.M. Klomp, C.P.F. Terheggen, J.J. De Swart, Phys. Rev. C 49 (1994) 2950, http://dx.doi.org/10.1103/PhysRevC.49.2950.
[27] H.V. von Geramb, AIP Conf. Proc. 97 (1983) 44, http://dx.doi.org/10.1063/1.33973.
[28] L. Rikus, K. Nakano, H.V. Geramb, Nucl. Phys. A 414 (1984) 413, http://dx.doi.org/10.1016/0375-9474(84)90611-0.
[29] A. Etchegoyen, et al. OXBASH code, MSU-NSCL Report 524, 1985, unpublished.
[30] J.J. Kelly, L.E.A. code, University of Maryland, 1995, unpublished.
[31] M. Smith, et al., Phys. Rev. Lett. 101 (2008) 202501, http://dx.doi.org/10.1103/PhysRevLett.101.202501.
[32] I.J. Thompson, M.V. Zhukov, Phys. Rev. C 49 (1994) 1904, http://dx.doi.org/10.1103/PhysRevC.49.1904.
[33] J.P. Fernández-García, et al., Phys. Rev. Lett. 110 (2013) 142701, http://dx.doi.org/10.1103/PhysRevLett.110.142701.
[34] G. Blanchon, A. Bonaccorso, D. Brink, N.V. Mau, Nucl. Phys. A 791 (2007) 303, http://dx.doi.org/10.1016/j.nuclphysa.2007.04.014.
[35] M. Cavallaro, et al., Phys. Rev. Lett. 118 (2017) 012701, http://dx.doi.org/10.1103/PhysRevLett.118.012701.
[36] F.M. Nunes, I.J. Thompson, R.C. Johnson, Nucl. Phys. A 596 (1996) 171, http://dx.doi.org/10.1016/0375-9474(95)00398-3.
[37] E. Garrido, D.V. Fedorov, A. Jensen, Nucl. Phys. A 700 (2002) 117, http://dx.doi.org/10.1016/S0375-9474(01)01310-0.
[38] Y. Kikuchi, T. Myo, K. Katō, K. Ikeda, Phys. Rev. C 87 (2013) 034606, http://dx.doi.org/10.1103/PhysRevC.87.034606.
[39] H. Fortune, Phys. Lett. B 760 (2016) 577, http://dx.doi.org/10.1016/j.physletb.2016.07.033.
[40] A. Sanetullaev, et al., Phys. Lett. B 755 (2016) 481, http://dx.doi.org/10.1016/j.physletb.2016.02.060.
[41] S. Orrigo, H. Lenske, Phys. Lett. B 677 (2009) 214, http://dx.doi.org/10.1016/j.physletb.2009.05.024.

No nonminimally coupled massless scalar hair for spherically symmetric neutral black holes

Shahar Hod [a,b,*]

[a] The Ruppin Academic Center, Emeq Hefer 40250, Israel
[b] The Hadassah Academic College, Jerusalem 91010, Israel

ARTICLE INFO

Editor: M. Cvetič

ABSTRACT

We provide a remarkably compact proof that spherically symmetric neutral black holes cannot support static nonminimally coupled massless scalar fields. The theorem is based on causality restrictions imposed on the energy-momentum tensor of the fields near the regular black-hole horizon.

1. Introduction

The non-linearly coupled Einstein-scalar field equations have attracted the attention of physicists and mathematicians for more than five decades. Interestingly, the composed Einstein-scalar system is characterized by very powerful and elegant no-hair theorems [1–3] which rule out the existence of asymptotically flat black-hole solutions with regular event horizons that support various types of scalar (spin-0) matter configurations.

The early no-hair theorems of Chase [4] and Bekenstein [5] have ruled out the existence of regular black holes supporting static minimally coupled massless scalar field configurations. The no-hair theorems of Bekenstein [5] and Teitelboim [6] have excluded the existence of black-hole hair made of minimally coupled massive scalar fields [7–9]. Later no-hair theorems of Heusler [10] and Bekenstein [11] have ruled out the existence of neutral black-hole spacetimes supporting static matter configurations made of minimally coupled scalar fields with positive semidefinite self-interaction potentials.

The physically interesting regime of scalar fields nonminimally coupled to gravity has been investigated by several authors. In a very interesting paper, Mayo and Bekenstein [12] have proved that spherically symmetric charged black holes cannot support matter configurations made of charged scalar fields nonminimally coupled to gravity with generic values of the dimensionless coupling parameter ξ [the physical parameter ξ quantifies the nonminimal coupling of the field to gravity, see Eq. (4) below]. Intriguingly, the rigorous derivation of a no-hair theorem for *neutral* scalar fields nonminimally coupled to gravity seems to be a mathematically more challenging task. In particular, the important no-hair theorems of [12,13] can be used to rule out the existence of spherically symmetric scalar hairy configurations in the restricted physical regimes $\xi < 0$ and $\xi \geq 1/2$ [14].

The main goal of the present paper is to present a (remarkably compact) unified no-hair theorem for neutral massless scalar fields nonminimally coupled to gravity which is valid for *generic* values of the dimensionless coupling parameter ξ (in particular, below we shall extend the interesting no-scalar hair theorems of [12] and [13] to the physical regime of nonminimally coupled neutral scalar fields with $0 < \xi < 1/2$). Our theorem, to be proved below, is based on simple physical restrictions imposed by causality on the energy-momentum tensor of the fields near the regular horizon of the black-hole spacetime.

2. Description of the system

We consider a non-linear physical system composed of a neutral black hole of horizon radius r_H and a massless scalar field ψ with nonminimal coupling to gravity. The composed black-hole-scalar-field system is assumed to be static and spherically symmetric, in which case the spacetime outside the black-hole horizon is characterized by the curved line element [12] (we shall use natural units in which $G = c = 1$)

$$ds^2 = -e^{\nu}dt^2 + e^{\lambda}dr^2 + r^2(d\theta^2 + \sin^2\theta d\phi^2) , \qquad (1)$$

where $\nu = \nu(r)$ and $\lambda = \lambda(r)$ [here (t, r, θ, ϕ) are the Schwarzschild coordinates]. As explicitly proved in [12], regardless of the matter content of the curved spacetime, a non-extremal regular black hole is characterized by the near-horizon relations [15]

* Correspondence to: The Ruppin Academic Center, Emeq Hefer 40250, Israel.
E-mail address: shaharhod@gmail.com.

$$e^{-\lambda} = L \cdot x + O(x^2) \quad \text{where} \quad x \equiv \frac{r - r_H}{r_H} \quad ; \quad L > 0 \tag{2}$$

and

$$\lambda' r_H = -\frac{1}{x} + O(1) \quad ; \quad \nu' r_H = \frac{1}{x} + O(1) . \tag{3}$$

The curved black-hole spacetime is non-linearly and non-minimally coupled to a real massless scalar field ψ whose action is given by [12]

$$S = S_{EH} - \frac{1}{2} \int \left(\partial_\alpha \psi \partial^\alpha \psi + \xi R \psi^2\right) \sqrt{-g} d^4x , \tag{4}$$

where the dimensionless physical parameter ξ quantifies the strength of the nonminimal coupling of the field to gravity, $R(r)$ is the scalar curvature of the spacetime, and S_{EH} is the Einstein–Hilbert action. As explicitly shown in [12], in the near-horizon $x \ll 1$ region, R is given by the simple expression

$$R = \frac{4L - 2}{r_H^2} \cdot [1 + O(x)] . \tag{5}$$

From the action (4) one finds the characteristic differential equation [12]

$$\partial_\alpha \partial^\alpha \psi - \xi R \psi = 0 \tag{6}$$

for the nonminimally coupled scalar field. Using the metric components (1) of the curved black-hole spacetime, one can express the scalar radial equation in the form

$$\psi'' + \frac{1}{2}\left(\frac{4}{r} + \nu' - \lambda'\right)\psi' - \xi R e^{\lambda} \psi = 0 . \tag{7}$$

(Here a prime $'$ denotes a spatial derivative with respect to the radial coordinate r.)

The action (4) also yields the compact expressions [12]

$$T^t_t = e^{-\lambda} \frac{\xi(4/r - \lambda')\psi\psi' + (2\xi - 1/2)(\psi')^2 + 2\xi\psi\psi''}{1 - 8\pi\xi\psi^2} \tag{8}$$

and

$$T^t_t - T^\phi_\phi = e^{-\lambda} \frac{\xi(2/r - \nu')\psi\psi'}{1 - 8\pi\xi\psi^2} \tag{9}$$

for the components of the energy-momentum tensor. As explicitly proved in [12], regardless of the matter content of the theory, a regular hairy black-hole spacetime must be characterized by finite mixed components of the energy-momentum tensor:

$$\{|T^t_t|, |T^r_r|, |T^\theta_\theta|, |T^\phi_\phi|\} < \infty . \tag{10}$$

In addition, it was proved in [12] that causality requirements enforce the characteristic inequalities[1]

$$|T^\theta_\theta| = |T^\phi_\phi| \le |T^t_t| \ge |T^r_r| \tag{11}$$

[1] As explicitly shown by Bekenstein and Mayo [12], for spherically symmetric spacetimes one can write $\epsilon = -T^t_t - \sum_{i=1}^3 c_i^2 (T^t_t - T^i_i)$ and $j^\mu j_\mu = -(T^t_t)^2 - \sum_{i=1}^3 c_i^2 [(T^t_t)^2 - (T^i_i)^2]$, where $\epsilon \equiv T_{\mu\nu} u^\mu u^\nu$ and $j^\mu \equiv -T^\mu_\nu u^\nu$ are respectively the energy density and the Poynting vector according to a physical observer with a 4-velocity u^ν, and the coefficients $\{c_i\}_{i=0}^3$ are characterized by the normalization condition $-c_0^2 + \sum_{i=1}^3 c_i^2 = -1$ (this relation guarantees that $u^\mu u_\mu = -1$ [12]). For physically acceptable systems in which the transfer of energy is not superluminal, the energy density should be of the same sign as $-T^t_t$ and the Poynting vector should be non-spacelike ($j^\mu j_\mu \le 0$) for all observers [12] (that is, for all choices of the coefficients $\{c_i\}_{i=0}^3$). These physical requirements yield the characteristic energy conditions (11) [12].

on the components of the energy-momentum tensors of physically acceptable systems. Note that the relations [12]

$$\text{sgn}(T^t_t) = \text{sgn}(T^t_t - T^r_r) = \text{sgn}(T^t_t - T^\phi_\phi) \tag{12}$$

provide necessary conditions for the validity of the characteristic energy conditions (11).

3. The no-hair theorem for static nonminimally coupled massless scalar fields

In the present section we shall explicitly prove that a spherically symmetric non-extremal neutral black hole *cannot* support non-linear hair made of static nonminimally coupled massless scalar fields.

We start our proof with the scalar radial equation (7) which, in the near-horizon $x \ll 1$ region, can be written in the form [see Eqs. (2), (3), and (5)]

$$\frac{d^2\psi}{dx^2} + \frac{1}{x}\frac{d\psi}{dx} + \frac{\beta}{x}\psi = 0 \quad ; \quad \beta \equiv \xi(2 - 4L)/L . \tag{13}$$

The general mathematical solution of the ordinary differential equation (13) can be expressed in terms of the familiar Bessel functions (see Eq. 9.1.53 of [16])

$$\psi(x) = A \cdot J_0(2\beta^{1/2}x^{1/2}) + B \cdot Y_0(2\beta^{1/2}x^{1/2}) \quad \text{for} \quad x \ll 1 , \tag{14}$$

where $\{A, B\}$ are constants. Using Eqs. 9.1.8 and 9.1.12 of [16], one finds the asymptotic near-horizon behavior

$$\psi(x \to 0) = A \cdot [1 - \beta x + O(x^2)] + B \cdot [\pi^{-1} \ln(\beta x) + O(1)] \tag{15}$$

of the radial scalar function. Substituting Eqs. (2), (3), and (15) into Eq. (8) and taking cognizance of the energy condition (10) [12], one immediately realizes that the coefficient of the singular term in the asymptotic near-horizon solution (15) should vanish [17]:

$$B = 0 . \tag{16}$$

We therefore find that the nonminimally coupled scalar field is characterized by the near-horizon behavior

$$\psi(x \ll 1) = A \cdot J_0(2\beta^{1/2}x^{1/2}) . \tag{17}$$

Substituting (17) into (8) and (9) and using the near-horizon relations (2) and (3), one obtains the simple expressions

$$T^t_t = \xi \cdot \frac{L\psi\psi'}{r_H(1 - 8\pi\xi\psi^2)} \cdot [1 + O(x)] \tag{18}$$

and

$$T^t_t - T^\phi_\phi = -\xi \cdot \frac{L\psi\psi'}{r_H(1 - 8\pi\xi\psi^2)} \cdot [1 + O(x)] \tag{19}$$

for the components of the energy-momentum tensor in the near-horizon $x \ll 1$ region. We immediately find from (18) and (19) the near-horizon relation

$$T^t_t = -(T^t_t - T^\phi_\phi) , \tag{20}$$

in *contradiction* with the characteristic relation (12) imposed by causality on the energy-momentum components of physically acceptable systems.

4. Summary

In this compact analysis, we have proved that *if* a spherically symmetric neutral black hole can support non-linear configurations made of nonminimally coupled massless scalar fields, then in the near-horizon $(r - r_H)/r_H \ll 1$ region the energy momentum components of the fields must be characterized by the relation $T_t^t = -(T_t^t - T_\phi^\phi)$ [see Eq. (20)]. However, one realizes that this near-horizon behavior is in *contradiction* with the characteristic relation $\text{sgn}(T_t^t) = \text{sgn}(T_t^t - T_\phi^\phi)$ [see Eq. (12)] which, as explicitly proved in [12], is imposed by causality on the energy-momentum components of generic physically acceptable systems. Thus, there are no physically acceptable solutions for the eigenfunction of the external nonminimally coupled massless scalar fields except the trivial one, $\psi \equiv 0$.

We therefore conclude that spherically symmetric neutral black holes cannot support static configurations made of nonminimally coupled massless scalar fields with *generic* values of the dimensionless physical parameter ξ.

Acknowledgements

This research is supported by the Carmel Science Foundation. I would like to thank Yael Oren, Arbel M. Ongo, Ayelet B. Lata, and Alona B. Tea for helpful discussions.

References

[1] R. Ruffini, J.A. Wheeler, Phys. Today 24 (1971) 30.
[2] B. Carter, Black Holes, in: C. De Witt, B.S. De Witt (Eds.), Proceedings of 1972 Session of Ecole d'ete de Physique Theorique, Gordon and Breach, New York, 1973.
[3] J.D. Bekenstein, Phys. Today 33 (1980) 24.
[4] J.E. Chase, Commun. Math. Phys. 19 (1970) 276.
[5] J.D. Bekenstein, Phys. Rev. Lett. 28 (1972) 452.
[6] C. Teitelboim, Lett. Nuovo Cimento 3 (1972) 326.
[7] It is worth mentioning that recent studies [8,9] of the composed Einstein-scalar system have explicitly proved that spinning black holes can support stationary (rather than static) matter configurations made of massive scalar (bosonic) fields.
[8] S. Hod, Phys. Rev. D 86 (2012) 104026, arXiv:1211.3202;
S. Hod, Eur. Phys. J. C 73 (2013) 2378, arXiv:1311.5298;
S. Hod, Phys. Rev. D 90 (2014) 024051, arXiv:1406.1179;
S. Hod, Phys. Lett. B 739 (2014) 196, arXiv:1411.2609;
S. Hod, Class. Quantum Gravity 32 (2015) 134002, arXiv:1607.00003;
S. Hod, Phys. Lett. B 751 (2015) 177;
S. Hod, Class. Quantum Gravity 33 (2016) 114001;
S. Hod, Phys. Lett. B 758 (2016) 181, arXiv:1606.02306;
S. Hod, O. Hod, Phys. Rev. D 81 (2010) 061502, Rapid communication, arXiv:0910.0734;
S. Hod, Phys. Lett. B 708 (2012) 320, arXiv:1205.1872;
S. Hod, J. High Energy Phys. 01 (2017) 030, arXiv:1612.00014.
[9] C.A.R. Herdeiro, E. Radu, Phys. Rev. Lett. 112 (2014) 221101;
C.L. Benone, L.C.B. Crispino, C. Herdeiro, E. Radu, Phys. Rev. D 90 (2014) 104024;
C.A.R. Herdeiro, E. Radu, Phys. Rev. D 89 (2014) 124018;
C.A.R. Herdeiro, E. Radu, Int. J. Mod. Phys. D 23 (2014) 1442014;
Y. Brihaye, C. Herdeiro, E. Radu, Phys. Lett. B 739 (2014) 1;
J.C. Degollado, C.A.R. Herdeiro, Phys. Rev. D 90 (2014) 065019;
C. Herdeiro, E. Radu, H. Rúnarsson, Phys. Lett. B 739 (2014) 302;
C. Herdeiro, E. Radu, Class. Quantum Gravity 32 (2015) 144001;
C.A.R. Herdeiro, E. Radu, Int. J. Mod. Phys. D 24 (2015) 1542014;
C.A.R. Herdeiro, E. Radu, Int. J. Mod. Phys. D 24 (2015) 1544022;
P.V.P. Cunha, C.A.R. Herdeiro, E. Radu, H.F. Rúnarsson, Phys. Rev. Lett. 115 (2015) 211102;
B. Kleihaus, J. Kunz, S. Yazadjiev, Phys. Lett. B 744 (2015) 406;
C.A.R. Herdeiro, E. Radu, H.F. Rúnarsson, Phys. Rev. D 92 (2015) 084059;
C. Herdeiro, J. Kunz, E. Radu, B. Subagyo, Phys. Lett. B 748 (2015) 30;
C.A.R. Herdeiro, E. Radu, H.F. Rúnarsson, Class. Quantum Gravity 33 (2016) 154001;
C.A.R. Herdeiro, E. Radu, H.F. Rúnarsson, Int. J. Mod. Phys. D 25 (2016) 1641014;
Y. Brihaye, C. Herdeiro, E. Radu, Phys. Lett. B 760 (2016) 279;
Y. Ni, M. Zhou, A.C. Avendano, C. Bambi, C.A.R. Herdeiro, E. Radu, J. Cosmol. Astropart. Phys. 1607 (2016) 049;
M. Wang, arXiv:1606.00811.
[10] M. Heusler, J. Math. Phys. 33 (1992) 3497;
M. Heusler, Class. Quantum Gravity 12 (1995) 779.
[11] J.D. Bekenstein, Phys. Rev. D 51 (1995) R6608.
[12] A.E. Mayo, J.D. Bekenstein, Phys. Rev. D 54 (1996) 5059.
[13] A. Saa, Phys. Rev. D 53 (1996) 7377.
[14] It is worth mentioning that, for spherically symmetric (3 + 1)-dimensional black holes, Saa [13] has also demonstrated the absence of scalar hair $\psi(r)$ in the nonminimal coupling regime $0 < \xi < 1/6$ with $8\pi\psi^2(r) < 1/\xi$ or with $1/\xi < 8\pi\psi^2(r) < [6\xi(1/6 - \xi)]^{-1}$ and in the nonminimal coupling regime $\xi > 1/6$ with $8\pi\psi^2(r) \neq 1/\xi$. In the present paper we shall provide a unified no-hair theorem for nonminimally coupled scalar fields which is valid for generic values of the dimensionless coupling parameter ξ and for generic values of the scalar field eigenfunction $\psi(r)$.
[15] As shown in [12], the expansion coefficient is given by $L \equiv 1 + 8\pi T_t^t(r_H) r_H^2 > 0$, where $-T_t^t(r_H)$ is the energy density of the matter fields at the black-hole horizon.
[16] M. Abramowitz, I.A. Stegun, Handbook of Mathematical Functions, Dover Publications, New York, 1970.
[17] It is worth noting that, for $\beta = 0$, one finds from (13) the asymptotic near-horizon functional behavior $\psi(x \to 0) = A \cdot \ln(x) + B$, where $\{A, B\}$ are constants. Substituting this expression into (8) and taking cognizance of the energy condition (10) [12], one immediately realizes that, for $A \neq 0$, this near-horizon scalar function is physically unacceptable. One therefore concludes that, for physically acceptable spacetimes, the scalar eigenfunction is strictly constant. Substituting $\psi = B$ into Eqs. (8) and (9), one finds that the components of the energy momentum tensor are identically zero. Thus, the constant B has no influence on physical quantities and one may therefore take $B = 0$ without loss of generality.

Onset of η-nuclear binding in a pionless EFT approach

N. Barnea [a], B. Bazak [b], E. Friedman [a], A. Gal [a,*]

[a] Racah Institute of Physics, The Hebrew University, 91904 Jerusalem, Israel
[b] IPNO, CNRS/IN2P3, Univ. Paris-Sud, Université Paris-Saclay, F-91406 Orsay, France

ARTICLE INFO

Editor: W. Haxton

Keywords:
Few-body systems
Mesic nuclei
$\not\pi$EFT calculations

ABSTRACT

ηNNN and $\eta NNNN$ bound states are explored in stochastic variational method (SVM) calculations within a pionless effective field theory (EFT) approach at leading order. The theoretical input consists of regulated NN and NNN contact terms, and a regulated *energy dependent* ηN contact term derived from coupled-channel models of the $N^*(1535)$ nucleon resonance. A self consistency procedure is applied to deal with the energy dependence of the ηN subthreshold input, resulting in a weak dependence of the calculated η-nuclear binding energies on the EFT regulator. It is found, in terms of the ηN scattering length $a_{\eta N}$, that the onset of binding $\eta\,^3$He requires a minimal value of Re $a_{\eta N}$ close to 1 fm, yielding then a few MeV η binding in $\eta\,^4$He. The onset of binding $\eta\,^4$He requires a lower value of Re $a_{\eta N}$, but exceeding 0.7 fm.

1. Introduction

The ηN s-wave interaction near threshold, $E_{\rm th}(\eta N) = 1487$ MeV, is attractive as realized first by coupling to the πN channel [1] and subsequently confirmed, e.g. [2], by coupling the ηN channel to the entire set of meson–baryon channels with 0^- octet mesons and $\frac{1}{2}^+$ octet baryons, thereby generating dynamically the $N^*(1535)$ S_{11} resonance. The size of the resulting ηN energy dependent s-wave attraction, however, is strongly model dependent with values of the real part of the ηN scattering length as low as 0.2 fm [2] and up to nearly 1 fm in the K-matrix model of Green and Wycech (GW) [3] and in the recent Giessen coupled channels study [4]. Following the work of Ref. [1] it was soon realized that η-nuclear quasibound states might exist [5,6] with widths determined by the scale of the imaginary part of the ηN scattering length. This imaginary part, due mostly to $\eta N \to \pi N$, is small in all models, between 0.2 to 0.3 fm. Nevertheless, no η-nuclear quasibound state has ever been established beyond doubt [7].

Recent optical-model calculations of such bound states [8,9] using several energy-dependent ηN model amplitudes are summarized in Refs. [10,11]. Whereas the appearance of η-nuclear bound states is robust in any of these ηN interaction models, the value of mass number A at which binding begins is model dependent. Thus, the relatively strong ηN attraction in model GW even admits in such calculations a $1s_\eta$ bound state in ^{4}He, with as low binding energy as 1.2 MeV and width of 2.3 MeV [11] calculated using a static ^{4}He density. Unfortunately, the η-nucleus optical model approach is not trustable for as light nuclei as ^{4}He, and genuine few-body calculations are required.

Photon- and hadron-induced reactions on nuclear targets provide useful constraints on possible η bound states in very light nuclei, where according to a recent review [12] the most straightforward interpretation of the data is that ηd is unbound, $\eta\,^3$He is nearly or just bound, and $\eta\,^4$He is bound. Our previous few-body ηNN and ηNNN calculations [13], using the Minnesota [14] and Argonne AV4' [15] NN potentials, agree with this conjecture as far as the ηd and $\eta\,^3$He systems are concerned. A similar conclusion for $\eta\,^3$He has been reached recently by evaluating the $pd \to \eta\,^3$He near-threshold reaction [16]. And a recent WASA-at-COSY search for a possible $\eta\,^4$He bound state in the $dd \to\,^3$He$N\pi$ reaction placed upper limits of a few nb on the production of a near-threshold bound state [17]. On the theoretical side, no precise few-body calculation of $\eta NNNN$ bound-state has ever been reported for $\eta\,^4$He.[1]

The present work reports for the first time on precise few-body $\eta NNNN$ calculations in which the Stochastic Variational Method (SVM) is applied to η plus few-nucleon Hamiltonians constructed in Leading Order (LO) within a Pionless Effective Field Theory. While this $\not\pi$EFT approach has been applied before to few-nucleon

* Corresponding author.
E-mail address: avragal@savion.huji.ac.il (A. Gal).

[1] A very recent preprint by Fix and Kolesnikov [18] reports on few-body calculations of the $\eta\,^3$He and $\eta\,^4$He scattering lengths, concluding that these systems are unbound.

systems, e.g. [19,20], and more recently in lattice-nuclei calculations [21,22], it is extended here to include constituent pseudoscalar mesons for which pion exchange with nucleons is forbidden by parity conservation of the strong interactions. In particular, the single ηN contact term required in LO is provided by the ηN s-wave scattering amplitude $F_{\eta N}(E_{sc})$ at a subthreshold energy E_{sc} derived self consistently within the few-body calculation, as practised in our previous work [13]. This is demonstrated in two ηN interaction models, GW [3] and CS (Cieplý–Smejkal [23]), which exhibit strong energy dependence of $F_{\eta N}(E)$ arising from the proximity of the $N^*(1535)$ resonance. The results of the few-body calculations reported in the present work suggest that the onset of binding $\eta\,^3$He requires a value of Re$\,a_{\eta N}$ close to 1 fm, whereas the onset of binding $\eta\,^4$He requires a somewhat weaker ηN interaction with Re$\,a_{\eta N}$ exceeding 0.7 fm.

2. Methodology

Here we outline the methodology of the present work, including the few-body SVM used, the choice of NN, NNN and ηN $\not\pi$EFT regulated contact terms, and the self consistent treatment of the energy-dependent ηN term.

2.1. SVM calculations

SVM calculations were introduced in the mid seventies to few-body nuclear problems [24], and used extensively with correlated Gaussian bases since the mid 1990s [25]. The SVM was benchmarked together with six other few-body methods in calculating the ^{4}He binding energy [26]. Correlated Gaussian trial wavefunctions in this method are written as

$$\Psi = \sum_k c_k \mathcal{A}\left(\left[\mathcal{Y}_L^k(\hat{\mathbf{x}})\chi_S^k\right]_{JM} \xi_{TT_z}^k \exp(-\frac{1}{2}\mathbf{x}^T A_k \mathbf{x})\right) \quad (1)$$

where the summation index k runs with linear variational parameters c_k on all possible values of the total spin S and the total orbital angular momentum L, as well as on all possible intermediate coupling schemes, χ_S and ξ_T stand for spin and isospin functions of the N-particle system, respectively, $\mathcal{Y}_L$ is the orbital part of Ψ formed by coupling successively spherical harmonics in the $(N-1)$ relative coordinates of which the vector $\mathbf{x}$ is made, and $\mathcal{A}$ antisymmetrizes over nucleons. The matrix A_k introduces $N(N-1)/2$ nonlinear variational parameters which are chosen stochastically. For a comprehensive review see Ref. [27].

2.2. Pionless EFT nuclear interactions

Here we follow a $\not\pi$EFT approach at LO. To this order the nuclear interaction consists of two-body and three-body contact (zero-range) terms,

$$V_2(ij) = \left[c_S^\Lambda \frac{1}{4}(1-\boldsymbol{\sigma}_i\cdot\boldsymbol{\sigma}_j) + c_T^\Lambda \frac{1}{4}(3+\boldsymbol{\sigma}_i\cdot\boldsymbol{\sigma}_j)\right]\delta_\Lambda(r_{ij}), \quad (2)$$

$$V_3(ijk) = d_3^\Lambda\,\delta_\Lambda(r_{ij}, r_{jk}), \quad (3)$$

where these contact terms are smeared by using normalized-to-one Gaussians with a regulating momentum-space scale (cut-off) parameter Λ:

$$\delta_\Lambda(r_{ij}) = \left(\frac{\Lambda}{2\sqrt{\pi}}\right)^3 \exp\left(-\frac{\Lambda^2}{4}r_{ij}^2\right),$$
$$\delta_\Lambda(r_{ij}, r_{jk}) = \delta_\Lambda(r_{ij})\delta_\Lambda(r_{jk}), \quad (4)$$

with $\delta_\Lambda(r_{ij})$ in the zero-range limit $\Lambda\to\infty$ becoming a Dirac $\delta(\mathbf{r}_{ij})$ function. For a given value of the scale parameter Λ, the two-body low-energy constants (LEC) c_S^Λ and c_T^Λ are fitted to the $S=0$ pn scattering length and to the $S=1$ deuteron binding energy, respectively, and the three-body LEC d_3^Λ is fitted to the ^{3}H binding energy. Following Ref. [28] a small corrective proton–proton contact term, with LEC c_{pp}^Λ, is introduced together with the Coulomb interaction between protons to reproduce the ^{3}He binding energy. These two-body and three-body LECs are listed in Ref. [28] where c_{pp}^Λ was found to effectively adjust c_S^Λ by less than 0.1% over the full Λ-range tested in the present work. The ^{4}He calculated binding energy $B(^4\text{He})$ provides then a check on how reasonable this LO $\not\pi$EFT version is. This is demonstrated in Table 1 for four representative values of Λ. The calculated values of $B(^4\text{He})$ depend only moderately on the scale parameter Λ, exhibiting renormalization scale invariance by approaching in the limit $\Lambda\to\infty$ a finite value 27.8±0.2 MeV which compares well with $B_{\text{exp}}(^4\text{He}) = 28.3$ MeV, despite the fact that only LO contributions are accounted for in this $\not\pi$EFT version.

Table 1
^{4}He binding energies $B(^4\text{He})$ (in MeV) in LO $\not\pi$EFT SVM calculations, with LECs fitted to NN and NNN low-energy data [28]. The $\Lambda\to\infty$ limit of $B(^4\text{He})$ was evaluated by using higher values of the scale Λ than listed here.

Λ (fm^{-1})	2	4	6	8	$\to\infty$	exp.
$B(^4\text{He})$	22.4	22.9	24.2	25.1	27.8±0.2	28.3

2.3. Pionless EFT ηN interactions

Parity conservation forbids pion exchange in the ηN interaction, suggesting thereby that a $\not\pi$EFT approach may be justified. With spin and isospin zero for the η meson, a single ηN contact term is all one needs at LO. Below we derive the corresponding LEC from the ηN s-wave scattering amplitude $F_{\eta N}(E)$ calculated in two meson–baryon coupled-channel interaction models, GW [3] and CS [23], and shown in Fig. 1. Whereas the GW model used in our previous work [13] is an on-shell K-matrix model that considers $\eta N \leftrightarrow \pi N$ coupling, the CS model is a meson–baryon multi-channel chirally motivated model in which the ηN interaction is extremely short ranged and practically momentum independent below the momentum breakdown scale specified in the next paragraph. Both models capture the main features of the underlying $N^*(1535)$ resonance which peaks about 50 MeV above the ηN threshold energy $E_{\text{th}} = 1487$ MeV and generates considerable energy dependence of $F_{\eta N}$ near threshold. In particular, both real and imaginary parts of $F_{\eta N}(E)$, which at threshold are given by the scattering lengths (in fm)

$$a_{\eta N}^{\text{GW}} = 0.96 + i0.26 \qquad a_{\eta N}^{\text{CS}} = 0.67 + i0.20, \quad (5)$$

decrease monotonically in these models upon going into the subthreshold region while displaying considerable model dependence.

The ηN scattering lengths listed above are of order 1 fm or less, much smaller than the NN scattering lengths whose large size justifies the use of $\not\pi$EFT in light nuclei. For ηN interactions as weak as implied by this size of $a_{\eta N}$, and with no ηN bound or virtual state expected, the ηN scattering length alone does not provide a meaningful criterion of fitting into an EFT approach. Alternatively, we estimate $p_\eta R \lesssim \frac{\pi}{2}$ for the momentum p_η of a weakly bound η-nuclear state in a square well of radius R. With $R = 2$ fm or a bit larger for the He isotopes, we get $p_\eta \lesssim 150$ MeV/c (≈0.76 fm^{-1}). Since the lowest-mass allowed meson exchange in pseudoscalar meson interaction with octet baryons is owing to vector mesons, with a typical mass of $m_\rho = 770$ MeV, this range of η-nuclear momenta can be accommodated comfortably within the nuclear LO $\not\pi$EFT approach of the preceding subsection. The $\not\pi$EFT small parameter associated with

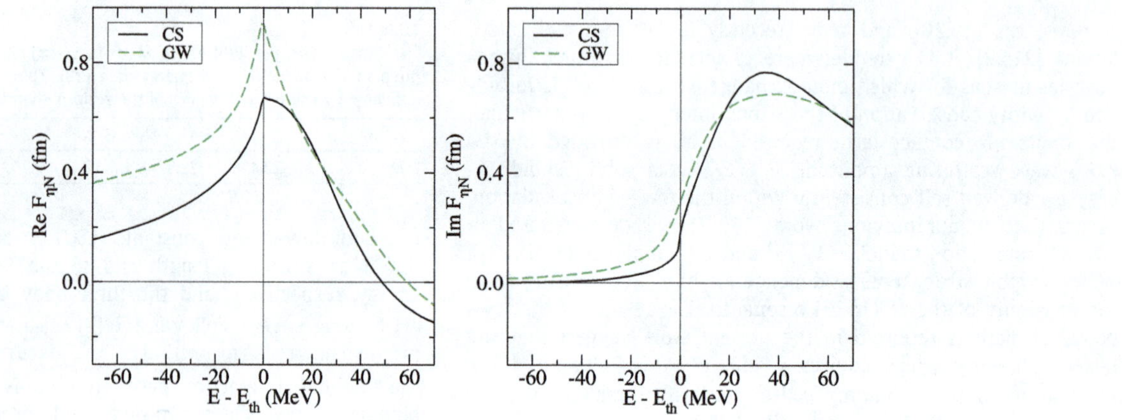

Fig. 1. Real (left panel) and imaginary (right panel) parts of the ηN cm s-wave scattering amplitude $F_{\eta N}(E)$ as a function of $E - E_{\rm th}$, with $E = \sqrt{s}$ the total ηN cm energy, in the GW [3] and CS [23] meson–baryon coupled-channel interaction models. The vertical line marks the ηN threshold energy $E_{\rm th}$. Figure adapted from Ref. [11].

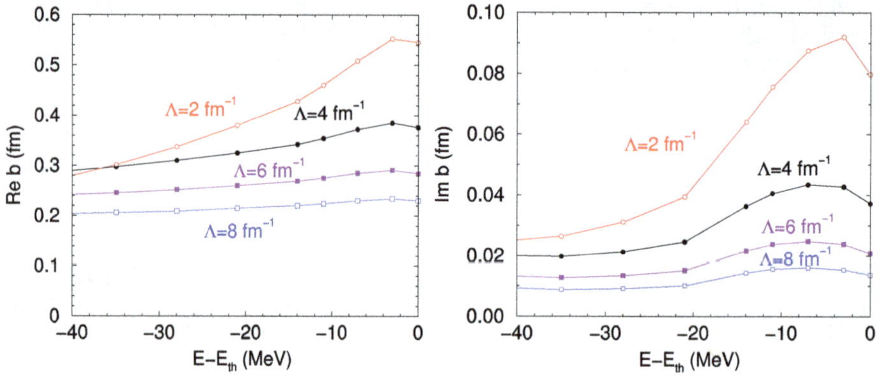

Fig. 2. Real (left panel) and imaginary (right panel) parts of the strength parameter $b^{\Lambda}_{\eta N}(E)$ of the ηN effective potential (6) at subthreshold energies $E < E_{\rm th}$ for four values of the scale (cut-off) parameter Λ, all of which result in the same scattering amplitude $F^{\rm GW}_{\eta N}$ [3] shown in Fig. 1.

this momentum breakdown scale of $Q^{\rho}_{\rm high} \approx m_{\rho} = 3.9$ fm^{-1} is given by $(p_{\eta}/Q^{\rho}_{\rm high})^2 \approx 0.04$.

As in previous work [13], and in order to account for the energy dependence inherent in the meson–baryon coupled channel dynamical generation of the $N^*(1535)$ resonance, we construct energy-dependent local potentials $v_{\eta N}(E)$ that produce the ηN energy dependent scattering amplitude $F_{\eta N}(E)$ below threshold in models GW and CS. For a given ηN interaction model, the on-shell scattering amplitude $F_{\eta N}(E)$ serves as a single datum to which LO $\not\pi$EFT regulated contact terms of the form

$$v_{\eta N}(E;r) = c^{\Lambda}_{\eta N}(E)\,\delta_{\Lambda}(r), \quad c^{\Lambda}_{\eta N}(E) = -\frac{4\pi}{2\mu_{\eta N}}\,b^{\Lambda}_{\eta N}(E), \tag{6}$$

are fitted. Here δ_{Λ} is a regulating normalized-to-one Gaussian with scale parameter Λ, as per Eq. (4), and $c^{\Lambda}_{\eta N}(E)$ is an energy dependent LEC conveniently related through the ηN reduced mass $\mu_{\eta N}$ to a strength function $b^{\Lambda}_{\eta N}(E)$ of length dimension. The specific value $c^{\Lambda}_{\eta N}(E_{\rm sc})$ of this LEC for a given cut-off Λ is determined self consistently in the η-nuclear SVM calculation as detailed in the next subsection. By using the *same* value of Λ in all NN, NNN and ηN regulating Gaussians we reach a consistent extension of the nuclear $\not\pi$EFT to a combined η-nuclear $\not\pi$EFT approach. In order to study the renormalization scale invariance of our few-body η-nuclear results, as shown for the purely nuclear case of $B(^4{\rm He})$ in Table 1, we discuss below $\not\pi$EFT calculations done for several representative values of the scale parameter, $\Lambda = 2, 4, 6, 8$ fm^{-1}. The last two values, clearly, exceed the momentum breakdown scale $Q^{\rho}_{\rm high} \approx 3.9$ fm^{-1} of the underlying $N^*(1535)$ resonance model for the ηN interaction, or even more so the lower momentum breakdown scale $q^{\rho}_{\rm high} \approx 3.0$ fm^{-1} set by excitation of vector meson degrees of freedom absent in the underlying $N^*(1535)$ dynamical models considered here, such as the ρ meson produced at threshold in the strong pion exchange reaction $\pi N \to \rho N$ with $p^{\rho}_{\rm th} = 586$ MeV/c. Finally, the model dependence of the LO ηN contact term introduced by studying two quite different ηN interaction models, GW and CS, leaves little motivation to go at present beyond LO. Hence, discussion of higher orders in $\not\pi$EFT is left to future work.

For a given value of Λ, the subthreshold values of the complex strength parameter $b^{\Lambda}_{\eta N}(E)$ in Eq. (6) were fitted to the complex phase shifts derived from the subthreshold scattering amplitudes $F_{\eta N}(E)$ in models GW and CS. The resulting values of the strength parameter $b^{\Lambda}_{\eta N}(E)$ for ηN subthreshold energies in model GW, shown in Fig. 2, fall off monotonically for both real and imaginary parts in going deeper below threshold, except for small kinks near threshold that reflect the threshold cusp of Re $F_{\eta N}(E)$ at $E_{\rm th}$ in Fig. 1. Similar curves for $b^{\Lambda}_{\eta N}(E)$ are obtained in model CS, with values smaller uniformly for both real and imaginary parts than model GW yields, in accordance with the relative strength of the generating scattering amplitudes $F_{\eta N}(E)$ shown in Fig. 1. We note that increasing Λ leads to weaker strengths $b^{\Lambda}_{\eta N}(E)$ and also to a weaker energy dependence. Inspecting Fig. 2, one also notes that Im $b^{\Lambda}_{\eta N}(E) \ll$ Re $b^{\Lambda}_{\eta N}(E)$, which justifies treating Im $v_{\eta N}$ perturbatively in the present calculations.

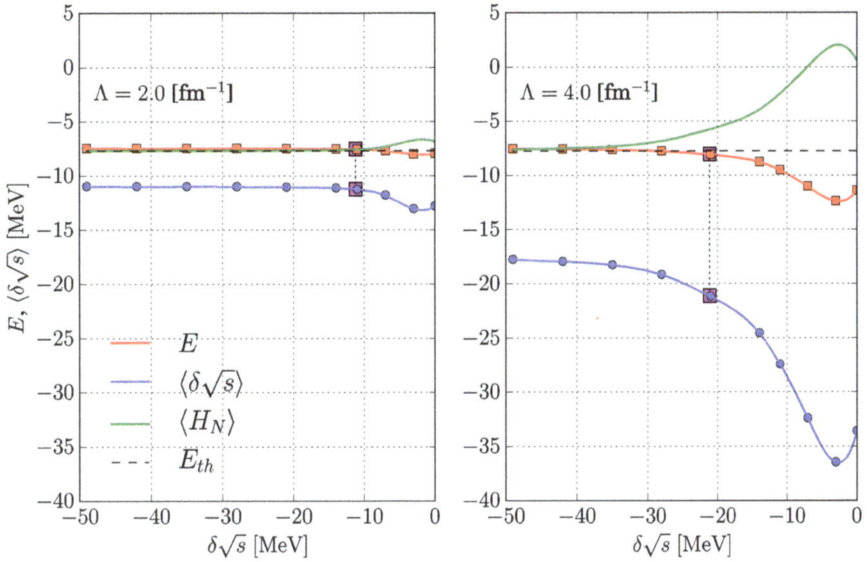

Fig. 3. Calculated $\eta\,^3$He bound-state energies E (squares) and expectation values $\langle\delta\sqrt{s}\rangle$ (circles) from Eq. (7), using LO $\not\pi$EFT NN and NNN potentials, and the ηN potentials $v^{\rm GW}_{\eta N}(E)$ for two values of the scale parameter Λ, as a function of the input energy shift $\delta\sqrt{s}$ used for the energy argument of $v^{\rm GW}_{\eta N}(E)$. The dashed vertical line marks the self consistent values of E and $\langle\delta\sqrt{s}\rangle$. The dashed horizontal line marks the ^{3}He core g.s. energy serving as threshold for a bound η, and the curve above it shows the squeezed core energy $\langle H_N\rangle$.

2.4. Energy dependence

To determine the ηN subthreshold energy at which $v_{\eta N}(E)$ is evaluated as input to the η-nuclear few-body calculations reported below, we denote the shift away from threshold by $\delta\sqrt{s}\equiv\sqrt{s}-\sqrt{s_{\rm th}}$, expressing it in terms of output expectation values [13]:

$$\langle\delta\sqrt{s}\rangle = -\frac{B}{A} - \beta_N\frac{1}{A}\langle T_N\rangle + \frac{A-1}{A}E_\eta - \xi_A\beta_\eta\left(\frac{A-1}{A}\right)^2\langle T_\eta\rangle\,, \tag{7}$$

where $\beta_{N(\eta)}\equiv m_{N(\eta)}/(m_N+m_\eta)$, $\xi_A\equiv Am_N/(Am_N+m_\eta)$, T_N and T_η are the nuclear and η kinetic energy operators evaluated in terms of internal Jacobi coordinates, with $T=T_N+T_\eta$ the total intrinsic kinetic energy of the system, B is the total binding energy of the η-nuclear few-body system and $E_\eta=\langle\Psi|(H-H_N)|\Psi\rangle$, where H_N is the Hamiltonian of the purely nuclear part in its own cm frame, and the total Hamiltonian H is evaluated in the overall cm frame. The imaginary, absorptive part of the ηN interaction is suppressed in this discussion. Noting that $(A-1)\langle T_{N:N}\rangle$ in Eq. (7) of Ref. [13] equals $\langle T_N\rangle$ here, Eq. (7) coincides with the former equation apart from a kinematical factor ξ_A introduced here to make correspondence with the η-nuclear, last Jacobi coordinate with which T_η is associated. Requiring that the expectation value $\langle\delta\sqrt{s}\rangle$ on the l.h.s. of Eq. (7), as derived from the solution of the Schroedinger equation, agrees with the input value $\delta\sqrt{s}$ for $v_{\eta N}(E)$, this equation defines a self-consistency cycle in our few-body η-nuclear calculations. Since each one of the four terms on the r.h.s. of Eq. (7) is negative, the self consistent energy shift $\delta\sqrt{s_{\rm sc}}$ is also negative, with size exceeding a minimum nonzero value obtained from the first two terms in the limit of vanishing η binding. Eq. (7) in the limit $A\gg 1$ coincides with the nuclear-matter expression used in Refs. [8,9] for calculating η-nuclear quasibound states.

Fig. 3 demonstrates how the self consistency requirement works in actual calculations. The three curves plotted in each panel are obtained by interpolating a sequence of calculated $\eta\,^3$He bound-state energies (squares) and the corresponding expectation values $\langle\delta\sqrt{s}\rangle$ (circles) from Eq. (7) for $A=3$, as a function of the input $\delta\sqrt{s}$ to the energy argument $E_{\rm th}+\delta\sqrt{s}$ of $v^{\rm GW}_{\eta N}$, for two choices of the momentum scale parameter $\Lambda=2,4$ fm^{-1}. The dashed vertical line marks the self consistent value of $\delta\sqrt{s}$ at which the outcome bound-state energy $E(\eta\,^3{\rm He})$ is evaluated, and the dashed horizontal line marks the ^{3}He core energy $E(^3{\rm He})$. Note that the self consistent value $E_{\rm sc}(\eta\,^3{\rm He})$ is *higher* than $E(^3{\rm He})$ in the left panel for $\Lambda=2$ fm^{-1}, while it is *lower* than $E(^3{\rm He})$ in the right panel for $\Lambda=4$ fm^{-1}. This means that, correspondingly, $\eta\,^3$He is slightly unbound for $\Lambda=2$ fm^{-1} while slightly bound for $\Lambda=4$ fm^{-1}. We note furthermore that for threshold values $v^{\rm GW}_{\eta N}(E_{\rm th})$, i.e. $\delta\sqrt{s}=0$, $\eta\,^3$He is bound in both cases (and also if the often used but unfortunately unfounded self consistency requirement [29] $\delta\sqrt{s}=E_\eta$ is imposed). Finally, the upper curve in Fig. 3 shows the expectation value $\langle H_N\rangle$ of the nuclear core energy, which is bounded from below by the ^{3}He core energy $E(^3{\rm He})$ marked by the dashed horizontal line.

3. Results and discussion

Separation energies $B_\eta\equiv B(\eta\,^A{\rm He})-B(^A{\rm He})$ (often called η binding energies) of the $\eta\,^A$He isotopes with $A=3,4$ were calculated self consistently in the SVM using LO $\not\pi$EFT NN and NNN regulated contact terms introduced in Eqs. (2)–(4), and a regulated ηN energy dependent contact term specified by Eq. (6), with scale parameters $\Lambda=2,4,6,8$ fm^{-1}. Two ηN coupled-channels models were used, GW [3] and CS [23]. The CS ηN interaction was found by far too weak to bind $\eta\,^3$He, and only by a fraction of MeV short of binding $\eta\,^4$He. The binding energies B_η were evaluated using real Hamiltonians in which Im $v_{\eta N}$ was disregarded. Restoring Im $v_{\eta N}$ in optical model calculations was found particularly important for near-threshold bound states, lowering their calculated B_η by 0.2±0.1 MeV. The η-nuclear widths Γ_η were calculated with wavefunctions $\Psi_{\rm g.s.}$ generated by these real Hamiltonians:

$$\Gamma_\eta = -2\,\langle\,\Psi_{\rm g.s.}\,|\,{\rm Im}\,V_\eta\,|\,\Psi_{\rm g.s.}\,\rangle\,. \tag{8}$$

Here, V_η sums over all pairwise ηN interactions. Since $|{\rm Im}\,V_\eta|\ll|{\rm Re}\,V_\eta|$, this is a reasonable approximation.

Binding energies B_η and widths Γ_η resulting from these self consistent calculations are listed in Table 2 for the ηN poten-

Table 2
η binding energies and widths (MeV) in the He isotopes from SVM calculations using ηN potentials $v^{GW}_{\eta N}(E)$ with scale parameters $\Lambda = 2, 4$ fm^{-1}, together with the corresponding self consistent values of the downward energy shift (in MeV) $\delta\sqrt{s_{sc}}$. The values of $\Gamma(\eta\,^3$He) shown here outdate the erroneous, too large widths listed in Ref. [13].

Λ (fm^{-1})	$\eta\,^3$He			$\eta\,^4$He		
	$\delta\sqrt{s_{sc}}$	B_η	Γ_η	$\delta\sqrt{s_{sc}}$	B_η	Γ_η
2	−11.2	−0.16	0.24	−16.5	−0.15	0.34
4	−21.1	+0.30	1.46	−32.2	+1.54	2.82

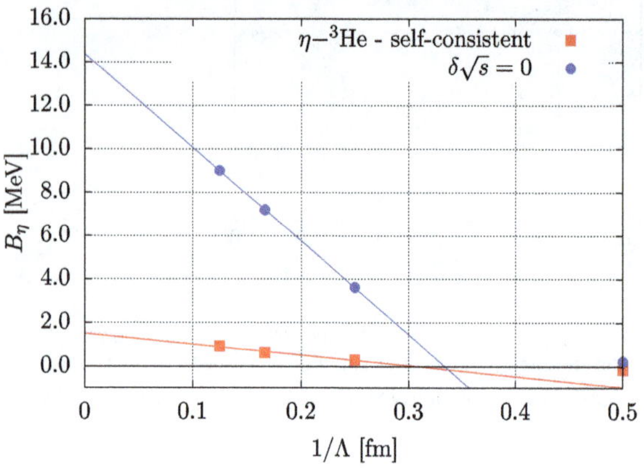

Fig. 4. $B_\eta(\eta\,^3$He) as a function of Λ^{-1}, calculated using ηN potentials $v^{GW}_{\eta N}(E)$ with scale parameters (from right to left) $\Lambda = 2, 4, 6, 8$ fm^{-1}. Squares (red) denote self consistent calculations, circles (blue) denote calculations with threshold values of the ηN interaction. Linear extrapolation to a point-like interaction, $\Lambda \to \infty$, is marked by straight lines. (For interpretation of the references to color in this figure legend, the reader is referred to the web version of this article.)

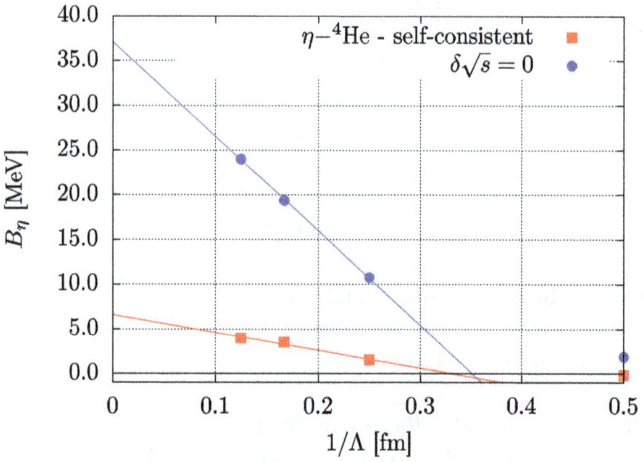

Fig. 5. Same as in Fig. 4, but for $\eta\,^4$He instead of $\eta\,^3$He.

tials $v^{GW}_{\eta N}(E)$ with scale parameters $\Lambda = 2, 4$ fm^{-1}, values for which our self consistency procedure was demonstrated in Fig. 3. Higher values of Λ exceed by far the momentum breakdown scale q^{ρ}_{high} introduced in our previous work [13] for $N^*(1535)$ resonance meson–baryon models in which the ηN scattering amplitude $F_{\eta N}$ is determined. Taken literally, this would mean that the GW ηN interaction hardly binds $\eta\,^3$He, if at all, and is likely to bind slightly $\eta\,^4$He, with B_η of order 1 MeV.

Sequences of calculated values of η binding energy B_η using ηN potentials $v^{GW}_{\eta N}(E)$ are shown for $\eta\,^3$He and $\eta\,^4$He in Figs. 4 and 5, respectively, as a function of Λ^{-1}. The figures demonstrate that the larger Λ, the larger is the resulting η binding energy B_η in spite of a similar increase in the value of $-\delta\sqrt{s_{sc}}$ in self consistent calculations which implies a weaker ηN potential strength $b(E_{sc})$. The dependence of B_η on Λ is weak for $\eta\,^3$He and moderate for $\eta\,^4$He in these calculations. For $\Lambda \geq 4$ fm^{-1}, B_η varies linearly in Λ^{-1}, with an average error of 50 keV for $\eta\,^3$He and 300 keV for $\eta\,^4$He, and with twice these errors upon extrapolating $\Lambda \to \infty$. In contrast, for calculations done at the ηN threshold, i.e. $\delta\sqrt{s} = 0$, the resulting values of B_η shown in the figures depend strongly on Λ with almost perfect linear dependence on Λ^{-1} for $\Lambda \geq 4$ fm^{-1}.

Interestingly, Figs. 4 and 5 also suggest that B_η assumes a *finite* value $B^{\Lambda\to\infty}_\eta$ in the limit of point ηN interaction, with no need to introduce a stabilizing three-body ηNN repulsive contact term analogous to the NNN contact term d^λ_3 of Eq. (3). This might be related to the weakness of the ηN interaction, with no ηN and ηNN bound states, compared to the much stronger NN interaction which generates NN and NNN bound states. In the latter case, a Thomas collapse of the NNN bound state is averted in $\not\pi$EFT by promoting a non-derivative NNN contact term from N^{2}LO to LO.

4. Conclusion

To summarize, we have presented genuine few-body SVM calculations of ηNNN ($\eta\,^3$He) bound states and, for the first time, also $\eta NNNN$ ($\eta\,^4$He) bound states, using LO $\not\pi$EFT interactions where the ηN interaction contact term was derived in coupled channels studies of the $N^*(1535)$ nucleon resonance. Special care was taken of the energy dependence of the input ηN subthreshold scattering amplitude by using a self consistency procedure. The present results exhibit renormalization scale invariance of the calculated η binding energies, without having to introduce a repulsive ηNN contact interaction. For physically motivated values of Λ, the onset of $\eta\,^3$He binding occurs for Re$\,a_{\eta N}$ close to 1 fm, as in model GW [3], consistently with our previous hyperspherical-basis ηNNN calculations [13]. The onset of $\eta\,^4$He binding requires a lower value of Re $a_{\eta N}$, exceeding however 0.7 fm; it is comfortably satisfied in model GW but not in model CS [23]. Further dedicated experimental searches for $\eta\,^4$He bound states are desirable in order to confirm the recent negative report from WASA-at-COSY [17] which, taken at face value, implies that Re $a_{\eta N} \lesssim 0.7$ fm. Similar results and conclusions hold valid in SVM calculations using non-EFT realistic nuclear models [14,15] augmented by the same ηN interaction models used here, and will be reported elsewhere.

Acknowledgements

We thank Jiří Mareš and Martin Schaefer for useful discussions on related matters. This work was supported in part (NB) by the Israel Science Foundation grant 1308/16, in part (NB, BB) by Pazi Fund grants, and in part (EF, AG) by the EU initiative FP7, HadronPhysics3, under the SPHERE and LEANNIS cooperation programs.

References

[1] R.S. Bhalerao, L.C. Liu, Phys. Rev. Lett. 54 (1985) 865.
[2] T. Waas, N. Kaiser, W. Weise, Phys. Lett. B 379 (1996) 34.
[3] A.M. Green, S. Wycech, Phys. Rev. C 71 (2005) 014001.
[4] V. Shklyar, H. Lenske, U. Mosel, Phys. Rev. C 87 (2013) 015201.
[5] Q. Haider, L.C. Liu, Phys. Lett. B 172 (1986) 257.
[6] L.C. Liu, Q. Haider, Phys. Rev. C 34 (1986) 1845.
[7] C. Wilkin, EPJ Web Conf. 130 (2016) 01007.
[8] E. Friedman, A. Gal, J. Mareš, Phys. Lett. B 725 (2013) 334.
[9] A. Cieplý, E. Friedman, A. Gal, J. Mareš, Nucl. Phys. A 925 (2014) 126.
[10] A. Gal, E. Friedman, N. Barnea, A. Cieplý, J. Mareš, D. Gazda, Acta Phys. Pol. B 45 (2014) 673.
[11] J. Mareš, N. Barnea, A. Cieplý, E. Friedman, A. Gal, EPJ Web Conf. 130 (2016) 03006.
[12] B. Krusche, C. Wilkin, Prog. Part. Nucl. Phys. 80 (2015) 43.
[13] N. Barnea, E. Friedman, A. Gal, Phys. Lett. B 747 (2015) 345.
[14] D.R. Thompson, M. LeMere, Y.C. Tang, Nucl. Phys. A 286 (1977) 53.
[15] R.B. Wiringa, S.C. Pieper, Phys. Rev. Lett. 89 (2002) 182501.

[16] J.J. Xie, W.H. Liang, E. Oset, P. Moskal, M. Skurzok, C. Wilkin, Phys. Rev. C 95 (2017) 015202.
[17] P. Adlarson, et al., WASA-at-COSY Collaboration, Nucl. Phys. A 959 (2017) 102.
[18] A. Fix, O. Kolesnikov, arXiv:1703.06591.
[19] U. van Kolck, Nucl. Phys. A 645 (1999) 273.
[20] P.F. Bedaque, H.-W. Hammer, U. van Kolck, Nucl. Phys. A 676 (2000) 357.
[21] N. Barnea, L. Contessi, D. Gazit, F. Pederiva, U. van Kolck, Phys. Rev. Lett. 114 (2015) 052501.
[22] J. Kirscher, N. Barnea, D. Gazit, F. Pederiva, U. van Kolck, Phys. Rev. C 92 (2015) 054002.
[23] A. Cieplý, J. Smejkal, Nucl. Phys. A 919 (2013) 46.
[24] V.I. Kukulin, V.M. Krasnopol'sky, J. Phys. G 3 (1977) 795.
[25] K. Varga, Y. Suzuki, Phys. Rev. C 52 (1995) 2885.
[26] H. Kamada, et al., Phys. Rev. C 64 (2001) 044001.
[27] Y. Suzuki, K. Varga, Stochastic Variational Approach to Quantum Mechanical Few-Body Problems, Springer-Verlag, Berlin, 1998.
[28] J. Kirscher, E. Pazy, J. Drachman, N. Barnea, arXiv:1702.07268.
[29] C. García-Recio, T. Inoue, J. Nieves, E. Oset, Phys. Lett. B 550 (2002) 47.

27

On the mass of the world-sheet 'axion' in $SU(N)$ gauge theories in $3+1$ dimensions

Andreas Athenodorou [a,b], Michael Teper [c,*]

[a] Computation-based Science and Technology Research Center, The Cyprus Institute, 20 Kavafi Str., Nicosia 2121, Cyprus
[b] Department of Physics, University of Cyprus, POB 20537, 1678 Nicosia, Cyprus
[c] Rudolf Peierls Centre for Theoretical Physics, University of Oxford, 1 Keble Road, Oxford OX1 3NP, UK

ARTICLE INFO

Editor: A. Ringwald

ABSTRACT

There is numerical evidence that the world sheet action of the confining flux tube in $D=3+1$ $SU(N)$ gauge theories contains a massive excitation with 0^- quantum numbers whose mass shows some decrease as one goes from $SU(3)$ to $SU(5)$. Moreover it has been shown that the natural coupling of this pseudoscalar has a topological interpretation making it natural to call it the world-sheet 'axion'. Recently it has been pointed out that if the mass of this 'axion' vanishes as $N \to \infty$ then it becomes possible for the world sheet theory to be integrable in the planar limit. In this paper we perform lattice calculations of this 'axion' mass from $SU(2)$ to $SU(12)$, which allows us to make a controlled extrapolation to $N=\infty$ and so test this interesting possibility. We find that the 'axion' does not in fact become massless as $N \to \infty$. So if the theory is to possess planar integrability then it must be some other world sheet excitation that becomes massless in the planar limit.

1. Introduction

The spectrum and world sheet action of confining flux tubes in $D=3+1$ $SU(N)$ gauge theories is now known to be, for the most part, remarkably simple. That is to say, the spectrum is very close to the spectrum of the light cone quantisation of the bosonic string theory [1], which is only consistent in $D=26$ and $D=3$, and which we shall refer to as the Nambu–Goto spectrum. (We restrict ourselves here to flux tubes that wind around one of the spatial tori, so that there is no Coulomb interaction, and no extra boundary terms to the world sheet action.) The remarkable simplicity obtained in lattice calculations (see [2] and references therein) is now well understood theoretically. Long flux tubes can be very accurately described by the established series of universal terms in the world sheet action [3] (see also [4]) while shorter flux tubes can be well understood from the near-integrability of the world-sheet theory in this limit [5–7]. The latter framework provides a powerful way to translate the observed energies into world sheet S-matrix elements and, where appropriate, into extra fields and interactions in the world sheet action. In particular one ground state in [2] showed large deviations from the simple Nambu–Goto spectrum, in a way that suggested that it might consist of a massive pseudoscalar world-sheet particle on the background flux tube. The analysis in [6] using the formalism of [5] shows quite convincingly that this is indeed the case, and that the mass can be read off from the excitation energy above the absolute ground state to a good approximation. Moreover the natural coupling of this pseudoscalar has a topological interpretation [5,6], making it natural to call it the world-sheet 'axion'. Recently [7] these authors have pointed out that the $N=\infty$ world-sheet theory might be integrable, but only if it possesses at least one massless mode in addition to the usual massless 'phonons' associated with the string's spontaneous breaking of the bulk translation symmetry. They also noted that the observed decrease in the lattice estimates of the axion mass when one goes from $SU(3)$ to $SU(5)$ [2] raises the very interesting possibility that the axion mass might decrease to zero as $N \to \infty$ thus providing the extra massless mode needed for integrability. Locating a place for integrability in the planar limit of $SU(N)$ gauge theories has the potential to provide some analytic control over these theories and so is an exciting possibility. This motivates the present paper in which we perform lattice calculations of the axion mass for larger N, so as to see whether this mass vanishes or not in the planar limit.

* Corresponding author.
E-mail addresses: a.athenodorou@cyi.ac.cy (A. Athenodorou), mike.teper@physics.ox.ac.uk (M. Teper).

2. Lattice calculation

2.1. Lattice setup

The lattice calculations in this paper are essentially a direct continuation of the lattice calculations in [2]. Here we briefly recall that our lattice field variables are $SU(N)$ matrices, U_l, residing on the links l of a periodic $L_x \times L_y \times L_z \times L_t$ lattice, with lattice spacing a. The Euclidean path integral is

$$Z = \int \mathcal{D}U \exp\{-\beta S[U]\}, \tag{1}$$

where $\mathcal{D}U$ is the Haar measure and we use the standard plaquette action,

$$\beta S = \beta \sum_p \left\{1 - \frac{1}{N}\mathrm{ReTr}U_p\right\} \quad ; \quad \beta = \frac{2N}{g^2}. \tag{2}$$

Here U_p is the ordered product of link matrices around the plaquette p. We write $\beta = 2N/g^2$, where g^2 becomes the continuum coupling when $a \to 0$. Monte Carlo calculations are performed using a standard Cabibbo–Marinari heat bath [8] plus over-relaxation algorithm.

2.2. Calculating flux tube energies

We calculate the energy of a flux tube that winds around the periodic space–time in the x direction and which has length $l = aL_x$. The details of the calculation are exactly the same as described at some length in [2]. Here it is useful to recall that such a flux tube has the following relevant quantum numbers: spin, J, around the axis; a parity, $P_{\shortparallel}$, arising from $x \to -x$ reflections supplemented by charge conjugation; the $D = 2+1$ parity, $P_\perp$, arising from $(y,z) \to (-y,z)$. We set the momenta along and transverse to the flux tube to zero in this paper.

As described in detail in [2] our operators $\phi(t)$ are essentially Wilson lines that wind around the x-torus and are summed over y and z, as well as the starting point in x, so as to have zero momentum. These loops are decorated with various deviations from the direct path so as to allow us to produce operators $\phi_q(t)$ with various non-trivial quantum numbers q. We extract ground states from the asymptotic decay of the correlators of such operators, $\langle \phi_q^\dagger(t)\phi_q(0)\rangle \propto \exp\{-E_q(l)t\}$ as $t \to \infty$ where $E_q(l)$ is the ground state energy of the flux tube with the quantum numbers q. An important constraint is that the statistical error is roughly independent of t, so the exponentially decreasing 'signal' can rapidly disappear into the 'noise'. That is to say we need $aE_q(l)$ to be small for a reliable calculation.

In this paper we shall focus on the quantum numbers $J^{P_{\shortparallel}P_\perp} = 0^{++}$ and $J^{P_{\shortparallel}P_\perp} = 0^{--}$. From the 0^{++} ground state we extract the string tension, while from the difference between the two ground state energies we can extract the 'axion' mass using

$$M_A \simeq E_{0^{--}}(l) - E_{0^{++}}(l) \tag{3}$$

as long as l is not too large. (If l is large, then the lightest 0^{--} state will be the one with massless phonons rather than with a massive axion [2].) This is, of course, an approximation but in practice it is a rather good one: using eqn. (3) with the spectra in [2] would give $M_A/\sqrt{\sigma} \sim 1.9 - 2.0$, while the correct value [5,6] is $M_A/\sqrt{\sigma} \sim 1.85(3)$.

2.3. Calculational strategy

Ideally we would wish to repeat our calculation in [2] for a range of values of N that is large enough for us to have confidence in our $N \to \infty$ extrapolation. Since the cost of pure gauge calculations increases roughly as $\propto N^3$, this would require substantially larger computer resources than employed in [2] where our main calculations were for $SU(3)$, with only some for $SU(5)$. So we shall follow a more limited strategy here which entails the risk of extra systematic errors, which we will need to address in detail later on in this paper.

We will perform calculations for $N \le 12$ which should be an adequate range, given that the leading large-N correction should be $\propto 1/N^2$. However to reduce the computational cost we will work at a larger lattice spacing than in [2] since this means that we can use lattices that are smaller in lattice units. We will of course need to check (in Section 3) that the ensuing lattice spacing corrections are in fact negligible. Of course if a is larger, then so is the energy $aE(l)$ in lattice units, which means it will be harder to extract the energy reliably from the correlators, as discussed in Section 2.2. To compensate for this we perform the calculation for small values of l where the value of $aE(l)$ will be modest for those states where $E(l)$ decreases with decreasing l when l is small. This is the case for the absolute ground state but is in general not the case for excitations of this ground state because the 'phonon' momenta that usually provide the excitation of the flux tube increase roughly as $\propto 1/l$ as l decreases. Fortunately, the 0^{--} excited state we are interested in has no phonons and is a counterexample to this. So, as we shall see below, we are able to calculate the ground state energies of the 0^{++} and 0^{--} flux tubes with reasonable accuracy. (Although we will, unfortunately, not be able to calculate the energies of other excited states.) Of course we need to use a value of l where the extraction of the axion mass from the difference between the 0^{++} and 0^{--} flux tube energies is justified, and this we demonstrate in Section 3, where we shall also see that it is enough to perform calculations for a single value of l rather than a range of values. All these restrictions on our calculations will allow us to perform a reasonably accurate calculation even for this large range of N.

In addition to the above caveats there is a more general problem with lattice calculations at large N. This is the rapid loss of ergodicity as N increases in exploring fields with different topological charge Q. On a periodic lattice a change in Q requires a fluctuation that starts as a zero action gauge singularity around some hypercube, then under the action of the Monte Carlo update deforms into a dislocation, then grows into an instanton with a small core and then gradually grows into an instanton with a typically sized core. With such local Monte Carlo changes the instanton has to pass through a stage where it is small on physical length scales albeit not small on lattice scales. However at physically small scales one can do a semiclassical estimate and, as is well known [9], the probability of small instantons is exponentially suppressed in N. (Albeit with some caveats [10].) Thus the probability of changing Q in a Monte Carlo is also suppressed exponentially with N. This has long been known [11–13]. The question, then, is whether such a freezing of Q has damaging implications for the calculations in this paper. This is a particularly relevant question in our context because our 'axion' arises from a topological interaction in the world-sheet action which can be induced by a θ term in the bulk $SU(N)$ gauge theory [14]. This is the issue we address in detail in Section 4.

3. Results

We perform calculations in $SU(N)$ lattice gauge theories with $N = 2, 3, 4, 5, 6, 7, 8, 10, 12$. These calculations are performed at a lattice spacing that is (nearly) constant in physical units with $a\sqrt{\sigma} \simeq 0.300 \pm 0.001$, as we see from Table 1. We also keep the

Table 1
Ground state energies of flux tubes of length $l = aL_x$ with $J^{P_{\parallel}P_{\perp}} = 0^{++}$ and 0^{--} quantum numbers, and also the string tension. For lattices, groups and couplings shown.

Gauge Group	β	$L_x \times L_y \times L_z \times L_t$	$a\sqrt{\sigma}$	$aE_{0^{++}}$	$aE_{0^{--}}$
$SU(2)$	2.360	$8 \times 12 \times 12 \times 16$	0.29944(77)	0.5716(38)	1.290(14)
$SU(3)$	5.825	$8 \times 12 \times 12 \times 16$	0.29733(67)	0.5613(33)	1.157(25)
		$10 \times 12 \times 12 \times 16$	0.2987(13)	0.7806(81)	1.357(52)
$SU(4)$	10.70	$8 \times 12 \times 12 \times 16$	0.30007(75)	0.5747(37)	1.159(21)
$SU(5)$	17.00	$8 \times 12 \times 12 \times 16$	0.29951(57)	0.5720(28)	1.138(29)
$SU(6)$	24.71	$8 \times 12 \times 12 \times 16$	0.29809(72)	0.5650(35)	1.089(26)
$SU(7)$	33.825	$8 \times 12 \times 12 \times 16$	0.29935(63)	0.5712(31)	1.094(17)
$SU(8)$	44.355	$8 \times 12 \times 12 \times 16$	0.29915(98)	0.5702(48)	1.103(27)
		$8 \times 16 \times 16 \times 20$	0.30047(83)	0.5767(41)	1.080(23)
		$8 \times 18 \times 18 \times 24$	0.30007(89)	0.5747(44)	1.105(20)
		$8 \times 24 \times 24 \times 32$	0.30031(51)	0.5759(25)	1.090(21)
$SU(10)$	69.617	$8 \times 12 \times 12 \times 16$	0.30078(55)	0.5782(27)	1.121(27)
		$8 \times 16 \times 16 \times 20$	0.30037(77)	0.5762(38)	1.110(23)
		$8 \times 18 \times 18 \times 24$	0.30049(43)	0.5768(21)	1.097(11)
		$8 \times 24 \times 24 \times 32$	0.30086(77)	0.5786(38)	1.109(20)
$SU(12)$	100.50	$8 \times 12 \times 12 \times 16$	0.30090(67)	0.5788(33)	1.120(26)
		$8 \times 16 \times 16 \times 20$	0.30045(59)	0.5766(29)	1.088(23)
		$8 \times 18 \times 18 \times 24$	0.30082(53)	0.5784(26)	1.063(18)
		$8 \times 24 \times 24 \times 32$	0.30112(59)	0.5799(29)	1.109(27)

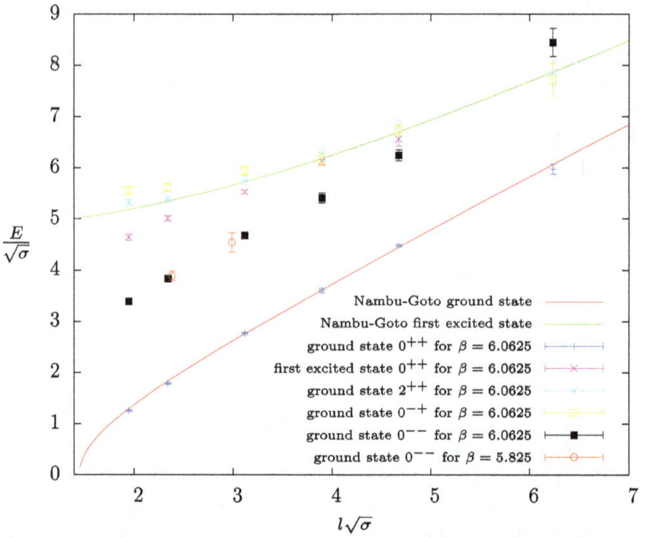

Fig. 1. Energies of the low-lying flux tube states with the quantum numbers shown, versus the length of the flux tube. Results for $SU(3)$ at $\beta = 6.0635$ taken from [2], compared to our new results for the $J^{P_{\parallel}P_{\perp}} = 0^{--}$ flux tube at $\beta = 5.825$ (○). (For interpretation of the references to colour in this figure, the reader is referred to the web version of this article.)

same the length $l = 8a$ of the winding flux tube, i.e. $l\sqrt{\sigma} \simeq 2.40$ in physical units.

We first need to check that using this relatively coarse a does not lead to significant lattice spacing corrections. To do that we focus on $SU(3)$, where we have an additional calculation using $l = 10a$, and we compare our results to those we obtained in [2] at a smaller value of a, corresponding to $a\sqrt{\sigma} \simeq 0.195$ (and which was itself checked in [2] against results obtained on an even finer lattice with $a\sqrt{\sigma} \simeq 0.129$). The comparison is displayed in Fig. 1 and we see that our values of the 0^{--} flux tube energy are perfectly consistent with those at the smaller lattice spacing. Given that the leading lattice spacing corrections should be $\propto a^2\sigma$, which decreases by a factor of ~ 2.4 between these two calculations, it appears that any such corrections must be negligible in our calculation.

The second thing we learn from Fig. 1 is that our choice of $l\sqrt{\sigma} \simeq 2.40$ for the calculation of the energy gap $\Delta E(l) = E_{0^{--}}(l) - E_{0^{++}}(l)$ is in the range where $\Delta E(l)$ is approximately independent of l, and so provides a good estimate of the 'axion' mass using $M_A \simeq \Delta E(l)$.

We will assume that our above checks for $SU(3)$ carry over to our calculations at other values of N, which is certainly reasonable for $N > 3$.

A further check one needs to perform is whether our transverse lattice size is large enough. To check this we show in Fig. 2 how the ground state energies of the 0^{++} and 0^{--} flux tubes vary with the volume, using the values for $N = 8, 10, 12$ in Table 1. We see that there is no visible volume dependence, demonstrating that our initial choice of a $8 \times 12 \times 12 \times 16$ lattice is entirely adequate, at least for these larger values of N.

An important final issue that still needs to be addressed has to do with the rapid loss of topological ergodicity as one increases N. This will be dealt with in detail in Section 4 where we will argue that this poses no obstacle to our calculations.

We can now turn to our estimates of the 'axion' mass, M_A, using eqn. (3). This is shown as a function of N in Fig. 3, where we have averaged the values obtained on the four different volumes for $N = 8, 10, 12$. We fit these values with a linear function of $1/N^2$, which is the expected leading large-N correction [15], giving

$$\frac{M_A}{\sqrt{\sigma}} = 1.713(14) + \frac{2.74(7)}{N^2} \quad ; \quad \chi^2/n_{dof} = 1.12 \quad N \in [2, 12]. \tag{4}$$

As shown, the fit has an entirely acceptable χ^2 per degree of freedom, so we do not need to include any higher order terms in $1/N^2$. (No doubt we would need to do so if our calculations had much smaller statistical errors.) This fit is shown in Fig. 3, and while one sees that $M_A/\sqrt{\sigma}$ does indeed decrease with increasing N, it is equally clear that the 'axion' mass does not vanish as $N \to \infty$.

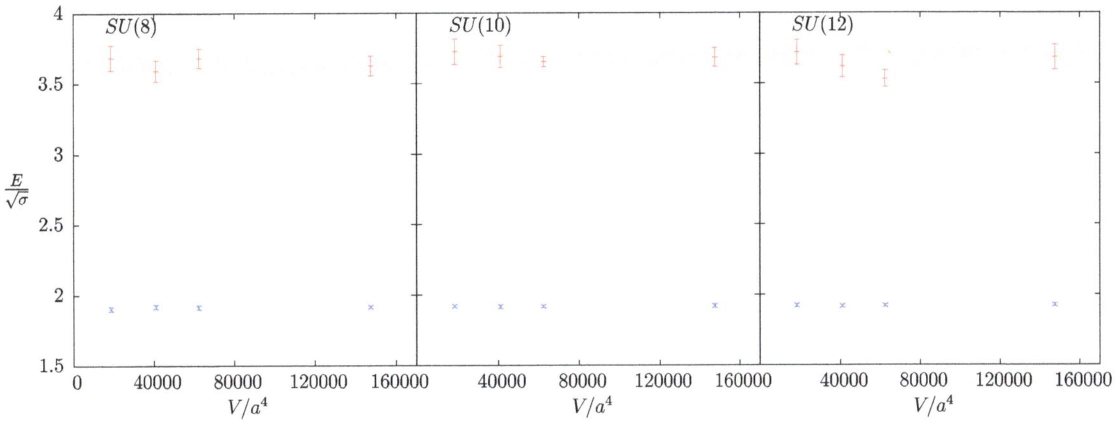

Fig. 2. The volume dependence of the energies of the $J^{P_{\parallel} P_{\perp}} = 0^{++}$ (×, blue) and 0^{--} (+, red) flux tube ground states. The bands correspond to the energy levels obtained with the largest volume i.e. 8 × 24 × 24 × 32. (For interpretation of the references to colour in this figure legend, the reader is referred to the web version of this article.)

Table 2
Various energy scales associated with the bulk gauge theory, together with their large-N extrapolation. Glueball quantum number labels are J^{PC}.

Group	$M_{0^{++}}/\sqrt{\sigma}\vert_{\beta_a}$	$M_{0^{++}}/\sqrt{\sigma}\vert_{cont}$	$(M_{0^{-+}} - M_{0^{++}})/\sqrt{\sigma}$	$(M_{0^{++*}} - M_{0^{++}})/\sqrt{\sigma}$
$SU(2)$	3.46(6)	3.78(7)		
$SU(3)$	2.97(6)	3.55(7)	2.40(10)	2.06(7)
$SU(4)$	2.82(6)	3.36(6)	2.22(9)	2.20(8)
$SU(5)$	2.81(7)	3.38(16)	2.15(10)	2.22(9)
$SU(6)$	2.67(7)	3.25(9)	2.26(10)	2.18(8)
$SU(8)$	2.68(8)	3.55(12)	2.14(10)	2.16(10)
$SU(\infty)$	2.62(5)	3.29(6)	2.11(9)	2.23(8)

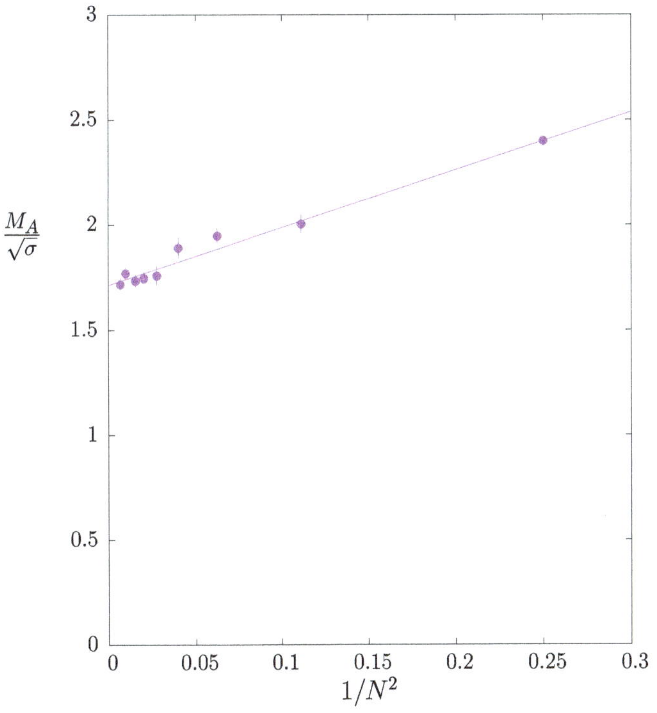

Fig. 3. The 'axion' mass, using eqn. (3), in units of the string tension, versus $1/N^2$ together with a linear extrapolation in $1/N^2$ to $N = \infty$.

As an aside, it is interesting to compare our above values for M_A to other characteristic energy scales in $SU(N)$ gauge theories, particularly as $N \to \infty$. We do this in Table 2. Note that our labelling of the quantum numbers of a glueball state is the conventional J^{PC}. The first column gives estimates of the mass of the lightest scalar glueball, $M_{0^{++}}$, in units of the string tension as calculated at the same values $\beta = \beta_a$ as our 'axion' mass. (The scalar mass is the mass gap and is from [16] for $N = 2, 3, 4, 6, 8$ and from [11] for $N = 5$. In some cases we need to interpolate from neighbouring β values.) The second column in Table 2 lists the continuum values of the scalar mass [16,11]. In a naive constituent model for the lightest scalar glueball, half its mass provides a rough estimate for the mass of a 'constituent' gluon (rather like half the ρ mass providing a rough estimate for the mass of a 'constituent' quark). The comparison with the 'axion' mass is complicated by the large lattice spacing corrections to the mass gap visible at $\beta = \beta_a$ in Table 2 while, as we have seen in Fig. 1, the 'axion' mass does not show significant lattice corrections. If we compare M_A to half the continuum mass gap we see that the values are quite similar at $N = \infty$, but much less so if we use the mass gap at $\beta = \beta_a$. In the third column of Table 2 we list the value of the gap between the lightest scalar and pseudoscalar glueballs.[1] (The masses are from [17], and have been obtained at values of β that are slightly larger than our β_a values. We expect that the gap would increase slightly if one were to extrapolate to β_a.) In a model for glueballs in which they are composed of closed loops of (fundamental) flux, the pseudoscalar could be due to the excitation of an 'axion' mode on the flux loop of the scalar glueball, so that $M_{0^{-+}} - M_{0^{++}} \simeq M_A$. Of course a 0^{-+} can also arise from the naive oscillations and rotations of the flux tube. However in a simple model [18] one finds in the case of $SU(2)$ (see Fig. 3b of [19]) that if one chooses the short distance cut-off parameter f of the model [19] at a natural value of $f \sim 1$ then while that reproduces the $M_{0^{-+}}$ and $M_{0^{++}}$ masses it is at the cost of losing the observed splitting between the 2^{++} and 2^{-+}. If we want to split the latter states then we require a larger value of f that leads to a mass

[1] We thank the referee for encouraging us to look at this quantity.

for the 0^{-+} that is too large. So in this latter case the idea that the relatively light observed 0^{-+} is in fact the lightest 0^{++} with an 'axion' excitation is not implausible in principle. Although the $N \to \infty$ value of the splitting listed in Table 2 is somewhat larger than the value of M_A, given the rough approximations involved, it is surely consistent with this general possibility. Finally in the fifth column we list the gap between the ground and first excited scalar glueballs. (Again from [17].) This is again a typical low-lying excitation energy of the bulk theory, but one which has no obvious link to the world-sheet 'axion'. The fact that it is very similar to the gap between the 0^{-+} and the 0^{++} suggests that the most that one can claim firmly from these comparisons is that the 'axion' mass is similar to typical excitation energies of the low-lying glueball spectrum.

4. Topology and non-ergodicity

As we explained earlier the usual local Monte Carlo algorithms rapidly lose the ability to change the topological charge Q of a lattice gauge field as N is increased. This suppression is related to the suppression of small instantons. If ρ is the size of the instanton and $\lambda(\rho) = g^2(\rho)N$ is the 't Hooft coupling on the scale of ρ, then the density of instantons is $D(\rho) \propto \exp\{-8\pi^2 N/\lambda(\rho)\}$ once ρ is small enough for this semi-classical calculation to be accurate [9]. (There are significant qualifications [10] that we do not enter into here.) Now, a change in Q is accompanied by an instanton growing from $\rho \sim a$ to $\rho \sim 1/\Lambda$ where $\Lambda \sim \sqrt{\sigma}$ is the physical length scale of the gauge theory. (Or the reverse process.) So this change will be suppressed by some factor $\propto \exp\{-8\pi^2 N/\lambda(\rho)\}$ with $\rho \sim O(a)$. Since $\lambda(\rho \sim a)$ decreases as a decreases, one way to delay (in N) this suppression is to work at a value of a that is not very small. In fact this is what we have done in this paper, in part for other reasons. However this only delays the onset of the problem, and eventually one needs to confront it.

An important point is that since our theory has a non-zero mass gap, the value of an observable will only depend on the topological fluctuations within a finite neighbourhood of that observable, i.e. within a distance $\sim 1/\Lambda$. So the effect on an observable of the total topological charge Q being frozen at some constant value can be made arbitrarily small by making the space–time volume sufficiently large. So a direct method to check whether an observable is affected by this freezing is simply to calculate it on a range of ever larger volumes to see if it changes. This is in fact the main reason that we calculated the flux tube energies on four different volumes for our largest three values of N in Table 1. The smallest volume of the four is the 'standard' volume we use for $2 \le N \le 7$. In physical units the space–time volume orthogonal to the flux tube is already a substantial $\sim \{3.6/\sqrt{\sigma}\}^2 \times 4.8/\sqrt{\sigma}$. The largest volume is a much larger $\sim \{7.2/\sqrt{\sigma}\}^2 \times 9.6/\sqrt{\sigma}$. Nonetheless as we see in Fig. 2 there is no sign of any change in the relevant flux tube energies as we increase the volume, strongly suggesting that the topological freezing is already unimportant on our 'standard' volume.

While such direct tests are the most convincing, it is interesting to see what is actually happening to the topological charge at these larger values of N. Since the method of calculation is fairly standard, we only briefly summarise it. We recall that the topological charge Q is the integral over Euclidean space–time of a topological charge density, $Q(x)$, which can be expressed in terms of the field strengths as $32\pi^2 Q(x) = \epsilon_{\mu\nu\rho\sigma}\mathrm{Tr}\{F_{\mu\nu}(x)F_{\rho\sigma}(x)\}$. If on the lattice we replace $F_{ij}(x)$ by the plaquette $U_{ij}(x)$ then we obtain a lattice topological charge density $a^4 Q_L(x)$ such that $Q_L(x) \to Q(x)$ for smooth fields. At finite β this lattice measure receives both additive and multiplicative renormalisation, which can be removed

Table 3
Fluctuations of the total topological charge, Q, and the integral of the absolute value of the topological charge density, $|Q(x)|$, on various lattice volumes for the $N = 8, 10, 12$.

Gauge Group	Lattice	$\langle Q_I^2\rangle$	$\langle \sum_x \lvert Q_I(x)\rvert \rangle$
$SU(8)$	$8 \times 12 \times 12 \times 16$	4.16(96)	2.55(13)
	$8 \times 16 \times 16 \times 20$	4.08(72)	5.23(15)
	$8 \times 18 \times 18 \times 24$	9.88(1.74)	8.48(17)
	$8 \times 24 \times 24 \times 32$	27.4(4.0)	20.08(19)
$SU(10)$	$8 \times 12 \times 12 \times 16$	1.09(21)	2.22(6)
	$8 \times 16 \times 16 \times 20$	2.61(73)	5.43(13)
	$8 \times 18 \times 18 \times 24$	7.86(1.96)	8.21(17)
	$8 \times 24 \times 24 \times 32$	7.66(1.10)	19.76(17)
$SU(12)$	$8 \times 12 \times 12 \times 16$	0.07(4)	2.09(6)
	$8 \times 16 \times 16 \times 20$	0.35(11)	5.22(12)
	$8 \times 18 \times 18 \times 24$	0.47(13)	8.23(10)
	$8 \times 24 \times 24 \times 32$	0.48(10)	19.47(15)

by smoothening the lattice fields in various ways. We shall employ 'cooling' [20] which consists of performing 'Monte Carlo'-type sweeps with the difference that one locally minimises the action. Under this process Q will be quasi-stable, since instantons are minima of the continuum action. $Q(x)$ on the other hand will be gradually deformed as one performs more cooling sweeps. (For example, neighbouring instantons and anti-instantons can reduce their action by gradually annihilating each other.) We refer to [11, 13,20] for much more detailed discussions.

In our calculation we perform 40 cooling sweeps on a sample of 40 or 80 lattice fields that should be mutually independent for standard physical observables. From experience we expect that this amount of cooling will leave the total charge Q unaffected, except for the possible disappearance of very small instantons. But at larger N there will be almost none of these. (That is, after all, why Q freezes in the Monte Carlo ensemble.) Each of these lattice fields was the last of a separate Monte Carlo sequence used for the calculation of the flux tube energies, and each started from a near-frozen starting lattice field with $Q = 0$. So if the ergodicity in Q were to be seriously suppressed, then we would expect to find the fluctuations of Q around $Q = 0$ to be suppressed. We list in Table 3 the values of $\langle Q^2\rangle$ that we obtain. We recall that Q is the difference between the number of instantons, n_+, and the number of anti-instantons, n_-, and since the correlation length is finite, and since $\langle n_+\rangle = \langle n_-\rangle \propto V$, we will have $\langle Q^2\rangle \propto V$ once the space–time volume V is large enough. Indeed in the dilute gas approximation one can readily show that $\langle Q^2\rangle = \langle n_+\rangle + \langle n_-\rangle = 2\langle n_+\rangle$. So it is usual to define the topological susceptibility $\chi_t = \langle Q^2\rangle/V$, and to express it in physical units, e.g. $\chi_t^{\frac{1}{4}}/\sqrt{\sigma}$. If we take our largest volume in Table 3 for $SU(8)$ we find $\chi_t^{\frac{1}{4}}/\sqrt{\sigma} \simeq 0.389(14)$ which is entirely consistent with the values at smaller N listed in Table 11 of [11] and Table 2 of [13]. That is to say, there is no visible suppression in the fluctuations of Q implying that we still have adequate ergodicity in Q in $SU(8)$ at $\beta = 44.355$. This is in fact no surprise since we chose this value of $a(\beta)$ for our calculations in the expectation, based on earlier calculations, that this would indeed be the case for $SU(8)$. More interesting is $SU(10)$ and $SU(12)$. Since $a\sqrt{\sigma}$ is the same as for $SU(8)$, we would expect that values of $\langle Q^2\rangle$ to be very similar for $N = 8, 10, 12$ if there continues to be adequate ergodicity in Q. However what we observe in Table 3 is a dramatic suppression of $\langle Q^2\rangle$ as we increase N to $N = 10$ and then $N = 12$. Clearly Q is indeed freezing for our highest values of N, despite our rather coarse value of $a(\beta)$.

As an important aside we remind the reader that while changing Q is necessarily associated with fields that contain small instantons (as described above) and so will be suppressed at large N, one can produce an instanton anti-instanton pair as a normal unsuppressed long-distance fluctuation. As one updates the field the separation between such pairs can grow so that even if $Q = 0$ for the total volume V, we expect that in any sufficiently large subvolume $\tilde{V}$ with $V \gg \tilde{V}$ the integrated topological charge $\tilde{Q} = \sum_{x \in \tilde{V}} Q(x)$ will have a restricted susceptibility $\langle \tilde{Q}^2 \rangle / \tilde{V}$ that equals the true susceptibility $\langle Q^2 \rangle / V$, when the latter is calculated with 'infinite' statistics so as to overcome any partial non-ergodicity in Q.

So, as remarked earlier, the relevant question here is whether the topological fluctuations in a subvolume that is large enough to contain the flux tube physics of interest, are significantly suppressed by the observed freezing of Q on the total volume for $N = 10, 12$. Answering this question directly is certainly possible but would require calculations that take us beyond the scope of this paper. However we do perform a step in that direction. This is provided by our calculation of $Q_{abs} \equiv \sum_x |Q(x)|$, whose average values are listed in Table 3. In the dilute gas approximation one can easily show that $\langle Q_{abs} \rangle = \langle (n_+ + n_-) \rangle = \langle Q^2 \rangle$. Of course the cooling will tend to decrease Q_{abs} because of the gradual annihilation of nearby (anti)instanton pairs, and in any case in reality the 'gas' is surely not dilute. Nonetheless the approximate equality in Table 3 between $\langle Q_{abs} \rangle$ and $\langle Q^2 \rangle$ for $SU(8)$, suggests that this argument has an approximate validity, and that $\langle Q_{abs} \rangle$ does give us a measure of the local density of topological fluctuations. If we now compare to the values of $\langle Q_{abs} \rangle$ listed in Table 3 for $SU(10)$ and $SU(12)$, we see that they are almost exactly the same. We take this as some evidence that the observed onset of a serious non-ergodicity in Q at our largest values of N will not have a significant impact on the topological fluctuations in the relevant subvolume $\tilde{V}$ as long as $\tilde{V} \ll V$. For our largest volume, $V \simeq \{2.4/\sqrt{\sigma}\} \times \{7.2/\sqrt{\sigma}\}^2 \times 9.6/\sqrt{\sigma}$, this inequality is presumably well satisfied.

5. Conclusions

In this paper we provided a calculation of the world-sheet 'axion' mass in $D = 3 + 1$ $SU(N)$ gauge theories, using the difference between the energies of the lightest $J^{P_{\parallel} P_{\perp}} = 0^{++}$ and 0^{--} flux tubes to estimate that mass. Although our calculations were at a fixed and rather coarse value of the lattice spacing, comparisons with earlier $SU(3)$ calculations reassured us that the estimates are reliable. Our calculations covered a much larger range of N than before, so as to allow a convincing large-N extrapolation. This required us to address the known rapid loss of lattice Monte Carlo ergodicity in the topological charge, when N becomes large. We addressed this directly by performing our calculations on a range of space–time volumes, and also by calculating the topological charge density, all of which strongly suggests that the clearly visible loss of this ergodicity does not have a significant impact on our results.

Our unambiguous conclusion is that the world sheet 'axion' has a finite non-zero mass at $N = \infty$, and that its mass is similar to typical excitation energies of the low-lying glueball spectrum. (Such as the gap between the lightest scalar and pseudoscalar, or the lightest two scalars, or indeed the 'constituent' gluon mass defined as half the mass gap.) That is to say, it cannot play the role of the extra massless world-sheet mode that is needed if the $N = \infty$ world sheet theory is to be integrable [7]. It needs to be stressed however that the currently available flux tube calculations [2] are incomplete and do not provide accurate calculations of all quantum numbers. Such calculations will require a basis of flux tube operators that is more extensive than that employed in [2] and until such a calculation is completed the possibility of some extra massless world-sheet modes at $N = \infty$ certainly cannot be excluded.

Acknowledgements

We would like to thank Sergei Dubovsky, Raphael Flauger and Victor Gorbenko for encouraging this investigation and for discussions on various aspects of this project. In addition AA acknowledges Krzysztof Cichy for discussions on topological aspects of this work. Furthermore we are indebted to participants at both the recent Flux Tubes conference held at the Perimeter Institute (in May 2015) and the earlier Confining Flux Tubes and Strings conference held at ECT, Trento (in July 2010). We are grateful to these institutions for hosting these very productive meetings. AA has been partially supported by an internal program of the University of Cyprus under the name of BARYONS. In addition, AA acknowledges the hospitality of the Cyprus Institute where part of this work was carried out. The numerical computations were carried out on the computing cluster in Oxford Theoretical Physics.

References

[1] P. Goddard, J. Goldstone, C. Rebbi, C. Thorn, Quantum dynamics of a massless relativistic string, Nucl. Phys. B 56 (1973) 109;
J. Arvis, The exact Q anti-Q potential in Nambu string theory, Phys. Lett. B 127 (1983) 106.
[2] A. Athenodorou, B. Bringoltz, M. Teper, Closed flux tubes and their string description in D=3+1 SU(N) gauge theories, JHEP 1102 (2011) 030, arXiv:1007.4720.
[3] O. Aharony, Z. Komargodski, The effective theory of long strings, JHEP 1305 (2013) 118, arXiv:1302.6257;
O. Aharony, E. Karzbrun, On the effective action of confining strings, JHEP 0906 (2009) 012, arXiv:0903.1927;
S. Dubovsky, R. Flauger, V. Gorbenko, Effective string theory revisited, JHEP 1209 (2012) 044, arXiv:1203.1054.
[4] M. Lüscher, Peter Weisz, String excitation energies in SU(N) gauge theories beyond the free-string approximation, JHEP 0407 (2004) 014, arXiv:hep-th/0406205;
J.M. Drummond, Universal subleading spectrum of effective string theory, arXiv:hep-th/0411017;
M. Lüscher, Symmetry breaking aspects of the roughening transition in gauge theories, Nucl. Phys. B 180 (1981) 317;
M. Lüscher, K. Symanzik, P. Weisz, Anomalies of the free loop wave equation in the WKB approximation, Nucl. Phys. B 173 (1980) 365.
[5] S. Dubovsky, R. Flauger, V. Gorbenko, Flux tube spectra from approximate integrability at low energies, J. Exp. Theor. Phys. 120 (3) (2015) 399, arXiv:1404.0037;
S. Dubovsky, R. Flauger, V. Gorbenko, Solving the simplest theory of quantum gravity, JHEP 1209 (2012) 133, arXiv:1205.6805.
[6] S. Dubovsky, R. Flauger, V. Gorbenko, Evidence for a new particle on the worldsheet of the QCD flux tube, Phys. Rev. Lett. 111 (6) (2013) 062006, arXiv:1301.2325.
[7] S. Dubovsky, V. Gorbenko, Towards a theory of the QCD string, JHEP 1602 (2016) 022, arXiv:1511.01908;
P. Cooper, S. Dubovsky, V. Gorbenko, A. Mohsen, S. Storace, Looking for integrability on the worldsheet of confining strings, JHEP 1504 (2015) 127, arXiv:1411.0703.
[8] N. Cabibbo, E. Marinari, A new method for updating SU(N) matrices in computer simulations of gauge theories, Phys. Lett. B 119 (1982) 387.
[9] E. Witten, Instantons, the quark model, and the 1/N expansion, Nucl. Phys. B 149 (1979) 285.
[10] M. Teper, Instantons and the 1/N expansion, Z. Phys. C 5 (1980) 233.
[11] B. Lucini, M. Teper, SU(N) gauge theories in four dimensions: exploring the approach to N = infinity, JHEP 0106 (2001) 050, arXiv:hep-lat/0103027.
[12] L. Del Debbio, H. Panagopoulos, P. Rossi, E. Vicari, Spectrum of confining strings in SU(N) gauge theories, JHEP 0201 (2002) 009, arXiv:hep-th/0111090.
[13] B. Lucini, M. Teper, U. Wenger, Topology of SU(N) gauge theories at T=0 and T=Tc, Nucl. Phys. B 715 (2005) 461, arXiv:hep-lat/0401028.
[14] P. Mazur, V. Nair, Strings in QCD and theta vacua, Nucl. Phys. B 284 (1987) 146.
[15] G. 't Hooft, A planar diagram theory for strong interactions, Nucl. Phys. B 72 (1974) 461.
[16] B. Lucini, M. Teper, U. Wenger, Glueballs and k-strings in SU(N) gauge theories: calculations with improved operators, JHEP 0406 (2004) 012, arXiv:hep-lat/0404008.
[17] B. Lucini, A. Rago, E. Rinaldi, Glueball masses in the large N limit, JHEP 1008 (2010) 119, arXiv:1007.3879.

[18] N. Isgur, J. Paton, A flux tube model for hadrons in QCD, Phys. Rev. D 31 (1985) 2910.
[19] T. Moretto, M. Teper, Glueball spectra of SU(2) gauge theories in three-dimensions and four-dimensions: a comparison with the Isgur–Paton flux tube model, arXiv:hep-lat/9312035.
[20] M. Teper, Instantons in the quantized SU(2) vacuum: a lattice Monte Carlo investigation, Phys. Lett. B 162 (1985) 357;
M. Teper, The topological susceptibility in SU(2) lattice gauge theory: an exploratory study, Phys. Lett. B 171 (1986) 86;
D. Smith, M. Teper, Topological structure of the SU(3) vacuum, Phys. Rev. D 58 (1998) 014505, arXiv:hep-lat/9801008.

On the stability of non-supersymmetric supergravity solutions

Ali Imaanpur *, Razieh Zameni

Department of Physics, School of sciences, Tarbiat Modares University, P.O. Box 14155-4838, Tehran, Iran

ARTICLE INFO

Editor: N. Lambert

ABSTRACT

We examine the stability of some non-supersymmetric supergravity solutions that have been found recently. The first solution is $AdS_5 \times M_6$, for M_6 an stretched CP^3. We consider breathing and squashing mode deformations of the metric, and find that the solution is stable against small fluctuations of this kind. Next we consider type IIB solution of $AdS_2 \times M_8$, where the compact space is a $U(1)$ bundle over $N(1,1)$. We study its stability under the deformation of M_8 and the 5-form flux. In this case we also find that the solution is stable under small fluctuation modes of the corresponding deformations.

1. Introduction

Selecting a stable solution among the many candidate supergravity solutions is a major problem in any Kaluza–Klein compactification. One way to guarantee the stability is to demand that the solution preserve a portion of supersymmetry [1–3]. In the absence of supersymmetry, on the other hand, it is difficult to conclude whether a particular solution is stable. In fact, one needs to examine the stability under small perturbations in all possible directions of the potential. Moreover, even if a solution is stable under such small perturbations, there still remains the question of stability under nonperturbative effects [4]. Finding non-supersymmetric stable solutions, however, becomes important if we are to construct realistic phenomenological models in which supersymmetry is spontaneously broken.

Freund–Rubin solutions can be divided into two main classes depending on whether or not the compact space encompasses (electric) fluxes [5,6]. When the flux has components only along the AdS direction, it has been observed that the majority of solutions either preserve supersymmetry (and hence stable), or at least are perturbatively stable. For solutions that support flux in the compact direction (Englert type), however, supersymmetry is often broken. They are in fact suspected to be unstable, though, the direct computation of mass spectrum and determination of stability is more involved. Englert type solution of $AdS_4 \times S^7$, for instance, was shown to be unstable [7], and this was further generalized to seven dimensional spaces which admit at least two Killing spinors [8]. Pope–Warner solution is another non-supersymmetric example which supports flux in the compact direction [9], and was proved to be unstable much later [10]. Englert type solutions, in spite of their possible instability, have played a key role in studying the holographic superconductors. By employing similar techniques that we use in this paper, domain wall solutions were found that interpolate between the Englert type and the skew-whiffed solutions. The domain wall solutions were then used to describe holographic superconductor phase diagrams [11].

The stability of Freund–Rubin type geometries of the form $AdS_p \times M_q$, where AdS_p is anti-de Sitter spacetime and M_q a compact manifold, has also increasingly been studied after the discovery of the AdS/CFT correspondence [12]. Stability is important for understanding a possible dual conformal field theory (CFT) description. For stable solutions, the spectrum of the masses directly yields the dimensions of certain operators in such a CFT. Unstable solutions can still have a dual CFT description but the physics is different [13]. Since the curvature of AdS is negative, not all the tachyonic modes lead to instability. In fact, scalars with $m^2 < 0$ may also appear if their masses are not below a bound set by the curvature scale of AdS [3].

Recently, some new non-supersymmetric compactifying solutions of eleven-dimensional supergravity and type IIB supergravity have been found [14,15]. Specifically, the eleven-dimensional supergravity solution consists of $AdS_5 \times M_6$, where for M_6 there are two possible choices. For the first solution M_6 is CP^3 with the standard Fubini-Study metric, which was derived and studied in [16], and it was further shown that is perturbatively stable [17]. For the second solution S^2 fibers of CP^3 are slightly stretched with respect to the base manifold. Type IIB solution, on the other hand, is $AdS_2 \times M_8$, where M_8 is a $U(1)$ bundle over $N(1,1)$. All these

* Corresponding author.
E-mail addresses: aimaanpu@modares.ac.ir (A. Imaanpur), r.zameni@modares.ac.ir (R. Zameni).

solutions have fluxes in the compact direction, they break supersymmetry and therefore it is important to know whether they are stable.

It is also interesting to see how these new solutions might arise from near horizon geometries of some particular brane configurations. This would then lead us to the construction of the CFT duals [18]. For the eleven-dimensional supergravity solutions, first notice that the compact manifold admits a nontrivial 2-cycle over which we can wrap branes. Therefore, one way to get the AdS_5 factor is to construct a Ricci flat cone over the compact manifold and then consider fractional 3-branes (wrapped M5-branes over the 2-cycles) in the orthogonal directions placed at the tip of the cone. The near horizon geometry of this brane configuration would be $AdS_5 \times M_6$. Similarly, for the type IIB case, since M_8 is Einstein and admits nontrivial 3-cycles, we can construct a Ricci flat cone over it, and then put fractional D0-branes (D3-branes wrapped over 3-cycles of M_8) at the tip of the cone. Therefore we expect $AdS_2 \times M_8$ solution to arise as the near horizon limit of this D0-brane configuration.

In this paper we examine the stability of solutions under small perturbations of the metric. For getting consistent equations of motion on AdS, however, we also need to introduce deformations of the fluxes. Here we follow an approach which is close to that of [19,20]. For compactification to AdS_5 the metric deformations correspond to the breathing and squashing modes. Including the deformation of the 4-form flux would correspond to three massive mode excitations on the AdS space. In type IIB case, however, the bundle structure of the compact manifold allows a more general deformation, which, in turn, results in seven massive mode excitations. Apart from deriving the mass spectrum of small fluctuations, our approach has the advantage of providing us with a set of consistent reduced equations on AdS space, so that any solution to these equations can be uplifted to a supergravity solution in eleven or ten dimensions.

2. Stability of $AdS_5 \times CP^3$ compactification

In this section we consider the solution $AdS_5 \times M_6$, where M_6 is CP^3 written as an S^2 bundle over S^4 [14], and study its stability under small perturbations. We start by deforming the metric along the fiber and the base by some unknown scalar functions on AdS_5. To get consistent reduced equations we see that the 4-form flux also needs to be deformed. After deriving the curvature tensor of the metric we write the supergravity equations of motion, and then linearize the equations around the known solutions. This allows us to read the mass of the small fluctuations corresponding to those deformations. If the mass squared falls in the Breitenlohner–Freedman range then the solution is stable against such perturbations.

To begin with, let us take the eleven dimensional spacetime to be the direct product of a 5 and 6-dimensional spaces,

$$ds_{11}^2 = ds_{AdS_5}^2 + ds_6^2. \tag{1}$$

For the 6-dimensional space the metric reads

$$\begin{aligned} ds_6^2 &= d\mu^2 + \frac{1}{4}\sin^2\mu\Sigma_i^2 + \lambda^2(d\theta - \sin\phi A_1 + \cos\phi A_2)^2 \\ &\quad + \lambda^2\sin^2\theta\big(d\phi - \cot\theta(\cos\phi A_1 + \sin\phi A_2) + A_3\big)^2, \end{aligned} \tag{2}$$

with λ the squashing parameter, and

$$A_i = \cos^2\frac{\mu}{2}\Sigma_i, \tag{3}$$

$$d\Sigma_i = -\frac{1}{2}\epsilon_{ijk}\Sigma_j \wedge \Sigma_k. \tag{4}$$

This is an S^2 bundle over S^4, and for $\lambda^2 = 1$ we get the Fubini–Study metric on CP^3.

To discuss the stability, we deform the metric as follows:

$$\begin{aligned} d\bar{s}^2 &= e^{2A(x)}g_{\alpha\beta}dx^\alpha dx^\beta \\ &\quad + e^{2B(x)}\Big(d\mu^2 + \frac{1}{4}\sin^2\mu\Sigma_j^2\Big) \\ &\quad + e^{2C(x)}(d\theta - \sin\phi A_1 + \cos\phi A_2)^2 \\ &\quad + e^{2C(x)}\sin^2\theta\big(d\phi - \cot\theta(\cos\phi A_1 + \sin\phi A_2) + A_3\big)^2, \end{aligned} \tag{5}$$

where $g_{\alpha\beta}$ is the AdS_5 metric, and $A(x)$, $B(x)$, and $C(x)$ are arbitrary scalar functions on AdS_5. In fact, $B(x)$ and $C(x)$ correspond to what is usually called the breathing and the squashing mode deformations. We choose the following vielbein basis

$$\begin{aligned} &\bar{e}^\alpha = e^{A(x)}e^\alpha && \alpha = \bar{0}, \bar{1}, \bar{2}, \bar{3}, \bar{4} \\ &\bar{e}^0 = e^{B(x)}e^0 && \\ &\bar{e}^i = e^{B(x)}e^i && i = 1, 2, 3 \\ &\bar{e}^a = e^{C(x)}e^a && a = 5, 6, \end{aligned} \tag{6}$$

where the indices $\alpha, \beta, \ldots$ indicate the 5d spacetime coordinates, and the rest are related to the 6-dimensional space, and

$$\begin{aligned} &e^0 = d\mu, \qquad e^i = \frac{1}{2}\sin\mu\Sigma_i, \\ &e^5 = \lambda(d\theta - \sin\phi A_1 + \cos\phi A_2), \\ &e^6 = \lambda\sin\theta\big(d\phi - \cot\theta(\cos\phi A_1 + \sin\phi A_2) + A_3\big). \end{aligned} \tag{7}$$

Evaluation of the Ricci tensor of this deformed metric yields

$$\begin{aligned} \bar{R}_{\alpha\beta} &= e^{-2A}\big\{R_{\alpha\beta} - \nabla^2 A\delta_{\alpha\beta} + 4\partial_\beta B\partial_\alpha(A - B) \\ &\quad + 2\partial_\beta C\partial_\alpha(A - C)\big\}, \end{aligned} \tag{8}$$

$$\bar{R}_{ij} = \big(3e^{-2B} - e^{2(C-2B)} - e^{-2A}\nabla^2 B\big)\delta_{ij}, \tag{9}$$

$$\bar{R}_{ab} = \big(e^{-2C} + e^{2(C-2B)} - e^{-2A}\nabla^2 C\big)\delta_{ab}. \tag{10}$$

Next, as in [14], we want to write a similar ansatz for the gauge field strength. However, since we have perturbed the metric with some scalar functions on AdS space we must add an extra term for consistency. Further, it is easier first to write the Hodge dual ansatz as follows

$$\bar{*}_{11}F_4 = \bar{\epsilon}_5 \wedge \big(\alpha(x)e^{56} + \gamma(x)K\big) + \bar{*}_5 d\eta \wedge Im\Omega, \tag{11}$$

where we have defined,

$$R_1 = \sin\phi\big(e^{01} + e^{23}\big) - \cos\phi\big(e^{02} + e^{31}\big), \tag{12}$$

$$\begin{aligned} R_2 &= \cos\theta\cos\phi\big(e^{01} + e^{23}\big) \\ &\quad + \cos\theta\sin\phi\big(e^{02} + e^{31}\big) - \sin\theta\big(e^{03} + e^{12}\big), \end{aligned} \tag{13}$$

$$\begin{aligned} K &= \sin\theta\cos\phi\big(e^{01} + e^{23}\big) + \sin\theta\sin\phi\big(e^{02} + e^{31}\big) \\ &\quad + \cos\theta\big(e^{03} + e^{12}\big), \end{aligned} \tag{14}$$

$$Re\Omega = R_1 \wedge e^5 + R_2 \wedge e^6, \tag{15}$$

$$Im\Omega = R_1 \wedge e^6 - R_2 \wedge e^5, \tag{16}$$

$$\omega_4 = e^0 \wedge e^1 \wedge e^2 \wedge e^3. \tag{17}$$

As $F_4 \wedge F_4 = 0$ (see (21)), the Maxwell equation reads

$$d\bar{*}_{11}F_4 = \bar{\epsilon}_5 \wedge (\alpha - \gamma) \wedge Im\Omega + d\bar{*}_5 d\eta \wedge Im\Omega = 0, \tag{18}$$

where we used [14],

$$de^{56} = Im\Omega, \quad dK = -Im\Omega, \quad dIm\Omega = 0. \tag{19}$$

Hence, Maxwell equation implies

$$\nabla^2\eta = \gamma(x) - \alpha(x), \tag{20}$$

where $\nabla^2 = \bar{*}_5 d\bar{*}_5 d$. Changing the basis through (6), from (11) we see that

$$\begin{aligned} \bar{*}_{11} F_4 = \bar{\epsilon}_5 \wedge \big(\alpha e^{-2C(x)} \bar{e}^{56} + \gamma e^{-2B(x)} \overline{K}\big) \\ + e^{-2B(x)-C(x)} \bar{*}_5 d\eta \wedge \overline{Im\Omega}, \end{aligned} \tag{21}$$

where bar indicates barred basis in (6). Therefore, for F_4 we find

$$\begin{aligned} F_4 = -\alpha(x) e^{-2C(x)} \overline{\omega}_4 - \gamma(x) e^{-2B(x)} \overline{K} \wedge \bar{e}^{56} \\ + e^{-2B(x)-C(x)} d\eta \wedge \overline{Re\Omega}. \end{aligned} \tag{22}$$

Let us now check the Bianchi identity $dF_4 = 0$. Since $d\omega_4 = 0$, $dRe\Omega = 4\omega_4 - 2e^{56} \wedge K$, and also $Im\Omega \wedge K = Im\Omega \wedge e^{56} = 0$, the Bianchi identity requires

$$\begin{aligned} -d\big(\alpha e^{4B(x)-2C(x)}\big) = 4d\eta, \\ d\big(\gamma e^{2C(x)}\big) = 2d\eta. \end{aligned} \tag{23}$$

The above equations, in turn, imply

$$\gamma(x) = -\frac{1}{2}\alpha(x) e^{4B(x)-4C(x)} + \frac{1}{2}\alpha_0\big(e^{-2C_1} + 2e^{2C_1}\big) e^{-2C(x)}, \tag{24}$$

where α_0 and C_1 are two constants. Using (20) and (24), the equation of motion for α reads

$$\nabla^2\alpha = -4\alpha\nabla^2 B + 2\alpha\nabla^2 C - 4[\gamma - \alpha] e^{-4B+2C}, \tag{25}$$

as we will later expand around constant solutions, here we have dropped quadratic derivative terms.

Next, let us turn to the Einstein equations [14]:

$$R_{MN} = \frac{1}{12} F_{MPQR} F_N^{PQR} - \frac{1}{3.48} g_{MN} F_{PQRS} F^{PQRS}, \tag{26}$$

where $M, N, P, \ldots = 0, 1, \ldots, 10$. With ansatz (22), we can calculate the right hand side of the above equations:

$$\bar{R}_{ij} = \left(\frac{1}{3}\alpha^2 e^{-4C(x)} + \frac{1}{6}\gamma^2 e^{-4B(x)}\right)\delta_{ij}, \tag{27}$$

$$\bar{R}_{ab} = \left(-\frac{1}{6}\alpha^2 e^{-4C(x)} + \frac{2}{3}\gamma^2 e^{-4B(x)}\right)\delta_{ab}. \tag{28}$$

Using (9) and (10) on the LHS of the above equations yields

$$3e^{-2B} - e^{2(C-2B)} - e^{-2A}\nabla^2 B = \frac{1}{3}\alpha^2 e^{-4C} + \frac{1}{6}\gamma^2 e^{-4B}, \tag{29}$$

$$e^{-2C} + e^{2(C-2B)} - e^{-2A}\nabla^2 C = -\frac{1}{6}\alpha^2 e^{-4C} + \frac{2}{3}\gamma^2 e^{-4B}. \tag{30}$$

Combining (24), (25), (29) and (30), we get three equations which have a constant solution $\alpha_0^2 = 4$, $C_1 = B_1 = 0$ corresponding to the first solution in [14] with $\lambda^2 = 1$. To study the small fluctuations, we expand around this constant solution and only keep the linear terms to get

$$\begin{aligned} \nabla^2\alpha &= 22\alpha + \frac{176}{3}B - \frac{56}{3}C, \\ \nabla^2 B &= -\alpha + \frac{10}{3}B + \frac{14}{3}C, \\ \nabla^2 C &= 2\alpha + \frac{52}{3}B + \frac{8}{3}C. \end{aligned} \tag{31}$$

Our next task is to find the mass spectrum. This is easily found by diagonalizing the mass matrix appearing on the RHS of equations (31). The mass spectrum reads

$$\mathbb{M} = diag(-2, 12, 18). \tag{32}$$

To see whether the first mode is stable, we need to invoke the Breitenlohner–Freedman (BF) stability bound on AdS_{d+1} which requires

$$m^2 \geq m_{BF}^2 = -\frac{d^2}{4}, \tag{33}$$

for the mode to be stable. For AdS_5 we need to have $m^2 \geq -4$, so we conclude that the solution is stable against all three fluctuation modes. This conclusion agrees with the result of [17] who proved the stability of this particular solution by analyzing the spectrum of forms on CP^3.

The squashed solution in [14] with $\lambda^2 = 2$, on the other hand, here corresponds to a solution with $\alpha_0^2 = 4$, $B_1 = 0$, and $e^{2C_1} = 2$. Expanding and linearizing the three equations (25), (29) and (30) around this solution we find

$$\begin{aligned} \nabla^2\alpha &= 13\alpha - \frac{40}{3}B + \frac{232}{3}C, \\ \nabla^2 B &= -\frac{1}{4}\alpha + \frac{16}{3}B - \frac{1}{3}C, \\ \nabla^2 C &= \frac{1}{2}\alpha + \frac{16}{3}B + \frac{35}{3}C. \end{aligned} \tag{34}$$

Diagonalizing the mass spectrum we find

$$\mathbb{M} = diag(3, 9, 18), \tag{35}$$

which is clearly stable.

3. Stability of type IIB compactifications to AdS_2

Another solution that we would like to study its stability is a compactification of type IIB theory on a $U(1)$ bundle over $N(1,1)$. Let us start by taking the following seven-dimensional metric of $N(1,1)$ [15]:

$$\begin{aligned} ds^2_{N(1,1)} = d\mu^2 + \frac{1}{4}\sin^2\mu\big(\Sigma_1^2 + \Sigma_2^2 + \cos^2\mu\,\Sigma_3^2\big) \\ + \lambda^2(d\theta - \sin\phi A_1 + \cos\phi A_2)^2 \\ + \lambda^2 \sin^2\theta\big(d\phi - \cot\theta(\cos\phi A_1 + \sin\phi A_2) + A_3\big)^2 \\ + \tilde{\lambda}^2(d\tau - A)^2, \end{aligned} \tag{36}$$

where λ and $\tilde{\lambda}$ are the squashing parameters, and

$$\begin{aligned} A_1 = \cos\mu\,\Sigma_1, \qquad A_2 = \cos\mu\,\Sigma_2, \\ A_3 = \frac{1}{2}\big(1 + \cos^2\mu\big)\Sigma_3, \end{aligned} \tag{37}$$

and,

$$A = \cos\theta d\phi + \sin\theta(\cos\phi A_1 + \sin\phi A_2) + \cos\theta A_3. \tag{38}$$

Note that the base manifold admits a closed 2-form, the Kähler form:

$$J = \frac{1}{4}da = \frac{1}{4}d\big(\sin^2\mu\,\Sigma_3\big) = e^{03} - e^{12}, \tag{39}$$

so that $dJ = 0$. Therefore, we can construct a $U(1)$ bundle over $N(1,1)$ as follows

$$ds_8^2 = ds^2_{N(1,1)} + \hat{\lambda}^2(dz - a)^2, \tag{40}$$

with $\hat{\lambda}$ measuring the scale of new $U(1)$ fiber.

To discuss the small fluctuations, as in previous section, we perturb the metric by scalar functions on AdS_2 as follows:

$$
\begin{aligned}
d\bar{s}^2_{10} &= e^{2A(x)}ds^2_{AdS_5} \\
&+ e^{2B(x)}\left(d\mu^2 + \frac{1}{4}\sin^2\mu(\Sigma_1^2 + \Sigma_2^2 + \cos^2\mu\Sigma_3^2)\right) \\
&+ e^{2C(x)}(d\theta - \sin\phi A_1 + \cos\phi A_2)^2 \\
&+ e^{2C(x)}\left(\sin^2\theta\left(d\phi - \cot\theta(\cos\phi A_1 + \sin\phi A_2) + A_3\right)^2\right) \\
&+ e^{2E(x)}(d\tau - A)^2 + e^{2D(x)}(dz - a)^2. \qquad (41)
\end{aligned}
$$

Let us choose the following basis,

$$
\begin{aligned}
&\bar{e}^\alpha = e^{A(x)}e^\alpha \qquad \alpha = \tilde{0}, \tilde{1} \\
&\bar{e}^i = e^{B(x)}e^i \qquad i = 0, 1, 2, 3 \\
&\bar{e}^a = e^{C(x)}e^a \qquad a = 5, 6 \\
&\bar{e}^7 = e^{E(x)}e^7, \qquad \bar{e}^8 = e^{D(x)}e^8, \qquad (42)
\end{aligned}
$$

where,

$$
\begin{aligned}
&e^0 = d\mu, \qquad e^1 = \frac{1}{2}\sin\mu\Sigma_1, \\
&e^2 = \frac{1}{2}\sin\mu\Sigma_2, \qquad e^3 = \frac{1}{2}\sin\mu\cos\mu\Sigma_3, \\
&e^5 = (d\theta - \sin\phi A_1 + \cos\phi A_2), \\
&e^6 = \sin\theta\left(d\phi - \cot\theta(\cos\phi A_1 + \sin\phi A_2) + A_3\right), \\
&e^7 = d\tau - A, \qquad e^8 = dz - a. \qquad (43)
\end{aligned}
$$

Now in terms of the barred basis the Ricci tensor is diagonal and reads,

$$
\begin{aligned}
\bar{R}_{ij} &= (6e^{-2B} - 4e^{2(C-2B)} - 2e^{2(E-2B)} \\
&\quad - 8e^{2(D-2B)} - e^{-2A}\nabla^2 B)\delta_{ij}, \\
\bar{R}_{ab} &= \left(4e^{2(C-2B)} - \frac{1}{2}e^{2(E-2C)} + e^{-2C} - e^{-2A}\nabla^2 C\right)\delta_{ab}, \\
\bar{R}_{77} &= 4e^{2(E-2B)} + \frac{1}{2}e^{2(E-2C)} - e^{-2A}\nabla^2 E, \\
\bar{R}_{88} &= 16e^{2(D-2B)} - e^{-2A}\nabla^2 D. \qquad (44)
\end{aligned}
$$

To write the self-dual 5-form field strength ansatz, we follow the prescription presented in [15]. However, as the metric is deformed by scalar functions we need to add some extra terms. Let us start by writing the following 5-form

$$
\begin{aligned}
\omega_5 &= (\alpha e^{-4B-D}\bar{\omega}_4 + \beta e^{-2B-2C-D}\bar{K} \wedge \bar{e}^{56} \\
&+ \gamma e^{-2B-C-E-D}\bar{e}^7 \wedge \overline{Im\Omega}) \wedge \bar{e}^8 \\
&+ \xi e^{-2B-2C-E}\bar{J} \wedge \bar{e}^{567} + e^{-2B-C-D}d\eta_1 \wedge \overline{Re\Omega} \wedge \bar{e}^8 \\
&+ e^{-2C-E-D}d\eta_2 \wedge \bar{e}^{56} \wedge \bar{e}^{78} \\
&+ e^{-2B-E-D}d\eta_3 \wedge \bar{K} \wedge \bar{e}^{78}, \qquad (45)
\end{aligned}
$$

with $\alpha, \beta, \gamma, \xi, \eta_1, \eta_2$, and η_3 are now taken to be scalar functions over spacetime and barred basis are defined as in (42). Requiring ω_5 to be closed, we get

$$
\begin{aligned}
&d\xi - 4d\eta_2 = 0, \\
&-d\gamma - 2d\eta_2 + d\eta_3 = 0, \\
&d\alpha - 8d\eta_1 - 4d\eta_3 = 0, \\
&d\beta + 2d\eta_1 - 2d\eta_2 - d\eta_3 = 0, \qquad (46)
\end{aligned}
$$

where for $N(1,1)$ we used $de^{56} = 2Im\Omega$, $dK = -Im\Omega$, and $dRe\Omega = 8\omega_4 - 2e^{56} \wedge K$ (these are different from the ones in previous section as the bases are different). Solving the above equations for α we get

$$
\alpha = -4\beta + 6\xi + 8\gamma + 4\beta_0 - 6\xi_0 - 8\gamma_0. \qquad (47)
$$

Now, taking the Hodge dual of (45) (with $\epsilon_{\tilde{0}\tilde{1}01235678} = 1$, where $\tilde{0}$ and $\tilde{1}$ refer to AdS_2 coordinates), we find

$$
\begin{aligned}
\bar{*}_{11}\omega_5 &= (-\alpha e^{-4B-D}\bar{e}^{567} - \beta e^{-2B-2C-D}\bar{K} \wedge \bar{e}^7) \wedge \bar{\epsilon}_2 \\
&+ \left(\gamma e^{-2B-C-E-D}\overline{Re\Omega} - \xi e^{-2B-2C-E}\bar{J} \wedge \bar{e}^8\right) \wedge \bar{\epsilon}_2 \\
&+ e^{-2B-C-D}\bar{*}d\eta_1 \wedge \overline{Im\Omega} \wedge \bar{e}^7 + e^{-2C-E-D}\bar{*}d\eta_2 \wedge \bar{\omega}_4 \\
&+ e^{-2B-E-D}\bar{*}d\eta_3 \wedge \bar{K} \wedge \bar{e}^{56}. \qquad (48)
\end{aligned}
$$

Requiring the above 5-form to be closed results to the following equations

$$
\begin{aligned}
&2\alpha e^{-4B+2C+E-D} + \beta e^{-2C-D+E} + 2\gamma e^{-D-E} \\
&\quad - e^{2C-E-D}\nabla^2\eta_3 = 0, \\
&4\beta e^{-2C-D+E} - 8\gamma e^{-D-E} + 8\xi e^{-2C-E+D} \\
&\quad - e^{4B-2C-E-D}\nabla^2\eta_2 = 0, \\
&2\alpha e^{-4B+2C+E-D} - \beta e^{-2C-D+E} - e^{E-D}\nabla^2\eta_1 = 0. \qquad (49)
\end{aligned}
$$

Using (46) and (49), we can find three equations

$$
\begin{aligned}
\nabla^2\xi &= 16\beta e^{-4B+2E} - 32\gamma e^{-4B+2C} + 32\xi e^{-4B+2D}, \\
\nabla^2\gamma &= 2\alpha e^{-4B+2E} + \beta(e^{-4C+2E} - 8e^{-4B+2E}) \\
&\quad + \gamma(2e^{-2C} + 16e^{-4B+2C}) - 16\xi e^{-4B+2D}, \\
\nabla^2\beta &= 2\alpha(e^{-4B+2E} - 2e^{-4B+2C}) \\
&\quad + \beta(e^{-4C+2E} + 2e^{-2C} + 8e^{-4B+2E}) \\
&\quad + 2\gamma(e^{-2C} - 8e^{-4B+2C}) + 16\xi e^{-4B+2D}. \qquad (50)
\end{aligned}
$$

With these constraints on scalar functions we can now write down the ansatz for the self-dual 5-form:

$$
\bar{F}_5 = \bar{*}_{11}\omega_5 + \omega_5 \qquad (51)
$$

which satisfies the equation of motion $d * F_5 = 0$.

Let us now consider the Einstein equations, taking the dilaton and axion to be constant, in the Einstein frame we have [15]:

$$
\begin{aligned}
R_{MN} &= \frac{1}{4.4!}\left(F_{MPQRS}F_N^{PQRS} - \frac{1}{10}F_{PQRSL}F^{PQRSL}g_{MN}\right) \\
&+ \frac{e^{-\phi}}{4}\left(H_{MPQ}H_N^{PQ} - \frac{1}{12}H_{PQR}H^{PQR}g_{MN}\right) \\
&+ \frac{e^{\phi}}{4}\left(F_{MPQ}F_N^{PQ} - \frac{1}{12}F_{PQR}F^{PQR}g_{MN}\right). \qquad (52)
\end{aligned}
$$

Using (44) and (51), the Einstein equations reduce to the following equations:

$$
\begin{aligned}
&6e^{-2B} - 4e^{2(C-2B)} - 2e^{2(E-2B)} - 8e^{2(D-2B)} \\
&\quad - e^{-2A}\nabla^2 B = \frac{\alpha^2}{4}e^{-8B-2D}, \\
&4e^{2(C-2B)} - \frac{1}{2}e^{2(E-2C)} + e^{-2C} - e^{-2A}\nabla^2 C \\
&\quad = -\frac{\alpha^2}{4}e^{-8B-2D} + \frac{\beta^2}{2}e^{-4B-4C-2D} + \frac{\xi^2}{2}e^{-4B-4C-2E},
\end{aligned}
$$

$$4e^{2(E-2B)}+\frac{1}{2}e^{2(E-2C)}-e^{-2A}\nabla^2 E$$
$$=-\frac{\alpha^2}{4}e^{-8B-2D}-\frac{\beta^2}{2}e^{-4B-4C-2D}$$
$$+\frac{\xi^2}{2}e^{-4B-4C-2E}+\gamma^2 e^{-2E-4B-2C-2D},$$
$$16e^{2(D-2B)}-e^{-2A}\nabla^2 D$$
$$=\frac{\alpha^2}{4}e^{-8B-2D}+\frac{\beta^2}{2}e^{-4B-4C-2D}$$
$$-\frac{\xi^2}{2}e^{-4B-4C-2E}+\gamma^2 e^{-2E-4B-2C-2D}. \quad (53)$$

The solution found in [15] corresponds to the following constant solution of equations (50) and (53):

$$\alpha_0=\frac{3}{2},\quad \beta_0=\frac{3}{16},\quad \gamma_0=-\frac{3}{16},$$
$$\xi_0=-\frac{3}{8},\quad e^{2C_1}=e^{2E_1}=\frac{1}{4},\quad e^{2D_1}=\frac{3}{16}. \quad (54)$$

To discuss the stability of this solution, we linearize equations (50) and (53) about the above solution to get,

$$\nabla^2 B=24B-2C-E+3D+16\beta-24\xi-32\gamma,$$
$$\nabla^2 C=-4B+26C+5E-3D-32\beta+48\xi+32\gamma,$$
$$\nabla^2 E=-4B+10C+21E-3D+48\xi+64\gamma,$$
$$\nabla^2 D=12B-6C-3E+21D-48\xi,$$
$$\nabla^2\beta=3B-\frac{9}{2}C+\frac{15}{4}E-\frac{9}{4}D+16\beta,$$
$$\nabla^2\xi=3C+\frac{3}{2}E-\frac{9}{2}D+4\beta+6\xi-8\gamma,$$
$$\nabla^2\gamma=-3B-\frac{3}{2}C+\frac{9}{4}E+\frac{9}{4}D+16\gamma. \quad (55)$$

As in previous section, we can diagonalize the RHS of (55) to find the mass spectrum:

$$\mathbb{M}=diag(2.14, 2.14, 12, 24, 29.85, 29.85, 30), \quad (56)$$

with all the eigenvalues positive, we conclude that the solution (54) is stable against small fluctuations.

4. Conclusions

In this paper we examined the stability of some recently found non-supersymmetric solutions of ten and eleven dimensional supergravity. We perturbed the metric and the form fluxes by some space–time dependent scalar functions so that to reduce the equations of motion consistently on AdS space. We then linearized these equations around solutions corresponding to those of [14] and [15]. For the compactification of the form $AdS_5\times M_6$, we found that the two solutions, with squashing parameters of $\lambda^2=1$ and $\lambda^2=2$, are both stable against the kind of small fluctuations that correspond to the breathing and the squashing modes. This result is in agreement with the analysis of [17] who proved the stability in the case of $\lambda^2=1$ using a different approach. For type IIB solution of $AdS_2\times M_8$, on the other hand, we observed that there are more modes that can be consistently excited on AdS space. We derived the equations of motion of these modes, and by linearizing them around the solution of [15] showed that this solution is also stable.

We showed that the solutions of [14,15] are stable against some particular small perturbations of the metric and the fluxes. However, to complete the proof of stability one needs to consider more general perturbations and study their spectrum. Moreover, having proved that a solution is perturbatively stable there still remains to check the solution against nonperturbative instabilities [4]. Recently, the authors of [21] showed that there is an instanton solution which destabilizes the $AdS_5\times CP^3$ solution, and hence concluded that it is nonperturbatively unstable. It is therefore interesting to see whether similar instanton solutions exist for $AdS_2\times M_8$.

The method we used in this paper led us to a set of consistent reduced equations on AdS space. Consequently, a solution of these reduced equations can be uplifted to derive new eleven-dimensional or type IIB solutions. Therefore, apart from deriving the mass spectrum of small fluctuations, our approach can also be useful in searching for new supergravity solutions. In particular, it is interesting to look for domain wall solutions which interpolate between different vacua.

References

[1] E. Witten, Search for a realistic Kaluza–Klein theory, Nucl. Phys. B 186 (1982) 412.
[2] G.W. Gibbons, C.M. Hull, N.P. Warner, The stability of gauged supergravity, Nucl. Phys. B 218 (1983) 173.
[3] P. Breitenlohner, D.Z. Freedman, Positive energy in anti-desitter backgrounds and gauged extended supergravity, Phys. Lett. B 115 (1982) 197.
[4] E. Witten, Instability of the Kaluza–Klein vacuum, Nucl. Phys. B 195 (1982) 481.
[5] P.G. Freund, M.A. Rubin, Dynamics of dimensional reduction, Phys. Lett. B 97 (1980) 233.
[6] M.J. Duff, B.E.W. Nilsson, C.N. Pope, Kaluza–Klein supergravity, Phys. Rep. 130 (1986) 1.
[7] B. Biran, P. Spindel, Instability of the parallelized seven sphere: an eleven-dimensional approach, Phys. Lett. B 141 (1984) 181.
[8] D.N. Page, C.N. Pope, Instabilities in Englert type supergravity solutions, Phys. Lett. B 145 (1984) 333.
[9] C.N. Pope, N.P. Warner, An SU(4) invariant compactification of $d=11$ supergravity on a stretched seven sphere, Phys. Lett. B 150 (1985) 352.
[10] N. Bobev, N. Halmagyi, K. Pilch, N.P. Warner, Supergravity instabilities of non-supersymmetric quantum critical points, Class. Quantum Gravity 27 (2010) 235013, arXiv:1006.2546 [hep-th].
[11] J.P. Gauntlett, J. Sonner, T. Wiseman, Quantum criticality and holographic superconductors in M-theory, J. High Energy Phys. 1002 (2010) 060, arXiv:0912.0512 [hep-th].
[12] J.M. Maldacena, The Large N limit of superconformal field theories and supergravity, Int. J. Theor. Phys. 38 (1999) 1113, Adv. Theor. Math. Phys. 2 (1998) 231, arXiv:hep-th/9711200.
[13] A. Adams, E. Silverstein, Closed string tachyons, AdS/CFT, and large N QCD, Phys. Rev. D 64 (2001) 086001, arXiv:hep-th/0103220.
[14] A. Imaanpur, New compactifications of eleven dimensional supergravity, Class. Quantum Gravity 30 (2013) 065021, arXiv:1205.1349 [hep-th].
[15] A. Imaanpur, Type IIB flux compactifications on twistor bundles, Phys. Lett. B 729 (2014) 45–49, arXiv:1309.5773v2 [hep-th].
[16] C.N. Pope, P. van Nieuwenhuizen, Compactifications of $d=11$ supergravity on Kahler manifolds, Commun. Math. Phys. 122 (1989) 281.
[17] J.E. Martin, H.S. Reall, On the stability and spectrum of non-supersymmetric AdS(5) solutions of M-theory compactified on Kahler–Einstein spaces, J. High Energy Phys. 0903 (2009) 002, arXiv:0810.2707 [hep-th].
[18] I.R. Klebanov, E. Witten, Superconformal field theory on three-branes at a Calabi–Yau singularity, Nucl. Phys. B 536 (1998) 199, arXiv:hep-th/9807080.
[19] D.N. Page, Classical stability of round and squashed seven spheres in eleven-dimensional supergravity, Phys. Rev. D 28 (1983) 2976.
[20] J.P. Gauntlett, S. Kim, O. Varela, D. Waldram, Consistent supersymmetric Kaluza–Klein truncations with massive modes, J. High Energy Phys. 0904 (2009) 102, arXiv:0901.0676 [hep-th].
[21] H. Ooguri, L. Spodyneiko, New Kaluza–Klein instantons and decay of AdS vacua, arXiv:1703.03105 [hep-th].

Overlaps after quantum quenches in the sine-Gordon model

D.X. Horváth [a,b,*], G. Takács [a,b]

[a] MTA-BME "Momentum" Statistical Field Theory Research Group, Budafoki út 8, 1111 Budapest, Hungary
[b] Department of Theoretical Physics, Budapest University of Technology and Economics, Budafoki út 8, 1111 Budapest, Hungary

ARTICLE INFO

Editor: N. Lambert

ABSTRACT

We present a numerical computation of overlaps in mass quenches in sine-Gordon quantum field theory using truncated conformal space approach (TCSA). To improve the cut-off dependence of the method, we use a novel running coupling definition which has a general applicability in free boson TCSA. The numerical results for the first breather overlaps are compared with the analytic continuation of a previously proposed analytical Ansatz for the initial state in a related sinh-Gordon quench, and good agreement is found between the numerical data and the analytical prediction in a large energy range.

1. Introduction

One of the most challenging problems in contemporary physics is the understanding of dynamical and relaxation phenomena in closed quantum systems out of equilibrium. Motivated by both theoretical interest and experimental relevance, recent studies led to a series of interesting discoveries such as the experimental observation of the lack of thermalization in integrable systems [1–4]. To explain the stationary state of integrable quantum systems, the concept of the generalized Gibbs ensemble (GGE) was proposed [5], and recently experimentally confirmed [6]. It also turned out that the GGE was generally incomplete when only including the well-known local conserved charges [7,8], and its completion made necessary the inclusion of novel quasi-local charges [9,10]. Adding to this the unconventional, often ballistic nature of quantum transport [11,12] or the confinement effects in the spread of correlations in non-integrable systems [13] indeed, a remarkable range of exotic behavior has emerged in recent years.

A paradigmatic framework for non-equilibrium dynamics is provided by quantum quenches [14], in which the initial state (which is typically the ground state of some pre-quench Hamiltonian H_0) is subject to evolution driven by a post-quench Hamiltonian H, which is obtained from H_0 by instantaneously changing some parameters of the system. For the purpose of computing the time evolution it is useful to know the overlaps, i.e. the amplitudes of the post-quench excitations in the initial state. Indeed, in the case of integrable post-quench dynamics, knowledge of these overlaps often enables the determination of steady state properties, and even the time evolution [15–20]. However, the determination of the overlaps is generally a very difficult task. When both the pre-quench and post-quench theories are non-interacting, the overlaps can be determined using the Bogoliubov transformation linking the pre- and post-quench excitation modes, but in genuinely interacting integrable models there are only few cases in which the overlaps are explicitly known. These cases mostly include spin chains and the Lieb–Liniger model [21–27].

Quantum field theories are known to provide universal descriptions of statistical models and many-body systems, valid at long distances, and therefore quantum quenches in field theories are interesting, especially in the quest for universal characteristics and behavior under quantum quenches. In massive relativistic integrable quantum field theories there exists a number of efficient approaches to the quench dynamics, which depend on the assumption that the initial state $|\Psi(0)\rangle$ can be written in a squeezed vacuum form in terms of post-quench Zamolodchikov–Faddeev creation operators $Z_a^\dagger(\vartheta)$ for asymptotic particle states and the post-quench vacuum $|0\rangle$

$$|\Psi(0)\rangle = \mathcal{N} \exp \int \frac{d\vartheta}{2\pi} K_{a,b}(\vartheta) Z_a^\dagger(-\vartheta) Z_b^\dagger(\vartheta)|0\rangle \,, \tag{1.1}$$

which is just the analogue of the Bogoliubov solution for free theories. The above form of the initial state is equivalent to the statement that the multi-particle creation amplitudes factorize into products of independent single pair creation amplitudes. This is obviously reminiscent of the factorization property of scattering in integrable quantum field theories [28], which justifies calling this class of quenches "integrable". Such a form of the initial state enables the application of methods based on thermodynamic Bethe Ansatz (TBA) [15–17], form factor based spectral expansions [29,

* Corresponding author.
E-mail address: esoxluciuslinne@gmail.com (D.X. Horváth).

30] or semi-classical approach [18]. However, even within the class of integrable quenches no exact solutions are known for the overlap functions $K_{a,b}$ apart from non-interacting quantum field theory models.

In the present work a numerical method is provided that is able to determine the overlaps in mass quenches in the sine-Gordon model with the same pre- and post-quench interaction strength, using the framework of the Truncated Conformal Space Approach (TCSA), originally introduced in [34] and extended to the sine-Gordon model in [35]. Recently an Ansatz for the overlaps was proposed for the quench from a massive free boson to an interacting sinh-Gordon model [31,32], which has already been used to obtain predictions for steady state expectation values [33]. Keeping in mind, that the initial state in [31,32] is the ground state of the free bosonic theory, whereas in our method the initial state is a ground state of an interacting model, we compare the analytical continuation of the Ansatz to sine-Gordon theory with the numerical overlaps for pairs of B_1 breathers, identify the energy ranges in which they match and hence the Ansatz provides a valid description for the B_1 overlaps, and discuss the origin of deviations, when they occur.

2. Overlaps in quantum field theory quenches

2.1. The sinh-Gordon quench overlaps and their continuation to sine-Gordon

The work [32] considered a quench from a massive free bosonic theory with particle mass m_0 to the massive sinh-Gordon theory

$$\mathcal{A}=\int d^2x\left(\frac{1}{2}\partial_\mu\Phi\partial^\mu\Phi-\frac{\mu^2}{g^2}\cosh g\Phi\right), \tag{2.1}$$

with coupling g and physical particle mass m (which is equal to μ in the classical limit). For these quenches it was argued that the initial state can be cast into the exponential form

$$|\Psi(0)\rangle=\mathcal{N}\exp\left\{\int\frac{d\vartheta}{2\pi}K(\vartheta)Z^\dagger(-\vartheta)Z^\dagger(\vartheta)\right\}|0\rangle\,. \tag{2.2}$$

However, properly demonstrating that the initial state has the form (1.1) in terms of post-quench asymptotic particle states is far from straightforward, and has only been possible in the non-interacting case (also in some interacting quenches in spin chains and Bose gases, where the exact overlaps are known and factorize in the thermodynamic limit [21–25,27]).

A class of states which has the exponential form is given by so-called integrable boundary states introduced in [36]; however, they cannot be considered physical initial states since they are not normalizable. As shown in [14,37] it is possible to construct proper initial states in terms of a boundary state $|B\rangle$ using the form

$$|\Psi(0)\rangle=e^{-\sum\tau_iQ_i}|B\rangle$$

where the Q_i are local conserved charges. Assuming that (as usual) the one-particle states are eigenstates of the Q_i, this obviously preserves the exponential form of the state, but it is hard to identify the physical quench (i.e. the pre-quench Hamiltonian H_0) that results in this state for a particular choice of the real parameters τ_i.

For quenches starting from a large initial mass in the sinh-Gordon field theory arguments in favor of the exponential form were advanced in [31], and even the following Ansatz was proposed for the function K:

$$K(\vartheta)=K_{free}(\vartheta)K_D(\vartheta)\,, \tag{2.3}$$

where $K_{free}(\vartheta)$ is given by

$$K_{free}(\vartheta)=\frac{E_0(\vartheta)-E(\vartheta)}{E_0(\vartheta)+E(\vartheta)},$$

$$E(\vartheta)=m\cosh\vartheta,\quad E_0(\vartheta)=\sqrt{m_0^2+m^2\sinh^2\vartheta}\,,$$

and is identical (up to a sign) with the Bogoliubov amplitude in a mass quench $m_0\longrightarrow m$ within a free bosonic model, while $K_D(\vartheta)$ is the amplitude of the Dirichlet boundary ($\Phi=0$) state in sinh-Gordon theory[1]:

$$K_D(\vartheta)=i\tanh(\vartheta/2)$$
$$\times\frac{\cosh(\vartheta/2-i\pi B/8)\sinh(\vartheta/2+i\pi(B+2)/8)}{\sinh(\vartheta/2+i\pi B/8)\cosh(\vartheta/2-i\pi(B+2)/8)},$$

$$B(g)=\frac{2g^2}{8\pi+g^2}\,. \tag{2.4}$$

In the follow-up work [32] it was shown that provided the initial state contains only multiple particle states composed of pairs with opposite momenta, extensivity of the charges guaranteeing integrability leads to factorization of multi-pair amplitudes and therefore results in an exponential form of the state, the only undetermined parameter being the pair creation amplitudes $K_{a,b}(\vartheta)$. However, the pair structure itself remains mainly an assumption supported only by some heuristic arguments [32].

Furthermore an infinite integral equation hierarchy was derived that determines (at least in principle) the full form of the initial state in terms of the post-quench multi-particle states for the quenches from a free massive boson to the sinh-Gordon model, and it was further shown that the simple Ansatz (2.3) was a very good numerical solution of the lowest member of the hierarchy provided the exponential form of the initial state was assumed. In addition, the next member of the hierarchy was used for a numerical test of the factorization assumption itself, which worked well within the limitations of the numerics.

Continuing to imaginary couplings $g=i\beta$ results in sine-Gordon theory

$$\mathcal{A}=\int d^2x\left(\frac{1}{2}\partial_\mu\Phi\partial^\mu\Phi+\frac{\mu^2}{\beta^2}\cos\beta\Phi\right), \tag{2.5}$$

and it is useful to introduce $\xi=\beta^2/(8\pi-\beta^2)=-B/2$. The fundamental excitations are a doublet of soliton/antisoliton of mass M. In the attractive regime ($\xi<1$) the spectrum also contains breathers B_r (soliton–antisoliton bound states) with masses $m_r=2M\sin r\pi\xi/2$ with r a positive integer less than ξ^{-1}. Due to integrability, the exact factorized S matrix is also known [28]. Under the analytic continuation to imaginary couplings the sinh-Gordon particle corresponds to the first breather B_1, which can be supported both by perturbation theory and the correspondence between the respective S matrix amplitudes. As a result, form factors of local operators and reflection factors containing only the first breather B_1 are also identical to the corresponding sinh-Gordon quantities under the same analytic continuation, which are all known in the model.

Here we consider sine-Gordon quenches which correspond to abruptly changing the soliton mass $M_0\to M$ while leaving the interaction strength ξ unaltered in the Hamiltonian H associated with (2.5). Note that under the analytic continuation this is related to a mass quench within sinh-Gordon theory with a fixed coupling g, while the Ansatz (2.3) was obtained for a quench from a free boson to sinh-Gordon theory. However, provided the interaction in the initial Hamiltonian does not play a significant role, we can ex-

[1] K_D can be obtained by analytic continuation from the first breather boundary amplitude in sine-Gordon theory which was obtained in [38].

pect that an analytic continuation

$$K_{B_1B_1}(\vartheta) = \frac{E_0(\vartheta) - E(\vartheta)}{E_0(\vartheta) + E(\vartheta)} K_D(\vartheta) \tag{2.6}$$

$$K_D(\vartheta) = i\tanh\left(\frac{\vartheta}{2}\right) \frac{\cosh\left(\frac{\vartheta}{2} + \frac{i\pi\xi}{4}\right)\sinh\left(\frac{\vartheta}{2} + \frac{i\pi(1-\xi)}{4}\right)}{\sinh\left(\frac{\vartheta}{2} - \frac{i\pi\xi}{4}\right)\cosh\left(\frac{\vartheta}{2} - \frac{i\pi(1-\xi)}{4}\right)}$$

gives a good approximation to the first breather pair creation amplitude in the sine-Gordon mass quench. We shall return to the issue of the initial interaction later when discussing the numerical results. Note that the amplitude depends only on the quench mass ratio, which is the same for each particle species since ξ is fixed, so we substituted the first breather mass by the soliton mass.

2.2. Overlaps in finite volume

In our numerical calculation we consider the system in a finite volume L with periodic boundary conditions, therefore we briefly recall the theory of finite size dependence of boundary state amplitudes, worked out in [39]. To keep the formulas short, we consider only one species of particles as the generalization to more than one species is rather obvious. Denoting the pre-quench ground state by $|B\rangle$, the most general expansion in terms of the post-quench eigenstates is

$$|B\rangle = |0\rangle + \sum_{n=1}^{\infty} \int \prod_{i=1}^{n} \frac{d\vartheta_i}{2\pi} K_n(\vartheta_1, ...\vartheta_n)\delta\left(\sum_{i=1}^{n} m\sinh\vartheta_i\right) \times |\vartheta_1, ..., \vartheta_n\rangle , \tag{2.7}$$

while in finite volume one obtains

$$|B\rangle_L = |0\rangle_L + \sum_{n=1}^{\infty} \sum_{I_1,...I_n}' N_n K_n(\vartheta_1^*, ...\vartheta_n^*)|I_1, ..., I_n\rangle_L , \tag{2.8}$$

where $\{\vartheta_i^*\}$ are the solutions of the Bethe–Yang equations, i.e. the system

$$Q_i = mL\sinh\vartheta_i + \sum_{j=1,\neq i}^{n} \delta(\vartheta_i - \vartheta_j) = 2\pi I_i \;, i = 1, ..., n, \tag{2.9}$$

where $\delta(\vartheta) = -i\ln S(\vartheta)$ is the phase-shift corresponding to the two particle S-matrix $S(\vartheta)$, m is the physical mass of the particle and I_i are the quantum numbers that characterize the finite volume states, with the prime meaning that only zero momentum states are included. In [39] the N_n functions were explicitly determined

$$N_n(\vartheta_1^*, ...\vartheta_n^*) = \frac{\sqrt{\rho_n(\vartheta_1^*, ...\vartheta_n^*)}}{\bar\rho_{n-1}(\vartheta_1^*, ...\vartheta_{n-1}^*)} + O(e^{-\mu' L}) , \tag{2.10}$$

where ρ_n is the density of states given by the Bethe–Yang Jacobi determinant [40]

$$\rho_n = \det\left\{\frac{\partial Q_k}{\partial\vartheta_j}\right\}_{j,k=1,...,n}$$

whereas $\bar\rho_{n-1}$ is the so-called reduced density of states which takes into account momentum conservation and is computed as the Jacobian

$$\bar\rho_{n-1} = \det\left\{\frac{\partial \bar Q_k}{\partial\vartheta_j}\right\}_{j,k=1,...,n-1}$$

of the constrained Bethe–Yang equations

$$\bar Q_i = mL\sinh\vartheta_i + \sum_{j=1,\neq i}^{n-1} \delta(\vartheta_i - \vartheta_j) + \delta(\vartheta_i - \tilde\vartheta) = 2\pi I_i \, , \tag{2.11}$$

$$i = 1, ..., n-1 \, , \quad \tilde\vartheta = -\sinh^{-1}\Big(\sum_{i=1}^{n-1}\sinh\vartheta_i\Big)$$

Formula (2.10) is exact to all orders in the inverse volume L^{-1} as indicated by correction terms that decay exponentially with the volume with some characteristic scale μ' (cf. [40]).

For the case when the expansion of the initial state in terms of the post-quench eigenstates only contains paired states

$$|B\rangle = |0\rangle + \sum_{n=1}^{\infty} \int \prod_{i=1}^{n} \frac{d\vartheta_i}{2\pi} K_n(\vartheta_1, ...\vartheta_n)| -\vartheta_1, \vartheta_1 ..., -\vartheta_n, \vartheta_n\rangle \, , \tag{2.12}$$

the appropriate constrained Bethe–Yang equations are

$$\bar Q_i^p = mL\sinh\vartheta_i + \delta(2\vartheta_i) + \sum_{j\neq i} \delta(\vartheta_i - \vartheta_j) + \delta(\vartheta_i + \vartheta_j) = 2\pi I_i \, , \quad i = 1, ..., n \, , \tag{2.13}$$

with solution $\{\vartheta_i^*\}$, and the finite volume expansion is

$$|B\rangle_L = |0\rangle_L + \sum_{n=1}^{\infty} \sum_{I_1,...I_n} N_n(\vartheta_1^*, ...\vartheta_n^*)K_n(\vartheta_1^*, ...\vartheta_n^*) \times |-I_1, I_1 ..., -I_n, I_n\rangle_L \, , \tag{2.14}$$

with the N_n functions [39]

$$N_n(\vartheta_1^*, ...\vartheta_n^*) = \frac{\sqrt{\rho_{2n}(-\vartheta_1^*, \vartheta_1^* ... -\vartheta_n^*, \vartheta_n^*)}}{\bar\rho_n^p(\vartheta_1^*, ...\vartheta_n^*)} + O(e^{-\mu' L}) \, ,$$

$$\bar\rho_n^p = \det\left\{\frac{\partial \bar Q_k^p}{\partial\vartheta_j}\right\}_{j,k=1,...,n} \, . \tag{2.15}$$

3. Overlaps from TCSA

3.1. TCSA for the sine-Gordon mass quench

We now turn to studying sine-Gordon mass quenches in truncated conformal space approach (TCSA), following the ideas in [41] which applied a similar approach to Ising field theory. For sine-Gordon, TCSA consists of representing the model as a compactified free massless boson conformal field theory (CFT) perturbed by a relevant operator, with the Hamiltonian

$$H = \int dx \frac{1}{2} :(\partial_t\Phi)^2 + (\partial_x\Phi)^2: -\frac{\lambda}{2}\int dx\,(V_1 + V_{-1}) \tag{3.1}$$

$$V_a =: e^{ia\beta\Phi} :$$

where the semicolon denotes normal ordering in terms of the massless scalar field modes. In a finite volume L, the spectrum of the free boson CFT is discrete and can be truncated to a finite subspace by introducing an upper cut-off e_{cut} in terms of the eigenvalue of the dilatation operator (which gives the energy in conformal units). Physical energies and volumes can be expressed in units of the soliton mass using the relation

$$\lambda = \frac{2\Gamma(\Delta)}{\pi\Gamma(1-\Delta)}\left(\frac{\sqrt{\pi}\Gamma\left(\frac{1}{2-2\Delta}\right)M}{2\Gamma\left(\frac{\Delta}{2-2\Delta}\right)}\right)^{2-2\Delta}, \quad \Delta = \frac{\beta^2}{8\pi} \tag{3.2}$$

so that the dimensionless volume variable and Hamiltonian can be defined as $l = ML$ and $h = H/M$, respectively. For more details on the sine-Gordon TCSA the interested reader is referred to [35].

The initial state corresponds to the ground state of the same Hamiltonian (3.1) with λ replaced by λ_0 corresponding to M_0. When considering the post-quench evolution in dimensionless volume $l = ML$, implementing the quench means using the ground state computed in the rescaled volume $l_0 = M_0 l/M$ [41].

The cut-off dependence of TCSA can be (partially) eliminated using renormalization group methods [42–44]. Here we used a modified version of the running coupling prescription in [45]. We can write an effective Hamiltonian in the form

$$H_{eff} = \int dx \frac{1}{2} : (\partial_t \Phi)^2 + (\partial_x \Phi)^2 : + \lambda_0 \mathbb{I} + \frac{\lambda_1}{2}\int dx\,(V_1 + V_{-1}) + \frac{\lambda_2}{2}\int dx\,(V_2 + V_{-2})$$

where we included counter terms generated at leading order according to the fusion rules $V_a V_b \sim V_{a+b}$. Introducing the dimensionless couplings

$$\tilde{\lambda}_a = \frac{\lambda_a L^{2-2h_a}}{(2\pi)^{1-2h_a}}, \quad h_a = \frac{a^2\beta^2}{8\pi}$$

the running couplings $\tilde{\lambda}_i$ are determined by the RG equations

$$\tilde{\lambda}_c(n) - \tilde{\lambda}_c(n-1) = \frac{1}{2n - d_0(l)} \sum_{a,b} \tilde{\lambda}_a(n)\tilde{\lambda}_b(n) C^c_{ab} \frac{n^{2h_{abc}-2}}{\Gamma(h_{abc})^2} \times (1 + O(1/n)) \tag{3.3}$$

where n is the cut-off expressed in conformal levels, C^c_{ab} is the operator product coefficient, $h_{abc} = h_a + h_b - h_c$ and $d_0(l)$ is the vacuum scaling function (cf. [45]). At the lowest order it is only necessary to run the couplings λ_0 and λ_2, from their starting values $\lambda_0 = 0$ and $\lambda_2 = 0$ at $n = \infty$.

The couplings must be run following (3.3) down from $n = \infty$ to the appropriate value of n_{cut} corresponding to the given cut-off e_{cut}. It must be taken into account that the $c = 1$ Hilbert space is spanned by Fock modules $\mathcal{F}_a$ created from the vacuum by V_a and the Hamiltonian is block-diagonal in terms of the Fock modules, symbolically:

$$\begin{pmatrix} H_0+\mathbb{I} & V_1 & V_2 & & & & \\ V_{-1} & H_0+\mathbb{I} & V_1 & V_2 & & & \\ V_{-2} & V_{-1} & H_0+\mathbb{I} & V_1 & V_2 & & \\ & \ddots & \ddots & \ddots & \ddots & \ddots & \\ & & V_{-2} & V_{-1} & H_0+\mathbb{I} & V_1 & V_2 \\ & & & V_{-2} & V_{-1} & H_0+\mathbb{I} & V_1 \\ & & & & V_{-2} & V_{-1} & H_0+\mathbb{I} \end{pmatrix}$$

and the eventual value of n_{cut} depends on the block one considers. Namely, when computing the coefficient of the block V_2 between $\mathcal{F}_a$ and $\mathcal{F}_{a+2}$, the intermediate states in the OPE $V_1 V_1 \sim V_2$ are from $\mathcal{F}_{a+1}$ which determines the level n_{cut} appropriate for the given block, and similarly for V_{-2} between $\mathcal{F}_a$ and $\mathcal{F}_{a-2}$ n_{cut} is fixed from $\mathcal{F}_{a-1}$. For the identity term between $\mathcal{F}_a$ and $\mathcal{F}_a$ there are two possible intermediate modules $\mathcal{F}_{a\pm 1}$, so the identity coupling must be split into two pieces $\lambda_{0\pm}$, each of them running down to the appropriate n_{cut} determined by the highest level in $\mathcal{F}_{a\pm 1}$.

The block-dependent running coupling corresponds to including a non-local counter term. The fact that such counter terms are necessary was noted in [46]; they account for $1/n$ corrections in the running coupling. In the sine-Gordon there is a large $1/n$ effect resulting from the fact that the cut-off level is heavily module dependent, ranging from e_{cut} in Fock module $\mathcal{F}_0$ to 0 for the Fock modules with the largest indices $\mathcal{F}_{\pm a_{max}}$. The consistency of this scheme was verified by numerically checking the cut-off dependence of the 15 lowest-lying levels in the TCSA spectrum, which proved to be negligible with this method.

3.2. The B_1–B_1 pair amplitude

Now we turn to numerical results for the B_1–B_1 pair amplitudes and compare them with the infinite volume prediction (2.6). The first task is to identify states corresponding to B_1–B_1 pairs in the numerical spectrum of the post-quench Hamiltonian. Solving the constrained Bethe–Yang system (2.13) one obtains the possible rapidities from which the energy levels can be computed. However, apart from the lowest lying levels, the identification is not feasible by merely comparing the TCSA and the Bethe–Yang energies due the density of the TCSA spectrum. This issue can be overcome by supplementing the energy selection procedure with a comparison of the finite volume form factors of the fields V_1 and V_2 obtained using the formalism developed in [40], to the TCSA matrix elements (for an exposition of how this works in sine-Gordon theory cf. [47]). As the form factors depend sensitively on the particle content of the state, the identification can unambiguously be performed.

Having identified the proper states in the set of numerical eigenstates the numerical overlaps can be obtained from their scalar product with the initial state, divided by the vacuum overlap to eliminate the normalization factor $\mathcal{N}$ in (2.2). As the TCSA matrix elements of the perturbing operator are real numbers, all the numerically computed eigenvectors are also real, corresponding to a specific convention for the phases of the post-quench eigenstates. Therefore the phase of the overlap function $K(\vartheta)$ is absent from the data, so after normalizing the TCSA overlap values with the inverse of (2.15) we compare their modulus to the value obtained from (2.6). This comparison is shown in Fig. 1 for a few of the quenches we considered; the conclusion is that it works well except in the low energy range, and that both the free particle and the Dirichlet parts of the analytic formula (2.6) are important.

Deviations in the low energy range can be attributed to two sources. First, the initial state is different from the free massive vacuum for which (2.6) (or more precisely, its sinh-Gordon counterpart (2.3)) was obtained. However, the difference is the presence of a relevant perturbation in the pre-quench Hamiltonian, which affects most the low-lying modes due to its relevance. Second, when modeling the finite size effects in Section 2 we used a formalism that neglects exponential corrections in the volume, which normally affect the lower lying states more. Unfortunately, it is not easy to separate these effects, and so we cannot say anything more definite about the low-energy behavior. However, the analytically continued solution (2.6) definitely provides a good description of the amplitudes in the mid-to-high energy range.

3.3. Amplitudes for 4 B_1 particles and factorization

Once the amplitude $K(\vartheta)$ is pinned down, all higher overlaps are determined by the exponential form of the state. This entails

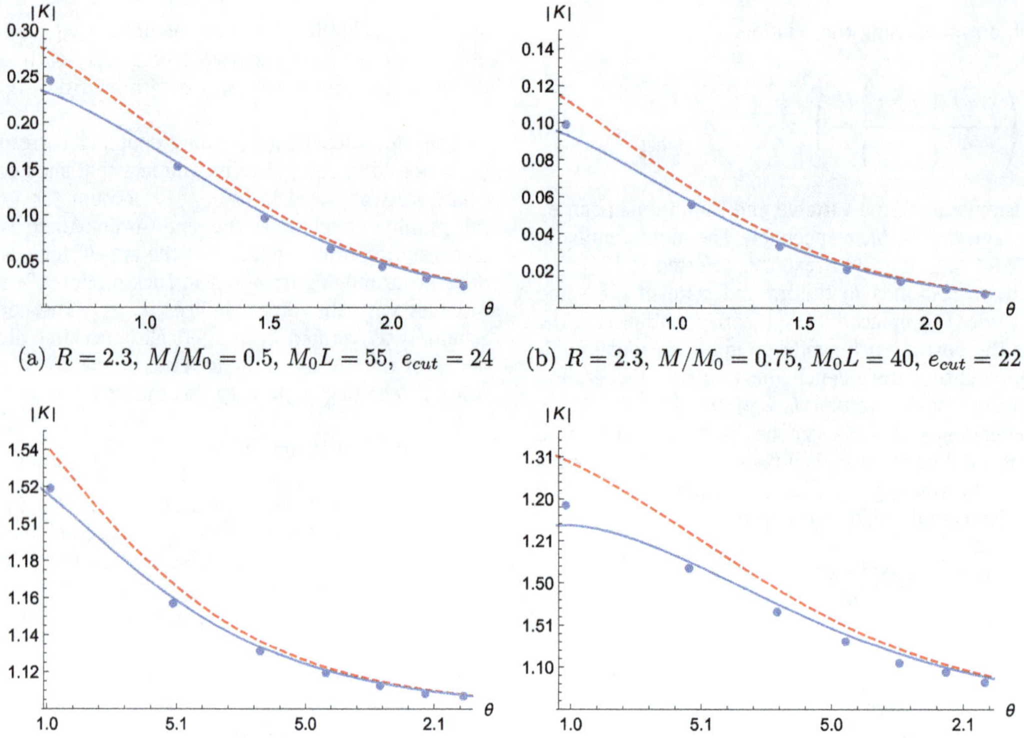

(a) $R = 2.3$, $M/M_0 = 0.5$, $M_0L = 55$, $e_{cut} = 24$ (b) $R = 2.3$, $M/M_0 = 0.75$, $M_0L = 40$, $e_{cut} = 22$

(c) $R = 2.3$, $M/M_0 = 1.5$, $M_0L = 22$, $e_{cut} = 24$ (d) $R = 2.0$, $M/M_0 = 0.5$, $M_0L = 50$, $e_{cut} = 24$

Fig. 1. The pair amplitude for some mass quenches in sine-Gordon theory. The sine-Gordon coupling β is parametrized as $\beta = \sqrt{4\pi}/R$. The blue (continuous) curves correspond to the sine-Gordon Ansatz (2.6), and the red (dashed) ones to the free theory solutions. (For interpretation of the references to color in this figure legend, the reader is referred to the web version of this article.)

Table 1
Overlaps for 4–B_1 paired states $|-\vartheta_1^*, \vartheta_1^*, -\vartheta_2^*, \vartheta_2^*\rangle$. The sine-Gordon coupling β is parametrized as $\beta = \sqrt{4\pi}/R$. To eliminate differences in phase conventions of energy eigenstates the modulus of the overlaps is reported.

Rapidities $\vartheta_1^*, \vartheta_2^*$	BY energy	TCSA energy	Normalized overlap	Factorized prediction
(a) $R = 2.3$, $M/M_0 = 0.5$, $M_0L = 55$, $e_{cut} = 24$				
{0.671828, 1.44047}	1.13089	1.13133	0.0255928	0.0244265
{0.651971, 1.72849}	1.34668	1.34742	0.0168602	0.0162507
{1.19428, 1.70726}	1.51794	1.51918	0.0108541	0.0108043
{0.642841, 1.95028}	1.56712	1.56853	0.0117727	0.0113951
{0.637471, 2.1315}	1.73764	1.79245	0.0083998	0.0083603
(b) $R = 2.0$, $M/M_0 = 0.5$, $M_0L = 50$, $e_{cut} = 24$				
{0.549607, 1.22608}	1.33758	1.33916	0.0296001	0.0278547
{0.524061, 1.51576}	1.56955	1.57217	0.0139780	0.0195048
{1.03577, 1.48932}	1.74274	1.74686	0.0142756	0.0149960
{0.512645, 1.73741}	1.80847	1.81308	0.0137404	0.0141252
{1.00938, 1.72497}	1.98019	1.98813	0.0130837	0.0108970
(c) $R = 2.3$, $M/M_0 = 0.75$, $M_0L = 40$, $e_{cut} = 22$				
{0.618879, 1.36023}	1.60354	1.60489	0.00454695	0.00344512
{0.599521, 1.64559}	1.89691	1.89943	0.00292898	0.00219769
{1.12225, 1.62472}	2.12314	2.12725	0.00176750	0.00132404
{0.590723, 1.866}	2.19785	2.20287	0.00203194	0.00150384
{1.10091, 1.85633}	2.42300	2.43139	0.00128401	0.00091002

the factorization property which states that states which do not have an exclusive pair structure in terms of particles have zero overlap, and for paired states the overlap is just equal to the product of individual pair state overlaps.

Another prediction from factorization is that the overlaps for paired 4–B_1 states is the product of pair overlaps. This is also consistent with the TCSA data as shown in Table 1. For the quenches in sub-tables (a) and (b), the overlaps are large enough so that one can observe a quantitative agreement between the predictions of (2.6) and the TCSA results. For the example in sub-table (c), the overlap is too small to be measured and the agreement is only qualitative.

The question is whether this constitutes a non-trivial test of overlap factorization? Note that when the quench is small factorization is expected to be valid to a very good approximation. A small quench means that the average energy density $\mathcal{E}$ after the quench satisfies

$$\mathcal{E} = \frac{1}{L}\left(\langle\Psi(0)|H|\Psi(0)\rangle - \langle 0|H|0\rangle\right) \ll m_1^2$$

with respect to the mass of the lightest particle m_1. In such a case the density of even the lightest pairs is so small that the average distance d between pairs is much larger than the correlation length m_1^{-1}. Since the interactions are suppressed by the distance as $e^{-m_1 d}$, the multi-pair amplitudes are expected to factorize irrespective of integrability when the quench is small.

We evaluated $\mathcal{E}$ for all the quenches for which we could produce reliable TCSA data and found that d was at least an order of magnitude larger than m_1^{-1}, therefore all observed deviations from factorization are expected to be TCSA related (either truncation errors or unmodeled finite size effects). Indeed, when testing the overlaps for non-paired 4-B_1 states, they proved to be an order of magnitude smaller than the overlaps for paired states, and were of the same order as the deviations between the prediction (2.6) and the measured two-particle overlap, which is consistent with factorization.

4. Conclusions

In this paper we studied mass quenches in the sine-Gordon integrable quantum field theory in the attractive regime, in particular, we numerically determined the two-particle overlaps for the B_1 breathers in the finite volume theory with a periodic boundary condition. Making use of the Ansatz (2.3) proposed in [31,32] for quenches from the free bosonic theory to the interacting sinh-Gordon theory and the well-known analytic continuation between the sine- and sinh-Gordon theories, the numerical overlaps were compared with the Ansatz. The numerical data points and the theoretical curve were found to match very well in the middle and high energy range, with some quantitative deviations in the low energy part which can be attributed to initial state interactions and finite size effects. These results also confirm the consistency of the derivation of the sinh-Gordon Ansatz, which relies on the assumption that the initial state contains only multiple particle states composed of pairs with opposite momenta, which lacks rigorous justification at this moment.

For the numerical determination of the overlaps the truncated conformal space approach was used. To improve upon the usual renormalization group treatment of TCSA [42–44], we added non-local counter terms that dominate the next order corrections in the inverse energy cut-off. Whereas in general, the construction of such non-local terms is difficult, in sine-Gordon TCSA it is easy to implement this correction and the accuracy of the numerical spectrum can be substantially improved.

As a closing remark, it has to be mentioned that the quantum sine-Gordon theory is a very interesting model in its own right, attracting a lot of attention due to its theoretical tractability and experimental relevance. Quenches in sine-Gordon theory have recently become realizable in experimental setups, describing the evolution of the relative phase of trapped and coupled condensates of cold atoms [48]. The knowledge of the pair amplitudes in (1.1) is crucial for the computation of steady state one- and two-point functions by currently available techniques [18,30], and the method presented in this paper provides a direct way to the numerical determination of the pair overlaps. Indeed, in addition to the B_1 overlaps presented here, our method can be used to extract pair amplitudes for higher breathers and soliton–antisoliton pairs. In this work we refrained from reporting the corresponding numerical data, since at present we have no theoretical description for them. The understanding of these overlaps, which is important for a full description of sine-Gordon quenches, is relegated to future works.

Acknowledgements

The authors are grateful to Márton Kormos and Tibor Rakovszky for their contributions in an early stage of this work and for useful discussions. This research was supported by the Momentum grant LP2012-50 of the Hungarian Academy of Sciences and by the K2016 grant no. 119204 of the research agency NKFIH.

References

[1] T. Kinoshita, T. Wenger, D.S. Weiss, Nature 440 (2006) 900–903.
[2] S. Trotzky, Y.-A. Chen, A. Flesch, I.P. McCulloch, U. Schollwöck, J. Eisert, I. Bloch, Nat. Phys. 8 (2012) 325–330, arXiv:1101.2659.
[3] M. Cheneau, P. Barmettler, D. Poletti, M. Endres, P. Schauss, T. Fukuhara, C. Gross, I. Bloch, C. Kollath, S. Kuhr, Nature 481 (2012) 484–487, arXiv:1111.0776.
[4] M. Gring, M. Kuhnert, T. Langen, T. Kitagawa, B. Rauer, M. Schreitl, I. Mazets, D.A. Smith, E. Demler, J. Schmiedmayer, Science 337 (2012) 1318–1322, arXiv:1112.0013.
[5] M. Rigol, V. Dunjko, V. Yurovsky, M. Olshanii, Phys. Rev. Lett. 98 (2007) 050405, arXiv:cond-mat/0604476.
[6] T. Langen, S. Erne, R. Geiger, B. Rauer, T. Schweigler, M. Kuhnert, W. Rohringer, I.E. Mazets, T. Gasenzer, J. Schmiedmayer, Science 348 (2015) 207–211, arXiv:1411.7185.
[7] B. Wouters, J. De Nardis, M. Brockmann, D. Fioretto, M. Rigol, J.-S. Caux, Phys. Rev. Lett. 113 (2014) 117202, arXiv:1405.0172.
[8] B. Pozsgay, M. Mestyán, M.A. Werner, M. Kormos, G. Zaránd, G. Takács, Phys. Rev. Lett. 113 (2014) 117203, arXiv:1405.2843.
[9] E. Ilievski, J. De Nardis, B. Wouters, J.-S. Caux, F.H.L. Essler, T. Prosen, Phys. Rev. Lett. 115 (2015) 157201, arXiv:1507.02993.
[10] E. Ilievski, M. Medenjak, T. Prosen, L. Zadnik, J. Stat. Mech. 1606 (2016) 064008, arXiv:1603.00440.
[11] J. Sirker, R.G. Pereira, I. Affleck, Phys. Rev. Lett. 103 (2009) 216602, arXiv:0906.1978.
[12] T. Prosen, Phys. Rev. Lett. 106 (2011) 217206, arXiv:1103.1350.
[13] M. Kormos, M. Collura, G. Takács, P. Calabrese, Nat. Phys. 13 (2017) 246–249, arXiv:1604.03571.
[14] P. Calabrese, J. Cardy, Phys. Rev. Lett. 96 (2006) 136801, arXiv:cond-mat/0601225;
P. Calabrese, J. Cardy, J. Stat. Mech. 0706 (2007) P06008, arXiv:0704.1880.
[15] B. Pozsgay, J. Stat. Mech. 1101 (2011) P01011, arXiv:1009.4662.
[16] D. Fioretto, G. Mussardo, New J. Phys. 12 (2010) 055015, arXiv:0911.3345.
[17] J.-S. Caux, F.H.L. Essler, Phys. Rev. Lett. 110 (2013) 257203, arXiv:1301.3806.
[18] M. Kormos, G. Zaránd, Phys. Rev. E 93 (2016) 062101, arXiv:1507.02708;
C.P. Moca, M. Kormos, G. Zaránd, Semi-semiclassical theory of quantum quenches in one dimensional systems, arXiv:1609.00974.
[19] Andrea De Luca, Gabriele Martelloni, Jacopo Viti, Phys. Rev. A 91 (2015) 021603, arXiv:1409.8482.
[20] J. De Nardis, L. Piroli, J.-S. Caux, J. Phys. A 48 (2015) 43FT01, arXiv:1505.03080.
[21] K.K. Kozlowski, B. Pozsgay, J. Stat. Mech. 1205 (2012) P05021, arXiv:1201.5884.
[22] J.D. Nardis, B. Wouters, M. Brockmann, J.-S. Caux, Phys. Rev. A 89 (2014) 033601, arXiv:1308.4310.
[23] B. Pozsgay, J. Stat. Mech. 1406 (2014) P06011, arXiv:1309.4593.
[24] M. Brockmann, J. De Nardis, B. Wouters, J.-S. Caux, J. Phys. A 47 (2014) 145003, arXiv:1401.2877.
[25] M. Brockmann, J. Stat. Mech. 1405 (2014) P05006, arXiv:1402.1471.
[26] M. Brockmann, J. De Nardis, B. Wouters, J.-S. Caux, J. Phys. A 47 (2014) 345003, arXiv:1403.7469.
[27] L. Piroli, P. Calabrese, J. Phys. A 47 (2014) 385003, arXiv:1407.2242.
[28] A.B. Zamolodchikov, Al.B. Zamolodchikov, Ann. Phys. 120 (1979) 253–291.
[29] D. Schuricht, F.H.L. Essler, J. Stat. Mech. 1204 (2012) P04017, arXiv:1203.5080.
[30] B. Bertini, D. Schuricht, F.H.L. Essler, J. Stat. Mech. 1410 (2014) P10035, arXiv:1405.4813.
[31] S. Sotiriadis, G. Takács, G. Mussardo, Phys. Lett. B 734 (2014) 52–57, arXiv:1311.4418.
[32] D.X. Horváth, S. Sotiriadis, G. Takács, Nucl. Phys. B 902 (2016) 508–547, arXiv:1510.01735.
[33] B. Bertini, L. Piroli, P. Calabrese, J. Stat. Mech. 1606 (2016) 063102, arXiv:1602.08269.
[34] V.P. Yurov, A.B. Zamolodchikov, Int. J. Mod. Phys. A 5 (1990) 3221–3246.
[35] G. Feverati, F. Ravanini, G. Takács, Phys. Lett. B 430 (1998) 264–273, arXiv:hep-th/9803104.
[36] S. Ghoshal, A. Zamolodchikov, Int. J. Mod. Phys. A 9 (1994) 3841–3886, arXiv:hep-th/9306002.
[37] J. Cardy, J. Stat. Mech. 1602 (2016) P023103, arXiv:1507.07266.
[38] S. Ghoshal, Int. J. Mod. Phys. A 9 (1994) 4801–4810, arXiv:hep-th/9310188.
[39] M. Kormos, B. Pozsgay, J. High Energy Phys. 1004 (2010) 112, arXiv:1002.2783.

[40] B. Pozsgay, G. Takács, Nucl. Phys. B 788 (2008) 167–208, arXiv:0706.1445.
[41] T. Rakovszky, M. Mestyán, M. Collura, M. Kormos, G. Takács, Nucl. Phys. B 911 (2016) 805–845, arXiv:1607.01068.
[42] G. Feverati, K. Graham, P.A. Pearce, G.Z. Tóth, G. Watts, A renormalization group for TCSA, arXiv:hep-th/0612203.
[43] R.M. Konik, Y. Adamov, Phys. Rev. Lett. 98 (2007) 147205, arXiv:cond-mat/0701605.
[44] P. Giokas, G. Watts, The renormalization group for the truncated conformal space approach on the cylinder, arXiv:1106.2448.
[45] M. Lencsés, G. Takács, J. High Energy Phys. 1509 (2015) 146, arXiv:1506.06477.
[46] M. Hogervorst, S. Rychkov, B.C. van Rees, Phys. Rev. D 91 (2015) 025005, arXiv:1409.1581.
[47] G. Fehér, G. Takács, Nucl. Phys. B 852 (2011) 441–467, arXiv:1106.1901.
[48] T. Schweigler, V. Kasper, S. Erne, B. Rauer, T. Langen, T. Gasenzer, J. Berges, J. Schmiedmayer, On solving the quantum many-body problem, arXiv:1505.03126.

Photon mass via current confinement

Vivek M. Vyas [a,*], Prasanta K. Panigrahi [b]

[a] Raman Research Institute, Sadashivnagar, Bengaluru, 560 080, India
[b] Department of Physical Sciences, Indian Institute of Science Education and Research Kolkata, Mohanpur, 741 246, India

ARTICLE INFO

Editor: B. Grinstein

Keywords:
Photon mass
Current confinement

ABSTRACT

A parity invariant theory, consisting of two massive Dirac fields, defined in three dimensional space–time, with the confinement of a certain current is studied. It is found that the electromagnetic field, when coupled minimally to these Dirac fields, becomes massive owing to the current confinement. It is seen that the origin of photon mass is not due to any kind of spontaneous symmetry breaking, but only due to current confinement.

1. Introduction

Field theories in three dimensional space time have been a subject of intense study since a couple of decades now. There are several reasons which make such field theories interesting. Firstly often the theories in lower dimensions are simpler than their higher dimensional counterparts. Secondly, it offers new structures like possibility of gauge invariant mass term for gauge field in the form of Chern–Simons term in the action. Interestingly, it was recently found that planar QED with a tree level Chern–Simons term admits a photon which is composite [1]. Theories with Chern–Simons term are found to play an important role in physics of quantum Hall effect and anyonic superconductors [2–6]. Models which exhibit dynamical mass generation and spontaneous chiral symmetry breaking have been constructed and extensively studied [7–11]. In recent years, with the discovery of graphene [12] and topological insulators [13] there has been a renewed interest in the study of lower dimensional field theories.

Colour confinement is one of the still not well understood aspect of QCD. One of the main hindrance is the fact that the low energy dynamics in such theory becomes non-perturbative, which makes dealing with them difficult. To circumvent this difficulty, there have been attempts to assume colour confinement from the beginning and work subsequently to see if one can get some idea about the dynamics of non-Abelian gauge fields [14–16]. In a remarkable paper by Srinivasan and Rajasekaran, it was shown that by assuming quark confinement it was possible to get QCD out of it [16]. Confinement has also been studied in theories defined in three dimensional space–time. It was shown by Polyakov that compact planar QED exhibits charge confinement [17]. While the case of non-compact QED was studied by Grignani et al. [18].

In this paper, it is shown that an assumption of confinement of a certain current gives rise to the photon mass. The theory consider here consists of two species of free Dirac fermions living on the plane, defined such that the theory is even under parity. These fermions are minimally coupled to the photon field. It is found that although the photons in the theory are massive, there is no spontaneous symmetry breaking. It is also shown that when such a theory is defined over a manifold with finite boundary, then there exist massless particles living on the boundary.

In the following section the model is introduced and its various features are discussed. Section 3 deals with the effective action of photon and its mass. Section 4 deals with the case when the theory lives on a manifold with a finite boundary, followed by a brief summary.

2. The model

The Lagrangian describing two species of massive Dirac fermions living in $2+1$ dimensional space–time reads:

$$\mathscr{L}_D = \bar{\psi}_+(i\gamma_+^\mu \partial_\mu - m)\psi_+ + \bar{\psi}_-(i\gamma_-^\mu \partial_\mu - m)\psi_-. \quad (1)$$

Here, gamma matrices are defined for ψ_+ field as $\gamma_+^0 = \sigma_3, \gamma_+^1 = i\sigma_1$ and $\gamma_+^2 = i\sigma_2$. Gamma matrices for ψ_- field are also same as ψ_+ except for γ^2, which is defined as $\gamma_+^2 = -\gamma_-^2$. This deliberate difference in choice of gamma matrices ensures that the

* Corresponding author.
E-mail addresses: vivekv@rri.res.in (V.M. Vyas), pprasanta@iiserkol.ac.in (P.K. Panigrahi).

Lagrangian is even under parity. It is known that, unlike four dimensional space-time, in the three dimensional world parity transformation is defined by reflecting one of the space axis, say Y axis, $(x, y) \to (x, -y)$. Instead of working with two spinor fields $\psi_\pm$, one can work in a reducible representation by defining $\Psi = (\psi_+, \psi_-)^T$, with $\beta = \gamma^0 = \mathbf{1} \otimes \sigma_3$, $\alpha_1 = \mathbf{1} \otimes \sigma_1$ and $\alpha_2 = \sigma_3 \otimes \sigma_2$, so that the above Lagrangian now reads:

$$\mathscr{L}_D = \bar{\Psi}(i\gamma^\mu \partial_\mu - m)\Psi,$$

where $\gamma_{1,2} = \beta\alpha_{1,2}$. Under parity operation, Ψ transforms as $\mathscr{P}\Psi(x, y, t)\mathscr{P}^{-1} = (\sigma_1 \otimes \mathbf{1})\Psi(x, -y, t)$. It can be checked that under this peculiar parity transformation, above Lagrangian remains even despite of having a mass term [19].

As it stands, apart from above mentioned parity transformation, the Lagrangian of this theory is invariant under two independent continuous rigid transformations:

$$\psi_+(r) \to e^{-i\theta}\psi_+(r),\ \psi_-(r) \to e^{-i\theta}\psi_-(r); \tag{2}$$

$$\psi_+(r) \to e^{-i\lambda}\psi_+(r),\ \psi_-(r) \to e^{i\lambda}\psi_-(r). \tag{3}$$

Here θ and λ are continuous real parameters. These being continuous symmetry operations, give rise to conserved currents as per the Noether theorem:

$$\partial_\mu(j_+^\mu + j_-^\mu) = 0 \text{ and } \partial_\mu(j_+^\mu - j_-^\mu) = 0,$$

where $j^\mu(r) = \bar{\psi}(r)\gamma^\mu\psi(r)$. It turns out that under parity, current $J^\mu = j_+^\mu + j_-^\mu = \bar{\Psi}\gamma^\mu\Psi$ transforms as a vector[1]: $\mathscr{P}J^\mu(x, y, t)\mathscr{P}^{-1} = \Lambda^\mu{}_\nu J^\nu(x, -y, t)$, whereas $\tilde{J}^\mu = j_+^\mu - j_-^\mu = -i\bar{\Psi}\gamma^\mu(\sigma_3 \otimes \sigma_3)\Psi$, transforms as a pseudovector: $\mathscr{P}\tilde{J}^\mu(x, y, t)\mathscr{P}^{-1} = \tilde{J}^\mu(x, -y, t)$. Since the physical photon field transforms as a vector under parity operation: $\mathscr{P}A^\mu(x, y, t)\mathscr{P}^{-1} = \Lambda^\mu{}_\nu A^\nu(x, -y, t)$, its coupling with the current $j_+^\mu + j_-^\mu$ preserves parity while making the symmetry transformation (2) local.

In this paper, we are interested in looking at the physical consequences if the current $j_+^\mu - j_-^\mu$ is confined. As pointed out by Kugo and Ojima in the context of QCD, and further discussed at length in Ref. [20], that the statement of colour charge confinement can be accurately stated as the absence of charge bearing states in the physical sector of the Hilbert space: $Q_{colour}|phys\rangle = 0$. In what follows, we shall work with a stronger condition than the Kugo-Ojima condition, and demand that the physical space of the theory described by Lagrangian (1) should not have any states which carry $(j_+^\mu - j_-^\mu)$ current, that is: $(j_+^\mu - j_-^\mu)|phys\rangle = 0$.[2] This shall be referred to as *current confinement condition* henceforth. Since we are demanding a priori that this current confinement condition should hold, it ought to be understood as a constraint. There exists a well known powerful technique to implement such a constraint using what is called the Lagrange multiplier (auxiliary) field [21]. One postulates the existence of a Lagrange multiplier field which is such that its only appearance in the action is via its coupling to the constraint condition. Thus the equation of motion corresponding to this field, obtained by demanding that the functional variation of the action with respect to this field be zero, is simply the constraint condition. It is worth pointing out that such Lagrange multiplier fields have no dynamics of their own, in the sense that there are no terms in the action comprising of spatial or temporal derivatives of these fields to begin with, and their sole purpose of existence is to ensure implementation of the constraint.

Thus by enlarging the degree of freedom in the theory by an additional field, one ensures that the constraint condition gets neatly embedded into the action, and hence into the dynamics of the theory. In our case the Lagrange multiplier Bose field is a_μ, which is meant to implement the constraint $(j_+^\mu - j_-^\mu)$, will only couple to it so that the Lagrangian (1) gets an additional term:

$$\mathscr{L} = \bar{\psi}_+(i\gamma_+^\mu\partial_\mu - m)\psi_+ + \bar{\psi}_-(i\gamma_-^\mu\partial_\mu - m)\psi_- + a_\mu\left(j_+^\mu - j_-^\mu\right). \tag{4}$$

Note that the equation of motion for a_μ field: $\frac{\delta S}{\delta a_\mu} = 0$, gives the constraint $j_+^\mu - j_-^\mu = 0$. In order to preserve parity, a_μ field has to be a pseudovector owing to its coupling with pseudovector current. Thus a_μ can in general be written as curl of some vector field χ: $a_\mu = \epsilon_{\mu\nu\lambda}\partial^\nu\chi^\lambda$, and can not have a contribution that can be written as a gradient of some scalar field. This asserts that a_μ can not be a gauge field, since a gauge field under a gauge transformation transforms as a vector $\partial_\mu\Lambda$, which is not consistent with the pseudovector nature of a_μ. Further note that since a_μ is curl of χ_μ, it immediately follows that its divergence vanishes: $\partial_\mu a^\mu = 0$.

In functional integral formulation of quantum field theory, generating functional is an object of central importance, which for this theory reads[3]:

$$Z[\eta_\pm, \bar{\eta}_\pm] = N\int \mathscr{D}[\bar{\psi}_\pm, \psi_\pm, a_\mu]\, e^{iS},$$

$$\text{where } S = \int d^3x\left(\mathscr{L} + \bar{\eta}_\pm\psi_\pm + \bar{\psi}_\pm\eta_\pm\right).$$

Here η and $\bar{\eta}$ are external sources which are coupled to Fermi fields $\bar{\psi}$ and ψ respectively.

Before we proceed with the details of the quantum theory, it is worth pointing out that if one functionally integrates a_μ in the above generating functional, one immediately obtains the current confinement condition $\delta(j_+^\mu - j_-^\mu)$, since a_μ appears linearly in the action. This clearly shows that in the quantum theory as well, the Lagrange multiplier field a_μ is properly implementing the current confinement constraint.

Since the Lagrangian (4) of the theory is invariant under transformation (3), requirement that the generating functional of the theory should also be invariant under (3), that is $\delta Z = 0$, is not unreasonable. Interestingly, it will be seen that this will give rise to Ward–Takahashi identities amongst various n-point functions in this theory and lead to non-trivial consequences. Demanding that Z be invariant under infinitesimal version of transformation (3) means $\delta Z = 0$, which can be written as:

$$\int \mathscr{D}[\bar{\psi}_\pm, \psi_\pm, a_\mu]\,(\delta S)\, e^{iS} = 0.$$

This can further be simplified to read:

$$\int \mathscr{D}[\bar{\psi}_\pm, \psi_\pm, a_\mu]\left(\mp\bar{\eta}_\pm(x)\psi_\pm(x) \pm \bar{\psi}_\pm(x)\eta_\pm(x)\right) e^{iS} = 0. \tag{5}$$

In terms of the generating functional of connected diagrams $W[\bar{\eta}_\pm, \eta_\pm] = -i\ln Z[\bar{\eta}_\pm, \eta_\pm]$, equation (5) becomes:

$$\bar{\eta}_+(x)\frac{\delta W}{\delta\bar{\eta}_+(x)} - \bar{\eta}_-(x)\frac{\delta W}{\delta\bar{\eta}_-(x)} - \eta_+(x)\frac{\delta W}{\delta\eta_+(x)} + \eta_-(x)\frac{\delta W}{\delta\eta_-(x)} = 0. \tag{6}$$

[1] Λ is diagonal matrix $\Lambda = \mathrm{diag}(1, 1, -1)$.

[2] The physical space here stands for the set of states in the vector space of the theory, which do not have negative norm [21]. In case when the negative normed states are altogether absent, then the condition $(j_+^\mu - j_-^\mu)|phys\rangle = 0$ holds for the whole of Hilbert space and hence becomes an operator condition $(j_+^\mu - j_-^\mu) = 0$.

[3] Since the current in this theory couples directly to a_μ, it is treated as a dynamical variable instead of χ_μ. Such a treatment is advocated by Hagen in Ref. [22].

It is often convenient to work with the effective action $\Gamma[\bar{\psi}_{\pm}, \psi_{\pm}]$ which is defined to be Legendre transform of $W[\bar{\eta}_{\pm}, \eta_{\pm}]$: $W[\bar{\eta}_{\pm}, \eta_{\pm}] = \Gamma[\bar{\psi}_{\pm}, \psi_{\pm}] + \int d^3x\,(\bar{\eta}_{\pm}\psi_{\pm} + \bar{\psi}_{\pm}\eta_{\pm})$, so that equation (6) reads:

$$\frac{\delta\Gamma}{\delta\psi_+(x)}\psi_+(x) - \frac{\delta\Gamma}{\delta\psi_-(x)}\psi_-(x) - \frac{\delta\Gamma}{\delta\bar{\psi}_+(x)}\bar{\psi}_+(x) + \frac{\delta\Gamma}{\delta\bar{\psi}_-(x)}\bar{\psi}_-(x) = 0. \tag{7}$$

This is the master equation from which one can get Ward–Takahashi identities connecting various vertex functions, by taking appropriate derivatives. The two-point function for $+$ species of fermions $iS_F(x-y) = \langle T\left(\psi_+(x)\bar{\psi}_+(y)\right)\rangle$, in terms of Γ is given by:

$$S_F^{-1}(y-x) = \left.\frac{\delta^2\Gamma}{\delta\psi(x)\delta\bar{\psi}(y)}\right|_{\psi=\bar{\psi}=0.}$$

Taking functional derivative of the master equation (7), once by $\bar{\psi}_+(y)$ followed by once with $\psi_+(z)$, one obtains following Ward–Takahashi identity for fermion Greens function:

$$\delta(x-z)S_F^{-1}(y-x) = \delta(x-y)S_F^{-1}(x-z). \tag{8}$$

This implies that $S_F^{-1}(y-x) = \delta(x-y)\int d^3z\, S_F^{-1}(x-z)$, whose only solution is $S_F^{-1}(y-x) \propto \delta(y-x)$. Above identity is very powerful, since it has allowed for an exact determination of propagator in this interacting theory. Exactly similar identity would also hold for propagator of $-$ species of fermions. It is worth mentioning, that this model is one of the rare cases where full propagator of this theory is known without any approximation. Presence of a physically observable particle in a theory, manifests as poles of propagator in momentum space. In our case, as is clearly evident, the propagator is regular everywhere in momentum space, which means that the Dirac fermion in our theory is not a propagating mode. This is particularly surprising since we started with a free Dirac theory with a constraint condition on currents, and it appears that condition is severe enough to not allow free fermion propagation.

In the absence of Dirac fermions, it is a natural to inquire about the particle excitations in this theory. In order to answer this question, it is instructive to study the four-point function in this theory. Apart from a trivial non-propagating solution discussed above, assuming validity of translational invariance, it can be shown that the Ward–Takahashi identity for four point function admits a solution of the kind: $\langle T\left(\psi_+(x_1)\psi_+(x_2)\bar{\psi}_+(y_1)\bar{\psi}_+(y_2)\right)\rangle \propto \delta(x_1-y_1)\delta(x_2-y_2)f(x_1-x_2)$, where f is some function of (x_1-x_2). This means that this Ward–Takahashi identity allows for propagation of composite operator $\psi(x)\bar{\psi}(y)|_{x=y}$, which describes charge neutral excitations consisting of fermion–antifermion bound states. It is worth mentioning, that the absence of fermions as elementary excitations and occurrence of bound states in a constrained theory like above, also appeared in a model of colour confinement proposed by Rajasekaran and Srinivasan [16]. Interestingly, they showed that quarks and gluons (which appeared as bound states) did not propagate and were confined, whereas mesons (colour neutral bound states of quarks) were propagating excitation in their model.

3. Electromagnetic response

In this section we focus our attention on the electromagnetic response of the theory. Lagrangian (4) with the minimal coupling of fermion fields to the photon field is given by:

$$\mathscr{L} = \bar{\psi}_+(i\partial\!\!\!/_+ - m + a\!\!\!/ + A\!\!\!/)\psi_+ + \bar{\psi}_-(i\partial\!\!\!/_- - m - a\!\!\!/ + A\!\!\!/)\psi_- - \frac{1}{4}F_{\mu\nu}F^{\mu\nu}. \tag{9}$$

In order to find the response of the theory under the influence of photon field, it is imperative that the photon field be treated as an external field. Terms involving ghosts and gauge fixing, which are absent in the above Lagrangian, have been incorporated by appropriate modification of measure $\mathscr{D}[A_\mu]$. In order to take into account effects due to quantum corrections, which arise from virtual fermion loop excitation, one needs to find out the effective action by integrating out fermion field. The effective action up to quadratic terms in fields, obtained using derivative expansion of fermion determinant [23,24] reads:

$$\mathscr{L}_{eff} = -\frac{1}{12\pi|m|}f_{\mu\nu}f^{\mu\nu} - \frac{m}{2\pi|m|}\epsilon^{\mu\nu\rho}\left(A_\mu f_{\nu\rho} + a_\mu F_{\nu\rho}\right) - \frac{1}{4}F_{\mu\nu}F^{\mu\nu}. \tag{10}$$

It can be shown that, in the limit of large m this approximation is valid and higher order terms can be neglected. As is evident, a_μ did not have a kinetic term to start with, but fermion loops have made it dynamical. Further, A_μ and a_μ fields are coupled by a mixed Chern–Simons term, which has a topological nature [25–29]. In other words, this implies that a_μ field has now become electromagnetically charged due to presence of virtual fermion cloud around it, with current being given by $J^\mu = \epsilon^{\mu\nu\rho}\partial_\nu a_\rho$, which is conserved off shell by construction. It is interesting to note that in this effective Lagrangian, the pseudovector field a_μ is coupled to dual of $F_{\mu\nu}$ so that the effective action is even under parity.

We started with a theory consisting of two species of massive Dirac fermions, with the assumption of current confinement. The current confinement being an independent condition, in the sense that it is not a consequence of the equations of motion of the theory, was understood as a constraint. We employed a judicious way of implementing this constraint using the Lagrange multiplier field a_μ, which essentially does book keeping of the constraint. Even though constraint condition is stated in terms of fermion fields, it can not be viewed in isolation since the Dirac fields are coupled to the photon field. Thus even after integrating out the fermion fields, the effects of current confinement condition survive and manifest as coupling between a_μ and A_μ in (10). Since the role of a_μ is only to ensure implementation of the constraint, it is imperative to integrate it out to see the effect of current confinement on the dynamics of the photon field. On integrating out a_μ field from Lagrangian (10), one arrives at an effective action for electromagnetic field:

$$\mathscr{L}_{eff} = -\frac{1}{4}F_{\mu\nu}F^{\mu\nu} + \frac{3|m|}{\pi}F_{\mu\nu}\frac{1}{\partial^2}F^{\mu\nu}. \tag{11}$$

As is evident, interaction with a_μ field has induced gauge invariant mass $M = \frac{12|m|}{\pi}$ for the physical electromagnetic field [30]. One may wonder that the differential operator $\frac{1}{\partial^2}$ in the Lagrangian may compromise locality and causality. However it has been shown in Ref. [30] that both of these features are intact. The action of $\frac{1}{\partial^2}$ on a function becomes transparent by going over to Fourier space,[4] that is:

[4] Alternatively, one may also consider the action of this differential operator in terms of convolution by a suitable Greens function $G(x)$, subject to the appropriate boundary conditions of the problem. The Greens function $G(x)$ is defined to solve: $\partial^2 G(x) = \delta(x)$. With the knowledge of boundary conditions, formally this can be inverted: $G(x) = \frac{1}{\partial_x^2}\delta(x)$, so that $\frac{1}{\partial_x^2}f(x) = \int dy\,\frac{1}{\partial_x^2}\delta(x-y)f(y) = \int dy\,G(x-y)f(y)$.

$$\frac{1}{\partial^2} f(x) = \int \frac{d^3p}{(2\pi)^3} \frac{-1}{p^2} e^{-ipx} \tilde{f}(p), \tag{12}$$

where $\tilde{f}(p)$ is Fourier transform of $f(x)$. Occurrence of such terms in action have been long known, for example it is known to appear in the effective action of Schwinger model when one integrates out fermions, as also in the action of two dimensional gravity theory studied by Polyakov [30].

It is known that in the planar world there exists Chern–Simons Lagrangian:

$$\mathcal{L} = -\frac{1}{4} F_{\mu\nu} F^{\mu\nu} + \frac{M}{2} \epsilon_{\mu\nu\lambda} A^{\mu} \partial^{\nu} A^{\lambda} \tag{13}$$

which is also gauge invariant and describes massive photon field [26]. However unlike Lagrangian (11) the mass term in this theory has a topological origin, and the theory evidently violates parity.

It may be tempting to believe that the photon mass term in the theory occurs because of some kind of spontaneous symmetry breaking and associated Anderson–Higgs mechanism. However, note that the theory is invariant under two kinds of continuous rigid transformations, which are generated by two conserved charges $Q_{1,2} = \int d^2x : j^0_+ \pm j^0_- :$. Vacuum expectation value of conserved current $\langle vac | j^{\mu}_{\pm}(x) | vac \rangle$ can be written as [31]:

$$\begin{aligned} \langle j^{\mu}_{\pm} \rangle &= \lim_{x \to y} \langle \bar{\psi}_{a,\pm}(y) \gamma^{\mu}_{ab,\pm} \psi_{b,\pm}(x) \rangle \\ &= -i \lim_{x \to y} \mathrm{tr}\, \gamma^{\mu}_{\pm} S_{F,\pm}(x-y), \end{aligned} \tag{14}$$

where *tr* stands for trace over Dirac induces. Since $S_{F,\pm}(x-y) = \text{const.} \times \delta(x-y)$ in this theory, one finds that $\langle vac | j^{\mu}_{\pm} | vac \rangle = 0$. This straightforwardly implies that, the charges $Q_{1,2}$ annihilate the vacuum $Q_{1,2} | vac \rangle = 0$ in this theory. This emphatically shows that there is no spontaneous symmetry breaking whatsoever, and that the current confinement is responsible for the photon mass.

4. Boundary theory

In above discussions we have assumed that the theory lives on two dimensional manifold whose boundary lies at the infinity, and further all the fields in discussion were assumed to decay sufficiently quickly so that surface terms in the action contribute negligibly. In this section we shall consider the case when the boundary is finite, in which case it may not be possible to ignore contribution due to the surface terms.

As noted above, low energy effective action describing the dynamics of low energy electronic excitation, subject to the current constraint, coupled to electromagnetic field is given by (10):

$$\begin{aligned} \mathcal{L} = &-\frac{1}{4} F_{\mu\nu} F^{\mu\nu} - \frac{1}{12\pi |m|} f_{\mu\nu} f^{\mu\nu} \\ &- \frac{m}{2\pi |m|} \epsilon^{\mu\nu\rho} \left(A_{\mu} f_{\nu\rho} + a_{\mu} F_{\nu\rho} \right). \end{aligned}$$

Note that the last mixed Chern–Simons term is not invariant under local gauge transformation: $A_{\mu} \to A_{\mu} + \partial_{\mu} \Lambda$, where Λ is some regular function of x. As a result, the change in action is given by:

$$\delta S_{CS} = \left(\frac{sgn(m)}{2\pi} \right) \int d^3x \, \epsilon^{\mu\nu\rho} \partial_{\mu} \left(\Lambda f_{\nu\rho} \right).$$

Above volume integral can be converted to a surface integral, defined on closed boundary of the manifold, to give an action:

$$\delta S_{CS} = \left(\frac{sgn(m)}{2\pi} \right) \int_B d^2x \, \epsilon^{\mu\nu} \Lambda f_{\mu\nu}.$$

This term, as it stands, is not gauge invariant, and is defined on the boundary, which encloses the bulk. Gauge invariance of any given theory, is a statement that, the theory is constrained, and possesses redundant variables. We observe that, our theory to start with was gauge invariant at classical level. One loop corrections arising out of fermion loops, generate Chern–Simons term, which exhibits gauge noninvariance. Because, our theory to start with was gauge invariant, and hence constrained, consistency demands that quantum(corrected) theory should also respect the imposed constraints, and hence should be gauge invariant. The occurrence of above gauge noninvariance, simply implies that one is only looking at one particular sector of the theory, and there exists other dynamical sector, whose dynamics is such that it compensates with the one above to render the total theory gauge invariance. Following Ref. [5], we demand that there must exist a corresponding gauge theory living on the boundary, defined such that it contributes a gauge noninvariant term of exactly opposite character and hence cancels the one written above. The simplest term, living on boundary, that obeys above condition is:

$$S_B = \frac{-sgn(m)}{2\pi} \int_B d^2x \theta \epsilon^{\mu\nu} f_{\mu\nu},$$

where $\theta(x,t)$ is a Bose field, which transforms as $\theta \to \theta + \Lambda$ under a gauge transformation. In general, this scalar field would be dynamical, and with a gauge invariant kinetic term, the boundary action reads:

$$S_B = \int_B d^2x \left[c \left(\partial_{\mu}\theta - A_{\mu} \right)^2 - \frac{sgn(m)}{2\pi} \theta \epsilon^{\mu\nu} f_{\mu\nu} \right].$$

Owing to its peculiar transformation gauge transformation property, a quadratic mass term for θ is not gauge invariant. Hence, in a gauge theory framework like this, θ field remains massless. Since the coupling of θ field with a field is anomalous, it turns out that the chiral current in this quantum theory is no longer conserved.

5. Conclusion

In this paper, we have shown that, in a parity invariant theory of two free Dirac fields living on a plane, confinement of current $j^{\mu}_{+} - j^{\mu}_{-}$ gives rise to the photon mass. To the best of our knowledge this is the only model in which current confinement paves the way to the gauge boson mass. A unique feature of this mechanism of photon mass generation is that there is no kind of spontaneous symmetry breaking involved. It is found that in case when the theory is defined on a manifold with a boundary, consistency implies the existence of massless particles on the boundary. It would be interesting to investigate whether it is possible to have a composite photon from confinement, as was seen in planar QED with a tree level Chern–Simons term [1]. Further, it is believed that the connection between gauge boson mass, compositeness and current confinement, as seen in this theory, could have some implications in the theory of strong interactions – QCD. Work along these lines is in progress and shall be published in due course.

Acknowledgements

The authors would like to thank the anonymous referee for his/her constructive comments which has helped the authors again a better understanding of this theory. The authors also thank Prof. V. Srinivasan for several useful discussions. VMV thanks Dr. R. P. Singh and Physical Research Laboratory, Ahmedabad for their hospitality while this work was being done.

References

[1] K. Abhinav, P. Panigrahi, Quantum and thermal fluctuations and pair-breaking in planar QED, J. High Energy Phys. 2016 (3) (2016) 1.

[2] N. Nagaosa, Quantum Field Theory in Condensed Matter Physics, Springer Verlag, 1999.

[3] P.K. Panigrahi, R. Ray, B. Sakita, Effective Lagrangian for a system of nonrelativistic fermions in $2+1$ dimensions coupled to an electromagnetic field: application to anyonic superconductors, Phys. Rev. B 42 (7) (1990) 4036.

[4] A. Gangopadhyaya, P.K. Panigrahi, Anyonic superconductivity in a modified large-U Hubbard model, Phys. Rev. B 44 (17) (1991) 9749.

[5] F. Wilczek, Fractional Statistics and Anyon Superconductivity, World Scientific Publ. Co., Inc., 1990.

[6] R. Laughlin, The relationship between high-temperature superconductivity and the fractional quantum Hall effect, Science 242 (4878) (1988) 525.

[7] T. Appelquist, M. Bowick, D. Karabali, L. Wijewardhana, Spontaneous chiral-symmetry breaking in three-dimensional QED, Phys. Rev. D 33 (12) (1986) 3704.

[8] T. Appelquist, M. Bowick, D. Karabali, L. Wijewardhana, Spontaneous breaking of parity in $(2+1)$-dimensional QED, Phys. Rev. D 33 (12) (1986) 3774–3776.

[9] G. Semenoff, L. Wijewardhana, Dynamical mass generation in 3d four-fermion theory, Phys. Rev. Lett. 63 (24) (1989) 2633–2636.

[10] T. Appelquist, D. Nash, L. Wijewardhana, Critical behavior in $(2+1)$-dimensional QED, Phys. Rev. Lett. 60 (25) (1988) 2575–2578.

[11] T. Appelquist, M. Bowick, E. Cohler, L. Wijewardhana, Chiral-symmetry breaking in $2+1$ dimensions, Phys. Rev. Lett. 55 (17) (1985) 1715–1718.

[12] K. Novoselov, A. Geim, S. Morozov, D. Jiang, Y. Zhang, S. Dubonos, I. Grigorieva, A. Firsov, Electric field effect in atomically thin carbon films, Science 306 (5696) (2004) 666.

[13] X.-L. Qi, S.-C. Zhang, Topological insulators and superconductors, Rev. Mod. Phys. 83 (2011) 1057–1110.

[14] T. Eguchi, New approach to collective phenomena in superconductivity models, Phys. Rev. D 14 (10) (1976) 2755.

[15] K. Kikkawa, Quantum corrections in superconductor models, Prog. Theor. Phys. 56 (3) (1976) 947.

[16] G. Rajasekaran, V. Srinivasan, Generation of gluons from quark confinement, Pramana 10 (1) (1978) 33.

[17] A. Polyakov, Quark confinement and topology of gauge theories, Nucl. Phys. B 120 (3) (1977) 429.

[18] G. Grignani, G. Semenoff, P. Sodano, Confinement-deconfinement transition in three-dimensional QED, Phys. Rev. D 53 (1996) 7157–7161.

[19] C.R. Hagen, Parity conservation in Chern–Simons theories and the anyon interpretation, Phys. Rev. Lett. 68 (1992) 3821–3825.

[20] N. Nakanishi, I. Ojima, Covariant Operator Formalism of Gauge Theories and Quantum Gravity, World Scientific, 1990.

[21] M. Henneaux, C. Teitelboim, Quantization of Gauge Systems, Princeton University Press, 1992.

[22] C. Hagen, Action-principle quantization of the antisymmetric tensor field, Phys. Rev. D 19 (8) (1979) 2367.

[23] K.S. Babu, A. Das, P. Panigrahi, Derivative expansion and the induced Chern–Simons term at finite temperature in $2+1$ dimensions, Phys. Rev. D 36 (12) (1987) 3725.

[24] A. Das, S. Panda, Temperature-dependent anomalous statistics, J. Phys. A 25 (5) (1992) L245.

[25] S. Deser, R. Jackiw, S. Templeton, Three-dimensional massive gauge theories, Phys. Rev. Lett. 48 (15) (1982) 975.

[26] S. Deser, R. Jackiw, S. Templeton, Topologically massive gauge theories, Ann. Phys. 140 (2) (1982) 372.

[27] R. Jackiw, S. Templeton, How super-renormalizable interactions cure their infrared divergences, Phys. Rev. D 23 (10) (1981) 2291.

[28] J.F. Schonfeld, A mass term for three-dimensional gauge fields, Nucl. Phys. B 185 (1) (1981) 157.

[29] W. Siegel, Unextended superfields in extended supersymmetry, Nucl. Phys. B 156 (1) (1979) 135.

[30] V.M. Vyas, V. Srinivasan, A gauge theory of massive spin one particles, Int. J. Theor. Phys. 55 (5) (2016) 2610–2620.

[31] K. Fujikawa, H. Suzuki, Path Integrals and Quantum Anomalies, Clarendon Press, 2004.

Quantum non-equilibrium effects in rigidly-rotating thermal states

Victor E. Ambruş

Department of Physics, West University of Timişoara, Bd. Vasile Pârvan 4, Timişoara, 300223, Romania

ARTICLE INFO

Editor: M. Cvetič

Keywords:
Rigidly-rotating thermal states
Landau frame
Beta frame
Dirac field
Klein–Gordon field
Dirichlet boundary conditions

ABSTRACT

Based on known analytic results, the thermal expectation value of the stress-energy tensor (SET) operator for the massless Dirac field is analysed from a hydrodynamic perspective. Key to this analysis is the Landau decomposition of the SET, with the aid of which we find terms which are not present in the ideal SET predicted by kinetic theory. Moreover, the quantum corrections become dominant in the vicinity of the speed of light surface (SOL). While rigidly-rotating thermal states cannot be constructed for the Klein–Gordon field, we perform a similar analysis at the level of quantum corrections previously reported in the literature and we show that the Landau frame is well-defined only when the system is enclosed inside a boundary located inside or on the SOL. We discuss the relevance of these results for accretion disks around rapidly-rotating pulsars.

1. Introduction

In relativistic fluid dynamics, global thermal equilibrium can be attained if the product βu^μ between the inverse local temperature β and the four-velocity u^μ of the flow satisfies the Killing equation [1–5]. A special property of thermal equilibrium is that the stress-energy tensor (SET) $T^{\mu\nu}_{\rm eq} = (E+P)u^\mu u^\nu + P g^{\mu\nu}$ corresponds to that of an ideal fluid of energy density E and pressure P [2,6–8].[1] In this letter, we will show that a quantum field theory (QFT) computation of the SET for rigidly-rotating thermal states (RRTS) contains non-ideal terms, as well as corrections to E which become important near the speed of light surface (SOL). We discuss the relevance of these corrections in the context of an astrophysical application.

2. Kinetic theory analysis

In space–times with axial symmetry, RRTS in thermal equilibrium can be described using the Killing vector corresponding to rotations about the z axis, i.e., $\beta u = \beta_0(\partial_t + \Omega\partial_\varphi)$, where Ω is the angular velocity of the rotating state [7]. On Minkowski space, the particle four-flow $N^\mu_{\rm eq}$ and stress-energy tensor $T^{\mu\nu}_{\rm eq}$ corresponding to RRTS are given by:

$$N^\mu_{\rm eq} = n u^\mu, \qquad T^{\mu\nu}_{\rm eq} = (E+P)u^\mu u^\nu + P g^{\mu\nu}, \tag{1}$$

while β and $u = u^\mu\partial_\mu$ are given by:

$$\beta = \gamma^{-1}\beta_0, \qquad u = \gamma(\partial_t + \Omega\partial_\varphi), \tag{2}$$

where γ is the Lorentz factor of a co-rotating observer at distance ρ from the z axis:

$$\gamma = \left(1-\rho^2\Omega^2\right)^{-1/2}. \tag{3}$$

The Killing vector βu becomes null on the SOL, where $\rho\Omega \to 1$ and co-rotating observers travel at the speed of light. From Eq. (2), it can be seen that the temperature β^{-1} diverges as the SOL is approached. The energy density E for massless particles obeying Fermi–Dirac (F–D) and Bose–Einstein (B–E) statistics is given by [6]:

$$E_{\rm F-D} = \frac{7\pi^2\gamma^4}{60\beta_0^4}, \qquad E_{\rm B-E} = \frac{\pi^2\gamma^4}{30\beta_0^4}, \tag{4}$$

while $P = E/3$. Since E and P diverge as inverse powers of the distance to the SOL, RRTS are well-defined only up to the SOL. While such divergent states clearly cannot occur in nature, rigid rotation can be induced in astrophysical systems, such as accretion disks around rapidly-rotating neutron stars or magnetars, where the intense magnetic field can lock charged particles into rigid rotation.[2]

E-mail address: victor.ambrus@e-uvt.ro.

[1] We use Planck units with $c = \hbar = k_B = 1$, while the metric signature is $(-,+,+,+)$.

[2] In such systems, various mechanisms prevent the violation of special relativity [9].

We investigate the role of quantum corrections in such systems in Sec. 6.

3. Stress-energy tensor decompositions

Before discussing the quantum analogue of Eqs. (4), the tools necessary to analyse the SET in out of equilibrium states must be introduced. The main difficulty comes due to the equivalence between mass and energy in special relativity, which makes the distinction between the velocity u^μ and the heat flux q^μ ambiguous. For a general (time-like) choice of u^μ, N^μ can be decomposed as [10]:

$$N^\mu = n u^\mu + V^\mu, \tag{5}$$

where $n = -u_\mu N^\mu$ and the flow of particles in the local rest frame (LRF) V^μ is given by:

$$V^\mu = \Delta^\mu{}_\nu N^\nu \tag{6}$$

In the above, $\Delta^{\mu\nu} = u^\mu u^\nu + g^{\mu\nu}$ is the projector on the hypersurface orthogonal to u^μ. The decomposition of the SET reads:

$$T^{\mu\nu} = E u^\mu u^\nu + (P + \varpi)\Delta^{\mu\nu} + W^\mu u^\nu + W^\nu u^\mu + \pi^{\mu\nu}, \tag{7}$$

where the dynamic pressure ϖ, flow of energy in the LRF W^μ and shear stress $\pi^{\mu\nu}$, together with V^μ, represent non-equilibrium terms. The quantities on the right hand side of Eq. (7) can be obtained through:

$$E = u^\mu u^\nu T_{\mu\nu}, \quad P + \varpi = \frac{1}{3}\Delta^{\mu\nu} T_{\mu\nu}, \quad W^\mu = -\Delta^{\mu\nu} u^\lambda T_{\nu\lambda},$$

$$\pi^{\mu\nu} = \left(\Delta^{\mu\lambda}\Delta^{\nu\sigma} - \frac{1}{3}\Delta^{\mu\nu}\Delta^{\lambda\sigma}\right) T_{\lambda\sigma}, \tag{8}$$

For a massless fluid, $\varpi = 0$. The heat flux q^μ is defined as [10]:

$$q^\mu = W^\mu - \frac{E+P}{n} V^\mu. \tag{9}$$

In the Eckart (particle) frame [2,8,11], u_e^μ is defined as the unit vector parallel to N^μ. Observers in the LRF of the Eckart frame see a flow of energy ($W_e^\mu = q_e^\mu$) and no flow of particles ($V_e^\mu = 0$). Since N^μ cannot be obtained using the QFT approach considered in this paper, the Eckart velocity u_e^μ cannot be defined. Hence, we will not consider the Eckart frame further in this paper.

In the Landau (energy) frame [2,8,12], $u^\mu \equiv u_L^\mu$ is defined as the eigenvector of $T^\mu{}_\nu$ corresponding to the (real, positive) Landau energy density E_L:

$$T^\mu{}_\nu u_L^\nu = -E_L u_L^\mu, \tag{10}$$

such that $W_L^\mu = 0$, which implies that there is no energy flux in the LRF. Since $V_L^\mu = -\frac{n_L}{E_L + P_L} q_L^\mu$ is in general non-zero, an observer in the LRF of the Landau frame will detect a non-vanishing particle flux.

Finally, the β-frame (thermometer frame) for the case of rigid rotation is defined with respect to [4]:

$$u_\beta = \gamma(\partial_t + \Omega\partial_\varphi). \tag{11}$$

A special property of the β-frame is that the LRF temperature is highest compared to the temperature measured with respect to any other frame [4]. In general, V_β^μ and W_β^μ do not vanish, such that the β-frame is a mixed particle-energy frame [13]. Due to the simplicity of its construction, we will start the analysis of the quantum SET with respect to the β-frame.

4. Klein–Gordon field

We now analyse the construction of RRTS from a QFT perspective. A first surprise comes from the analysis of the RRTS of the Klein–Gordon field: in the unbounded Minkowski space, there exist modes which have a non-vanishing Minkowski energy ω (i.e., with respect to the static Hamiltonian $H_s = i\partial_t$), while their co-rotating energy $\widetilde{\omega} = \omega - \Omega m$, measured with respect to the rotating Hamiltonian $H_r = i(\partial_t + \Omega\partial_\varphi)$, vanishes. For such modes, the Bose–Einstein density of states factor $(e^{\beta\widetilde{\omega}} - 1)^{-1}$ diverges, yielding divergent thermal expectation values (t.e.v.s) at every point in the space-time [14,15]. The kinetic theory result (4) is clearly unaffected by this vanishing co-rotating energy modes catastrophy. Indeed, the problematic modes are no longer present in the QFT formulation if the system is enclosed within a boundary placed inside or on the SOL [15,16]. Furthermore, a recent perturbative QFT analysis allows the computation of quantum corrections to the kinetic theory SET [17], which we will analyse in detail in what follows. For completeness, we present an analysis of the connection between these perturbative results and the non-perturbative QFT approach in Appendix A.

Substituting the results in Table III of Ref. [17] into Eq. (34) in Ref. [17] yields the following β-frame (2) decomposition of the SET:

$$E_\beta = \frac{\pi^2\gamma^4}{30\beta_0^4} + \frac{\Omega^2\gamma^6}{36\beta_0^2}, \qquad W_\beta = \frac{\Omega^3\gamma^7}{18\beta_0^2}\left(\rho^2\Omega\partial_t + \partial_\varphi\right),$$

$$\pi_\beta^{\mu\nu} = \frac{\Omega^2\gamma^6}{54\beta_0^2}\begin{pmatrix} \gamma^2 - 1 & 0 & \Omega\gamma^2 & 0 \\ 0 & 1 & 0 & 0 \\ \Omega\gamma^2 & 0 & \rho^{-2}\gamma^2 & 0 \\ 0 & 0 & 0 & -2 \end{pmatrix}, \tag{12}$$

where $\omega = \Omega\gamma^2\partial_z$, $a = \rho\Omega^2\gamma^2\partial_\rho$ and $\gamma = \beta_0^2\Omega^3\gamma^3(\rho^2\Omega\partial_t + \partial_\varphi)$ were used in Eq. (34) of Ref. [17]. Compared to the kinetic theory result (1), the quantum SET contains non-vanishing contributions in the form of the non-ideal terms W^μ and $\pi^{\mu\nu}$. Moreover, the second term in E_β (12) represents a correction to $E_{\text{B-E}}$ (4) which becomes dominant in the vicinity of the SOL due to the γ^6 factor.

The construction of the Landau frame yields:

$$E_L = \frac{E_\beta}{3} - \frac{1}{2}\hat{\boldsymbol{W}}_\beta \cdot \boldsymbol{\pi}_\beta \cdot \hat{\boldsymbol{W}}_\beta + \sqrt{\left(\frac{2E_\beta}{3} + \frac{1}{2}\hat{\boldsymbol{W}}_\beta \cdot \boldsymbol{\pi}_\beta \cdot \hat{\boldsymbol{W}}_\beta\right)^2 - W_\beta^2}, \tag{13}$$

$$u_L^\mu = \sqrt{\frac{E_L + \frac{1}{3}E_\beta + \hat{\boldsymbol{W}}_\beta \cdot \boldsymbol{\pi}_\beta \cdot \hat{\boldsymbol{W}}_\beta}{2(E_L - \frac{1}{3}E_\beta + \frac{1}{2}\hat{\boldsymbol{W}}_\beta \cdot \boldsymbol{\pi}_\beta \cdot \hat{\boldsymbol{W}}_\beta)}} \times \left(u_\beta^\mu + \frac{W_\beta^\mu}{E_L + \frac{1}{3}E_\beta + \hat{\boldsymbol{W}}_\beta \cdot \boldsymbol{\pi}_\beta \cdot \hat{\boldsymbol{W}}_\beta}\right), \tag{14}$$

where $W_\beta^2 = \rho^2\Omega^6\gamma^{12}/324\beta_0^4 \geq 0$, $\hat{\boldsymbol{W}}_\beta \equiv W_\beta/\sqrt{W_\beta^2}$ and $\hat{\boldsymbol{W}}_\beta \cdot \boldsymbol{\pi}_\beta \cdot \hat{\boldsymbol{W}}_\beta = \Omega^2\gamma^6/54\beta_0^2$. Surprisingly, the Landau frame is well-defined only for $0 \leq \rho\Omega \leq (\rho\Omega)_{\text{lim}}$, where

$$(\rho\Omega)_{\text{lim}} = \sqrt{x^2 + x + 1} - x, \qquad x = \frac{5}{4\pi^2}(\beta_0\Omega)^2. \tag{15}$$

When $\rho\Omega > (\rho\Omega)_{\text{lim}}$, E_L is no longer real. It can be seen that $(\rho\Omega)_{\text{lim}}$ decreases from 1 at $\beta_0\Omega = 0$ (large temperatures or slow rotation) down to 0.5 as $\beta_0\Omega \to \infty$.

We are again forced to regard the RRTS of the Klein–Gordon field as ill-defined. The natural question to ask is whether the problem with defining the Landau frame persists when the system

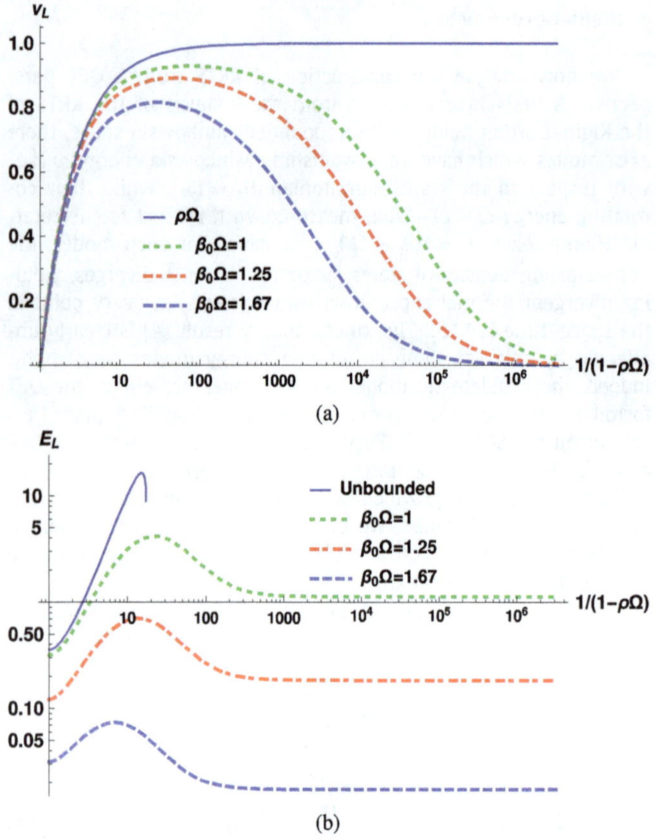

Fig. 1. (a) Landau velocity $v_L = \rho u_L^{\varphi}/u_L^0$ and (b) Landau energy E_L for massless Klein–Gordon particles enclosed inside a cylinder located on the SOL ($R = \Omega^{-1}$). The continuous curve in (a) shows the velocity $\rho\Omega$ for the case of rigid rotation, while in (b) it corresponds to the Landau energy (13) in the unbounded case. This curve is interrupted when E_L becomes complex and the Landau frame is no longer well-defined.

is enclosed inside a boundary. Following Ref. [15], we construct the Landau frame for the case when the system is enclosed inside a cylinder of radius $R = \Omega^{-1}$ (i.e. placed on the SOL), on which Dirichlet boundary conditions are imposed. Fig. 1 shows that the Landau frame is well defined arbitrarily close to the boundary, where the Landau velocity $v_L = \rho u_L^{\varphi}/u_L^0$ decreases to 0 due to the boundary conditions. It can also be seen in Fig. 1 that both v_L and E_L increase monotonically as β_0 is increased. Fig. 1(b) also shows E_L for the unbounded Minkowski space (13) for the case when $\beta_0\Omega = 1$. The curve is interrupted at $\rho\Omega \simeq 0.942$, where E_L becomes complex.

5. Dirac field

The QFT analysis of the RRTS of the Dirac field is presented in Ref. [14]. The β-frame decomposition can be performed using u_β (2) for the components of the SET given in Eqs. (25c)–(25f) in Ref. [14], yielding:

$$E_\beta = \frac{7\pi^2\gamma^4}{60\beta_0^4} + \frac{\Omega^2}{24\beta_0^2}\left(4\gamma^6 - \gamma^4\right), \qquad (16a)$$

$$W_\beta = \frac{\Omega^3\gamma^7}{18\beta_0^2}(\rho^2\Omega\partial_t + \partial_\varphi), \qquad (16b)$$

while $P_\beta = E_\beta/3$ and $\pi_\beta^{\mu\nu} = 0$. It is remarkable that W_β^μ for the Dirac field (16b) has the same expression as that for the Klein–Gordon field (12). As in the case of the Klein–Gordon field, the first term in Eq. (16a) corresponds to $E_{\text{F-D}}$ (4), while the second term represents a quantum correction which dominates in the vicinity of the SOL. Fig. 2(a) demonstrates this behaviour and it can be seen that the correction increases when either Ω or β are increased.

The eigenvalue equation (10) can be solved analytically in terms of the Landau energy and velocity:

$$E_L = \frac{E_\beta}{3} + \sqrt{\frac{4E_\beta^2}{9} - W_\beta^2}, \qquad (17)$$

$$u_L^\mu = \sqrt{\frac{3E_L + E_\beta}{2(3E_L - E_\beta)}}\left(u_\beta^\mu + \frac{3W_\beta^\mu}{3E_L + E_\beta}\right). \qquad (18)$$

In contrast to the case of the Klein–Gordon field, the Landau frame is well-defined everywhere inside the SOL, since $4E_\beta^2/9W_\beta^2 > 1$ when $\rho\Omega < 1$. The ratio E_L/E_β decreases from 1 on the rotation axis down to $\frac{1}{3} + \frac{1}{\sqrt{3}}$ as the SOL is approached, where $W_\beta \to \frac{1}{3}E_\beta$. At fixed $\rho\Omega < 1$, E_L approaches E_β as either Ω or β are decreased, as confirmed in Fig. 2(b).

The Landau velocity $v_L = \rho u_L^{\varphi}/u_L^0 \ge \rho\Omega$ is compared to $\rho\Omega$ in Fig. 2(c). The difference $1 - \rho\Omega/v_L$ decreases to zero as the SOL is approached, while its value at the origin increases monotonically as $\beta_0\Omega$ is increased.

For completeness, we list below $\pi_L^{\mu\nu}$:

$$\pi_L^{\mu\nu} = \frac{2(E_\beta - E_L)(3E_L - 2E_\beta)}{3(3E_L - E_\beta)} u_\beta^\mu u_\beta^\nu + \frac{E_\beta - E_L}{3} g^{\mu\nu} - \frac{E_\beta - E_L}{3E_L - E_\beta}(u_\beta^\mu W_\beta^\nu + u_\beta^\nu W_\beta^\mu) - \frac{6E_L}{9E_L^2 - E_\beta^2} W_\beta^\mu W_\beta^\nu.$$

6. Astrophysical application

Let us now apply our results in the context of the millisecond pulsar PSR J1748-2446ad reported in Ref. [18]. Its pulse frequency is $\nu \simeq 716\,\text{Hz}$, such that the SOL is located at a distance $\rho_{\text{SOL}} = c/2\pi\nu \simeq 66.685\,\text{km}$ from the rotation axis. The typical surface temperature for a neutron star with characteristic age $\tau_c \ge 2.5 \times 10^7$ years is $T_s \simeq 10^5\,\text{K}$ [19]. Its radius is $r_s \lesssim 16\,\text{km}$ [18], such that the temperature on the rotation axis can be extrapolated as $T_0 = T_s/\gamma_s \simeq 9.7 \times 10^4\,\text{K}$. Let us now investigate the magnitude of the quantum corrections for massless Dirac fermions dragged into rigid rotation by the pulsars magnetic field ($B_{\text{surf}} \le 1.1 \times 10^9\,\text{G}$ [18]) by considering the following quantity:

$$\delta E_{\text{QFT}} = \frac{E_\beta}{E_{\text{F-D}}} - 1 = \frac{10}{7\pi^2}\left(\frac{h\nu}{K_B T_0}\right)^2\left(\gamma^2 - \frac{1}{4}\right) \simeq 1.8 \times 10^{-26}\left(\gamma^2 - \frac{1}{4}\right), \qquad (19)$$

where the appropriate units were reinserted. As pointed out in Ref. [17], the quantum correction is very small due to the presence of the Planck constant h. The value of γ at which $E_\beta = 2E_{\text{F-D}}$ is $\gamma \simeq 7.4 \times 10^{12}$, which would correspond for an electron to an energy of $m\gamma c^2 \simeq 3.8 \times 10^{18}\,\text{eV}$, comparable to cosmic rays energies. At such high values of γ, the distance to the SOL is of order $\sim 6 \times 10^{-22}\,\text{m}$, where the rotation of the accretion disk is most likely no longer rigid.

Since our analysis was performed at the level of massless fermions, it is worth mentioning that in the case of the pulsar PSR J1748-2446ad, the relativistic coldness [8] has the value $\zeta_0 = mc^2/k_B T_0 \simeq 6.1 \times 10^4$ in the case of electrons, while the ratio $mc^2/h\nu \simeq 1.7 \times 10^{17}$ also has a large value. These numbers indicate that the massless limit results presented in this paper may be inaccurate close to the rotation axis, where the properties of RRTS are heavily influenced by the value of m in both the kinetic theory

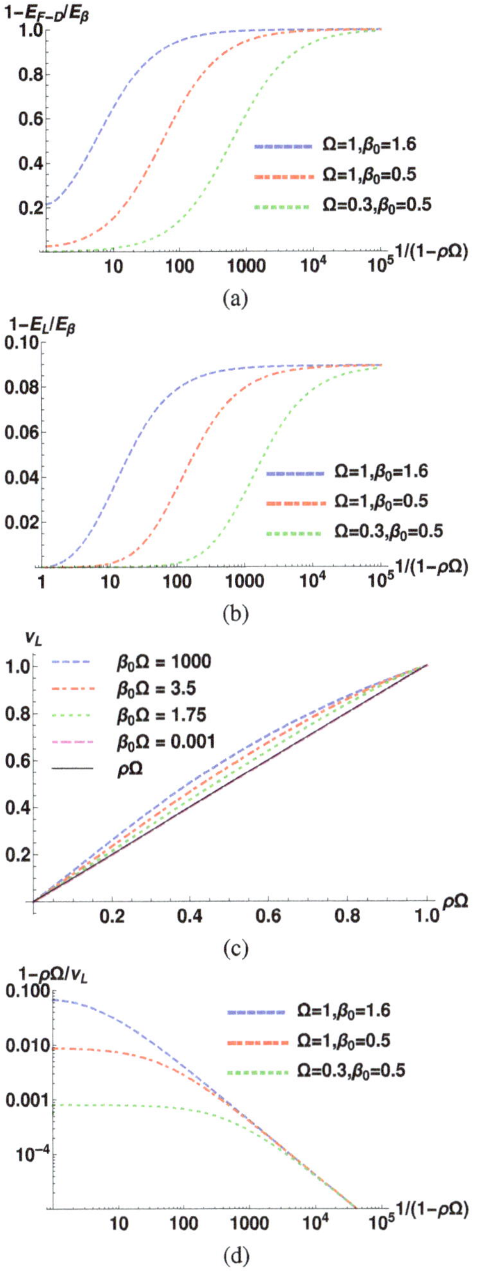

Fig. 2. (a) Comparison between the energy density obtained from kinetic theory $E_{\text{F-D}}$ (4) and the β-frame quantum energy density E_β (16a); (b) Comparison between the energy densities E_L (17) and E_β (16a); (c) Comparison between the Landau velocity $v_L = \rho u_L^\varphi / u_L^t$ and the velocity $\rho\Omega$ corresponding to rigid rotation; (d) Relative difference $1 - \rho\Omega/v_L$ between v_L and $\rho\Omega$.

[6] and in the QFT [14] approaches. Also in these latter references, it can be seen that the mass dependence disappears in the vicinity of the SOL, such that at $\gamma \sim 7.4 \times 10^{12}$, the particle constituents behave as though they were massless.

7. Conclusion

In summary, we investigated rigidly-rotating thermal states of massless Klein–Gordon and Dirac particles. In comparison to relativistic kinetic theory results, the QFT approach yields a non-ideal SET. An analysis of the quantum SET reveals the presence of quantum corrections to the energy density, as well as non-equilibrium terms such as the shear pressure tensor. These quantum terms become dominant as the speed of light surface (SOL) is approached. While for the Dirac field, the Landau frame can be defined everywhere up to the SOL, this is not so for the Klein–Gordon field, which we analysed based on the quantum corrections calculated in Ref. [17]. The Landau frame becomes everywhere well defined when the system is enclosed inside a boundary placed inside or on the SOL.

An evaluation of the order of magnitude of the quantum corrections in a realistic astrophysical system (i.e. for a millisecond pulsar) shows that for such systems, quantum corrections become important only at cosmic ray energies, in which case the rigid rotation must be maintained up to subnuclear distances from the SOL.

Acknowledgements

The author would like to thank Robert Blaga for preliminary discussions and for reading the manuscript. This work was supported by a grant of the Romanian National Authority for Scientific Research and Innovation, CNCS-UEFISCDI, project number PN-II-RU-TE-2014-4-2910.

Appendix A. QFT analysis of the Klein–Gordon field

It is well-known that the t.e.v. of the SET for the RRTS of the Klein–Gordon (KG) field is ill-defined throughout the whole space-time [14,15]. It is also known that this anomalous behaviour is due to modes which are not present once the system is enclosed within a boundary which excludes the space outside of the SOL [15,16,23]. Moreover, the kinetic theory treatment of the same system allows the SET to be computed uneventfully everywhere inside the SOL. Recently, quantum corrections to these kinetic theory results were reported in Ref. [17]. The purpose of this appendix is to bridge the gap between the perturbative analysis of Ref. [17] and the expressions obtained from the exact QFT approach.

The QFT analysis of the RRTS of the KG field can be performed following Refs. [14,15] by introducing co-rotating coordinates $x_r^\mu = (t_r, \rho_r, \varphi_r, z_r)$, defined via $\varphi_r = \varphi - \Omega t$, such that:

$$ds^2 = -\gamma_r^{-2} dt_r^2 + 2\rho_r^2 \Omega \, dt_r \, d\varphi_r + d\rho_r^2 + \rho_r^2 d\varphi_r^2 + dz_r^2. \tag{A.1}$$

The KG field operator for scalar particles of mass μ can be expanded as:

$$\Phi(x_r) = \sum_{m_j=-\infty}^{\infty} \int_{\mu}^{\infty} \omega_j \, d\omega_j \int_{-p_j}^{p_j} dk_j \left[f_j(x_r) a_j + f_j^*(x_r) a_j^\dagger \right], \tag{A.2}$$

where $f_j(x_r) \equiv f_{\omega km}(x_r)$ are the mode solutions of the KG equation [14,15]:

$$f_{\omega km}(x_r) = \frac{1}{\sqrt{8\pi^2 |\omega|}} e^{-i\widetilde{\omega} t_r + im\varphi_r + ikz_r} J_m(q\rho_r). \tag{A.3}$$

In the above, $\widetilde{\omega} = \omega - \Omega m$ is the eigenvalue of the co-rotating Hamiltonian $H_r = i\partial_{t_r}$, while the transverse momentum q, longitudinal momentum k and Minkowski energy ω satisfy $\omega = \sqrt{q^2 + k^2 + \mu^2}$, with $p = \sqrt{\omega^2 - \mu^2}$ being the Minkowski momentum. The one-particle operators a_j and $a_j^\dagger$ satisfy the canonical commutation relations $[a_j, a_{j'}^\dagger] = \delta(j, j')$, where

$$\delta(j, j') = \delta_{m_j, m_{j'}} \delta(k_j - k_{j'}) \frac{\delta(\omega_j - \omega_{j'})}{|\omega_j|}. \tag{A.4}$$

Let us now consider the renormalised t.e.v. of the SET operator in the "new improved" [20] form corresponding to conformal coupling in Ref. [21]:

$$\langle : T_{\mu\nu} :\rangle_{\beta_0} = \langle : \frac{1}{3}\{\phi_{;\mu}, \phi_{;\nu}\} - \frac{1}{6}\{\phi, \phi_{;\mu\nu}\} - \frac{1}{6} g_{\mu\nu}(\phi^{;\lambda}\phi_{;\lambda} + \mu^2\phi^2) :\rangle_{\beta_0}, \tag{A.5}$$

where the colon indicates normal (Wick) ordering. The anticommutator $\{,\}$ was introduced to ensure operator symmetrisation. The above t.e.v. can be computed starting from [14,23]:

$$\langle : a_j a^\dagger_{j'} :\rangle_{\beta_0} = \frac{\delta(j, j')}{e^{\beta_0 \widetilde{\omega}_j} - 1}. \tag{A.6}$$

Introducing the notation G_{abc} through [22]:

$$G_{abc} = \frac{1}{\pi^2} \sum_{m=-\infty}^{\infty} \int_{\mu}^{\infty} \frac{d\omega}{e^{\beta_0 \widetilde{\omega}} - 1} \int_0^p dk\, \omega^a q^b m^c J_m^2(q\rho), \tag{A.7}$$

the t.e.v. of ϕ^2 and of the SET can be written as [22]:

$$\langle : \phi^2 :\rangle_{\beta_0} = \frac{1}{2} G_{000}, \tag{A.8}$$

$$\langle : T_{tt} :\rangle_{\beta_0} = \frac{1}{2} G_{200} + \frac{1}{24}\left(\frac{d^2}{d\rho^2} + \frac{1}{\rho}\frac{d}{d\rho}\right) G_{000},$$

$$\langle : T_{\rho\rho} :\rangle_{\beta_0} = \frac{1}{2} G_{020} - \frac{1}{2\rho^2} G_{002} + \frac{1}{8}\left(\frac{d^2}{d\rho^2} + \frac{5}{3\rho}\frac{d}{d\rho}\right) G_{000},$$

$$\langle : T_{\varphi\varphi} :\rangle_{\beta_0} = \frac{1}{2\rho^2} G_{002} - \frac{1}{24}\left(\rho^2\frac{d^2}{d\rho^2} + 3\rho\frac{d}{d\rho}\right) G_{000},$$

$$\langle : T_{zz} :\rangle_{\beta_0} = \frac{1}{2}(G_{200} - G_{020} - \mu^2 G_{000}) - \frac{1}{24}\left(\frac{d^2}{d\rho^2} + \frac{1}{\rho}\frac{d}{d\rho}\right) G_{000},$$

$$\langle : T_{t\varphi} :\rangle_{\beta_0} = -\frac{1}{2} G_{101}. \tag{A.9}$$

The functions G_{abc} (A.7) are clearly divergent due to the Bose–Einstein density of states factor $(e^{\beta_0\widetilde{\omega}} - 1)^{-1}$. In this section, we will present a procedure to isolate the regular part G^{reg}_{abc} of G_{abc} by splitting G_{abc} as follows:

$$G_{abc} = G^{\text{reg}}_{abc} + G^{\infty}_{abc}, \tag{A.10}$$

where G^{∞}_{abc} absorbs the infinite part of G_{abc}. We will show that G^{reg}_{abc} leads to the corrections presented in Ref. [17].

The method that we will employ is analogous to that used in Ref. [14] for Dirac fermions, being based on expanding the Bose–Einstein density of states factor as follows [22]:

$$\frac{1}{e^{\beta_0(\omega - \Omega m)} - 1} = \sum_{n=0}^{\infty} \frac{(-\Omega)^n}{n!} m^n \frac{d^n}{d\omega^n}\left(\frac{1}{e^{\beta_0\omega} - 1}\right). \tag{A.11}$$

Since the left hand side of the above expression has a pole at $\omega = \Omega m$, the above expansion is not well defined when $\omega < \Omega m$. It is worth mentioning that the modes for which $\widetilde{\omega} < 0$ are no longer allowed when the system is enclosed inside a boundary placed inside or on the SOL [15,16]. Despite the fact that the modes with $\widetilde{\omega} < 0$ cannot be excluded from the mode sum in Eq. (A.7), we will show that the above procedure can still be used to recover the results in Ref. [17].

Substituting the expansion (A.11) into Eq. (A.7) yields:

$$G_{abc} = \frac{1}{\pi^2} \sum_{n=0}^{\infty} \frac{(-\Omega)^n}{n!} \int_{\mu}^{\infty} d\omega\, \omega^a \frac{d^n}{d\omega^n}(e^{\beta_0\omega} - 1)^{-1} \times \int_0^p dk\, q^b \sum_{m=-\infty}^{\infty} m^{n+c} J_m^2(q\rho). \tag{A.12}$$

The sum over m can be performed using the following formula:

$$\sum_{m=-\infty}^{\infty} m^{2n} J_m^2(z) = \sum_{j=0}^{n} \frac{\Gamma(j + \frac{1}{2})}{j!\sqrt{\pi}} a_{n,j} z^{2j}, \tag{A.13}$$

where the coefficients $a_{n,j}$ can be determined as follows:

$$a_{n,j} = \frac{1}{(2j)!} \lim_{\alpha\to 0} \frac{d^{2n}}{d\alpha^{2n}}\left(2\sinh\frac{\alpha}{2}\right)^{2j}, \tag{A.14}$$

such that $a_{n,j}$ vanishes when $j > n$. The following particular cases are required to evaluate Eqs. (A.8) and (A.9):

$$a_{j,j} = 1, \qquad a_{j+1,j} = \frac{1}{12} j(2j+1)(2j+2),$$

$$a_{j+2,j} = \frac{1}{1440} j(2j+1)(2j+2)(2j+3)(2j+4)(5j-1). \tag{A.15}$$

Furthermore, the integral over k in Eq. (A.12) can be performed using Eq. (A.11) in Ref. [14]:

$$\int_0^p dk\, q^\nu = \frac{\Gamma(\frac{\nu}{2} + 1)\sqrt{\pi}}{2\Gamma(\frac{\nu+1}{2} + 1)} p^{\nu+1}. \tag{A.16}$$

Let us apply the above procedure for G_{000}, which reduces to:

$$G_{000} = \frac{1}{\pi^2} \sum_{j=0}^{\infty} \frac{(\rho\Omega)^{2j}}{2j+1} \sum_{n=0}^{\infty} \frac{\Omega^{2n} a_{n+j,j}}{(2n+2j)!} \times \int_{\mu}^{\infty} d\omega\, p^{2j+1} \frac{d^{2n+2j}}{d\omega^{2n+2j}} (e^{\beta_0\omega} - 1)^{-1}. \tag{A.17}$$

In the massless case, $p = \omega$ and the integral over ω runs from 0 to ∞. Noting that:

$$\int_0^{\infty} d\omega\, \omega^{2j+1} \frac{d^{2n+2j}}{d\omega^{2n+2j}} (e^{\beta_0\omega} - 1)^{-1} = (2j+1)! \times \begin{cases} \dfrac{\pi^2}{6\beta_0^2}, & n = 0, \\ -\dfrac{1}{2} + \dfrac{1}{\beta_0} \lim\limits_{\omega\to 0} \omega^{-1}, & n = 1, \\ \dfrac{1}{\beta_0}(2n-2)! \lim\limits_{\omega\to 0} \omega^{-2n+1}, & n > 1. \end{cases} \tag{A.18}$$

It can be seen that the case $n = 0$ corresponds to G^{reg}_{000}. The first term $-\frac{1}{2}$ in the $n = 1$ piece represents a temperature-independent contribution (i.e. which survives in the limit of vanishing temperature, when $\beta_0 \to \infty$). This is the analogue of the spurious contributions highlighted in Ref. [14], which are induced due to the construction of the thermal state with respect to the Minkowski (static) vacuum (see the Iyer vs. Vilenkin discussion in Ref. [14]). The second term in the $n = 1$ piece and all further terms with $n > 1$ are divergent, being induced by the infrared divergence of the Bose–Einstein density of states factor:

$$\frac{1}{e^{\beta_0\omega} - 1} = \frac{1}{\beta_0\omega} - \frac{1}{2} + \text{odd, positive powers of } \beta_0\omega. \tag{A.19}$$

The result can be summarised as follows:

$$G^{\rm reg}_{000} = \frac{\gamma^2}{6\beta_0^2},$$

$$G^{\infty}_{000} = -\frac{\Omega^2\gamma^2}{24\pi^2}(\gamma^2 - 1) + \frac{\Omega^2}{\pi^2\beta_0}\sum_{j=0}^{\infty}(\rho\Omega)^{2j} \times \sum_{n=0}^{\infty}\frac{\Omega^{2n}(2j)!(2n)!a_{n+j+1,j}}{(2n+2j+2)!}\lim_{\omega\to 0}\omega^{-2n-1}. \tag{A.20}$$

After a similar analysis of the rest of the terms appearing in Eq. (A.9), the following regular contributions $\phi^2_{\rm reg}$ and $T^{\rm reg}_{\mu\nu}$ to $\langle :\phi^2:\rangle_{\beta_0}$ and $\langle : T_{\mu\nu} :\rangle_{\beta_0}$ can be obtained:

$$\phi^2_{\rm reg} = \frac{\gamma^2}{6\beta_0^2}, \tag{A.21}$$

$$T^{\rm reg}_{tt} = \frac{\pi^2\gamma^4}{90\beta_0^4}(4\gamma^2 - 1) + \frac{\Omega^2\gamma^6}{36\beta_0^2}(6\gamma^2 - 5),$$

$$T^{\rm reg}_{\rho\rho} = \frac{\pi^2\gamma^4}{90\beta_0^4} + \frac{\Omega^2\gamma^6}{36\beta_0^2},$$

$$\frac{1}{\rho^2}T^{\rm reg}_{\varphi\varphi} = \frac{\pi^2\gamma^4}{90\beta_0^4}(4\gamma^2 - 3) + \frac{\Omega^2\gamma^6}{36\beta_0^2}(6\gamma^2 - 5),$$

$$T^{\rm reg}_{zz} = \frac{\pi^2\gamma^4}{90\beta_0^4} - \frac{\Omega^2\gamma^6}{36\beta_0^2},$$

$$\frac{1}{\rho}T^{\rm reg}_{t\varphi} = -\rho\Omega\left[\frac{2\pi^2\gamma^6}{45\beta_0^4} + \frac{\Omega^2\gamma^6}{18\beta_0^2}(3\gamma^2 - 1)\right]. \tag{A.22}$$

Performing the β-frame decomposition with respect to u_β (2) on the above expressions yields Eqs. (12).

References

[1] S.R. de Groot, W.A. van Leeuwen, Ch.G. van Weert, Relativistic Kinetic Theory – Principles and Applications, North-Holland Publishing Company, Amsterdam, Netherlands, 1980.
[2] C. Cercignani, G.M. Kremer, The Relativistic Boltzmann Equation: Theory and Applications, Birkhäuser Verlag, Basel, Switzerland, 2002.
[3] F. Becattini, Phys. Rev. Lett. 108 (2012) 244502.
[4] F. Becattini, L. Bucciantini, E. Grossi, L. Tinti, Eur. Phys. J. C 75 (2015) 191.
[5] F. Becattini, Acta Phys. Pol. B 47 (2016) 1819.
[6] V.E. Ambruş, R. Blaga, Annals of West University of Timisoara - Physics 58 (2015) 89.
[7] V.E. Ambruş, I.I. Cotăescu, Phys. Rev. D 94 (2016) 085022.
[8] L. Rezzolla, O. Zanotti, Relativistic Hydrodynamics, Oxford University Press, Oxford, UK, 2013.
[9] D.L. Meier, Black Hole Astrophysics: The Engine Paradigm, Springer-Verlag, Berlin–Heidelberg, 2012.
[10] I. Bouras, E. Molnár, H. Niemi, Z. Xu, A. El, O. Fochler, C. Greiner, D.H. Rischke, Phys. Rev. C 82 (2010) 024910.
[11] C. Eckart, Phys. Rev. 58 (1940) 919.
[12] L.D. Landau, E.M. Lifshitz, Fluid Mechanics, 2nd ed., Pergamon Press, Oxford, UK, 1987.
[13] P. Ván, T.S. Biró, AIP Conf. Proc. 1578 (2014) 114.
[14] V.E. Ambruş, E. Winstanley, Phys. Lett. B 734 (2014) 296.
[15] G. Duffy, A.C. Ottewill, Phys. Rev. D 67 (2003) 044002.
[16] N. Nicolaevici, Class. Quantum Gravity 18 (2001) 5407.
[17] F. Becattini, E. Grossi, Phys. Rev. D 92 (2015) 045037.
[18] J.W.T. Hessels, S.M. Ransom, I.H. Stairs, P.C.C. Freire, V.M. Kaspi, F. Camilo, Science 311 (2006) 1901–1904.
[19] N.K. Glendenning, Compact Stars – Nuclear Physics, Particle Physics, and General Relativity, Springer-Verlag, New York, USA, 1997.
[20] P. Candelas, D. Deutsch, Proc. R. Soc. Lond. A 354 (1977) 79–99.
[21] N.D. Birrell, P.C.W. Davies, Quantum Fields in Curved Space, Cambridge University Press, Oxford, 1982.
[22] V.E. Ambruş, PhD thesis, 2014, eprint available at http://etheses.whiterose.ac.uk/id/eprint/7527.
[23] A. Vilenkin, Phys. Rev. D 21 (1980) 2260.

Quantum quench and scaling of entanglement entropy

Paweł Caputa [a,*], Sumit R. Das [b], Masahiro Nozaki [c], Akio Tomiya [d]

[a] *Center for Gravitational Physics, Yukawa Institute for Theoretical Physics, Kyoto University, Kyoto 606-8502, Japan*
[b] *Department of Physics and Astronomy, University of Kentucky, Lexington, KY 40506, USA*
[c] *Kadanoff Center for Theoretical Physics, University of Chicago, Chicago, IL 60637, USA*
[d] *Key Laboratory of Quark & Lepton Physics (MOE) and Institute of Particle Physics, Central China Normal University, Wuhan 430079, China*

ARTICLE INFO

Editor: M. Cvetič

ABSTRACT

Global quantum quench with a finite quench rate which crosses critical points is known to lead to universal scaling of correlation functions as functions of the quench rate. In this work, we explore scaling properties of the entanglement entropy of a subsystem in a harmonic chain during a mass quench which asymptotes to finite constant values at early and late times and for which the dynamics is exactly solvable. When the initial state is the ground state, we find that for large enough subsystem sizes the entanglement entropy becomes independent of size. This is consistent with Kibble–Zurek scaling for slow quenches, and with recently discussed "fast quench scaling" for quenches fast compared to physical scales, but slow compared to UV cutoff scales.

1. Introduction

The behavior of entanglement of a many-body system that undergoes a quantum quench has been a subject of great interest in recent times. When the quench is instantaneous (i.e. a sudden change of the hamiltonian), several results are known. Perhaps the best known result pertains to the entanglement entropy (EE) of a region of size l in a $1+1$ dimensional conformal field theory following a global instantaneous quench, $S_{EE}(l)$. As shown in [1], $S_{EE}(l)$ grows linearly in time till $t \approx l/2$ and then saturates to a constant value typical of a thermal state – a feature which has been studied extensively in both field theory and in holography. Generalizations of this result to conserved charges and higher dimensions have been discussed more recently [2–4]. The emphasis of these studies is to probe the *time evolution* of the entanglement entropy.

In physical situations, quantum quench has a finite rate, characterized by a time scale δt, that can vary from very small to very large. When the quench involves a critical point, universal scaling behavior has been found for correlation functions at *early* times. The most famous scaling appears for a global quench which starts from a massive phase with an initial gap m_g, crosses a critical point (chosen to be e.g. at time $t=0$) and ends in another massive phase. For *slow quenches* (large δt), it has been conjectured that quantities obey Kibble–Zurek scaling [5]: evidence for this has been found in several solvable models and in numerical simulations [6,7]. Such scaling follows from two assumptions. First, it is assumed that as soon as the initial adiabatic evolution breaks down at some time $-t_{KZ}$ (the Kibble–Zurek time) the system becomes roughly diabatic. Secondly, one assumes that the only length scale in the critical region is the instantaneous correlation length ξ_{KZ} at the time $t=-t_{KZ}$. This implies that, for example, one point functions scale as $\langle \mathcal{O}(t)\rangle \sim \xi_{KZ}^{-\Delta}$, where Δ denotes the conformal dimension of the operator $\mathcal{O}$ at the critical point. An improved conjecture involves scaling functions. For example, one and two point correlation functions are expected to be of the form [8–14]

$$\langle \mathcal{O}(t)\rangle \sim \xi_{KZ}^{-\Delta}\, F(t/t_{KZ})$$
$$\langle \mathcal{O}(\vec{x},t)\mathcal{O}(\vec{x}',t')\rangle \sim \xi_{KZ}^{-2\Delta} F\left[\frac{|\vec{x}-\vec{x}'|}{\xi_{KZ}}, \frac{(t-t')}{t_{KZ}}\right] \tag{1}$$

Some time ago, studies of slow quenches in AdS/CFT models have led to some insight into the origin of such scaling without making these assumptions [15].

For protocols in relativistic theories which asymptote to constant values at early times, one finds a different scaling behavior in the regime $\Lambda_{UV}^{-1} \ll \delta t \ll m_{phys}^{-1}$, where Λ_{UV} is the UV cutoff scale, and m_{phys} denotes any physical mass scale in the problem. For example,

$$\langle \mathcal{O}(t)\rangle \sim \delta t^{d-2\Delta} \tag{2}$$

where d is the space–time dimension. This "fast quench scaling" behavior was first found in holographic studies [16] and subse-

* Corresponding author.
E-mail address: caputa@nbi.dk (P. Caputa).

quently shown to be a completely general result in any relativistic quantum field theory [17]. The result follows from causality, and the fact that in this regime linear response becomes a good approximation. Finally, in the limit of an instantaneous quench, suitable quantities saturate as a function of the rate: for quench to a critical theory a rich variety of universal results are known in $1+1$ dimensions [18].

Much less is known about the behavior of entanglement and Renyi entropies *as functions of the quench rate* – a key ingredient of universality. This has been studied for the 1d Ising model (and generalizations) with a transverse field which depends *linearly* on time, $g(t) = 1 - \frac{t}{\tau_Q}$ [19,9,20]. The system is prepared in the instantaneous ground state at some initial time, crossing criticality at $t = 0$. The emphasis of [19] and [9,20] is on the slow regime, which means $\tau_Q \gg a$ where a is the lattice spacing, while [21] also studies smaller values of τ_Q. In particular, [19] and [21] studied the EE for half of a finite chain and found that the answer approaches $S_{EE} \sim \frac{1}{12}\log \xi_{KZ}$ after sufficiently slow quenches. This is consistent with the standard assumptions which lead to Kibble–Zurek scaling mentioned above. According to these assumptions, the system evolves adiabatically till $t = -t_{KZ}$ and enters a phase of diabatic evolution soon afterwards. Thus the state of the system at $t = 0$ is not far from the ground state of the instantaneous hamiltonian at $t = -t_{KZ}$. Furthermore when $\tau_Q \gg 1$ in lattice units, ξ_{KZ} is large, and the instantaneous state is close to criticality. In such a state, the entanglement entropy of a subregion of a large chain with N_A boundary points should obey an "area law" $\frac{c}{6} N_A \log(\xi_{KZ})$, where c is the central charge. When the subsystem is half space $N_A = 1$ and for the Ising model the central charge is $c = 1/2$. Similarly, [9,20] studied the EE of a subsystem of finite size l in an infinite 1d Ising model, with a transverse field linear in time, starting with the ground state at $t = -\infty$. The EE close to the critical point for $l \gg \xi_{KZ}$ was found to saturate to $S_{EE} = (\text{constant}) + \frac{1}{6}\log(\kappa(t)\xi_{KZ})$. The factor $\kappa(t)$ depends mildly on the time of measurement and $\kappa(-t_{KZ}) \approx 1$. Once again, this result is roughly that of a stationary system with correlation length ξ_{KZ}, as would be expected from Kibble–Zurek considerations. The factor $\kappa(t)$ is a correction to the extreme adiabatic–diabatic assumption. The paper [21] investigates an intermediate regime of fast quench (as described above). While this paper investigates scaling of S_{EE} as a function of quench rate in the slow regime, there is no similar analysis in the fast regime.

In this letter, we study entanglement entropy for a simple system: an infinite harmonic chain (i.e. a $1+1$ dimensional bosonic theory on a lattice) with a time dependent mass term which asymptotes to constant *finite* values at early and late times. We choose a mass function for which the quantum dynamics can be solved exactly. The use of such a protocol allows us to explore the whole range of quench rates, where the speed of quench is measured in units of the initial gap rather than the lattice scale. We compute the entanglement entropy for a subsystem of size l (in lattice units) in the middle of the quench and find that it scales in interesting ways as we change the quench rate. The dimensionless quantity which measures the quench timescale is $\Gamma_Q = m_0 \delta t$ where m_0 is the initial gap.

2. Our setup and quench protocols

The hamiltonian of the harmonic chain is given by

$$H = \frac{1}{2}\sum_{n=-\infty}^{\infty}\left[P_n^2 + (X_{n+1} - X_n)^2 + m^2(t)X_n^2\right] \tag{3}$$

where (X_n, P_n) are the usual canonically conjugate scalar field variables on an one dimensional lattice whose sites are labelled by the integer n. The mass term $m(t)$ is time dependent. All quantities are in lattice units. In terms of momentum variables X_k, P_k

$$X_n(t) = \int_{-\pi}^{\pi}\frac{dk}{2\pi}X_k(t)\,e^{ikn} \qquad P_n(t) = \int_{-\pi}^{\pi}\frac{dk}{2\pi}P_k(t)\,e^{ikn} \tag{4}$$

the equation of motion is given by

$$\frac{d^2X_k}{dt^2} + [4\sin^2(k/2) + m^2(t)]X_k = 0 \tag{5}$$

We are interested in functions $m(t)$ which asymptote to constant values m_0 at $t \to \pm\infty$, and pass through zero at $t = 0$. Let $f_k(t)$ be a solution of (5) which asymptotes to a purely positive frequency solution $\sim e^{-i\omega_0 t}/\sqrt{2\omega_0}$ at $t \to -\infty$, where

$$\omega_0^2 = 4\sin^2(k/2) + m_0^2. \tag{6}$$

A mode decomposition

$$X_k(t) = f_k(t)a_k + f_k^\star(t)a_{-k}^\dagger \tag{7}$$

with $[a_k, a_{-k'}^\dagger] = 2\pi\delta(k-k')$ can be then used to define the "in" vacuum by $a_k|0\rangle = 0$ for all k. The solutions $f_k(t)$ are chosen to satisfy the Wronskian condition $f_k(\dot{f}_k)^\star - (\dot{f}_k)f_k^\star = i$. The state $|0\rangle$ then denotes the Heisenberg picture ground state of the initial Hamiltonian. The normalized wavefunctional for the "in" vacuum state is given by

$$\Psi_0(X_k, t) = \prod_k \frac{1}{[\sqrt{2\pi}\,f_k^\star(t)]^{1/2}}\exp\left[\frac{1}{2}\left(\frac{\dot{f}_k(t)}{f_k(t)}\right)^\star X_k X_{-k}\right] \tag{8}$$

We will choose a quench protocol for a mass function for which the mode functions $f_k(t)$ can be solved exactly. The particular mass function we use is

$$m^2(t) = m_0^2\tanh^2(t/\delta t) \tag{9}$$

The corresponding mode functions are given by

$$f_k = \frac{1}{\sqrt{2\omega_0}}\frac{(2)^{i\omega_0\delta t}\cosh^{2\alpha}(t/\delta t)}{E'_{1/2}\tilde{E}_{3/2} - E_{1/2}\tilde{E}'_{3/2}} \times$$
$$[\tilde{E}'_{3/2}\,{}_2F_1(\tilde{a}, \tilde{b}, \frac{1}{2}; -\sinh^2(t/\delta t))$$
$$+E'_{1/2}\sinh(t/\delta t)\,{}_2F_1(\tilde{a} + \frac{1}{2}, \tilde{b} + \frac{1}{2}, \frac{3}{2}; -\sinh^2(t/\delta t))]$$

where we have defined

$$\begin{aligned}
\omega_0^2(k) &= 4\sin^2(k/2) + m_0^2 \\
\alpha &= \frac{1}{4}[1 + \sqrt{1 - 4m_0^2\delta t^2}] \\
\tilde{a} &= \frac{1}{4}[1 + \sqrt{1 - 4m_0^2\delta t^2}] + \frac{i}{2}\delta t\omega_0 \\
\tilde{b} &= \frac{1}{4}[1 + \sqrt{1 - 4m_0^2\delta t^2}] - \frac{i}{2}\delta t\omega_0 \\
E_{1/2} &= \frac{\Gamma(1/2)\Gamma(\tilde{b} - \tilde{a})}{\Gamma(\tilde{b})\Gamma(1/2 - \tilde{a})} \\
\tilde{E}_{3/2} &= \frac{\Gamma(3/2)\Gamma(\tilde{b} - \tilde{a})}{\Gamma(\tilde{b} + 1/2)\Gamma(1 - \tilde{a})} \\
E'_c &= E_c(\tilde{a} \leftrightarrow \tilde{b}).
\end{aligned} \tag{10}$$

3. Entanglement entropy

Our aim is to calculate the entanglement entropy for a subregion of the infinite chain consisting of l lattice points. This is most conveniently calculated by considering the $2l \times 2l$ matrix of correlators, and the symplectic matrix [4,22]

$$C = \begin{bmatrix} X_{mn}(t) & D_{mn}(t) \\ D_{mn}(t) & P_{mn}(t) \end{bmatrix} \quad J = \begin{bmatrix} 0 & I_{l\times l} \\ -I_{l\times l} & 0 \end{bmatrix}$$

where (m, n) denote lattice sites inside the subsystem and

$$X_{mn}(t) = \langle 0|X_m(t)X_n(t)|0\rangle, \quad P_{mn}(t) = \langle 0|P_m(t)P_n(t)|0\rangle,$$
$$D_{mn}(t) = \frac{1}{2}\langle 0|\{X_m(t), P_n(t)\}|0\rangle \tag{11}$$

The eigenvalues of the matrix iJC then occur in pairs $\pm\gamma_n(t)$ $n = 1, \cdots l$ where $\gamma_n(t) > 0$. The entanglement entropy S is given by

$$S = \sum_{n=1}^{l}\{[\gamma_n(t) + \frac{1}{2}]\log[\gamma_n(t) + \frac{1}{2}] - [\gamma_n(t) - \frac{1}{2}]\log[\gamma_n(t) - \frac{1}{2}]\} \tag{12}$$

In terms of the mode functions (10), the correlators (11) are given by the expressions

$$X_{mn}(t) = \int_{-\pi}^{\pi} \frac{dk}{2\pi}|f_k(t)|^2\cos(k|m-n|)$$

$$P_{mn}(t) = \int_{-\pi}^{\pi} \frac{dk}{2\pi}|\dot{f}_k(t)|^2\cos(k|m-n|)$$

$$D_{mn}(t) = \int_{-\pi}^{\pi} \frac{dk}{2\pi}\mathrm{Re}\left(f_k^*(t)\dot{f}_k(t)\right)\cos(k|m-n|) \tag{13}$$

The correlator $D_{mn}(t)$ is particularly important in the calculations which follow. This is because this quantity is exactly zero when the mass is time independent. For the same reason, this quantity vanishes in the leading order of the adiabatic expansion (i.e. when observables are replaced by their *static* answers with the instantaneous value of the mass), i.e. $f_k^{adia} = \frac{1}{\sqrt{2\omega_k(t)}}e^{-i\omega_k(t)t}$ with $\omega_k(t)^2 = 4\sin^2(k/2) + m^2(t)$.

The computation of correlation functions and the entanglement entropy then involves integrals over momenta k, which are computed numerically. As is well known, numerical integrals with oscillating functions in the integrand are tricky. Numerical methods reduce this to a summation of an alternating series, leading to slow convergence. To confirm our numerical integration, we check the convergence of the integration with varying precision. We perform all calculations with interval sizes $dk = 0.00001, 0.000001$.

4. Regimes of the quench rate

Before we examine these entropies, it is necessary to understand the various regimes of the dimensionless parameter $\Gamma_Q \equiv m_0\delta t$. This can be done by studying the scaling behavior of correlation functions. Since we are interested in the slow as well as the fast regime, we will stay close to the continuum limit where the dimensionless mass $m_0 a$ is small. In the following we express everything in lattice units. Then, Kibble Zurek scaling is expected to hold for these quantities for $\Gamma_Q \gg 1$, while we get the regime of fast quench when $\Gamma_Q \ll 1$, but $\delta t > 1$.

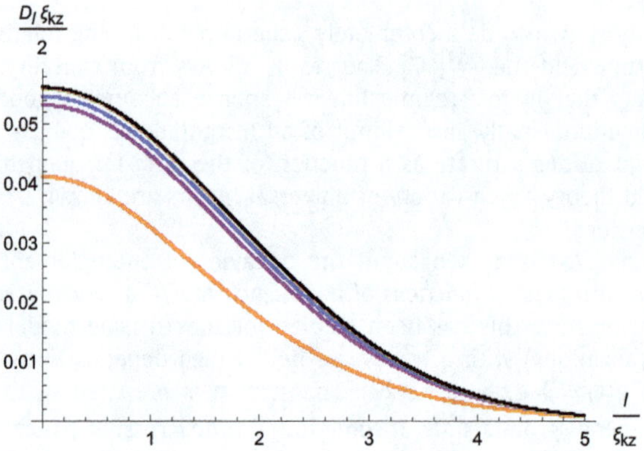

Fig. 1. (Color online.) Rescaled correlator $\frac{1}{2}\xi_{KZ}D_{mn}(t=0)$ as a function of $|m-n|$. The various values of Γ_Q, ξ_{KZ} are in different colors: On top of each other we see Red: ($\Gamma_Q = 1, \xi_{kz} = 400$) and Orange: ($\Gamma_Q = 1, \xi_{kz} = 500$), then Purple: ($\Gamma_Q = 5, \xi_{kz} = 400$) and Blue: ($\Gamma_Q = 10, \xi_{kz} = 400$) and again on top of each other: Green: ($\Gamma_Q = 100, \xi_{kz} = 400$), Pink: ($\Gamma_Q = 100, \xi_{kz} = 500$) and Black: ($\Gamma_Q = 500, \xi_{kz} = 500$).

For slow quench, the usual Landau criterion leads to a Kibble Zurek time and the instantaneous correlation length at this time to be $t_{KZ} = \xi_{KZ} = \sqrt{\delta t/m_0}$, and we expect e.g. the following leading answers at $t = 0$: $X_{mn}(0) \sim \xi_{KZ}^0 F_X(|m-n|/\xi_{KZ})$, $P_{mn}(0) \sim \xi_{KZ}^{-2}F_P(|m-n|/\xi_{KZ})$, $D_{mn}(0) \sim \xi_{KZ}^{-1}F_D(|m-n|/\xi_{KZ})$, where F_X, F_P, F_D are smooth functions. We indeed find this behavior for $\Gamma_Q \geq 100$. For example, Fig. 1 shows $\xi_{KZ}D_{mn}(0)$ as a function of $|m-n|$ for various values of Γ_Q. Clearly for large enough Γ_Q the points fall on top of each other, showing that the above scaling behavior holds. The results for $X_{mn}(0)$ and $\xi_{KZ}^2 P_{mn}(0)$ are also consistent with the above expectations.

In the fast quench regime $\Gamma_Q \ll 1$ the analysis of [17] leads to the following scaling relations for coincident correlators

$$X_{nn} \sim (\text{const}) + (\delta t)^2 \quad P_{nn} \sim (\text{const}) + (\delta t)^0 \quad D_{nn} \sim \delta t \tag{14}$$

We found this behavior for small $\Gamma_Q < 0.1$. For correlation functions X_{mn}, P_{mn}, D_{mn}, with the constant mass values subtracted, we expect these scalings to hold for large $|m-n| < \delta t < m_0^{-1}$, while for $\delta t < |m-n| < m_0^{-1}$ they should become independent of δt [17]. We indeed see this behavior for sufficiently small Γ_Q. This is most clearly seen for D_{mn} since the constant mass value of this quantity is exactly zero.

5. Scaling of entropies

Now that we have identified the slow and fast quench regimes, we go ahead and explore possible scaling properties of the entanglement entropy S_{EE}. This quantity is studied in detail at $t = 0$. (which is the middle of the quench) for both fast and slow quenches.

First, we consider the fast quench regime, $\Gamma_Q \ll 1$. We have computed the difference of the entanglement entropy at $t = 0$ from its value at $t = -\infty$,

$$\Delta S_A = S(t=0) - S(t=-\infty) \tag{15}$$

for various subsystem sizes l as a function of Γ_Q. Fig. 2 shows the results for $\xi_i = 600$ as a function of Γ_Q. When l is smaller than the initial correlation length $\xi_i = m_0^{-1}$ this quantity depends on l for a given Γ_Q. However for l sufficiently large compared to ξ_i we find that the l-dependence saturates.

Namely, the data for large l/ξ_i fits to the following result for small Γ_Q

$$\Delta S = c\,\Gamma_Q^2, \tag{16}$$

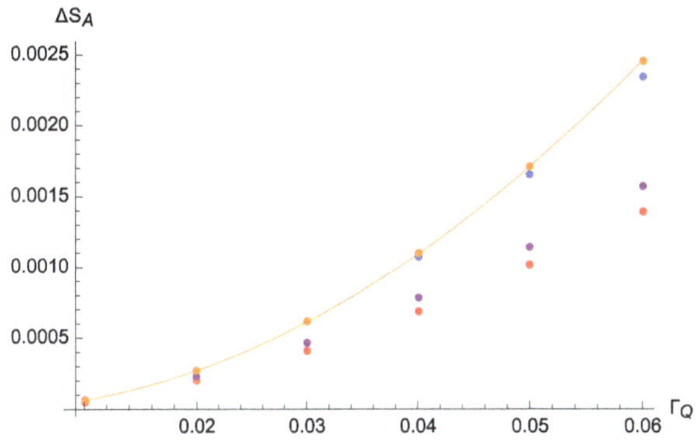

Fig. 2. (Color online.) ΔS_A at $t=0$ as a function of Γ_Q for fast quench. All the data are for $\xi_i=600$. Different colors correspond to different values of l. Red, Purple, Blue, Green and Orange have $l=5, 10, 100, 1000, 2000$ respectively. The data for $l=1000, 2000$ are almost on top of each other.

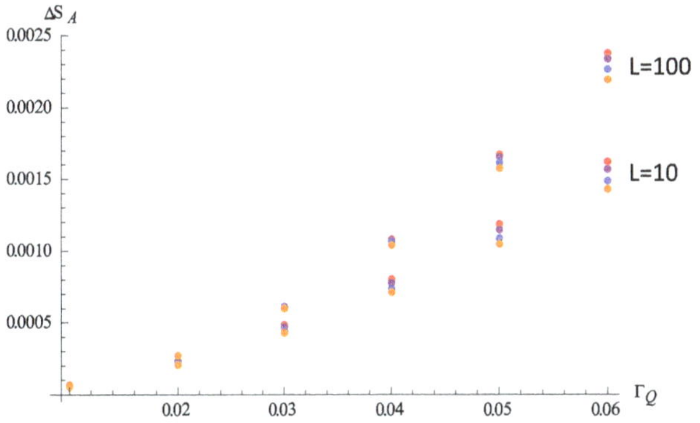

Fig. 3. (Color online.) ΔS_A at $t=0$ as a function of Γ_Q for fast quench. The data are for $l=10$ and $l=100$. Different colors correspond to different values of ξ_i. Red, Purple, Blue, and Orange have $\xi_i=500, 600, 800, 1000$ respectively.

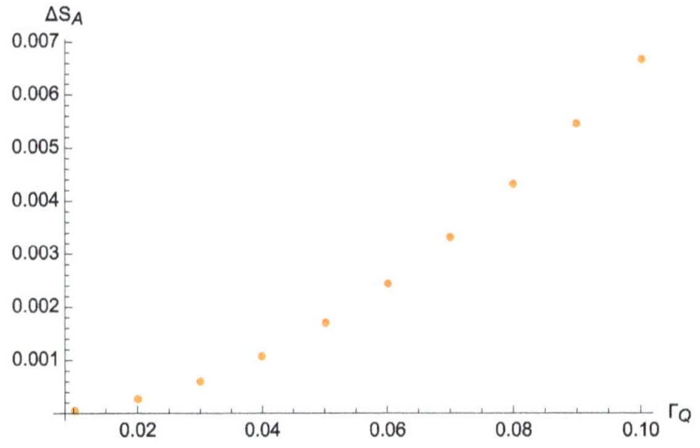

Fig. 4. (Color online.) ΔS_A at $t=0$ as a function of Γ_Q for fast quench. All the data are for $l=2000$. We plotted this for different values of ξ_i. Red, Purple, Blue, and Orange have $\xi_i=500, 600, 800, 1000$ respectively. However the data are basically on top of each other.

where the constant c is independent of l and ξ_i and approximately equal to 0.67.

In fact, for small l/ξ_i, ΔS_A does depend on ξ_i for a given Γ_Q and l, as shown in Figs. 3 and 4. These are plots of ΔS_A as a function of Γ_Q for a given l, and different values of ξ_i. Fig. 3 has $l=100$ with $\frac{l}{\xi_i}$ ranging from 0.1 to 0.2: the result has a clear dependence on ξ_i. Fig. 4 has $l=2000$ with $\frac{l}{\xi_i}$ ranging from 2.0 to 4.0 – the data for various values of ξ_i are right on top of each other.

This result (16) is in agreement with expectations based on [17]. In these papers it was argued that the fast quench scaling

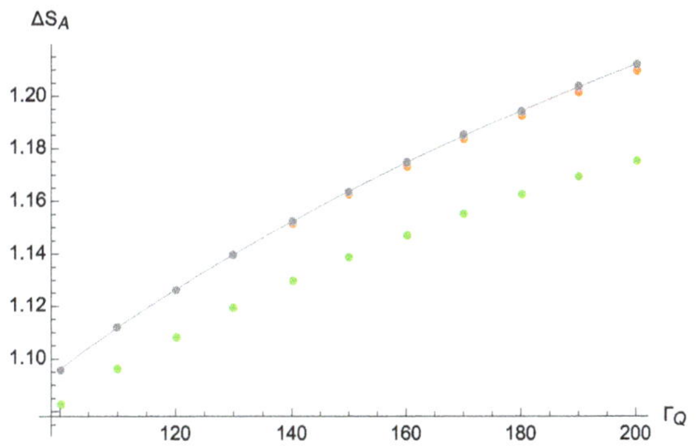

Fig. 5. (Color online.) ΔS_A at $t=0$ as a function of Γ_Q for slow quench. The data are for $\xi_i=40$ and $\xi_i=50$ for different values of l. However the points for these two different ξ_i are basically on top of each other. Green, orange, pink and grey correspond to $l=1000, 2000, 3000, 3500$. The $l=3000$ and $l=3500$ values are too close to be distingushable.

for one point functions follow from the fact that linear response becomes a good approximation in this regime. Even though the scaling behavior of the two point function is complicated (as mentioned above), one would expect that the δt dependence of the matrix elements of C would still be governed by perturbation theory. To lowest order this should be therefore proportional to m_0^2. In the regime of quench rates we are considering, the only other scale is δt. Since the entanglement entropy is dimensionless, we expect that the leading answer is proportional to $m_0^2\delta t^2$ which is Γ_Q^2. The fact that the result becomes independent of l is also expected since the measurement is made at $t=0$.

It would be interesting to gain further insight by using the perturbation theory for entanglement entropy (as e.g. developed in [23]).

Let us now consider quench rates slow enough to ensure that the correlation functions discussed above display standard Kibble Zurek scaling. Fig. 5 shows S_{EE} as a function of Γ_Q for a given m_0 various values of subsystem size l.

For large enough l we therefore find that ΔS_A is independent of l and fits the relation

$$\Delta S_A = (\text{constant}) + \frac{1}{6}\log(\Gamma_Q) \tag{17}$$

The constant in (17) is roughly 0.3. This result is consistent with the expectations from Kibble Zurek scenario. More precisely, the state of the system at $t=0$ is fairly close to the instantaneous ground state at $t=-t_{KZ}$, i.e. the ground state of the system with a fixed (time independent) correlation length ξ_{KZ}. For such a system the entanglement entropy should behave as $\frac{1}{3}\log(\xi_{KZ})$, independent of l for $l>\xi_{KZ}$. Since $\xi_{KZ}=\xi_i\sqrt{\Gamma_Q}$, this is $\frac{1}{3}\log(\xi_i)+\frac{1}{6}\log(\Gamma_Q)$. On the other hand, $S_{EE}(t=-\infty)$ is ground state entanglement entropy of the system at a fixed correlation length ξ_i and therefore behaves as $\frac{1}{3}\log(\xi_i)$ for $l\gg\xi_i$. Thus the KZ scenario predicts that when ΔS_A is expressed as a function of Γ_Q, the dependence on ξ_i should cancel – which is exactly as we find.

In conclusion, we have calculated the entanglement entropy of a subsystem in a harmonic chain in the presence of an exactly solvable mass quench and examined the dependence of this quantity at the middle of the quench on the quench rate. Our nonlinear quench protocol asymptotes to constant values at initial and final times: in this respect it differs significantly from the kind of protocols most commonly studied in the literature, e.g. couplings which behave linearly in time. For quench rates which are fast compared to the initial mass (but slow compared to the lattice scale) our results show, for the first time, that the fast quench scaling established for correlation functions in [17] extend to nonlocal quantities like entanglement entropy. For quench rates slow compared to the initial mass, our results are consistent with the predictions of Kibble–Zurek scenario, and with earlier results in the 1d Ising model.

Acknowledgements

We thank Shinsei Ryu, Marek Rams, Krishnendu Sengupta and Tadashi Takayanagi for discussions and comments on the manuscript. The work of PC is supported by the Simons Foundation through the "It from Qubit" collaboration. The work of S.R.D. is supported by a National Science Foundation grant NSF-PHY-1521045. The work of A.T. was supported in part by NSFC under grant number 11535012. A part of numerical computation in this work was carried out at the Yukawa Institute Computer Facility.

References

[1] P. Calabrese, J.L. Cardy, J. Stat. Mech. 0504 (2005) P04010, arXiv:cond-mat/0503393;
P. Calabrese, J. Cardy, J. Stat. Mech. 0706 (2007) P06008, arXiv:0704.1880 [cond-mat.stat-mech].
[2] P. Caputa, G. Mandal, R. Sinha, J. High Energy Phys. 1311 (2013) 052, arXiv:1306.4974 [hep-th].
[3] H. Liu, S.J. Suh, Phys. Rev. Lett. 112 (2014) 011601, arXiv:1305.7244 [hep-th];
H. Liu, S.J. Suh, Phys. Rev. D 89 (6) (2014) 066012, arXiv:1311.1200 [hep-th];
H. Casini, H. Liu, M. Mezei, J. High Energy Phys. 1607 (2016) 077, arXiv:1509.05044 [hep-th];
M. Mezei, arXiv:1612.00082 [hep-th].
[4] J.S. Cotler, M.P. Hertzberg, M. Mezei, M.T. Mueller, J. High Energy Phys. 1611 (2016) 166, arXiv:1609.00872 [hep-th].
[5] T.W.B. Kibble, J. Phys. A 9 (1976) 1387;
W.H. Zurek, Nature 317 (1985) 505.
[6] For a review, see A. Polkovnikov, K. Sengupta, A. Silva, M. Vengalattore, Rev. Mod. Phys. 83 (2011) 863, arXiv:1007.5331 [cond-mat.stat-mech].
[7] J. Dziarmaga, Adv. Phys. 59 (2010) 1063, arXiv:0912.4034 [cond-mat.quant-gas].
[8] W.H. Zurek, Phys. Rep. 276 (1996) 177.
[9] L. Cincio, J. Dziarmaga, M.M. Rams, W.H. Zurek, Phys. Rev. A 75 (2007) 052321, arXiv:cond-mat/0701768 [cond-mat.str-el].
[10] S. Deng, G. Ortiz, L. Viola, Europhys. Lett. 84 (2008) 67008.
[11] B. Damski, W.H. Zurek, New J. Phys. 11 (2009) 063014.
[12] J. Dziarmaga, M.M. Rams, New J. Phys. 12 (2010) 055007.
[13] M. Kolodrubetz, B.K. Clark, D.A. Huse, Phys. Rev. Lett. 109 (2012) 015701.
[14] A. Chandran, A. Erez, S.S. Gubser, S.L. Sondhi, Phys. Rev. B 86 (2012) 064304;
A. Chandran, F.J. Burnell, V. Khemani, S.L. Sondhi, J. Phys. Condens. Matter 25 (2013) 404214.
[15] P. Basu, S.R. Das, J. High Energy Phys. 1201 (2012) 103, arXiv:1109.3909 [hep-th];
For a review see S.R. Das, PTEP 2016, PTEP 2016 (12) (2016) 12C107, arXiv:1608.04407 [hep-th].
[16] A. Buchel, L. Lehner, R.C. Myers, A. van Niekerk, J. High Energy Phys. 1305 (2013) 067, arXiv:1302.2924 [hep-th];
A. Buchel, R.C. Myers, A. van Niekerk, Phys. Rev. Lett. 111 (2013) 201602, arXiv:1307.4740 [hep-th].
[17] S.R. Das, D.A. Galante, R.C. Myers, Phys. Rev. Lett. 112 (2014) 171601, arXiv:1401.0560 [hep-th];
S.R. Das, D.A. Galante, R.C. Myers, J. High Energy Phys. 1502 (2015) 167, arXiv:1411.7710 [hep-th];
S.R. Das, D.A. Galante, R.C. Myers, J. High Energy Phys. 1508 (2015) 073, arXiv:1505.05224 [hep-th];
S.R. Das, D.A. Galante, R.C. Myers, J. High Energy Phys. 1605 (2016) 164, arXiv:1602.08547 [hep-th].
[18] P. Calabrese, J.L. Cardy, Phys. Rev. Lett. 96 (2006) 136801, arXiv:cond-mat/0601225;
P. Calabrese, J. Cardy, arXiv:0704.1880 [cond-mat.stat-mech];
S. Sotiriadis, J. Cardy, Phys. Rev. B 81 (2010) 134305, arXiv:1002.0167 [quant-ph].
[19] F. Pollmann, S. Mukherjee, A.G. Green, J.E. Moore, Phys. Rev. E 81 (2010) 020101.
[20] A. Francuz, J. Dziarmaga, B. Gardas, W.H. Zurek, Phys. Rev. B 93 (7) (2016) 075134, arXiv:1510.06132 [cond-mat.stat-mech].
[21] E. Canovi, E. Ercolessi, P. Naldesi, L. Taddia, D. Vodola, Phys. Rev. B 89 (2014) 104303.
[22] K. Audenaert, J. Eisert, M.B. Plenio, R.F. Werner, Phys. Rev. A 66 (4) (2002) 042327, arXiv:quant-ph/0205025;
A. Botero, B. Reznik, Phys. Rev. A 70 (2004) 052329;
A. Coser, E. Tonni, P. Calabrese, J. Stat. Mech. 1412 (12) (2014) P12017, arXiv:1410.0900 [cond-mat.stat-mech].
[23] V. Rosenhaus, M. Smolkin, J. High Energy Phys. 1412 (2014) 179, arXiv:1403.3733 [hep-th];
V. Rosenhaus, M. Smolkin, J. High Energy Phys. 1502 (2015) 015, arXiv:1410.6530 [hep-th];
S. Leichenauer, M. Moosa, M. Smolkin, J. High Energy Phys. 1609 (2016) 035, arXiv:1604.00388 [hep-th].

Small-x asymptotics of the quark helicity distribution: Analytic results

Yuri V. Kovchegov [a], Daniel Pitonyak [b], Matthew D. Sievert [c,*]

[a] Department of Physics, The Ohio State University, Columbus, OH 43210, USA
[b] Division of Science, Penn State University-Berks, Reading, PA 19610, USA
[c] Theoretical Division, Los Alamos National Laboratory, Los Alamos, NM 87545, USA

ARTICLE INFO

Editor: J.-P. Blaizot

ABSTRACT

In this Letter, we analytically solve the evolution equations for the small-x asymptotic behavior of the (flavor singlet) quark helicity distribution in the large-N_c limit. These evolution equations form a set of coupled integro-differential equations, which previously could only be solved numerically. This approximate numerical solution, however, revealed simplifying properties of the small-x asymptotics, which we exploit here to obtain an analytic solution. We find that the small-x power-law tail of the quark helicity distribution scales as $\Delta q^S(x, Q^2) \sim \left(\frac{1}{x}\right)^{\alpha_h}$ with $\alpha_h = \frac{4}{\sqrt{3}}\sqrt{\frac{\alpha_s N_c}{2\pi}}$, in excellent agreement with the numerical estimate $\alpha_h \approx 2.31\sqrt{\frac{\alpha_s N_c}{2\pi}}$ obtained previously. We then verify this solution by cross-checking the predicted scaling behavior of the auxiliary "neighbor dipole amplitude" against the numerics, again finding excellent agreement.

1. Introduction

The small-x power-law behavior of parton distribution functions (PDFs) and hadronic structure functions at small Bjorken x is governed by quantum evolution equations which resum large logarithms of $\frac{1}{x}$. The most familiar of these is the linear Balitsky–Fadin–Kuraev–Lipatov (BFKL) equation [1,2] for the unpolarized structure functions F_1 and F_2 along with the quark and gluon PDFs at small x, which resums the single-logarithmic parameter $\alpha_s \ln \frac{1}{x} \sim 1$ (with α_s the strong coupling constant). The result of this resummation is a power-law growth at small x given by $F_1(x, Q^2) \sim q(x, Q^2) \sim \left(\frac{1}{x}\right)^{\alpha_P}$, with the leading-order (LO) exponent $\alpha_P = 1 + \frac{4\alpha_s N_c}{\pi} \ln 2$ known as the perturbative "Pomeron intercept" in the terminology of Regge theory. Here N_c is the number of colors.

The analogous small-x asymptotic behavior of the helicity PDFs $\Delta f(x, Q^2)$ and the polarized structure function $g_1(x, Q^2)$ has received much less attention than the unpolarized case. Early studies emphasized the role of exchanging polarized quarks [3–8] (the "Reggeon" in Regge theory), with important progress on the full polarized evolution made by Bartels, Ermolaev, and Ryskin [9,10]. Recently, we have derived the small-x evolution equations for the quark helicity PDFs $\Delta q^S(x, Q^2)$ and the polarized structure function $g_1(x, Q^2)$ [11,12] in the modern language of the dipole model. (In this Letter, we restrict our discussion to the flavor-singlet quark helicity distribution; for the non-singlet quark helicity distribution, see [12].) These helicity evolution equations, like the perturbative Reggeon evolution equations, resum the double-logarithmic parameter $\alpha_s \ln^2 \frac{1}{x} \sim 1$; they couple to both polarized quark and gluon exchange, and, in this respect, differ from the gluon-only unpolarized LO BFKL equation.

In general, the helicity evolution equations derived in [11, 12] form an infinite tower of operator equations analogous to the Balitsky hierarchy [13,14] for the unpolarized small-x evolution [14–21]. In both cases, the operator hierarchy closes in the large-N_c limit [14–17]. For helicity evolution, this still yields a pair of coupled integro-differential equations for the "polarized dipole amplitude" $G(x_{10}^2, z)$ and the auxiliary "neighbor dipole amplitude" $\Gamma(x_{10}^2, x_{21}^2, z)$ that must be solved to determine the power-law behavior at small x. (Here x_{ij}'s denote transverse sizes of dipoles and z is the softest longitudinal momentum fraction between the quark and antiquark in the dipole.) This asymptotic behavior of the polarized dipole $G(x_{10}^2, z) \sim (zs)^{\alpha_h}$ determines the corresponding small-x asymptotics of the helicity PDFs and the polarized structure function: $\Delta q^S(x, Q^2) \sim g_1(x, Q^2) \sim \left(\frac{1}{x}\right)^{\alpha_h}$, where we refer to the exponent α_h as the "helicity intercept" in analogy to the Pomeron intercept.

* Corresponding author.
E-mail addresses: kovchegov.1@osu.edu (Y.V. Kovchegov), dap67@psu.edu (D. Pitonyak), sievertmd@lanl.gov (M.D. Sievert).

In [22], we solved the large-N_c helicity evolution equations for α_h numerically, obtaining $\alpha_h \approx 2.31\sqrt{\frac{\alpha_s N_c}{2\pi}}$. We also found that such an intercept could lead to a significant enhancement of the contribution from the quark spin $\Delta\Sigma$ to the proton spin [22], which is not ruled out by current experimental data [23]. In the following Sections, we use an emergent scaling feature of this numerical solution, namely, that G depends only on a single combination of its arguments and not on each independently, to derive an analytic expression for α_h.

2. Solution of the large-N_c equations

In standard coordinates, the large-N_c helicity evolution equations read [11,12]

$$G(x_{10}^2, z) = G^{(0)}(x_{10}^2, z) + \frac{\alpha_s N_c}{2\pi} \int\limits_{\frac{1}{x_{10}^2 s}}^{z} \frac{dz'}{z'} \int\limits_{\frac{1}{z's}}^{x_{10}^2} \frac{dx_{21}^2}{x_{21}^2} \times \left[\Gamma(x_{10}^2, x_{21}^2, z') + 3G(x_{21}^2, z')\right], \tag{1a}$$

$$\Gamma(x_{10}^2, x_{21}^2, z') = G^{(0)}(x_{10}^2, z') + \frac{\alpha_s N_c}{2\pi} \int\limits_{\frac{1}{x_{10}^2 s}}^{z'} \frac{dz''}{z''} \times \int\limits_{\frac{1}{z''s}}^{\min\left\{x_{10}^2,\, x_{21}^2 \frac{z'}{z''}\right\}} \frac{dx_{32}^2}{x_{32}^2} \left[\Gamma(x_{10}^2, x_{32}^2, z'') + 3G(x_{32}^2, z'')\right], \tag{1b}$$

where x_{10}, x_{21}, x_{32} are the transverse sizes of various dipoles and z, z', z'' are longitudinal momentum fractions of the softest (anti-)quarks in the dipoles. Following [22], it is convenient to introduce the scaled logarithmic variables

$$\eta \equiv \sqrt{\frac{\alpha_s N_c}{2\pi}} \ln \frac{zs}{\Lambda^2}, \qquad s_{10} \equiv \sqrt{\frac{\alpha_s N_c}{2\pi}} \ln \frac{1}{x_{10}^2 \Lambda^2}, \tag{2a}$$

$$\eta' \equiv \sqrt{\frac{\alpha_s N_c}{2\pi}} \ln \frac{z's}{\Lambda^2}, \qquad s_{21} \equiv \sqrt{\frac{\alpha_s N_c}{2\pi}} \ln \frac{1}{x_{21}^2 \Lambda^2}, \tag{2b}$$

$$\eta'' \equiv \sqrt{\frac{\alpha_s N_c}{2\pi}} \ln \frac{z''s}{\Lambda^2}, \qquad s_{32} \equiv \sqrt{\frac{\alpha_s N_c}{2\pi}} \ln \frac{1}{x_{32}^2 \Lambda^2}, \tag{2c}$$

where Λ is an IR momentum cutoff and s is the center-of-mass-energy squared at which the helicity PDF is measured. In terms of these rescaled variables, the large-N_c helicity evolution equations are

$$G(s_{10}, \eta) = G^{(0)}(s_{10}, \eta) + \int\limits_{s_{10}}^{\eta} d\eta' \int\limits_{s_{10}}^{\eta'} ds_{21} \left[\Gamma(s_{10}, s_{21}, \eta') + 3\,G(s_{21}, \eta')\right], \tag{3a}$$

$$\Gamma(s_{10}, s_{21}, \eta') = G^{(0)}(s_{10}, \eta') + \int\limits_{s_{10}}^{\eta'} d\eta'' \int\limits_{\max\{s_{10}, s_{21}+\eta''-\eta'\}}^{\eta''} ds_{32} \times \left[\Gamma(s_{10}, s_{32}, \eta'') + 3\,G(s_{32}, \eta'')\right]. \tag{3b}$$

In the numerical solution of [22], two important features were observed in the asymptotic limit: (a) a negligible dependence on the choice of initial conditions $G^{(0)}$ and (b) the dependence of G only on the combination $\zeta \equiv \eta - s_{10}$, rather than on η and s_{10} separately (see Fig. 1). This scaling behavior also occurs in the neighbor dipole amplitude Γ (see Fig. 2) and sets in when $\zeta \gtrsim 1-2$. Let us therefore (a) trivially fix the initial conditions to $G^{(0)} = 1$ and (b) assume the $\eta - s_{10}$ scaling from the outset:

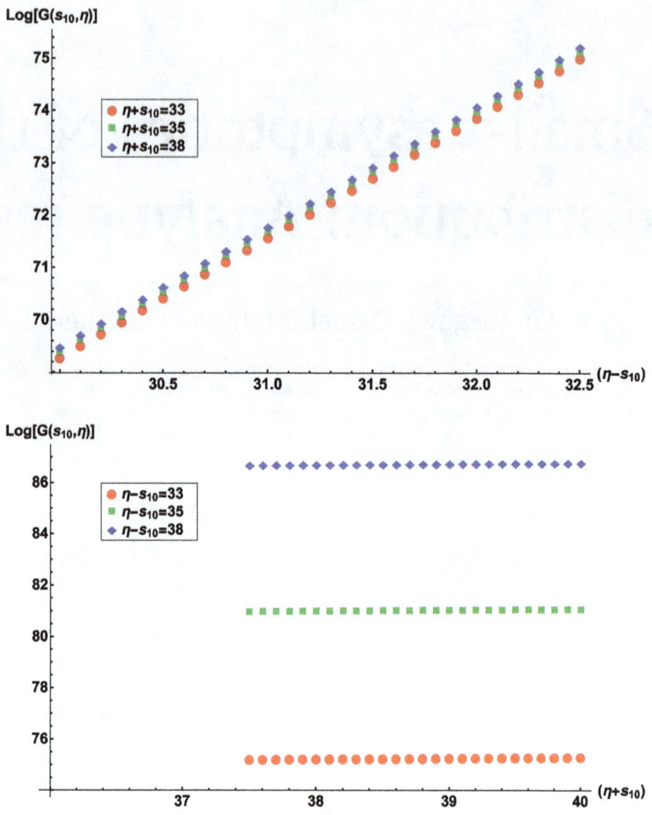

Fig. 1. Numerical solution of the scaled equations (3) as a function of $\eta - s_{10}$ for fixed $\eta + s_{10}$ (top panel) and as a function of $\eta + s_{10}$ for fixed $\eta - s_{10}$ (bottom panel). The grid parameters are $\eta_{max} = 40$, $\Delta\eta = 0.05$. One clearly sees that $\ln G$ has a linear dependence on $\eta - s_{10}$ and is independent of $\eta + s_{10}$.

$$G(s_{10}, \eta) = G(\eta - s_{10}), \tag{4a}$$

$$\Gamma(s_{10}, s_{21}, \eta') = \Gamma(\eta' - s_{10}, \eta' - s_{21}). \tag{4b}$$

Then Eqs. (3) become

$$G(\zeta) = 1 + \int\limits_0^{\zeta} d\xi \int\limits_0^{\xi} d\xi' \left[\Gamma(\xi, \xi') + 3\,G(\xi')\right], \tag{5a}$$

$$\begin{aligned}\Gamma(\zeta, \zeta') &= 1 + \int\limits_0^{\zeta'} d\xi \int\limits_0^{\xi} d\xi' \left[\Gamma(\xi, \xi') + 3\,G(\xi')\right] + \int\limits_{\zeta'}^{\zeta} d\xi \int\limits_0^{\zeta'} d\xi' \left[\Gamma(\xi, \xi') + 3\,G(\xi')\right] \\ &= G(\zeta') + \int\limits_{\zeta'}^{\zeta} d\xi \int\limits_0^{\zeta'} d\xi' \left[\Gamma(\xi, \xi') + 3\,G(\xi')\right],\end{aligned} \tag{5b}$$

with the boundary conditions

$$G(0) = 1, \quad \Gamma(\zeta', \zeta') = G(\zeta'). \tag{6}$$

Since $s_{10} \le s_{21}$, we consider $\Gamma(\zeta, \zeta')$ only in the range $\zeta > \zeta'$.

To solve (5), we first differentiate, obtaining

$$\partial_\zeta G(\zeta) = \int_0^\zeta d\xi' \left[\Gamma(\zeta, \xi') + 3\,G(\xi')\right], \tag{7a}$$

$$\partial_\zeta \Gamma(\zeta, \zeta') = \int_0^{\zeta'} d\xi' \left[\Gamma(\zeta, \xi') + 3\,G(\xi')\right], \tag{7b}$$

and we then introduce the Laplace transforms

$$G(\zeta) = \int \frac{d\omega}{2\pi i} e^{\omega\zeta}\, G_\omega, \qquad \Gamma(\zeta, \zeta') = \int \frac{d\omega}{2\pi i} e^{\omega\zeta'}\, \Gamma_\omega(\zeta), \tag{8a}$$

$$G_\omega = \int_0^\infty d\zeta\, e^{-\omega\zeta}\, G(\zeta), \qquad \Gamma_\omega(\zeta) = \int_0^\infty d\zeta'\, e^{-\omega\zeta'}\, \Gamma(\zeta, \zeta'). \tag{8b}$$

Note that $\Gamma(\zeta, \zeta')$ is mathematically well-defined for any $\zeta, \zeta' > 0$, although only $\zeta > \zeta'$ contributes to the evolution equations. Consider first the Laplace transform (8) of Eq. (7b),

$$\partial_\zeta \Gamma_\omega(\zeta) = \frac{1}{\omega}\left[\Gamma_\omega(\zeta) + 3\,G_\omega\right]. \tag{9}$$

This is just an ordinary differential equation in ζ, with the solution

$$\Gamma_\omega(\zeta) + 3\,G_\omega = e^{\frac{\zeta}{\omega}}\left[\Gamma_\omega(0) + 3\,G_\omega\right]; \tag{10}$$

substituting (10) back into (8) then gives

$$\Gamma(\zeta, \zeta') = \int \frac{d\omega}{2\pi i} e^{\omega\zeta'} \left\{ e^{\frac{\zeta}{\omega}}\left[\Gamma_\omega(0) + 3\,G_\omega\right] - 3\,G_\omega \right\}, \tag{11}$$

or, equivalently,

$$\Gamma(\zeta, \zeta') + 3\,G(\zeta') = \int \frac{d\omega}{2\pi i} e^{\omega\zeta' + \frac{\zeta}{\omega}} \left[\Gamma_\omega(0) + 3\,G_\omega\right]. \tag{12}$$

Using the second boundary condition in (6), Eq. (12) then fixes G, giving the general solution for G and Γ as

$$G(\zeta) = \frac{1}{4}\int \frac{d\omega}{2\pi i} e^{(\omega + \frac{1}{\omega})\zeta}\, H_\omega, \tag{13a}$$

$$\Gamma(\zeta, \zeta') = \int \frac{d\omega}{2\pi i} e^{\omega\zeta' + \frac{\zeta}{\omega}}\, H_\omega - \frac{3}{4}\int \frac{d\omega}{2\pi i} e^{(\omega + \frac{1}{\omega})\zeta'}\, H_\omega, \tag{13b}$$

where we have introduced the unknown function H_ω as

$$H_\omega \equiv \Gamma_\omega(0) + 3\,G_\omega. \tag{14}$$

It is useful to observe that, upon substituting Eq. (11) back into Eq. (7b), the consistency of the solution requires that

$$\int \frac{d\omega}{2\pi i} e^{\frac{\zeta}{\omega}} \frac{1}{\omega} H_\omega = 0. \tag{15}$$

Indeed, the ω contour in the Bromwich integral (8) runs parallel to the imaginary axis and to the right of all the poles of the integrand. Because the extra factor of $1/\omega$ in the integrand of Eq. (15) provides sufficient convergence at infinity, we can close the contour in the right half-plane, getting zero and confirming Eq. (15).

Finally, we can impose a further constraint on our results in Eqs. (13) by requiring them to also satisfy Eq. (7a). Plugging Eqs. (13) into Eq. (7a) and employing (15) gives the constraint

$$\int \frac{d\omega}{2\pi i} e^{\omega\zeta + \frac{\zeta}{\omega}} \left(\omega - \frac{3}{\omega}\right) H_\omega = 0. \tag{16}$$

It is convenient to define f_ω such that

$$H_\omega = \left(\frac{\omega}{\omega^2 - 3}\right) f_\omega \tag{17}$$

and expand f_ω in a Laurent series:

$$f_\omega = \sum_{n=-\infty}^{\infty} c_n\, \omega^n. \tag{18}$$

After expanding both f_ω with (18) and $e^{\zeta/\omega}$ in their respective series, we pick up the enclosed residues at $\omega = 0$ and obtain the constraint

$$\begin{aligned} 0 &= \int \frac{d\omega}{2\pi i} e^{\omega\zeta + \frac{\zeta}{\omega}} f_\omega \\ &= \sum_{n=-\infty}^{-1} c_n\, I_{-n-1}(2\zeta) + \sum_{n=0}^{\infty} c_n\, I_{n+1}(2\zeta) \\ &= c_{-1} I_0(2\zeta) + \sum_{n=1}^{\infty} (c_{n-1} + c_{-n-1}) I_n(2\zeta). \end{aligned} \tag{19}$$

Thus, we obtain $c_{-1} = 0$ and $c_n = -c_{-n-2}$ for $n \geq 0$, such that

$$f_\omega = \sum_{n=0}^{\infty} c_n \left(\omega^n - \frac{1}{\omega^{n+2}}\right). \tag{20}$$

However, we know that f_ω cannot contain large positive powers of ω, or else it would affect convergence at infinity and violate the consistency condition (15). Substituting (17) into (15) gives

$$0 = \int \frac{d\omega}{2\pi i} e^{\frac{\zeta}{\omega}} \frac{1}{\omega^2 - 3} f_\omega. \tag{21}$$

Taking $\zeta = 0$ for simplicity and using (20), we have

$$\begin{aligned} 0 &= \sum_{n=0}^{\infty} c_n \int \frac{d\omega}{2\pi i} \frac{1}{\omega^2 - 3}\left(\omega^n - \frac{1}{\omega^{n+2}}\right) \\ &= \sum_{n=1}^{\infty} c_n \int \frac{d\omega}{2\pi i} \frac{\omega^n}{\omega^2 - 3}, \end{aligned} \tag{22}$$

where for all sufficiently convergent integrals, we have closed the contour in the right-half plane and obtained zero. Therefore, the consistency condition (15) implies that $c_n = 0$ for $n \geq 1$, such that

$$H_\omega = c_0 \frac{\omega^2 - 1}{\omega\,(\omega^2 - 3)}. \tag{23}$$

The function (23) fixes the solution of the helicity evolution equations, giving for G in Eq. (13a)

$$G(\zeta) = \frac{c_0}{4}\int \frac{d\omega}{2\pi i} e^{\omega\zeta + \frac{\zeta}{\omega}} \frac{\omega^2 - 1}{\omega\,(\omega^2 - 3)}. \tag{24}$$

Using the first boundary condition in (6) in the limit $\zeta \to 0^+$ fixes the coefficient to $c_0 = 4$, after closing the contour in the left half-plane and collecting the residues at $\omega = 0, \pm\sqrt{3}$. Therefore, the complete asymptotic solution of the large-N_c helicity evolution equations is given by

$$G(\zeta) = \int \frac{d\omega}{2\pi i} e^{\omega\zeta + \frac{\zeta}{\omega}} \frac{\omega^2 - 1}{\omega\,(\omega^2 - 3)}, \tag{25a}$$

$$\Gamma(\zeta, \zeta') = 4\int \frac{d\omega}{2\pi i} e^{\omega\zeta' + \frac{\zeta}{\omega}} \frac{\omega^2 - 1}{\omega\,(\omega^2 - 3)} - 3\int \frac{d\omega}{2\pi i} e^{\omega\zeta' + \frac{\zeta'}{\omega}} \frac{\omega^2 - 1}{\omega\,(\omega^2 - 3)}. \tag{25b}$$

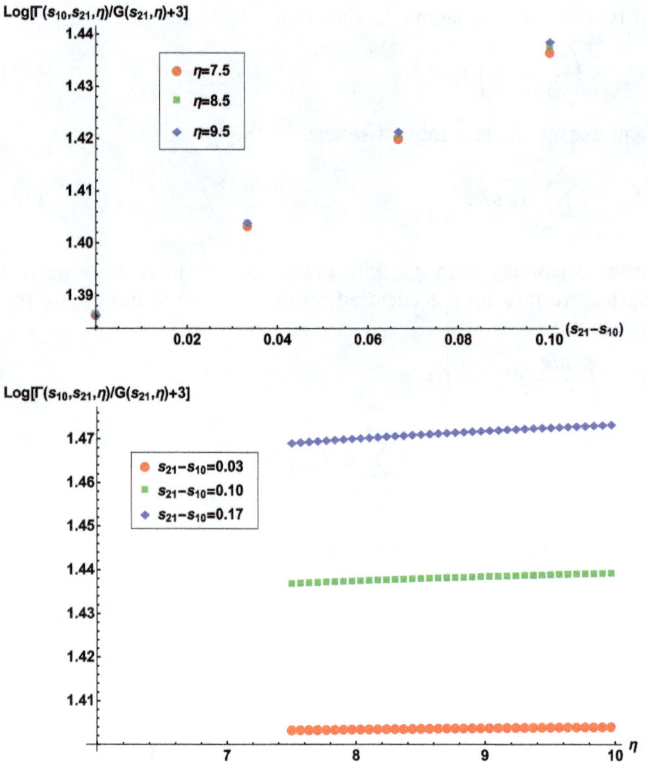

Fig. 2. Plot of the scaling ratio (28) in the numerical solution of [22] as a function of $s_{21} - s_{10}$ for fixed η (top panel) and as a function of η for fixed $s_{21} - s_{10}$ (bottom panel). The grid parameters are $\eta_{max} = 10$, $\Delta\eta = 0.03$.

The high-energy/small-x asymptotics of Eq. (25a), corresponding to $\zeta \sim \zeta' \gg 1$, are given by the right-most pole of the integrand at $\omega = +\sqrt{3}$. Keeping the contribution to (25) from this pole only, we obtain the final result

$$G(\zeta) \approx \frac{1}{3} e^{\frac{4}{\sqrt{3}}\zeta} \tag{26a}$$

$$\Gamma(\zeta, \zeta') \approx \frac{1}{3} e^{\frac{4}{\sqrt{3}}\zeta'} \left(4e^{\frac{\zeta-\zeta'}{\sqrt{3}}} - 3 \right) = G(\zeta') \left(4e^{\frac{\zeta-\zeta'}{\sqrt{3}}} - 3 \right). \tag{26b}$$

The asymptotic form of $G \sim e^{\frac{4}{\sqrt{3}}\zeta} \sim (zs)^{\alpha_h}$ in (26a) gives the analytic expression for the helicity intercept

$$\alpha_h = \frac{4}{\sqrt{3}} \sqrt{\frac{\alpha_s N_c}{2\pi}} \approx 2.3094 \sqrt{\frac{\alpha_s N_c}{2\pi}}, \tag{27}$$

in complete agreement with the numerical solution $\alpha_h \approx 2.31 \sqrt{\frac{\alpha_s N_c}{2\pi}}$ of [22]!

Finally, we note that (26b) makes a useful prediction for the form of Γ which can be straightforwardly tested against the existing numerical solution of [22]. In the units (2) used in the numerics, our analytic solution predicts that the ratio of Γ to G should scale as

$$\ln \left[\frac{\Gamma(s_{10}, s_{21}, \eta)}{G(s_{21}, \eta)} + 3 \right] = \ln 4 + \frac{1}{\sqrt{3}} (s_{21} - s_{10}). \tag{28}$$

This ratio, calculated in the numerical solution of [22], is plotted in Fig. 2, where we see excellent agreement with the features of (28). Qualitatively, no dependence of this ratio on η is seen, and the dependence on $s_{21} - s_{10}$ is linear. And even though we have not performed a detailed extrapolation from the discretized numerics to the continuum, we even see significant quantitative agreement with (28): the vertical intercept (in the top panel of Fig. 2) of 1.384 agrees fantastically with the expected $\ln 4 \approx 1.386$, and the slope of 0.517 is within 10% of the expected $\frac{1}{\sqrt{3}} \approx 0.577$. Indeed, if we perform a general fit of $\ln[\frac{\Gamma(s_{10}, s_{21}, \eta)}{G(s_{21}, \eta)} + 3]$ for $0 \le s_{10} \le s_{21} \le 0.10$ and $7.5 \le \eta \le 10$ to a function of the form $a s_{21} + b s_{10} + c\eta + d$, we find $a \approx -b \approx \frac{1}{\sqrt{3}}$ (within 10% accuracy) and $c \approx 0$, $d \approx \ln 4$ (with much greater accuracy). This preferred functional form is in excellent agreement with our analytic calculation (28). We also note that the numerics in [22] used scaling-violating initial conditions, so that the agreement seen here validates our claim of negligible dependence on the initial conditions. Thus, we can conclude with confidence that our analytic solution (26a) and helicity intercept (27) correctly reproduce the high-energy asymptotics of the numerical calculation in [22].

3. Conclusions

In this Letter, we have derived an analytic solution to the large-N_c helicity evolution equations (1) in the high-energy/small-x asymptotics. The central results are the solutions (26) for the polarized dipole amplitude G and the auxiliary neighbor dipole amplitude Γ, leading to the analytic expression for the helicity intercept (27). The key assumption which made such an analytic solution possible was the observation of emergent scaling behavior (4) as seen in the previous numerical solution of [22] (Fig. 1). It is interesting to observe that this scaling behavior is reminiscent of using $\ln k^-$ as the evolution variable rather than $\ln k^+$, as in [24]. We have checked our analytic results by comparing the predicted behavior of the auxiliary neighbor dipole amplitude Γ in Eq. (28) with the numerical solution in Fig. 2, finding excellent agreement.

Unfortunately, it is not clear whether the techniques used here can be extended to obtain an analytic solution of the helicity evolution equations in the large-$N_c \& N_f$ limit [11,12]. The addition of quark loops to the evolution kernel introduces terms which explicitly break the scaling property (4), similar to what was found for the Reggeon [8] (see also [24]). Therefore, we set aside the question of generalizing this approach to the large-$N_c \& N_f$ limit as a separate project, which we leave for future work.

Acknowledgements

The authors are greatly indebted to Edmond Iancu for his suggestion to look for a scaling solution to the helicity evolution equations and to Bin Wu for several helpful discussions of the involved integrals.

This material is based upon work supported by the U.S. Department of Energy, Office of Science, Office of Nuclear Physics under Award Number DE-SC0004286 (YK), within the framework of the TMD Topical Collaboration (DP), and DOE Contract No. DE-AC52-06NA25396 (MS). MS received additional support from the U.S. Department of Energy, Office of Science under the DOE Early Career Program Award No. 2012LANL7033 and the LANL LDRD program, project No. 20160183ER.

References

[1] E.A. Kuraev, L.N. Lipatov, V.S. Fadin, The Pomeranchuk singularity in non-Abelian gauge theories, Sov. Phys. JETP 45 (1977) 199–204.
[2] I. Balitsky, L. Lipatov, The Pomeranchuk singularity in quantum chromodynamics, Sov. J. Nucl. Phys. 28 (1978) 822–829.
[3] R. Kirschner, L. Lipatov, Double logarithmic asymptotics and Regge singularities of quark amplitudes with flavor exchange, Nucl. Phys. B 213 (1983) 122–148, http://dx.doi.org/10.1016/0550-3213(83)90178-5.

[4] R. Kirschner, Regge asymptotics of scattering amplitudes in the logarithmic approximation of QCD, Z. Phys. C 31 (1986) 135, http://dx.doi.org/10.1007/BF01559604.
[5] R. Kirschner, Regge asymptotics of scattering with flavor exchange in QCD, Z. Phys. C 67 (1995) 459–466, http://dx.doi.org/10.1007/BF01624588, arXiv:hep-th/9404158.
[6] R. Kirschner, Reggeon interactions in perturbative QCD, Z. Phys. C 65 (1995) 505–510, http://dx.doi.org/10.1007/BF01556138, arXiv:hep-th/9407085.
[7] S. Griffiths, D. Ross, Studying the perturbative Reggeon, Eur. Phys. J. C 12 (2000) 277–286, http://dx.doi.org/10.1007/s100529900240, arXiv:hep-ph/9906550.
[8] K. Itakura, Y.V. Kovchegov, L. McLerran, D. Teaney, Baryon stopping and valence quark distribution at small x, Nucl. Phys. A 730 (2004) 160–190, http://dx.doi.org/10.1016/j.nuclphysa.2003.10.016, arXiv:hep-ph/0305332.
[9] J. Bartels, B. Ermolaev, M. Ryskin, Nonsinglet contributions to the structure function g1 at small x, Z. Phys. C 70 (1996) 273–280, arXiv:hep-ph/9507271.
[10] J. Bartels, B.I. Ermolaev, M.G. Ryskin, Flavor singlet contribution to the structure function G(1) at small x, Z. Phys. C 72 (1996) 627–635, http://dx.doi.org/10.1007/s002880050285, arXiv:hep-ph/9603204.
[11] Y.V. Kovchegov, D. Pitonyak, M.D. Sievert, Helicity evolution at small x, J. High Energy Phys. 01 (2016) 072, http://dx.doi.org/10.1007/JHEP01(2016)072, arXiv:1511.06737.
[12] Y.V. Kovchegov, D. Pitonyak, M.D. Sievert, Helicity evolution at small x: flavor singlet and non-singlet observables, Phys. Rev. D 95 (1) (2017) 014033, http://dx.doi.org/10.1103/PhysRevD.95.014033, arXiv:1610.06197.
[13] I. Balitsky, Operator expansion for high-energy scattering, Nucl. Phys. B 463 (1996) 99–160, http://dx.doi.org/10.1016/0550-3213(95)00638-9, arXiv:hep-ph/9509348.
[14] I. Balitsky, Factorization and high-energy effective action, Phys. Rev. D 60 (1999) 014020, arXiv:hep-ph/9812311.
[15] I. Balitsky, Operator expansion for high-energy scattering, Nucl. Phys. B 463 (1996) 99–160, arXiv:hep-ph/9509348.
[16] Y.V. Kovchegov, Small-x F_2 structure function of a nucleus including multiple pomeron exchanges, Phys. Rev. D 60 (1999) 034008, arXiv:hep-ph/9901281.
[17] Y.V. Kovchegov, Unitarization of the BFKL pomeron on a nucleus, Phys. Rev. D 61 (2000) 074018, arXiv:hep-ph/9905214.
[18] J. Jalilian-Marian, A. Kovner, H. Weigert, The Wilson renormalization group for low x physics: gluon evolution at finite parton density, Phys. Rev. D 59 (1998) 014015, arXiv:hep-ph/9709432.
[19] J. Jalilian-Marian, A. Kovner, A. Leonidov, H. Weigert, The Wilson renormalization group for low x physics: towards the high density regime, Phys. Rev. D 59 (1998) 014014, arXiv:hep-ph/9706377.
[20] E. Iancu, A. Leonidov, L.D. McLerran, The renormalization group equation for the color glass condensate, Phys. Lett. B 510 (2001) 133–144, http://dx.doi.org/10.1016/S0370-2693(01)00524-X.
[21] E. Iancu, A. Leonidov, L.D. McLerran, Nonlinear gluon evolution in the color glass condensate. I, Nucl. Phys. A 692 (2001) 583–645, arXiv:hep-ph/0011241.
[22] Y.V. Kovchegov, D. Pitonyak, M.D. Sievert, Small-x asymptotics of the quark helicity distribution, Phys. Rev. Lett. 118 (5) (2017) 052001, http://dx.doi.org/10.1103/PhysRevLett.118.052001, arXiv:1610.06188.
[23] E.R. Nocera, E. Santopinto, Can sea quark asymmetry shed light on the orbital angular momentum of the proton?, arXiv:1611.07980.
[24] E. Iancu, J.D. Madrigal, A.H. Mueller, G. Soyez, D.N. Triantafyllopoulos, Resumming double logarithms in the QCD evolution of color dipoles, Phys. Lett. B 744 (2015) 293–302, http://dx.doi.org/10.1016/j.physletb.2015.03.068, arXiv:1502.05642.

Aspects of ultra-relativistic field theories via flat-space holography

Reza Fareghbal [a], Ali Naseh [b,*], Shahin Rouhani [c,b]

[a] Department of Physics, Shahid Beheshti University, G.C., Evin, Tehran 19839, Iran
[b] School of Particles and Accelerators, Institute for Research in Fundamental Sciences (IPM), P.O. Box 19395-5531, Tehran, Iran
[c] Department of Physics, Sharif University of Technology, P.O. Box 11365-9161, Tehran, Iran

ARTICLE INFO

Editor: N. Lambert

ABSTRACT

Recently it was proposed that asymptotically flat spacetimes have a holographic dual which is an ultra-relativistic conformal field theory. In this paper, we obtain the conformal anomaly for such a theory via the flat-space holography technique. Furthermore, using flat-space holography we obtain a C-function for this theory which is monotonically decreasing from the UV to the IR by employing the null energy condition in the bulk.

1. Introduction

It is an interesting question to extend the celebrated AdS/CFT correspondence [1] to the spacetimes other than AdS. One possible direction is exploration of holography for the asymptotically flat spacetimes. Since these spacetimes are given by taking zero cosmological constant limit from the asymptotically AdS spacetimes, it is of interest to look for the corresponding operation of flat-space limit in the boundary theory. It is proposed in papers [2] and [3] that the flat-space limit (zero cosmological constant limit) at the bulk side corresponds to taking ultra-relativistic limit of a dual CFT. The key point in this correspondence is the equivalence between symmetries of ultra-relativistic theory and asymptotic symmetries of asymptotically flat spacetimes at null infinity. These symmetries which are known as Bondi–Metzner–Sachs (BMS) symmetries [4] are infinite dimensional at three [5,6] and four dimensions [7].

The Flat/Ultra-relativistically contracted CFT correspondence can be used to get insight about the quantum nature of gravity in the asymptotically flat spacetimes or answer some questions about the ultra-relativistic theories by using the holographic calculations (see recent papers [8,9] and references therein). In this paper we want to use this duality and study some aspects of the ultra-relativistic theories. In particular we will zoom in on 2d ultra-relativistic field theories. This symmetry arises as a contraction of the Poincare symmetry, in the limit of vanishing velocity of light.

A related question in these field theories is to find a quantity which monotonically decreases with renormalization group (RG) flow. The manifestation of scale dependence in quantum field theory is the renormalization group (RG) flow of coupling constants with scale. It is however not clear why this irreversible flow (the name group should better read semi-group) should flow into a fixed point, while in principal, strange attractors are possible. That these pathological cases do not happen in physically relevant systems is interesting. In 2d this is due to Zamolodchikov's c-theorem [10], and in 4d due to a-theorem [11,12].[1] In other words the existence of a decreasing function on the RG flow enforces flow towards a fixed point (or limit cycle). There is an intuitive way of understanding the IR fixed point. In Wilson's renormalization group viewpoint [22], at any non-trivial energy scale, we simply integrate out the fast degrees of freedom. Eventually, we should not have any degrees of freedom left, in other words degrees of freedom reduce with scale, which is the essence of Cardy's formula [23].

It may be argued that for ultra-relativistic field theories, even if existence of fixed points cannot be proven, one is on safe grounds since in some limit the relativistic theory is approached and RG flow would be well behaved. However one expects that the existence of a c-theorem should also be provable in ultra-relativistic field theories. In this paper, using connection between ultra-relativistic field theory and asymptotically flat spacetimes, we propose a C-function which is monotonically decreasing from the UV to the IR. Our starting point is AdS/CFT and holographic cal-

* Corresponding author.
E-mail addresses: r_fareghbal@sbu.ac.ir (R. Fareghbal), naseh@ipm.ir (A. Naseh), rouhani@ipm.ir (S. Rouhani).

[1] Further discussions can be found in [13–21].

culation of C-function for a theory which is dual of asymptotically AdS spacetimes. We take the flat-space limit in this calculation and use the null energy condition in the bulk to find the C-function of the ultra-relativistic theory. We also use holographic method and calculate the trace anomaly of ultra-relativistic theory. The interesting point is that similar to the relativistic CFT, the conformal anomaly is related to the curvature of the spacetimes on which ultra-relativistic theory lives.

In section two we start with preliminaries and try to clarify the connection between ultra-relativistic contraction of CFT_2 and flat space limit of AdS_3 spacetimes. Section three is devoted to holographic calculations where we find conformal anomaly of the ultra-relativistic theory. In section four we propose a C-function for it.

2. Ultra-relativistic CFT and flat-space holography

In this paper we are interested in ultra-relativistic contraction of 2d CFTs, which is achievable if speed of light tends to zero. Another equivalent way of taking the ultra-relativistic limit of CFT is performed by scaling time; $t \to \epsilon t$ and taking $\epsilon \to 0$ limit [3]. If one starts with the generators of $SO(2,2)$ and contracts it by scaling of time, the final algebra is given by [3]

$$\begin{aligned}
&[E,P]=0, \quad [E,D]=E, \quad [E,B]=0, \quad [E,F]=0, \\
&[E,G]=-2B, \quad [P,D]=P, \quad [P,B]=-E, \quad [P,F]=-2B, \\
&[P,G]=2D, \quad [D,B]=0, \quad [D,F]=F, \quad [D,G]=G, \\
&[B,F]=0, \quad [B,G]=-F, \quad [F,G]=0.
\end{aligned} \tag{2.1}$$

The generators E, P, B, D, F, G are respectively energy, momentum, boost, dilation, acceleration and special conformal transformation (for their precise definitions see below (2.4)). Note that this algebra is isomorphic to Galilean Conformal Algebra (GCA) [24] simply by switching the space and time coordinates. A duality which seems to be peculiar to $1+1$ dimensions. The above algebra (2.1) has an affine extension with an infinite number of generators which is given by contraction of two copies of Virasoro algebra. More precisely, if we start with two copies of the conformal algebra with generators $l_n, \bar{l}_n$ and central charges $c, \bar{c}$ and define generators [3,24,25]

$$\begin{aligned}
M_n &= \lim_{\epsilon\to 0} \epsilon \left(l_n + \bar{l}_{-n}\right), \\
L_n &= \lim_{\epsilon\to 0} \left(l_n - \bar{l}_{-n}\right),
\end{aligned}$$

the final algebra is

$$\begin{aligned}
&[L_m, L_n] = (m-n)L_{m+n} + c_L m(m^2-1)\delta_{m+n,0}, \\
&[L_m, M_n] = (m-n)M_{m+n} + c_M m(m^2-1)\delta_{m+n,0},
\end{aligned} \tag{2.2}$$

where

$$\begin{aligned}
c_L &= \lim_{\epsilon\to 0} \frac{1}{12}(c-\bar{c}), \\
c_M &= \lim_{\epsilon\to 0} \frac{\epsilon}{12}(c+\bar{c}).
\end{aligned} \tag{2.3}$$

The generators of the global part of the algebra (2.2) are related to (2.1) by

$$\begin{aligned}
&L_0 = D, \quad L_1 = G, \quad L_{-1} = -P, \\
&M_0 = -B, \quad M_1 = F, \quad M_{-1} = E.
\end{aligned} \tag{2.4}$$

The interesting point is that (2.2) appears also in a completely different context. It is the asymptotic symmetry at the null infinity of the three dimensional asymptotically flat spacetimes [6]. Because of this, it was proposed in [2,3] that the holographic dual of asymptotically flat spacetimes is a field theory which has ultra-relativistic conformal symmetry. The dual ultra-relativistic field theory is given by contraction of CFT. Since according to AdS/CFT correspondence the parent CFT is dual to asymptotically AdS spacetimes, the contracted CFT corresponds to flat-space limit (zero cosmological constant limit) of the asymptotically AdS spacetimes. If one starts with AdS_3 spacetime given by

$$ds^2 = -\left(1+\frac{r^2}{\ell^2}\right)d\tau^2 + \frac{dr^2}{\left(1+\frac{r^2}{\ell^2}\right)} + r^2 d\phi^2, \tag{2.5}$$

the geometry which the dual CFT lives on it, is given by using conformal boundary of the AdS spacetime,

$$ds^2_{boundary} = \frac{r^2}{G^2}\left(-\frac{G^2}{\ell^2}d\tau^2 + G^2 d\phi^2\right) \tag{2.6}$$

where we have intentionally used Newton's constant in the conformal factor to make it dimensionless and also have a ℓ independent conformal factor. It is clear from (2.6) that the boundary CFT lives on a two dimensional flat spacetimes and its time coordinate, t, is related to the bulk as $t = \frac{G}{\ell}\tau$. Now taking the flat-space limit in the bulk side by sending $\ell \to \infty$ or $\epsilon = \frac{G}{\ell} \to 0$ (while keeping G fixed) corresponds to contraction of time in the boundary as $t \to \epsilon t$. Using this correspondence one can study some aspects of gravity in the three dimensional asymptotically flat spacetimes by using two dimensional ultra-relativistically contracted CFT. For example a quasi local stress tensor for asymptotically flat spacetimes is achievable if we start with energy momentum of a CFT and contract it [26].

To be more precise, let us consider energy–momentum tensor of a relativistic CFT, $T_{\mu\nu}$, which satisfies a relativistic conservation equation. The components of the energy–momentum tensor of the contracted CFT, $\tilde{T}_{\mu\nu}$ are given by [26]

$$\begin{aligned}
\tilde{T}_{++} + \tilde{T}_{--} &= \lim_{\epsilon\to 0} \epsilon\,(T_{++} + T_{--}) \\
\tilde{T}_{++} - \tilde{T}_{--} &= \lim_{\epsilon\to 0} (T_{++} - T_{--}) \\
\tilde{T}_{+-} &= \lim_{\epsilon\to 0} \epsilon\, T_{+-}
\end{aligned} \tag{2.7}$$

where $+$ and $-$ correspond to light-cone coordinates in both of the theories. The components of the stress tensor given by (2.7) can be used to calculate the charges of the asymptotic symmetries in the gravity side. The Poisson bracket of these charges has a central extension with a non-zero central charge c_M. The value of c_M is consistent with the one which is given by (2.3). Thus c_M which is defined by (2.3) has an appropriate definition in terms of the charges of the gravity theory.

3. Ultra-relativistic conformal anomaly

In this section we calculate the conformal anomaly of the ultra-relativistic conformal field theories using holography. The starting point for the holographic calculations is AdS/CFT and its standard dictionary for the relation between bulk and boundary quantities and then taking the flat-space limit. However, the flat-space limit or taking infinite radius limit of asymptotically AdS spacetimes is gauge dependent and is not well-defined for line elements written in the Fefferman–Graham coordinates. In order to have a well-defined flat limit all calculations will be done in the so called BMS gauge [27]. For the three-dimensional asymptotically locally AdS spacetimes the BMS gauge is given by,

$$ds^2 = \mathcal{A}(u,r,\phi)du^2 - 2e^{2\beta(u,\phi)}dudr + 2\mathcal{B}(u,r,\phi)dud\phi + r^2d\phi^2, \tag{3.1}$$

where for the Einstein gravity with negative cosmological constant,

$$S = \frac{1}{16\pi G}\int d^3x\sqrt{-g}\left(R + \frac{2}{\ell^2}\right), \tag{3.2}$$

$\mathcal{A}$ and $\mathcal{B}$ are determined as

$$\mathcal{A}(u,r,\phi) = -e^{4\beta(u,\phi)}\frac{r^2}{\ell^2} + \mathcal{M}(u,\phi),$$
$$\mathcal{B}(u,r,\phi) = -r\partial_\phi e^{2\beta(u,\phi)} + \mathcal{N}(u,\phi), \tag{3.3}$$

and

$$\partial_\phi\mathcal{M} - 2\partial_u\mathcal{N} + 4\mathcal{N}\partial_u\beta = 0$$
$$-8\partial_\phi\beta\partial_u\partial_\phi\beta - 4\partial_u\partial_\phi^2\beta + \frac{4}{\ell^2}\mathcal{N}\partial_\phi\beta + \frac{2}{\ell^2}\partial_\phi\mathcal{N} - e^{-4\beta}\partial_u\mathcal{M}$$
$$+4\mathcal{M}e^{-4\beta}\partial_u\beta = 0. \tag{3.4}$$

The quasi-local stress tensor $T_{\mu\nu}$ of the metric (3.1) are given by using the standard calculation of the holographic renormalization method [28,29]. According to AdS/CFT correspondence the components of $T_{\mu\nu}$ at the bulk correspond to expectation values of the energy–momentum tensor of the boundary CFT. These components are

$$T_{uu} = \frac{e^{4\beta}\left(\partial_\phi^2\beta + (\partial_\phi\beta)^2\right)}{4\pi\ell G} + \frac{\mathcal{M}}{16\pi\ell G},$$
$$T_{u\phi} = \frac{\mathcal{N}}{8\pi\ell G},$$
$$T_{\phi\phi} = \frac{\ell\mathcal{M}e^{-4\beta}}{16\pi G} - \frac{\ell(\partial_\phi\beta)^2}{4\pi G}. \tag{3.5}$$

Conformal boundary of the asymptotically locally AdS spacetimes (3.1) is given by

$$ds^2|_{C.B} = -\frac{G^2}{\ell^2}e^{4\beta(u,\phi)}du^2 + G^2d\phi^2 \tag{3.6}$$

where we have intentionally used G to make the conformal factor (which was used in (3.1) for defining conformal boundary) dimensionless. It is clear from (3.6) that the conformal boundary is non-flat. Thus the dual CFT must live on a non-flat spacetime. The Ricci scalar of the conformal boundary is

$$R_{C.B} = -\frac{4}{G^2}\left(\partial_\phi^2\beta + 2(\partial_\phi\beta)^2\right). \tag{3.7}$$

Using (3.5) and (3.7) we can calculate the trace of stress tensor as

$$T = g^{ij}T_{ij} = -\frac{\ell}{4\pi G^3}\left(\partial_\phi^2\beta + 2(\partial_\phi\beta)^2\right), \tag{3.8}$$

which can be written as

$$T = \frac{C}{24\pi}R_{C.B}, \tag{3.9}$$

where $C = 3\ell/2G$ is the Brown and Henneaux's central charge. The result (3.9) is the conformal anomaly of the boundary CFT.

We should note that the u dependence in function $\beta(u,\phi)$ is crucial to have conformal anomaly. For the case which β is only a function of ϕ coordinate we have

$$T = \Box\Phi(\phi), \tag{3.10}$$

where $\Box \equiv \nabla^i\nabla_i$ is defined by the metric (3.6) and $\Phi(\phi) = -\frac{C}{6\pi}\beta(\phi)$. Appearance of $\Box\Phi(\phi)$ on the right-hand side of (3.10) creates the opportunity to define an improved stress tensor:

$$\Theta_{ij} = T_{ij} + \left(\nabla_i\nabla_j - g_{ij}\Box\right)\Phi(\phi), \tag{3.11}$$

which is traceless. Therefore to have a non-trivial conformal anomaly (3.9), "u" dependence of the function β is crucial.

Now take flat-space limit in the above calculations and propose a holographic derivation for the ultra-relativistic conformal anomaly. The main steps have been done in paper [26] where the stress tensor near the null infinity of the asymptotically flat spacetimes $\tilde{T}_{ij}$ is related to the AdS counterpart T_{ij} as

$$\tilde{T}_{++} + \tilde{T}_{--} = \lim_{\frac{G}{\ell}\to 0}\frac{G}{\ell}(T_{++} + T_{--})$$
$$\tilde{T}_{++} - \tilde{T}_{--} = \lim_{\frac{G}{\ell}\to 0}(T_{++} - T_{--})$$
$$\tilde{T}_{+-} = \lim_{\frac{G}{\ell}\to 0}\frac{G}{\ell}T_{+-} \tag{3.12}$$

where $x^\pm = \frac{u}{\ell} \pm \phi$ for the asymptotically AdS solutions and $x^\pm = \frac{u}{G} \pm \phi$ for the asymptotically flat cases. We should note that in the BMS gauge, the asymptotically flat solutions are given also as (3.1). However, $\mathcal{A}$ and $\mathcal{B}$ are given by taking flat limit i.e. $G/\ell \to 0$, from the equations (3.3) and (3.4). Another point is that according to [26] the dual ultra-relativistic theory lives on a two dimensional spacetime which is given by taking flat limit from the conformal boundary of the AdS counterpart. For the current case the line element of the corresponding spacetime is given by

$$d\tilde{s}^2 = -e^{4\beta(u,\phi)}du^2 + G^2d\phi^2, \tag{3.13}$$

which is not flat and its Ricci scalar is the same as (3.7). It is clear that the standard definition of conformal infinity does not yield (3.13). However, it is possible to find (3.13) by an anisotropic scaling of the asymptotically flat metric. This method for definition of the conformal infinity is similar to [30] where the anisotropic scaling is proposed for definition of the conformal boundary of the non-relativistic theories.

Using (3.5) and (3.12) we have

$$\tilde{T}_{uu} = \frac{e^{4\beta}\left(\partial_\phi^2\beta + (\partial_\phi\beta)^2\right)}{4\pi G^2} + \frac{\mathcal{M}}{16\pi G^2},$$
$$\tilde{T}_{u\phi} = \frac{\mathcal{N}}{8\pi G^2},$$
$$\tilde{T}_{\phi\phi} = \frac{\mathcal{M}e^{-4\beta}}{16\pi} - \frac{(\partial_\phi\beta)^2}{4\pi}. \tag{3.14}$$

Finally we find

$$\tilde{T} = \tilde{g}^{ij}\tilde{T}_{ij} = \frac{1}{4\pi}c_M\tilde{R}, \tag{3.15}$$

where $\tilde{R}$ is the Ricci scalar of (3.13). This is one of the main results of the current paper.[2]

[2] We should note that taking flat limit from the stress tensor of asymptotically AdS spacetimes has been also studied in paper [31] which does not consider ultra-relativistic field theory as the dual of asymptotically flat spacetimes and our final result (3.15) is absent in that paper.

4. Ultra-relativistic c-theorem

In order to propose a holographic C-function for ultra-relativistic CFT, we first redo the holographic calculation for the asymptotically AdS spacetimes in the BMS gauge and then take the flat-space limit. The holographic proof of c-theorem in Fefferman–Graham gauge is provided in [32,33]. Note that the Fefferman–Graham gauge is not appropriate for taking the flat-space limit. Therefore in this section we firstly rederive that analysis in the BMS gauge which will be appropriate for taking the flat-space limit.

To see an RG flow in boundary field theory, we should add matter to the cosmological Einstein–Hilbert theory

$$S=\frac{1}{16\pi G}\int d^3x\sqrt{-g}\left(R+\frac{2}{\ell^2}\right)+S_{matter}, \tag{4.1}$$

where "matter" could be for example a scalar field. Note that without "matter" (4.1) admits the below radial solution (a solution which depends just on "r")

$$ds^2=-2dudr+r^2(-\frac{1}{l^2}d^2u+d^2x). \tag{4.2}$$

We demand that the boundary theory has 2d-Poincaré symmetry. The boundary theory lives on the hypersurface at constant "r". The most general possible 3d-bulk radial-metric in the BMS gauge which is consistent with requested symmetry can be written as

$$ds^2=-2dudr+e^{2B(r)}(-\frac{1}{l^2}d^2u+d^2x). \tag{4.3}$$

We propose the gravitational relativistic C-function in BMS gauge as

$$C(r)=c\,\frac{e^{-B(r)}}{B'(r)}, \tag{4.4}$$

which for that

$$C'(r)=-c\,\frac{e^{-B(r)}(B''(r)+B'(r)^2)}{B'(r)^2}. \tag{4.5}$$

Note that for the AdS$_3$ case ($B(r)=ln(r)$), the function $C(r)=c$ and $C'(r)=0$. The "c" is a constant but at fixed points (AdS solutions) it reduces to the central charge.

Moreover, according to the null-energy condition, for any null vector ξ^μ we must have

$$T^m_{\mu\nu}\xi^\mu\xi^\nu\geq 0, \tag{4.6}$$

where T^m is the matter stress-tensor. The matter stress-tensor for the theory (4.1) in geometry (4.3) has this structure

$$T^m_{\mu\nu}=\begin{pmatrix} EQ_{rr} & EQ_{ru} & 0\\ EQ_{ru} & EQ_{uu} & 0\\ 0 & 0 & EQ_{\phi\phi}\end{pmatrix}, \tag{4.7}$$

where

$$EQ_{rr}=-B''(r)-B'(r)^2,\qquad EQ_{uu}=\frac{e^{2B(r)}}{l^2}EQ_{ru},$$
$$EQ_{\phi\phi}=-e^{2B(r)}EQ_{ru},$$
$$EQ_{ru}=-\frac{1}{l^2}\left(e^{2B(r)}B''(r)+2e^{2B(r)}B'(r)^2-1\right).$$

Furthermore, by considering the vector ξ^μ as

$$\xi_\mu=\big(\xi_1(r,u,x),\xi_2(r,u,x),\xi_3(r,u,x)\big), \tag{4.8}$$

and solving $g^{\mu\nu}\xi_\mu\xi_\nu=0$, it is easy to see that the general null vector in geometry (4.3) has the form:

$$\xi^\mu=\left(\frac{\sqrt{\xi_2^2l^2-\xi_3^2}}{l},-\frac{(\xi_2 l+\sqrt{\xi_2^2l^2-\xi_3^2})l}{e^{2B(r)}},\frac{\xi_3}{e^{2B(r)}}\right). \tag{4.9}$$

Substituting the above null vector in (4.6) and using (4.7) gives

$$-\frac{(\xi_2^2l^2-\xi_3^2)}{l^2}\left(B''(r)+B'(r)^2\right)\geq 0. \tag{4.10}$$

Note that the first parenthesis in the above equation is positive because it also appears in the square root term of (4.9). By using this result in (4.5) and noting that $c>0$ for unitary theories, we see that $C'(r)$ is semipositive. Therefore by increasing "r", $C(r)$ also increases. Note that for the metric (4.2) the boundary lives at $r\to\infty$ which is the UV regime of the dual field theory.[3] Thus by moving from the boundary to the bulk in the gravity side ($r=\infty$ to $r=0$) which is equivalent to moving from the UV to the IR regime in the dual field theory, $C(r)$ has a monotonic behavior. This completes the proof of the monotonic behavior of the gravitational relativistic C-function (4.4) in the BMS gauge.

Let us see what happens in the flat space limit ($l\to\infty$). Firstly (4.3) in the flat space limit becomes

$$ds^2=-2dudr+e^{2B(r)}d^2x, \tag{4.11}$$

and (4.7) in the same limit becomes

$$T^m_{\mu\nu}=\begin{pmatrix} EQ_{rr} & 0 & 0\\ 0 & 0 & 0\\ 0 & 0 & 0\end{pmatrix}, \tag{4.12}$$

with

$$EQ_{rr}=-B''(r)-B'(r)^2. \tag{4.13}$$

Moreover the general null vector of the metric (4.11) is

$$\xi^\mu=\left(-\xi_2,-\frac{1}{2}\frac{\xi_3^2}{e^{2B(r)}\xi_2},\frac{\xi_3}{e^{2B(r)}}\right). \tag{4.14}$$

Substituting (4.14) and (4.12) in (4.6) results in

$$-\xi_2^2\left(B''(r)+B'(r)^2\right)\geq 0. \tag{4.15}$$

Furthermore, the flat space limit of (4.4) and (4.5) are

$$C(r)=6c_M\frac{e^{-B(r)}}{B'(r)}, \tag{4.16}$$

and

$$C'(r)=-6c_M\,\frac{e^{-B(r)}(B''(r)+B'(r)^2)}{B'(r)^2}, \tag{4.17}$$

respectively. By using (4.15) and assuming $c_M>0$ because of unitary reason [34], one can see that (4.17) is semipositive. The semipositivity of (4.17) guarantees that the proposed ultra-relativistic C-function (4.16) has a monotonic flow from the UV to the IR regime.

[3] One of the Killing vectors of AdS_3 in the BMS gauge is $\xi^\mu=(\lambda r,-\lambda u,-\lambda x)$. Therefore by moving to the boundary, the norm of the spatial coordinate (norm of the difference between zero and arbitrary point in the boundary) will decrease. This means moving to the UV regime of the dual field theory.

5. Discussion

In this paper we have derived a C-function for the ultra-relativistic CFT which decreases from the UV to the IR region. We have also derived the conformal anomaly by allowing the boundary to become curved whilst the bulk remains flat. In this case indeed the conformal anomaly appears and the trace of the energy–momentum tensor fails to vanish. We showed that the trace anomaly is proportional to the ultra-relativistic central charge c_M. The reason that central charge c_L does not appear in the final result is that for the ultra-relativistic field theories which we consider in this paper, c_L is actually zero. This happens because the bulk gravitational theory preserves parity. The non-vanishing c_L is achievable in the gravitational theories which are not parity invariant. For those gravitational theories, exploring the effect of c_L in the boundary conformal anomaly is interesting.

Moreover the appearance of anomaly in the trace of holographic stress tensor reflects the fact that in the process of holographic renormalization some counterterms are added which break conformal symmetry. In this paper we started with the variation of the on-shell action and those counterterms do not contribute and are not relevant. Note that these types of counterterms were found in the FG gauge in the renowned paper [35]. However the metric in FG gauge is not appropriate for the flat space limit. To see which types of counterterms are necessary to do the full analysis of holographic renormalization in the BMS gauge will be interesting.

Acknowledgements

We are indebted to Mahmoud Safari for his contributions in the early part of this project. R.F. would like to thank School of Particles and Accelerator of Institute for Research in Fundamental Sciences (IPM) for the research facilities.

References

[1] J.M. Maldacena, Int. J. Theor. Phys. 38 (1999) 1113;
J.M. Maldacena, Adv. Theor. Math. Phys. 2 (1998) 231;
E. Witten, Adv. Theor. Math. Phys. 2 (1998) 253;
S.S. Gubser, I.R. Klebanov, A.M. Polyakov, Phys. Lett. B 428 (1998) 105.

[2] A. Bagchi, Correspondence between asymptotically flat spacetimes and nonrelativistic conformal field theories, Phys. Rev. Lett. 105 (2010) 171601;
A. Bagchi, The BMS/GCA correspondence, arXiv:1006.3354 [hep-th].

[3] A. Bagchi, R. Fareghbal, BMS/GCA redux: towards flatspace holography from non-relativistic symmetries, J. High Energy Phys. 1210 (2012) 092, arXiv:1203.5795 [hep-th].

[4] H. Bondi, M.G. van der Burg, A.W. Metzner, Gravitational waves in general relativity. 7. Waves from axisymmetric isolated systems, Proc. R. Soc. Lond. A 269 (1962) 21;
R.K. Sachs, Gravitational waves in general relativity. 8. Waves in asymptotically flat space–times, Proc. R. Soc. Lond. A 270 (1962) 103;
R.K. Sachs, Asymptotic symmetries in gravitational theory, Phys. Rev. 128 (1962) 2851.

[5] A. Ashtekar, J. Bicak, B.G. Schmidt, Asymptotic structure of symmetry reduced general relativity, Phys. Rev. D 55 (1997) 669, arXiv:gr-qc/9608042.

[6] G. Barnich, G. Compere, Classical central extension for asymptotic symmetries at null infinity in three spacetime dimensions, Class. Quantum Gravity 24 (2007) F15, arXiv:gr-qc/0610130.

[7] G. Barnich, C. Troessaert, Symmetries of asymptotically flat 4 dimensional spacetimes at null infinity revisited, arXiv:0909.2617 [gr-qc].

[8] A. Bagchi, D. Grumiller, W. Merbis, Stress tensor correlators in three-dimensional gravity, arXiv:1507.05620 [hep-th].

[9] S.M. Hosseini, A. Veliz-Osorio, Gravitational anomalies, entanglement entropy, and flat-space holography, arXiv:1507.06625 [hep-th].

[10] A.B. Zamolodchikov, Irreversibility of the flux of the renormalization group in a 2D field theory, JETP Lett. 43 (1986) 730, Pis'ma Zh. Eksp. Teor. Fiz. 43 (1986) 565.

[11] M.A. Luty, J. Polchinski, R. Rattazzi, The a-theorem and the asymptotics of 4D quantum field theory, J. High Energy Phys. 1301 (2013) 152, http://dx.doi.org/10.1007/JHEP01(2013)152, arXiv:1204.5221 [hep-th].

[12] Z. Komargodski, A. Schwimmer, On renormalization group flows in four dimensions, J. High Energy Phys. 1112 (2011) 099, http://dx.doi.org/10.1007/JHEP12(2011)099, arXiv:1107.3987 [hep-th].

[13] A. Dymarsky, Z. Komargodski, A. Schwimmer, S. Theisen, On scale and conformal invariance in four dimensions, arXiv:1309.2921 [hep-th].

[14] J. Polchinski, Scale and conformal invariance in quantum field theory, Nucl. Phys. B 303 (1988) 226.

[15] V. Riva, J.L. Cardy, Scale and conformal invariance in field theory: a physical counterexample, Phys. Lett. B 622 (2005) 339, arXiv:hep-th/0504197.

[16] J.F. Fortin, B. Grinstein, A. Stergiou, Limit cycles and conformal invariance, J. High Energy Phys. 1301 (2013) 184, http://dx.doi.org/10.1007/JHEP01(2013)184, arXiv:1208.3674 [hep-th].

[17] K. Farnsworth, M. Luty, P. Prelipina, Scale invariance plus unitarity implies conformal invariance in four dimensions, arXiv:1309.4095 [hep-th].

[18] A. Bzowski, K. Skenderis, Comments on scale and conformal invariance in four dimensions, arXiv:1402.3208 [hep-th].

[19] Y. Nakayama, Phys. Rep. 569 (2015) 1, http://dx.doi.org/10.1016/j.physrep.2014.12.003, arXiv:1302.0884 [hep-th].

[20] A. Dymarsky, A. Zhiboedov, Scale-invariant breaking of conformal symmetry, J. Phys. A 48 (41) (2015) 41FT01, http://dx.doi.org/10.1088/1751-8113/48/41/41FT01, arXiv:1505.01152 [hep-th].

[21] A. Naseh, Scale vs conformal invariance from entanglement entropy, arXiv:1607.07899 [hep-th].

[22] K.G. Wilson, The renormalization group: critical phenomena and the Kondo problem, Rev. Mod. Phys. 47 (1975) 773.

[23] J.L. Cardy, Operator content of two-dimensional conformally invariant theories, Nucl. Phys. B 270 (1986) 186.

[24] A. Bagchi, R. Gopakumar, Galilean conformal algebras and AdS/CFT, J. High Energy Phys. 0907 (2009) 037, arXiv:0902.1385 [hep-th].

[25] A. Hosseiny, S. Rouhani, Affine extension of Galilean conformal algebra in $2+1$ dimensions, J. Math. Phys. 51 (2010) 052307, arXiv:0909.1203 [hep-th].

[26] R. Fareghbal, A. Naseh, Flat-space energy–momentum tensor from BMS/GCA correspondence, J. High Energy Phys. 1403 (2014) 005, arXiv:1312.2109 [hep-th].

[27] G. Barnich, A. Gomberoff, H.A. Gonzalez, The flat limit of three dimensional asymptotically anti-de Sitter spacetimes, Phys. Rev. D 86 (2012) 024020, arXiv:1204.3288 [gr-qc].

[28] V. Balasubramanian, P. Kraus, A stress tensor for anti-de Sitter gravity, Commun. Math. Phys. 208 (1999) 413, arXiv:hep-th/9902121.

[29] S. de Haro, S.N. Solodukhin, K. Skenderis, Holographic reconstruction of space-time and renormalization in the AdS/CFT correspondence, Commun. Math. Phys. 217 (2001) 595, arXiv:hep-th/0002230.

[30] P. Horava, C.M. Melby-Thompson, Anisotropic conformal infinity, Gen. Relativ. Gravit. 43 (2011) 1391, http://dx.doi.org/10.1007/s10714-010-1117-y, arXiv:0909.3841 [hep-th].

[31] R.N. Caldeira Costa, Aspects of the zero Λ limit in the AdS/CFT correspondence, Phys. Rev. D 90 (10) (2014) 104018, arXiv:1311.7339 [hep-th].

[32] D.Z. Freedman, S.S. Gubser, K. Pilch, N.P. Warner, Renormalization group flows from holography supersymmetry and a c theorem, Adv. Theor. Math. Phys. 3 (1999) 363, arXiv:hep-th/9904017.

[33] R.C. Myers, A. Sinha, Holographic c-theorems in arbitrary dimensions, J. High Energy Phys. 1101 (2011) 125, arXiv:1011.5819 [hep-th].

[34] A. Bagchi, S. Detournay, R. Fareghbal, J. Simón, Holography of 3D flat cosmological horizons, Phys. Rev. Lett. 110 (14) (2013) 141302, arXiv:1208.4372 [hep-th].

[35] M. Henningson, K. Skenderis, The holographic Weyl anomaly, J. High Energy Phys. 9807 (1998) 023, arXiv:hep-th/9806087.

Fermionic continuous spin gauge field in (A)dS space

R.R. Metsaev

Department of Theoretical Physics, P.N. Lebedev Physical Institute, Leninsky prospect 53, Moscow 119991, Russia

ARTICLE INFO

Editor: N. Lambert

Keywords:
Fermionic field
Continuous spin
Higher-spin field

ABSTRACT

Fermionic continuous spin field propagating in (A)dS space-time is studied. Gauge invariant Lagrangian formulation for such fermionic field is developed. Lagrangian of the fermionic continuous spin field is constructed in terms of triple gamma-traceless tensor-spinor Dirac fields, while gauge symmetries are realized by using gamma-traceless gauge transformation parameters. It is demonstrated that partition function of fermionic continuous spin field is equal to one. Modified de Donder gauge condition that considerably simplifies analysis of equations of motion is found. Decoupling limits leading to arbitrary spin massless, partial-massless, and massive fermionic fields are studied.

1. Introduction

In view of many interesting features of continuous spin gauge field theory this topic has attracted some interest in recent time [1–3]. For a list of references devoted to various aspects of continuous spin field, see Refs. [2,4,5]. It seems likely that some regime of the string theory can be related to continuous spin field theory (see, e.g., Refs. [6]). Other interesting feature of continuous spin field theory, which triggered our interest in this topic, is that the bosonic continuous spin field can be decomposed into an infinite chain of coupled scalar, vector, and totally symmetric tensor fields which consists of every field just once. We recall then that a similar infinite chain of scalar, vector and totally symmetric fields enters the theory of bosonic higher-spin gauge field in AdS space [7].

Supersymmetry plays important role in string theory and higher-spin gauge field theory. We expect that supersymmetry will also play important role in theory of continuous spin field. For a discussion of supersymmetry, we need to study bosonic and fermionic continuous spin fields. Lagrangian formulation of bosonic continuous spin field in flat space $R^{3,1}$ was studied in Ref. [2], while Lagrangian formulation of bosonic continuous spin field in flat space $R^{d,1}$ and $(A)dS_{d+1}$ space with arbitrary d was discussed in Ref. [3]. Lagrangian formulation of fermionic continuous spin field in flat space $R^{3,1}$ was obtained in Ref. [1]. So far Lagrangian formulation of fermionic continuous spin field in $(A)dS_{d+1}$ and $R^{d,1}$ with arbitrary d has not been discussed in the literature. Our major aim in this paper is to develop Lagrangian formulation of continuous spin fermionic field in flat space $R^{d,1}$ and $(A)dS_{d+1}$ space with arbitrary $d \geq 3$.[1] As by product, using our Lagrangian formulation, we compute partition functions of fermionic continuous spin fields in (A)dS and flat spaces and show that such partition functions are equal to 1. Considering various decoupling limits, we demonstrate how massless, partial-massless, and massive fermionic fields appear in the framework of Lagrangian formulation of fermionic continuous spin (A)dS field.

2. Lagrangian formulation of fermionic continuous spin field

Field content. To discuss gauge invariant and Lorentz covariant formulation of fermionic continuous spin field propagating in $(A)dS_{d+1}$ space we introduce the following set of Dirac complex-valued spin-$\frac{1}{2}$, spin-$\frac{3}{2}$, and tensor-spinor fields of the Lorentz $so(d,1)$ algebra

$$\psi^{a_1\ldots a_n\alpha}\,, \qquad n = 0, 1, \ldots, \infty\,, \tag{2.1}$$

where flat vector indices of the $so(d,1)$ algebra $a_1,\ldots,a_n$ run over $0,1,\ldots,d$, while α stands for spinor index. In what follows, the spinor indices will be implicit. In (2.1), fields with $n = 0$ and $n = 1$ are the respective Dirac spin-$\frac{1}{2}$ and spin-$\frac{3}{2}$ fields of the $so(d,1)$ algebra, while fields with $n \geq 2$ are the totally symmetric spin-$(n+\frac{1}{2})$ Dirac tensor-spinor fields of the $so(d,1)$ algebra.

[1] We agree with remarks of Authors in Refs. [1,2] that generalization of Lagrangian for continuous spin massless field in flat space $R^{3,1}$ to the case of $R^{d,1}$ with $d > 3$ is straightforward.

E-mail address: metsaev@lpi.ru.

Tensor–spinor fields $\psi^{a_1...a_n}$ (2.1) with $n \ge 3$ are considered to be triple gamma-traceless,

$$\gamma^a \psi^{abba_4...a_n} = 0\,, \qquad n = 3, 4, \ldots, \infty. \tag{2.2}$$

Dirac fields (2.1) subject to constraints (2.2) constitute a field content in our approach.

In order to obtain a gauge invariant description of the continuous spin fermionic field in an easy-to-use form, we introduce bosonic creation operators α^a, υ. Using the α^a, υ, we collect all fields appearing in (2.1) into a ket-vector $|\psi\rangle$,

$$|\psi\rangle = \sum_{n=0}^{\infty} \frac{\upsilon^n}{n!\sqrt{n!}} \alpha^{a_1} \ldots \alpha^{a_n} \psi^{a_1...a_n} |0\rangle\,. \tag{2.3}$$

We note that triple gamma-tracelessness constraint (2.2) can be represented as $\bar\alpha^2 \gamma\bar\alpha|\psi\rangle = 0$.

Lagrangian. Gauge invariant action and Lagrangian of the fermionic continuous spin field we found take the form

$$S = \int d^{d+1}x \mathcal{L}\,, \qquad \mathrm{i}\mathcal{L} = e\langle\psi|E|\psi\rangle\,, \tag{2.4}$$

$$E = E_{(1)} + E_{(0)}\,, \tag{2.5}$$

$$E_{(1)} = \not{D} - \alpha D\gamma\bar\alpha - \gamma\alpha\bar\alpha D + \gamma\alpha\not{D}\gamma\bar\alpha + \frac{1}{2}\gamma\alpha\alpha D\bar\alpha^2 + \frac{1}{2}\alpha^2\gamma\bar\alpha\bar\alpha D - \frac{1}{4}\alpha^2\not{D}\bar\alpha^2\,, \tag{2.6}$$

$$E_{(0)} = (1 - \gamma\alpha\gamma\bar\alpha - \frac{1}{4}\alpha^2\bar\alpha^2)e_1^\Gamma + (\gamma\alpha - \frac{1}{2}\alpha^2\gamma\bar\alpha)\bar e_1 + (\gamma\bar\alpha - \frac{1}{2}\gamma\alpha\bar\alpha^2)e_1\,, \tag{2.7}$$

$\langle\psi| \equiv (|\psi\rangle)^\dagger\gamma^0$, where $e = \det e_{\underline{m}}^a$ and $e_{\underline{m}}^a$ stands for vielbein in (A)dS space. In (2.6) and below, the notation $\not{D}$ is used for the Dirac operator in (A)dS space. A definition of quantities like αD, $\gamma\alpha$, α^2 may be found in Appendix. Quantities e_1^Γ, e_1, and $\bar e_1$ entering $E_{(0)}$ (2.7) are defined as

$$e_1^\Gamma = \frac{2\kappa_0}{2N_\upsilon + d - 1}\,, \qquad e_1 = e_\upsilon\bar\upsilon\,, \qquad \bar e_1 = -\upsilon e_\upsilon\,, \tag{2.8}$$

$$e_\upsilon = \Big(\frac{F_\upsilon}{(N_\upsilon + 1)(2N_\upsilon + d - 1)}\Big)^{1/2}\,, \qquad N_\upsilon = \upsilon\bar\upsilon\,, \tag{2.9}$$

$$F_\upsilon = \kappa_0^2 - \mu_0\big(N_\upsilon + \frac{d-1}{2}\big)^2 - \rho\big(N_\upsilon + \frac{d-1}{2}\big)^4\,, \tag{2.10}$$

$$\kappa_0 \text{ and } \mu_0 - \text{ dimensionfull parameters}, \tag{2.11}$$

$$\rho = -R^{-2} \text{ for AdS}; \quad \rho = 0 \text{ for flat}; \quad \rho = R^{-2} \text{ for dS}, \tag{2.12}$$

where the R stands for a radius of (A)dS space. The following remarks are in order.

i) Our Lagrangian depends on two arbitrary dimensionfull parameters κ_0, μ_0 and on ρ (2.12).

ii) The one-derivative operator $E_{(1)}$ (2.6) coincides with the one-derivative contribution to the Fang–Fronsdal operator that enters Lagrangian of fermionic massless field in (A)dS space.

Gauge symmetries. In order to describe gauge transformation of continuous spin field we use the following gauge transformation parameters:

$$\xi^{a_1...a_n\alpha}\,, \qquad n = 0, 1, \ldots, \infty\,, \tag{2.13}$$

where α stands for spinor index which will be implicit in what follows. In (2.13), gauge transformation parameters with $n = 0$ and $n = 1$ are the respective spin-$\frac{1}{2}$ and spin-$\frac{3}{2}$ Dirac fields of the Lorentz $so(d,1)$ algebra, while gauge transformation parameters with $n \ge 2$ are totally symmetric γ-traceless spin-$(n + \frac{1}{2})$ Dirac fields of the Lorentz $so(d,1)$ algebra,

$$\gamma^a \xi^{aa_2...a_n} = 0\,, \qquad n = 1, 2, \ldots, \infty\,. \tag{2.14}$$

In order to simplify the presentation of gauge transformation we use the oscillators α^a, υ and collect all gauge transformation parameters appearing in (2.13) into ket-vector $|\xi\rangle$ defined by the relation

$$|\xi\rangle = \sum_{n=0}^{\infty} \frac{\upsilon^{n+1}}{n!\sqrt{(n+1)!}} \alpha^{a_1} \ldots \alpha^{a_n} \xi^{a_1...a_n} |0\rangle\,. \tag{2.15}$$

In terms of the ket-vector $|\xi\rangle$, γ-tracelessness constraints (2.13) are represented as $\gamma\bar\alpha|\xi\rangle = 0$.

We now find that our Lagrangian (2.4)–(2.7) is invariant under a gauge transformation given by

$$\delta|\psi\rangle = G|\xi\rangle\,,$$

$$G = \alpha D - e_1 + \gamma\alpha\frac{e_1^\Gamma}{2N_\alpha + d - 1} - \alpha^2\frac{1}{2N_\alpha + d + 1}\bar e_1\,, \tag{2.16}$$

where the operators e_1^Γ, e_1, $\bar e_1$ entering derivative independent part of gauge transformation (2.16) are defined in (2.8)–(2.12).

Representation for gauge invariant Lagrangian (2.4)–(2.7) and the corresponding gauge transformation (2.16) is universal and valid for arbitrary theory of fermionic totally symmetric gauge (A)dS fields. Various theories of fermionic totally symmetric gauge (A)dS fields are distinguished only by explicit form of the operators e_1^Γ, e_1, $\bar e_1$ entering $E_{(0)}$ (2.7) and gauge transformation (2.16). This is to say that, operators E and G for fermionic totally symmetric massless, massive, conformal, and continuous spin fields in (A)dS depend on the oscillators α^a, $\bar\alpha^a$, the Dirac γ-matrices, and on the derivative D^a in the same way as operators E (2.5)–(2.7) and G (2.16). Namely, operators E and G for fermionic totally symmetric massless, massive, conformal, and continuous spin fields in (A)dS are distinguished only by the explicit form of the operators e_1^Γ, e_1, and $\bar e_1$. For the reader convenience we note that, for massless fields in $(A)dS_{d+1}$, the e_1^Γ, e_1, and $\bar e_1$ take the form

$$e_1^\Gamma = \sqrt{-\rho}\big(N_\upsilon + \frac{d-3}{2}\big)\,, \qquad e_1 = 0\,, \qquad \bar e_1 = 0\,, \quad \text{for massless fields in } (A)dS_{d+1}\,. \tag{2.17}$$

For spin-$(s + \frac{1}{2})$ and mass-m massive (A)dS field, the operators e_1^Γ, e_1, and $\bar e_1$ can be read from expressions (2.18)–(2.21) in Ref. [8].[2] For fermionic conformal fields in flat space, the operators e_1^Γ, e_1, and $\bar e_1$ can be read from expressions (3.31)–(3.34) in Ref. [13]. For fermionic conformal fields in (A)dS space, explicit expressions for the operators e_1^Γ, e_1, and $\bar e_1$ are still to be worked out.

3. (Ir)reducible classically unitary fermionic continuous spin field

Our Lagrangian (2.4) depends on the two parameters κ_0 and μ_0. In this Section, we find restrictions imposed on the κ_0 and μ_0 for irreducible and reducible classically unitary dynamical systems.

[2] In this paper, we use metric-like approach to gauge fields. Let us also mention frame-like and BRST approaches to gauge fields (see, e.g., Refs. [9,10]). It will be interesting to study fermionic continuous spin field by using frame-like and BRST approaches and establish their connection with a vector-superspace approach in Refs. [1,2,11]. Study of bosonic continuous spin field by using BRST method may be found in Ref. [12].

Let us start with our definition of classically unitary reducible and irreducible systems.

i) Lagrangian (2.4) is constructed out of complex-valued Dirac fields. In order for the action be hermitian the κ_0 (2.8) should be real-valued, while the quantity F_υ (2.10) should be positive for all eigenvalues $N_\upsilon = 0, 1, \ldots, \infty$. Introducing the notation

$$F_\upsilon(n) = \kappa_0^2 - \mu_0\big(n + \frac{d-1}{2}\big)^2 - \rho\big(n + \frac{d-1}{2}\big)^4,$$
$$F_\upsilon(n) \equiv F_\upsilon|_{N_\upsilon = n}, \qquad (3.1)$$

we note then that, depending on behaviour of the $F_\upsilon(n)$, we use the following terminology

$$F_\upsilon(n) \geq 0 \text{ for all } n = 0, 1, \ldots, \infty$$
$$\text{classically unitary system;} \qquad (3.2)$$
$$F_\upsilon(n) \neq 0 \text{ for all } n = 0, 1, \ldots, \infty,$$
$$\text{irreducible system;} \qquad (3.3)$$
$$F_\upsilon(n_r) = 0 \text{ for some } n_r \in 0, 1, \ldots, \infty,$$
$$\text{reducible system.} \qquad (3.4)$$

Relation (3.2) tells us that, if $F_\upsilon(n)$ (3.1) is positive for all n, then we will refer to fields (2.1) as classically unitary system. From (3.3), we learn that, if $F_\upsilon(n)$ (3.1) has no roots, then we will refer to fields (2.1) as irreducible system. For this case, Lagrangian (2.4) describes infinite chain of coupling fields (2.1). From (3.4), we learn that, if $F_\upsilon(n)$ (3.1) has roots, then we will refer to fields (2.1) as reducible system. For the reducible system, Lagrangian (2.4) and gauge transformation (2.16) are factorized and describe finite and infinite decoupled chains of fields.

From now on we assume that κ_0 is real-valued. Using definitions above-given in (3.2)–(3.4), we define (ir)reducible classically unitary systems as follows

$$F_\upsilon(n) > 0 \text{ for all } n = 0, 1, \ldots, \infty,$$
$$\text{irreducible classically unitary system;} \qquad (3.5)$$
$$F_\upsilon(n_r) = 0 \text{ for some } n_r \in 0, 1, \ldots, \infty,$$
$$F_\upsilon(n) > 0 \text{ for all } n = 0, 1, \ldots, \infty \text{ and } n \neq n_r$$
$$\text{reducible classically unitary system.} \qquad (3.6)$$

We now consider (ir)reducible classically unitary systems for flat, AdS, and dS spaces in turn.

Flat space, $\rho = 0$. Setting $\rho = 0$ in (3.1), it is easy to see that restrictions (3.5), (3.6) amount to the following restrictions on the parameters κ_0 and μ_0:

$$\mu_0 = 0, \qquad \kappa_0 \neq 0,$$
$$\text{for irreducible classically unitary system,} \qquad (3.7)$$
$$\mu_0 < 0, \qquad \kappa_0 - \text{ arbitrary },$$
$$\text{for irreducible classically unitary system,} \qquad (3.8)$$
$$\mu_0 = 0, \qquad \kappa_0 = 0,$$
$$\text{for reducible classically unitary system.} \qquad (3.9)$$

Below, we demonstrate that μ_0 is related to conventional mass parameter m as $\mu_0 = m^2$ (see (4.17) for $\rho = 0$). This implies that the cases $\mu_0 = 0$, $\mu_0 > 0$, and $\mu_0 < 0$ are associated with the respective massless, massive, and tachyonic fields. Note that, for the case of (3.9), we have $F_\upsilon(n) \equiv 0$ and this case describes a chain of conventional massless fields in flat space which consists of every spin just once. Case $\mu_0 = 0$ (3.7) describes massless continuous spin field. For such field in $R^{3,1}$, alternative representation for gauge invariant Lagrangian was obtained first in Ref. [1]. Case $\mu_0 < 0$ (3.8) describes tachyonic continuous spin field in flat space. Lagrangian gauge invariant description for fermionic massless continuous spin field in $R^{d,1}$, $d > 3$, and fermionic tachyonic continuous spin field in $R^{d,1}$, $d \geq 3$, has not been discussed in earlier literature. We expect that our continuous spin field with $\mu_0 < 0$ (3.8) is associated with tachyonic UIR of the Poincaré algebra. Tachyonic UIR of the Poincaré algebra are discussed, e.g., in Ref. [14].

Ignoring restrictions (3.2), (3.3), we find that restriction (3.4) leads to interesting reducible system. Namely, setting $\rho = 0$ in (3.1), we see that the equation $F_\upsilon(s) = 0$ gives

$$\kappa_0^2 = \big(s + \frac{d-1}{2}\big)^2 \mu_0 . \qquad (3.10)$$

Inserting $\rho = 0$ and κ_0^2 (3.10) into (3.1), we get

$$F_\upsilon(n) = (s-n)(s+n+d-1)\mu_0 . \qquad (3.11)$$

Now decomposing ket-vectors $|\psi\rangle$, (2.3) and $|\xi\rangle$ (2.15) as

$$|\psi\rangle = |\psi^{0,s}\rangle + |\psi^{s+1,\infty}\rangle , \qquad (3.12)$$
$$|\psi^{M,N}\rangle \equiv \sum_{n=M}^{N} \frac{\upsilon^n}{n!\sqrt{n!}} \alpha^{a_1} \ldots \alpha^{a_n} \psi^{a_1 \ldots a_n} |0\rangle , \qquad (3.13)$$
$$|\xi\rangle = |\xi^{0,s-1}\rangle + |\xi^{s,\infty}\rangle , \qquad (3.14)$$
$$|\xi^{M,N}\rangle \equiv \sum_{n=M}^{N} \frac{\upsilon^{n+1}}{n!\sqrt{(n+1)!}} \alpha^{a_1} \ldots \alpha^{a_n} \xi^{a_1 \ldots a_n} |0\rangle , \qquad (3.15)$$

and using $\rho = 0$ and F_υ as in (3.11), we verify that Lagrangian (2.4) and gauge transformation (2.16) are factorized,

$$\mathcal{L} = \mathcal{L}^{0,s} + \mathcal{L}^{s+1,\infty} , \qquad \mathrm{i}\mathcal{L}^{0,s} \equiv \langle \psi^{0,s} | E | \psi^{0,s} \rangle ,$$
$$\mathrm{i}\mathcal{L}^{s+1,\infty} \equiv \langle \psi^{s+1,\infty} | E | \psi^{s+1,\infty} \rangle , \qquad (3.16)$$
$$\delta |\psi^{0,s}\rangle = G |\xi^{0,s-1}\rangle , \qquad \delta |\psi^{s+1,\infty}\rangle = G |\xi^{s,\infty}\rangle , \qquad (3.17)$$

where E, G are given in (2.5), (2.16). Thus, for values of the parameters μ_0 and κ_0 given in (3.10), we see from (3.16), (3.17) that our Lagrangian and gauge transformations describe two decoupling fields $|\psi^{0,s}\rangle$ and $|\psi^{s+1,\infty}\rangle$. As κ_0 (3.10) is real-valued, we have $\mu_0 > 0$. Using the notation $\mu_0 = m^2$, we note that the field $|\psi^{0,s}\rangle$ describes a fermionic classically unitary spin-$(s + \frac{1}{2})$ mass-m massive field. For this case, $F_\upsilon(n) < 0$ (3.11) when $n = s+1, s+2, \ldots, \infty$ and this implies that the ket-vector $|\psi^{s+1,\infty}\rangle$ describes classically non-unitary fermionic field.

(A)dS space, $\rho \neq 0$. For (A)dS, our study of Eqs. (3.5), (3.6) is summarized as follows.

Statement 1. *For dS, equations (3.5), (3.6) do not have solutions.* This implies that continuous spin dS field is not realized neither as irreducible classically unitary system nor as reducible classically unitary system.[3]

Statement 2. For AdS, Eqs. (3.5) have solutions which we classify as Type IA,IB, II, and III solutions.

[3] Discussion of group theoretical aspects of quantum fields in dS space may be found in Refs. [15,16].

Type IA and IB solutions for AdS:

$$\mu_0 < \frac{1}{4}|\rho|(d-1)^2\,, \qquad \kappa_0 - \text{arbitrary}, \qquad IA; \tag{3.18}$$

$$\frac{1}{4}|\rho|(d-1)^2 \le \mu_0 \le \frac{1}{2}|\rho|(d-1)^2\,, \qquad \kappa_0^2 > \frac{(d-1)^2}{4}\Big(\mu_0 - \frac{(d-1)^2}{4}|\rho|\Big)\,, \qquad IB. \tag{3.19}$$

Type II solutions for AdS:

$$\mu_0 = 2|\rho|(\lambda_0 + \frac{d-1}{2})^2\,, \qquad \kappa_0^2 > |\rho|(\lambda_0 + \frac{d-1}{2})^4\,, \qquad \lambda_0 = 1, 2, \ldots, \infty\,. \tag{3.20}$$

Type III solutions for AdS:

$$\mu_0 = 2|\rho|(\lambda + \frac{d-1}{2})^2\,, \tag{3.21}$$

$$\kappa_0^2 > |\rho|(\lambda + \frac{d-1}{2})^4 - |\rho|\epsilon^2(\epsilon + 2\lambda_0 + d - 1)^2 \quad \text{for } 0 < \epsilon < \epsilon_r\,, \tag{3.22}$$

$$\kappa_0^2 > |\rho|(\lambda + \frac{d-1}{2})^4 - |\rho|(1-\epsilon)^2(\epsilon + 2\lambda_0 + d)^2\,, \quad \text{for } \epsilon_r < \epsilon < 1\,, \tag{3.23}$$

$$\kappa_0^2 > |\rho|(\lambda + \frac{d-1}{2})^4 - \frac{1}{4}|\rho|(2\lambda_0 + d)^2\,, \quad \text{for } \epsilon = \epsilon_r\,, \tag{3.24}$$

$$\lambda = \lambda_0 + \epsilon\,, \qquad 0 < \epsilon < 1\,, \qquad \lambda_0 = 0, 1, \ldots, \infty\,, \tag{3.25}$$

$$\epsilon_r \equiv \frac{1}{2}(\sqrt{(2\lambda_0 + d)^2 + 1} - 2\lambda_0 - d + 1)\,. \tag{3.26}$$

Note that type II solutions given in (3.20) are labelled by integer λ_0, while type III solutions given in (3.20)–(3.26) are labelled by ϵ, $0 < \epsilon < 1$, and integer λ_0 (3.25).

Statement 3. *For AdS, solution to Eqs. (3.6) is given by*

$$\kappa_0^2 = (s + \frac{d-1}{2})^2 m^2\,, \qquad \mu_0 \equiv m^2 + |\rho|(s + \frac{d-1}{2})^2\,, \tag{3.27}$$

$$|\rho|(s + \frac{d-3}{2})^2 < m^2 < |\rho|(s + \frac{d+1}{2})^2\,, \tag{3.28}$$

where, for the reader convenience, we introduce conventional mass parameter m (3.27).[4] Lagrangian (2.4) with κ_0 and μ_0 as in (3.27), (3.28) describes the reducible classically unitary system. Namely, let us decompose ket-vectors $|\psi\rangle$ (2.3) and $|\xi\rangle$ (2.15) as

$$|\psi\rangle = |\psi^{0,s}\rangle + |\psi^{s+1,\infty}\rangle\,, \tag{3.29}$$

$$|\xi\rangle = |\xi^{0,s-1}\rangle + |\xi^{s,\infty}\rangle\,, \tag{3.30}$$

where ket-vectors $|\psi^{M,N}\rangle$, $|\xi^{M,N}\rangle$ are defined in (3.13), (3.15). Then we can verify that Lagrangian (2.4) and gauge transformation (2.16) with κ_0 and μ_0 as in (3.27) are factorized as

$$\mathcal{L} = \mathcal{L}^{0,s} + \mathcal{L}^{s+1,\infty}\,, \qquad i\mathcal{L}^{0,s} \equiv e\langle\psi^{0,s}|E|\psi^{0,s}\rangle\,, \qquad i\mathcal{L}^{s+1,\infty} \equiv e\langle\psi^{s+1,\infty}|E|\psi^{s+1,\infty}\rangle\,, \tag{3.31}$$

$$\delta|\psi^{0,s}\rangle = G|\xi^{0,s-1}\rangle\,, \qquad \delta|\psi^{s+1,\infty}\rangle = G|\xi^{s,\infty}\rangle\,, \tag{3.32}$$

where E, G are given in (2.5), (2.16). In other words, $\mathcal{L}^{0,s}$ and $\mathcal{L}^{s+1,\infty}$ (3.31) are invariant under transformations governed by gauge transformation parameters $|\xi^{0,s-1}\rangle$ and $|\xi^{s,\infty}\rangle$ respectively.

The three Statements above-discussed are proved by noticing that $F_\upsilon(n)$ (3.1) has at most two roots. Thus we have to analyse the following three cases: 1) $F_\upsilon(n)$ has no roots; 2) $F_\upsilon(n)$ has one root; 3) $F_\upsilon(n)$ has two roots; We analyse these cases in turn.

i) Solution without roots of $F_\upsilon(n)$. Using (3.1), we note that equations (3.5) can alternatively be represented as

$$\kappa_0^2 > \max_{n=0,1,\ldots,\infty}\Big(\mu_0(n + \frac{d-1}{2})^2 + \rho(n + \frac{d-1}{2})^4\Big)\,. \tag{3.33}$$

We now see that, for dS space ($\rho > 0$), equation (3.33) has no solution. For AdS space ($\rho < 0$), we find that, depending on values of μ_0, equation (3.33) leads to Type IA, IB, II and III solutions.

ii) Solution with one root of $F_\upsilon(n)$. Using the notation $n_r = s$ for one root of $F_\upsilon(n)$, we see that Eqs. (3.6) amount to

$$F_\upsilon(s) = 0\,, \qquad F_\upsilon(n) > 0 \quad \text{for } n = 0, 1, \ldots, s-1, s+1, s+2, \ldots, \infty\,. \tag{3.34}$$

First we note that the equation $F_\upsilon(s) = 0$ leads to the following restrictions on μ_0 and κ_0

$$\kappa_0^2 = (s + \frac{d-1}{2})^2 m^2\,, \qquad \mu_0 \equiv m^2 - \rho(s + \frac{d-1}{2})^2\,. \tag{3.35}$$

Inserting μ_0, κ_0 (3.35) into (2.10), we cast the F_υ into the following form

$$F_\upsilon = (s - N_\upsilon)(s + N_\upsilon + d - 1)\Big(m^2 + \rho(N_\upsilon + \frac{d-1}{2})^2\Big)\,. \tag{3.36}$$

Lagrangian (2.4) with F_υ given in (3.36) describes reducible system of continuous spin field. This is to say that if we decompose $|\psi\rangle$ and $|\xi\rangle$ as in (3.29), (3.30) then we can check that Lagrangian (2.4) and gauge transformation (2.16) are factorized as in (3.31), (3.32).

Second, using μ_0, κ_0 (3.35) and considering inequalities $F_\upsilon(n) > 0$ in (3.34), we find the following restrictions on m^2:

$$|\rho|(s + \frac{d-3}{2})^2 < m^2 < |\rho|(s + \frac{d+1}{2})^2 \qquad \text{for AdS;} \tag{3.37}$$

$$-|\rho|\frac{(d-1)^2}{4} < m^2 < -|\rho|\infty\,, \qquad \text{for dS.} \tag{3.38}$$

We note that for the derivation of inequalities in (3.37), (3.38) it is convenient to use representation for F_υ given in (3.36). We note also that the left inequalities in (3.37), (3.38) are obtained by requiring $F_\upsilon(n) > 0$ for $n = 0, 1, \ldots, s-1$, while the right inequalities in (3.37), (3.38) are obtained by requiring $F_\upsilon(n) > 0$ for $n = s+1, \ldots, \infty$. It is easy to see that, for dS, inequalities (3.38) are inconsistent,[5] i.e., for dS, Eqs. (3.6) with one root $n_r = s$ do not have solutions. For AdS, we see that (3.35), (3.37) lead to (3.27), (3.28).

iii) Solution with two roots of $F_\upsilon(n)$. Denoting two roots of Eqs. (3.6) as s, S,

$$F_\upsilon(s) = 0\,, \qquad F_\upsilon(S) = 0\,, \qquad s \le S\,, \tag{3.39}$$

it is easy to see Eqs. (3.39) amount to the following relations

$$\kappa_0^2 = -\rho(s + \frac{d-1}{2})^2(S + \frac{d-1}{2})^2\,, \tag{3.40}$$

$$\mu_0 = -\rho(s + \frac{d-1}{2})^2 - \rho(S + \frac{d-1}{2})^2\,. \tag{3.41}$$

[4] For a finite component field, the m is defined from the Lagrangian, $i\mathcal{L} = e\psi_s(\not{\partial} + m)\psi_s + \ldots$ where dots stand for terms involving $\psi_{s'}$, $s' < s$, and for contributions which vanish while imposing the γ-tracelessness constraint.

[5] The left inequality in (3.38) tells us that m^2 should be bounded from below, while from the right inequality in (3.38) we learn that $m^2 = -\infty$. Note that the right inequality in (3.38) is obtained by requiring $F_\upsilon(n) > 0$ for $n = \infty$.

Inserting μ_0, κ_0 (3.40), (3.41) into (2.10), we find the following representation for F_υ:

$$F_\upsilon = -\rho(s - N_\upsilon)(s + N_\upsilon + d - 1)(S - N_\upsilon)(S + N_\upsilon + d - 1)\,. \tag{3.42}$$

Lagrangian (2.4) with F_υ as in (3.42) describes reducible system of continuous spin field. Namely let us decompose ket-vectors $|\psi\rangle$ (2.3) and $|\xi\rangle$ (2.15) as

$$|\psi\rangle = |\psi^{0,s}\rangle + |\psi^{s+1,S}\rangle + |\psi^{S+1,\infty}\rangle\,, \tag{3.43}$$

$$|\xi\rangle = |\xi^{0,s-1}\rangle + |\xi^{s,S-1}\rangle + |\xi^{S,\infty}\rangle\,, \tag{3.44}$$

where the ket-vectors $|\psi^{M,N}\rangle$, $|\xi^{M,N}\rangle$ are defined in (3.13), (3.15). Then we can verify that Lagrangian (2.4) and gauge transformation (2.16) with κ_0 and μ_0 as in (3.40), (3.41) are factorized,

$$\mathcal{L} = \mathcal{L}^{0,s} + \mathcal{L}^{s+1,S} + \mathcal{L}^{S+1,\infty}\,, \qquad \mathrm{i}\mathcal{L}^{M,N} \equiv e\langle\psi^{M,N}|E|\psi^{M,N}\rangle\,, \tag{3.45}$$

$$\delta|\psi^{0,s}\rangle = G|\xi^{0,s-1}\rangle\,, \qquad \delta|\psi^{s+1,S}\rangle = G|\xi^{s,S-1}\rangle\,,$$
$$\delta|\psi^{S+1,\infty}\rangle = G|\xi^{S,\infty}\rangle\,, \tag{3.46}$$

where E, G are given in (2.5), (2.16). In (3.45), (3.46), we assume that if $s = S$, then $|\psi^{s+1,s}\rangle \equiv 0$. Note that classically unitary system with $s = S$ is described by (3.27), (3.28) when $m^2 = |\rho|(s + \frac{d-1}{2})^2$. From (3.42), we learn that the ket-vector $|\psi^{0,s}\rangle$ describes classically unitary (non-unitary) fermionic massive spin-$(s + \frac{1}{2})$ field in AdS (dS), the ket-vector $|\psi^{s+1,S}\rangle$ describes classically non-unitary fermionic partial-massless spin-$(S+\frac{1}{2})$ field in (A)dS, while the ket-vector $|\psi^{S+1,\infty}\rangle$ describes classically unitary (non-unitary) fermionic infinite-component spin-$(S + \frac{3}{2})$ field in AdS (dS). Mass parameters of the fermionic fields $|\psi^{0,s}\rangle$ and $|\psi^{s+1,S}\rangle$ are given by

$$m^2 = -\rho\big(S + \frac{d-1}{2}\big)^2\,, \qquad \text{for } |\psi^{0,s}\rangle\,, \tag{3.47}$$

$$m^2 = -\rho\big(s + \frac{d-1}{2}\big)^2\,, \qquad \text{for } |\psi^{s+1,S}\rangle\,. \tag{3.48}$$

Mass parameter (3.48) can be represented in the form

$$m^2 = -\rho\big(S + \frac{d-3}{2} - k\big)^2\,, \qquad k \equiv S - s - 1\,, \quad \text{for } |\psi^{s+1,S}\rangle\,, \tag{3.49}$$

which tells that us that $|\psi^{s+1,S}\rangle$ describes a spin-$(S + \frac{1}{2})$ and depth-k partial-massless fermionic field. For $S = s + 1$ ($k = 0$), this field turns out to be spin-$(S + \frac{1}{2})$ massless fermionic field.[6]

4. Partition function and modified de Donder gauge condition

In this section, we are going to demonstrate that a partition function of the fermionic continuous spin field is equal to one. Namely, using gauge invariant Lagrangian (2.4), we find the following expression for the partition function Z of the fermionic continuous spin field

$$Z^{-1} = \prod_{n=0}^{\infty} Z_n^{-1}\,, \qquad Z_n^{-1} = \frac{\mathcal{D}_{n-1}(M_{n-1}^2)\mathcal{D}_{n-1}(M_{n-1}^2)}{\mathcal{D}_n(M_n^2)\mathcal{D}_{n-2}(M_{n-2}^2)}\,, \tag{4.1}$$

$$\mathcal{D}_n(M^2) = \sqrt{\det{}_n(-\not\Box + M^2)}\,, \qquad \not\Box \equiv \not D\not D\,, \tag{4.2}$$

$$M_n^2 \equiv \mu_0 + \rho\big(n + \frac{d-1}{2}\big)^2\,, \tag{4.3}$$

where, in (4.2), a determinant is computed on a space of the Lorentz $so(d,1)$ algebra spin-$(n+\frac{1}{2})$ Dirac field subject to γ-tracelessness constraint. Note also that in (4.1) we assume the convention $\mathcal{D}_{-2}(M^2) = 1$, $\mathcal{D}_{-1}(M^2) = 1$. From (4.1), we see immediately that the partition function of the fermionic continuous spin field is equal to one, $Z = 1$.

Alternatively, Z (4.1) can be cast into more convenient form by using general formula for the $\mathcal{D}_n(M^2)$ with arbitrary M^2 (4.2),

$$\mathcal{D}_n(M^2) = \mathcal{D}_n^\perp(M^2)\mathcal{D}_{n-1}(M^2 - \rho(2n + d - 2))\,, \tag{4.4}$$

where $\mathcal{D}_n^\perp(M^2)$ appearing in (4.4) takes the same form as in (4.2), with assumption that the determinant is computed on space of the gamma-traceless and divergence-free spin-$(n+\frac{1}{2})$ Dirac field of the Lorentz $so(d,1)$ algebra. For M_n^2 as in (4.3), relation (4.4) takes the form

$$\mathcal{D}_n(M_n^2) = \mathcal{D}_n^\perp(M_n^2)\mathcal{D}_{n-1}(M_{n-1}^2)\,. \tag{4.5}$$

Using (4.5), we see that Z_n^{-1} (4.1) can be represented as

$$Z_n^{-1} = \frac{\mathcal{D}_{n-1}^\perp(M_{n-1})}{\mathcal{D}_n^\perp(M_n)}\,. \tag{4.6}$$

Using (4.6), we find then, for fermionic continuous spin field in (A)dS and flat spaces, the same cancellation mechanism as for higher-spin fields in flat space [17]. Namely, from Z (4.1) and Z_n (4.6), we see the cancellation of determinant of the physical spin-$(n + \frac{1}{2})$ field and ghost determinant of spin-$(n + \frac{3}{2})$ field. Note that Z is equal to 1 without the use of special regularization procedure required for a computation of partition functions of higher-spin (A)dS gauge fields (see, e.g., Ref. [17]).

Representation for the partition function given in (4.1) can be obtained from Lagrangian (2.4) by using the well-known technique discussed in the earlier literature (see, e.g., Refs. [18–20]). Here, instead of a repetition of the well-known technicalities, we prefer to demonstrate how the expression for partition function of continuous spin fermionic field (4.1) leads to partition functions of massless, partial-massless and massive fermionic fields. Also we present our modified de Donder gauge for fermionic fields which allows us to obtain in a straightforward way the mass terms M_n^2 (4.3) entering the determinants in (4.1), (4.2).

Partial-massless and massless fields. As we said above, if κ_0 and μ_0 take values given in (3.40), (3.41), then field $|\psi^{s+1,S}\rangle$ appearing in (3.43) describes spin-$(S + \frac{1}{2})$, depth-$(S - s - 1)$ partial-massless field. Partition function for such field is obtained from (4.1) by considering contribution of Z_n with $n = s + 1, s + 2, \ldots, S$. Doing so, we get

$$Z_{\text{par-massl}}^{-1} = \prod_{n=s+1}^{S} Z_n^{-1}\,,$$

$$Z_{\text{par-massl}}^{-1} = \frac{\mathcal{D}_s(M_s^2)\mathcal{D}_{S-1}(M_{S-1}^2)}{\mathcal{D}_{s-1}(M_{s-1}^2)\mathcal{D}_S(M_S^2)}\,, \tag{4.7}$$

where Z_n entering (4.7) takes the same form as in (4.1), while M_n^2 is obtained by inserting μ_0 (3.41) into (4.3).

$$M_n^2 = \rho\big(n + \frac{d-1}{2}\big)^2 - \rho\big(s + \frac{d-1}{2}\big)^2 - \rho\big(S + \frac{d-1}{2}\big)^2\,. \tag{4.8}$$

Using (4.4) and M_n^2 (4.8), we verify relation (4.5). Using (4.5), we see that Z (4.7) takes the well-known form of partition function of spin-$(S + \frac{1}{2})$ and depth-$(S - s - 1)$ partial-massless field

[6] Using notation $k_{\rm CU}$ for the commonly used depth of partial-massless field we note that our k in (3.49) is related to $k_{\rm CU}$ as $k = k_{\rm CU} - 1$.

$$Z^{-1}_{\text{par-massl}} = \frac{\mathcal{D}^{\perp}_{S}(M^2_S)}{\mathcal{D}^{\perp}_{s}(M^2_s)}\,,$$

$$M^2_S = -\rho(S + \frac{d-1}{2})^2\,, \qquad M^2_s = -\rho(s + \frac{d-1}{2})^2\,, \tag{4.9}$$

where M^2_S, M^2_s given in (4.9) are obtained from (4.8). For $S = s+1$, a ket-vector $|\psi^{S,S}\rangle$ (3.43) describes massless field and Z (4.9) becomes the partition function of spin-$(S+\frac{1}{2})$ massless field

$$Z^{-1}_{\text{massl}} = \frac{\mathcal{D}^{\perp}_{S-1}(M^2_{S-1})}{\mathcal{D}^{\perp}_{S}(M^2_S)}\,,$$

$$M^2_{S-1} = -\rho(S + \frac{d-1}{2})^2 \qquad M^2_S = -\rho(S + \frac{d-3}{2})^2\,. \tag{4.10}$$

Massive field. As we noted, if κ_0 and μ_0 take values given in (3.35), then field $|\psi^{0,s}\rangle$ appearing in (3.29) describes spin-$(s+\frac{1}{2})$ and mass-m massive field. Partition function for such field is obtained from (4.1) by considering the contributions of Z_n with $n = 0, 1, \ldots, s$. Doing so, we get

$$Z^{-1}_{\text{massv}} = \prod_{n=0}^{s} Z_n^{-1}\,, \qquad Z^{-1}_{\text{massv}} = \frac{\mathcal{D}_{s-1}(M^2_{s-1})}{\mathcal{D}_s(M^2_s)}\,, \tag{4.11}$$

where Z_n entering (4.11) takes the same form as in (4.1), while M^2_n is obtained by inserting μ_0 (3.35) into (4.3),

$$M^2_n = m^2 - \rho(s-n)(s+n+d-1)\,. \tag{4.12}$$

Using (4.4) and M^2_n (4.12), we verify relation (4.5). Using then (4.5) for $n = s$, we find that Z (4.11) takes the well-known form of partition function for spin-$(s+\frac{1}{2})$ and mass-m massive fermionic field

$$Z_{\text{massv}} = \mathcal{D}^{\perp}_{s}(M^2_s)\,, \qquad M^2_s = m^2\,. \tag{4.13}$$

Modified de Donder gauge. Now we outline the derivation of the mass terms M^2_n (4.3) entering the determinants in (4.1), (4.2). To this end we propose to use a gauge condition which we refer to as modified de Donder gauge. A modified de Donder gauge is defined as follows

$$\bar{C}_{\text{mod}}|\psi\rangle = 0\,, \tag{4.14}$$

$$\bar{C}_{\text{mod}} \equiv \bar{C}_{\text{st}} - \frac{1}{2N_\alpha + d - 1}(\gamma\bar{\alpha} + \frac{1}{2}\gamma\alpha\bar{\alpha}^2)e_1^{\Gamma} + \frac{1}{2}\bar{\alpha}^2 e_1 - \frac{2N_\alpha + d - 3}{2N_\alpha + d - 1}\Pi^{[1,2]}_{\text{bos}}\bar{e}_1\,, \tag{4.15}$$

$$\bar{C}_{\text{st}} \equiv \bar{\alpha}D - \frac{1}{2}\alpha D\bar{\alpha}^2\,, \qquad \Pi^{[1,2]}_{\text{bos}} \equiv 1 - \alpha^2\frac{1}{2(2N_\alpha + d + 1)}\bar{\alpha}^2\,, \tag{4.16}$$

where the operators e_1^{Γ}, e_1, $\bar{e}_1$ are given in (2.8), (2.9). By analogy with Lagrangian (2.4) and gauge transformation (2.16), the representation for modified de Donder gauge given in (4.14)–(4.16) is universal and valid for arbitrary theory of fermionic totally symmetric (A)dS fields (see our discussion below Eq. (2.16)).

Using (4.14), we verify that first-order equations of motion obtained from Lagrangian (2.4) lead to the following second-order equations for $|\psi\rangle$[7]:

$$(\not{\Box} - M^2 + \rho\alpha^2\bar{\alpha}^2)|\psi\rangle = 0\,, \qquad M^2 \equiv \mu_0 + \rho(N_\upsilon + \frac{d-1}{2})^2\,, \tag{4.17}$$

where $\not{\Box}$ is given in (4.2). From (4.17), we see that, as we have promised, for flat space ($\rho = 0$), the parameter μ_0 can be interpreted as square of the conventional mass parameter m, $\mu_0 = m^2$.

Gauge-fixed equations of motion (4.17) are invariant under on-shell leftover gauge transformation that are obtained from (2.16) by the substitution $|\xi\rangle \to |\Xi\rangle$, where the $|\Xi\rangle$ satisfies the following equations of motion:

$$(\not{\Box} - M^2_{\Xi})|\Xi\rangle = 0\,, \qquad M^2_{\Xi} \equiv \mu_0 + \rho(N_\upsilon + \frac{d-3}{2})^2\,. \tag{4.18}$$

We note that ket-vector $|\Xi\rangle$ is obtained from (2.15) by the substitution $\xi^{a_1\ldots a_n} \to \Xi^{a_1\ldots a_n}$.

To cast equations (4.17) into decoupled form, we decompose triple γ-traceless ket-vector $|\psi\rangle$, $(\gamma\bar{\alpha})^3|\psi\rangle = 0$, into three γ-traceless ket-vectors $|\psi_I\rangle$, $|\psi_\Gamma\rangle$, and $|\psi_{II}\rangle$,

$$|\psi\rangle = |\psi_I\rangle + \gamma\alpha|\psi_\Gamma\rangle + \alpha^2|\psi_{II}\rangle\,, \qquad \gamma\bar{\alpha}|\psi_\tau\rangle = 0\,, \quad \tau = I, \Gamma, II\,, \tag{4.19}$$

$$|\psi_\tau\rangle = \sum_{n=0}^{\infty} \frac{\upsilon^{n+\lambda_\tau}}{n!\sqrt{(n+\lambda_\tau)!}}\alpha^{a_1}\ldots\alpha^{a_n}\psi_\tau^{a_1\ldots a_n}|0\rangle\,,$$

$$\lambda_I \equiv 0\,, \quad \lambda_\Gamma \equiv 1\,, \quad \lambda_{II} \equiv 2\,. \tag{4.20}$$

Inserting $|\psi\rangle$ (4.19) into (4.17), we obtain decoupled equations for $|\psi_\tau\rangle$, $\tau = I, \Gamma, II$:

$$(\not{\Box} - M^2_\tau)|\psi_\tau\rangle = 0\,, \qquad M^2_\tau = \mu_0 + \rho(N_\upsilon + \frac{d-1}{2} - \lambda_\tau)^2\,. \tag{4.21}$$

In terms of $\psi_\tau^{a_1\ldots a_n}$ and $\Xi^{a_1\ldots a_n}$, equations (4.21) and (4.18) take the following respective form

$$(\not{\Box} - M^2_n)\psi_\tau^{a_1\ldots a_n} = 0\,, \quad \tau = I, \Gamma, II\,, \qquad (\not{\Box} - M^2_n)\Xi^{a_1\ldots a_n} = 0\,, \tag{4.22}$$

$n = 0, 1, \ldots\infty$, where the mass terms M^2_n take the form given in (4.3). Thus we see that the mass terms entering determinants (4.1), (4.2) are indeed governed by M^2_n (4.3).

To conclude, we developed Lagrangian gauge invariant formulation for fermionic continuous spin field in (A)dS. We used our formulation to demonstrate that a partition function of the fermionic continuous spin field is equal to 1. Lagrangian descriptions of bosonic and fermionic continuous spin (A)dS fields obtained in Ref. [3] and in this paper provide possibility to discuss supersymmetric theories of continuous spin (A)dS fields. Recent interesting discussion of Lagrangian formulation for supermultiplets in AdS and flat spaces may be found in Refs. [24,25]. Needless to say that a problem of interacting continuous spin fields also deserves to be studied. Various light-cone gauge methods discussed in Refs. [26–31] might be helpful for this purpose. Use of Lorentz covariant methods (see, e.g., Refs. [32–36]) for study of interacting continuous spin fields could also be of some interest. Our modified de Donder gauge leads to simple equations of motion for fermionic fields. We think therefore that such gauge might be useful for studying various aspects of (A)dS field dynamics along the lines in Refs. [37,38].

Acknowledgements

This work was supported by the RFBR Grant No. 17-02-00317.

7 For the bosonic case, we expect that the results in Ref. [21] and Ref. [22] provide descriptions of bosonic continuous spin field at level of light-cone gauge equations of motion and unfolding equations of motion respectively. Interesting recent discussion of unfolded approach may be found in Ref. [23].

Appendix A. Notation

Vector indices of the Lorentz $so(d,1)$ algebra a,b,c run over values $0,1,\ldots,d$. Flat metric tensor η^{ab} is mostly positive. For simplicity, we drop the flat metric η^{ab} in scalar products. The vacuum $|0\rangle$, the creation operators α^a, υ, and the annihilation operators $\bar\alpha^a$, $\bar\upsilon$ are defined by the relations

$$[\bar\alpha^a,\alpha^b]=\eta^{ab}\,,\quad [\bar\upsilon,\upsilon]=1\,,\quad \bar\alpha^a|0\rangle=0\,,\quad \bar\upsilon|0\rangle=0\,,$$
$$\alpha^{a\dagger}=\bar\alpha^a\,,\qquad \upsilon^\dagger=\bar\upsilon\,. \qquad (A.1)$$

Throughout this paper, we refer to these operators as oscillators. For the Dirac γ-matrices we use the conventions $\{\gamma^a,\gamma^b\}=2\eta^{ab}$, $\gamma^{a\dagger}=\gamma^0\gamma^a\gamma^0$. On a space of ket-vector $|\psi\rangle$ (2.3), covariant derivative D^a is defined as $D^a=\eta^{ab}D_b$,

$$D_a\equiv e_a^{\underline{m}}D_{\underline{m}}\,,\qquad D_{\underline{m}}\equiv\partial_{\underline{m}}+\frac{1}{2}\omega_{\underline{m}}^{ab}M^{ab}\,,$$
$$M^{ab}\equiv\alpha^a\bar\alpha^b+\frac{1}{4}\gamma^a\gamma^b-(a\leftrightarrow b)\,, \qquad (A.2)$$

$\partial_{\underline{m}}=\partial/\partial x^{\underline{m}}$. In (A.2), the base manifold index $\underline{m}$ run over values $=0,1,\ldots,d$, the $D_{\underline{m}}$ denotes Lorentz covariant derivative, the $e_a^{\underline{m}}$ is inverse vielbein of $(A)dS_{d+1}$ space–time, the M^{ab} stands for a spin operator of the Lorentz algebra $so(d,1)$, and the $\omega_{\underline{m}}^{ab}$ denotes the Lorentz connection of $(A)dS_{d+1}$ space–time. Tensor–spinor field carrying the base manifold indices, $\psi^{\underline{m}_1\ldots\underline{m}_n}$, and tensor–spinor field carrying the flat indices, $\psi^{a_1\ldots a_n}$, are related as $\psi^{a_1\ldots a_n}\equiv e_{\underline{m}_1}^{a_1}\ldots e_{\underline{m}_n}^{a_n}\psi^{\underline{m}_1\ldots\underline{m}_n}$. Our conventions for scalar products are as follows

$$\alpha^2\equiv\alpha^a\alpha^a\,,\qquad \bar\alpha^2\equiv\bar\alpha^a\bar\alpha^a\,,$$
$$N_\alpha\equiv\alpha^a\bar\alpha^a\,,\qquad N_\upsilon\equiv\upsilon\bar\upsilon\,, \qquad (A.3)$$
$$\gamma\alpha\equiv\gamma^a\alpha^a\,,\qquad \gamma\bar\alpha\equiv\gamma^a\bar\alpha^a\,,$$
$$\alpha D\equiv\alpha^aD^a\,,\qquad \bar\alpha D\equiv\bar\alpha^aD^a\,,\qquad \not{D}\equiv\gamma^aD^a\,. \qquad (A.4)$$

References

[1] X. Bekaert, M. Najafizadeh, M.R. Setare, Phys. Lett. B 760 (2016) 320, arXiv:1506.00973 [hep-th].
[2] P. Schuster, N. Toro, Phys. Rev. D 91 (2015) 025023, arXiv:1404.0675 [hep-th].
[3] R.R. Metsaev, Phys. Lett. B 767 (2017) 458, arXiv:1610.00657 [hep-th].
[4] X. Bekaert, J. Mourad, J. High Energy Phys. 0601 (2006) 115, arXiv:hep-th/0509092.
[5] L. Brink, A.M. Khan, P. Ramond, X.z. Xiong, J. Math. Phys. 43 (2002) 6279, arXiv:hep-th/0205145.
[6] G.K. Savvidy, Int. J. Mod. Phys. A 19 (2004) 3171, arXiv:hep-th/0310085;
J. Mourad, Continuous spin particles from a string theory, arXiv:hep-th/0504118.
[7] M.A. Vasiliev, Phys. Lett. B 243 (1990) 378;
M.A. Vasiliev, Phys. Lett. B 567 (2003) 139, arXiv:hep-th/0304049.
[8] R.R. Metsaev, Phys. Lett. B 643 (2006) 205, arXiv:hep-th/0609029.
[9] K.B. Alkalaev, O.V. Shaynkman, M.A. Vasiliev, J. High Energy Phys. 0508 (2005) 069, arXiv:hep-th/0501108;
E.D. Skvortsov, Y.M. Zinoviev, Nucl. Phys. B 843 (2011) 559, arXiv:1007.4944 [hep-th].
[10] I.L. Buchbinder, V.A. Krykhtin, A.A. Reshetnyak, Nucl. Phys. B 787 (2007) 211, arXiv:hep-th/0703049;
A. Fotopoulos, M. Tsulaia, arXiv:0805.1346 [hep-th].
[11] V.O. Rivelles, Phys. Rev. D 91 (12) (2015) 125035, arXiv:1408.3576 [hep-th].
[12] A.K.H. Bengtsson, J. High Energy Phys. 1310 (2013) 108, arXiv:1303.3799 [hep-th].
[13] R.R. Metsaev, Conformal totally symmetric arbitrary spin fermionic fields, arXiv:1211.4498 [hep-th].
[14] X. Bekaert, N. Boulanger, arXiv:hep-th/0611263.
[15] E. Joung, J. Mourad, R. Parentani, J. High Energy Phys. 0608 (2006) 082, arXiv:hep-th/0606119;
E. Joung, J. Mourad, R. Parentani, J. High Energy Phys. 0709 (2007) 030, arXiv:0707.2907 [hep-th].
[16] T. Basile, X. Bekaert, N. Boulanger, J. High Energy Phys. 1705 (2017) 081, arXiv:1612.08166 [hep-th].
[17] M. Beccaria, A.A. Tseytlin, J. Phys. A 48 (27) (2015) 275401, arXiv:1503.08143 [hep-th].
[18] T. Creutzig, Y. Hikida, P.B. Ronne, J. High Energy Phys. 1202 (2012) 109, arXiv:1111.2139 [hep-th].
[19] A.A. Tseytlin, Nucl. Phys. B 877 (2013) 598, arXiv:1309.0785 [hep-th].
[20] A. Campoleoni, H.A. Gonzalez, B. Oblak, M. Riegler, J. High Energy Phys. 1604 (2016) 034, arXiv:1512.03353.
[21] R.R. Metsaev, Nucl. Phys. B 563 (1999) 295, arXiv:hep-th/9906217.
[22] D.S. Ponomarev, M.A. Vasiliev, Nucl. Phys. B 839 (2010) 466, arXiv:1001.0062 [hep-th].
[23] T. Basile, R. Bonezzi, N. Boulanger, J. High Energy Phys. 1704 (2017) 054, arXiv:1701.08645 [hep-th].
[24] I.L. Buchbinder, T.V. Snegirev, Y.M. Zinoviev, J. High Energy Phys. 1510 (2015) 148, arXiv:1508.02829.
[25] S.M. Kuzenko, M. Tsulaia, Nucl. Phys. B 914 (2017) 160, arXiv:1609.06910 [hep-th].
[26] A.K.H. Bengtsson, I. Bengtsson, L. Brink, Nucl. Phys. B 227 (1983) 31.
[27] R.R. Metsaev, Mod. Phys. Lett. A 6 (1991) 2411.
[28] R.R. Metsaev, Mod. Phys. Lett. A 6 (1991) 359.
[29] R.R. Metsaev, Nucl. Phys. B 759 (2006) 147, arXiv:hep-th/0512342.
[30] R.R. Metsaev, Nucl. Phys. B 859 (2012) 13, arXiv:0712.3526 [hep-th].
[31] A.K.H. Bengtsson, J. High Energy Phys. 1612 (2016) 134, arXiv:1607.06659 [hep-th];
D. Ponomarev, E.D. Skvortsov, J. Phys. A 50 (9) (2017) 095401, arXiv:1609.04655 [hep-th];
C. Sleight, M. Taronna, J. High Energy Phys. 1702 (2017) 095, arXiv:1609.00991 [hep-th].
[32] M.A. Vasiliev, Nucl. Phys. B 862 (2012) 341, arXiv:1108.5921 [hep-th];
E. Joung, M. Taronna, Nucl. Phys. B 861 (2012) 145, arXiv:1110.5918 [hep-th];
N. Boulanger, D. Ponomarev, E.D. Skvortsov, M. Taronna, Int. J. Mod. Phys. A 28 (2013) 1350162.
[33] R.R. Metsaev, Phys. Lett. B 720 (2013) 237, arXiv:1205.3131 [hep-th].
[34] A. Fotopoulos, M. Tsulaia, J. High Energy Phys. 1011 (2010) 086, arXiv:1009.0727 [hep-th];
M. Henneaux, G. Lucena Gómez, R. Rahman, J. High Energy Phys. 1208 (2012) 093, arXiv:1206.1048 [hep-th].
[35] R.R. Metsaev, Phys. Rev. D 77 (2008) 025032, arXiv:hep-th/0612279;
I.L. Buchbinder, V.A. Krykhtin, M. Tsulaia, Nucl. Phys. B 896 (2015) 1, arXiv:1501.03278.
[36] R. Manvelyan, K. Mkrtchyan, W. Ruehl, Phys. Lett. B 696 (2011) 410, arXiv:1009.1054 [hep-th];
M. Grigoriev, A.A. Tseytlin, J. Phys. A 50 (12) (2017) 125401, arXiv:1609.09381 [hep-th].
[37] D. Ponomarev, A.A. Tseytlin, J. High Energy Phys. 1605 (2016) 184, arXiv:1603.06273 [hep-th];
R. Manvelyan, K. Mkrtchyan, W. Ruhl, Nucl. Phys. B 803 (2008) 405, arXiv:0804.1211 [hep-th].
[38] D. Francia, J. Mourad, A. Sagnotti, Nucl. Phys. B 773 (2007) 203, arXiv:hep-th/0701163;
A. Fotopoulos, M. Tsulaia, J. High Energy Phys. 0910 (2009) 050, arXiv:0907.4061 [hep-th].

36

Analytical shear viscosity in hyperscaling violating black brane

Xiao-Mei Kuang [a,b], Jian-Pin Wu [c,*]

[a] Center for Gravitation and Cosmology, College of Physical Science and Technology, Yangzhou University, Yangzhou 225009, China
[b] Instituto de Física, Pontificia Universidad Católica de Valparaíso, Casilla 4059, Valparaíso, Chile
[c] Institute of Gravitation and Cosmology, Department of Physics, School of Mathematics and Physics, Bohai University, Jinzhou 121013, China

ARTICLE INFO

Editor: N. Lambert

ABSTRACT

In this letter, with the use of matching method, we investigate the shear viscosity in a non-relativistic boundary filed theory without hyperscaling symmetry, which is dual to a bulk charged hyperscaling violating black brane. By matching the solutions to the inner region and outer region at the matching region, we analytically obtain that the ratio of shear viscosity and the entropy density is alway $1/4\pi$ at zero temperature and finite temperatures. Our results satisfy the Kovtun–Starinets–Son (KSS) bound.

1. Introduction

Gauge/gravity duality is a beautiful approach to study the physics of strongly coupling sectors, because it connects a bulk gravitational theory and quantum field theory that lives in one less dimensions [1–3]. This allows us to explore the strongly coupled phenomena with the use of dual gravitational systems with weak coupling. In order to capture physics in a wider class of field theories, the duality has been generalized to the sectors beyond relativistic conformal symmetry. A remarkable generalization proposed in [4–11] is to consider the dual gravity with the metric

$$ds^2_{d+1} = u^{\frac{2\theta}{d-1}}\left(-\frac{1}{u^{2z}}dt^2 + \frac{1}{u^2}du^2 + \frac{1}{u^2}d\vec{x}^2\right), \tag{1}$$

which presents both a Lifshitz dynamical critical exponent z ($z \geq 1$) and a hyperscaling violating (HV) exponent θ. Under the scale-transformation $t \to \lambda^z t$, $x_i \to \lambda x_i$, $u \to \lambda u$, the metric transforms as $ds \to \lambda^{\theta/(d-1)}ds$, which breaks the scale-invariance. When $\theta = 0$, the above analysis recovers the known geometry with Lifshitz symmetry and it goes back to the pure AdS geometry when $z = 1$ and $\theta = 0$. Lots of extensive holographic study based on HV background have been present in [12–21] and references therein.

As the simplest implement of holography, AdS/CFT correspondence has been widely used in the study of hydrodynamic properties, such as transport coefficients of strongly coupled systems. Specially, it is found that the ratio of the shear viscosity (η) over the entropy density (s) has a universal value $1/4\pi$ in dual theories described by Einstein gravity [22,23], which has been extended into more general theories, see [24–27] and therein. It is then addressed in [28,29] that the value of ratio $1/4\pi$ gives a universal lower bound, namely the KSS bound, which should be satisfied by all sectors in nature. However, the violation of the viscosity bound has been studied in the presence of higher-derivative gravity corrections [30–40] and in anisotropic gauge/gravity dualities [41,42]. All the descriptions above are focused on finite temperature. Later, by borrowing the matching method proposed in [43] to study the holographic (non-)Fermi liquid, the transport coefficients including shear viscosity and electric conductivity were investigated at extremal AdS RN black hole with finite charge density [44,45]. At zero temperature, the ratio η/s of the boundary field theory is the same as that at the finite temperature boundary field theory. With the same method, the transport coefficients of field theory dual to AdS charged Gauss–Bonnet is performed in [46].

It will be interesting to explore the transport coefficients of hydrodynamic modes in wider boundary geometries like Eq. (1) accompanying with finite charge density, because it is more general than that discussed in AdS gravity. This means that one requires a charged black hole solution with the asymptotical behavior of Eq. (1) in the bulk theory. This kind of solution was firstly proposed in [7], which will be present in the next section.

The aim of this work is to disclose the shear viscosity of the field theory with finite charge density at any temperatures, dual to the bulk theory with hyperscaling violation found in [7]. We will use the matching method via comparing the solution of inner region and outer region at the matching region. Before computing the shear viscosity, we calculate the asymptotical solutions of the

* Corresponding author.
E-mail addresses: xmeikuang@gmail.com (X.-M. Kuang), jianpinwu@mail.bnu.edu.cn (J.-P. Wu).

perturbation modes by matching the solutions. The ratio of shear viscosity to entropy density keeps $1/4\pi$ both at zero and finite temperature, though the HV exponent θ explicitly contributes to the entropy density and shear viscosity. This means the hyperscaling violating in the system does not violate the KSS bound. Note that the shear viscosity via matching method in Lifshitz black hole without hyperscaling violation has been addressed in [47]. And more study of the shear viscosity of non-relativistic effective field theory in neutral case via various methods can be seen [48–52].

This paper is organized as follows. We briefly review the black brane solution at any temperature in HV theory in section 2. Using the matching method, we study the shear viscosity at zero and finite temperature in section 3. In both cases, the KSS bound is satisfied. The last section is the conclusion and discussion.

2. The charged HV black branes from Einstein–Dilaton–Maxwell theory

We start from Einstein–Dilaton–Maxwell (EDM) action in $3+1$ spacetime dimensions [53]

$$S_g = -\frac{1}{16\pi G}\int d^4x\sqrt{-g}\Big[R - \frac{1}{2}(\partial\phi)^2 + V(\phi) - \frac{1}{4}\left(e^{\lambda_1\phi}F^{\mu\nu}F_{\mu\nu} + e^{\lambda_2\phi}\mathcal{F}^{\mu\nu}\mathcal{F}_{\mu\nu}\right)\Big]. \tag{2}$$

The action contains two $U(1)$ gauge fields coupled to a dilaton field ϕ. The $U(1)$ field A with field strength $F_{\mu\nu}$ is required to have a charged solution, while the other gauge field $\mathcal{A}$ with field strength $\mathcal{F}_{\mu\nu}$ and the dilaton field are necessary to generate an anisotropic scaling. Here λ_1 and λ_2 are free parameters, which will be determined later. We can deduce the equations of motion for all the fields from the above action. The Einstein equation of motion for the metric is

$$\begin{aligned} R_{\mu\nu} - \frac{1}{2}Rg_{\mu\nu} &= \frac{1}{2}\partial_\mu\phi\partial_\nu\phi - \frac{V(\phi)}{2}g_{\mu\nu} \\ &\quad + \frac{1}{2}\Big[e^{\lambda_1\phi}(F_{\mu\rho}F^{\rho}_{\nu} - \frac{g_{\mu\nu}}{4}F^{\rho\sigma}F_{\rho\sigma}) \\ &\quad + e^{\lambda_2\phi}(\mathcal{F}_{\mu\rho}\mathcal{F}^{\rho}_{\nu} - \frac{g_{\mu\nu}}{4}\mathcal{F}^{\rho\sigma}\mathcal{F}_{\rho\sigma})\Big]. \end{aligned} \tag{3}$$

The equation of motion for the dilaton field is

$$\nabla^2\phi = -\frac{dV(\phi)}{d\phi} + \frac{1}{4}\left(\lambda_1 e^{\lambda_1\phi}F^{\mu\nu}F_{\mu\nu} + \lambda_2 e^{\lambda_2\phi}\mathcal{F}^{\mu\nu}\mathcal{F}_{\mu\nu}\right). \tag{4}$$

The Maxwell equations for the gauge fields are

$$\nabla_\mu\left(\sqrt{-g}e^{\lambda_2\phi}\mathcal{F}^{\mu\nu}\right) = 0, \tag{5}$$

$$\nabla_\mu\left(\sqrt{-g}e^{\lambda_1\phi}F^{\mu\nu}\right) = 0. \tag{6}$$

The potential $V(\phi)$ plays a very important role in obtaining a charged HV black brane. Following [7], we set $V(\phi) = V_0 e^{\gamma\phi}$ with γ and V_0 being free parameters. Then the analytical charged HV black brane solution is [7]

$$ds_4^2 = r^{-\theta}\left(-r^{2z}f(r)dt^2 + \frac{dr^2}{r^2 f(r)} + r^2(dx^2+dy^2)\right), \tag{7}$$

$$f = 1 - \left(\frac{r_h}{r}\right)^{2+z-\theta} + \frac{Q^2}{r^{2(z-\theta+1)}}\left[1 - \left(\frac{r_h}{r}\right)^{\theta-z}\right], \tag{8}$$

$$\mathcal{F}_{rt} = \sqrt{2(z-1)(2+z-\theta)}\, e^{\frac{2-\theta/2}{\sqrt{2(2-\theta)(z-1-\theta/2)}}\phi_0} r^{1+z-\theta}, \tag{9}$$

$$F_{rt} = Q\sqrt{2(2-\theta)(z-\theta)}\, e^{-\sqrt{\frac{z-1+\theta/2}{2(2-\theta)}}\phi_0} r^{-(z-\theta+1)}, \tag{10}$$

$$e^{\phi} = e^{\phi_0} r^{\sqrt{2(2-\theta)(z-1-\theta/2)}}. \tag{11}$$

Here, r_h is the radius of horizon satisfying $f(r_h) = 0$ and $Q = \frac{1}{16\pi G}\int e^{\lambda_1\phi}F_{rt}$ is the total charge of the black brane. All the parameters in the action can be determined by Lifshitz scaling exponent z and HV exponent θ

$$\begin{aligned} \lambda_1 &= \sqrt{\frac{2(z-1-\theta/2)}{2-\theta}}, \quad \lambda_2 = -\frac{2(2-\theta/2)}{\sqrt{2(2-\theta)(z-\theta/2-1)}}, \\ \gamma &= \frac{\theta}{\sqrt{2(2-\theta)(z-1-\theta/2)}}, \\ V_0 &= e^{\frac{-\theta\phi_0}{\sqrt{2(2-\theta)(z-1-\theta/2)}}}(z-\theta+1)(z-\theta+2). \end{aligned} \tag{12}$$

From Eqs. (9) and (10), we can obtain the gauge fields

$$\mathcal{A}_t = -\mu\left[1 - \left(\frac{r}{r_h}\right)^{2+z-\theta}\right] \quad \text{with } \mu = \frac{\sqrt{2(z-1)(2+z-\theta)}}{2+z-\theta}e^{\frac{2-\theta/2}{\sqrt{2(2-\theta)(z-1-\theta/2)}}\phi_0} r_h^{2+z-\theta}, \tag{13}$$

$$A_t = \mu\left[1 - \left(\frac{r_h}{r}\right)^{z-\theta}\right] \quad \text{with } \mu = Q\sqrt{\frac{2(2-\theta)}{z-\theta}}e^{-\sqrt{\frac{z-1+\theta/2}{2(2-\theta)}}\phi_0} r_h^{\theta-z}. \tag{14}$$

In addition, the Hawking temperature and entropy density are respectively

$$T = \frac{(2+z-\theta)r_h^z}{4\pi}\left[1 - \frac{(z-\theta)Q^2}{2+z-\theta}r_h^{2(\theta-z-1)}\right], \tag{15}$$

$$s = \frac{r_h^{2-\theta}}{4G}. \tag{16}$$

Before proceeding, we have to fix the valid region of the parameters z and θ. First, the background solution (7)–(11) are valid only if $z \geq 1$ and $\theta \geq 0$. Second, the condition $z - \theta \geq 0$ should be satisfied to make sure a well-defined chemical potential in the dual field theory. Third, it is obvious from equation (13) that $\theta < 2$. The null energy condition $(-\frac{\theta}{2}+1)(-\frac{\theta}{2}+z-1) \geq 0$ [7] gives us $\theta \leq 2(z-1)$. Thus, in this charged background, the range of the parameters is

$$\begin{cases} 0 \leq \theta \leq 2(z-1) & \text{for } 1 \leq z < 2\,, \\ 0 \leq \theta < 2 & \text{for } z \geq 2\,. \end{cases} \tag{17}$$

Furthermore, if we set $Q = \sqrt{\frac{2+z-\theta}{z-\theta}}r_h^{z-\theta+1}$, i.e., $\mu = \frac{\sqrt{2(2-\theta)(2+z-\theta)}}{z-\theta}r_h$, one gets the zero-temperature limit. Then the redshift factor $f(r)$ becomes

$$f(r)|_{T=0} = 1 - \frac{2(z-\theta+1)}{z-\theta}\left(\frac{r_h}{r}\right)^{z-\theta+2} + \frac{z-\theta+2}{z-\theta}\left(\frac{r_h}{r}\right)^{2(z-\theta+1)}. \tag{18}$$

Obviously, in the near horizon limit $r \to r_h$,

$$f(r)|_{T=0, r\to r_h} \simeq \frac{(z-\theta+1)(z-\theta+2)}{r_h^2}(r-r_h)^2. \tag{19}$$

Therefore, at the zero temperature, there exists the same near horizon geometry, $AdS_2 \times \mathbb{R}^2$, as that for RN-AdS background. Specially, we can define $u = r/r_h$ and change the coordinate via

$$u - 1 = \frac{\alpha}{\varsigma}, \tag{20}$$

with $\alpha = \frac{1}{(z-\theta+1)(z-\theta+2)r_h^z}$. Consequently, near the horizon, the metric is given by

$$ds^2 = r_h^{-\theta}\left[\frac{-dt^2+d\varsigma^2}{(z-\theta+1)(z-\theta+2)\varsigma^2}+r_h^2(dx^2+dy^2)\right], \tag{21}$$

with the curvature radius of AdS_2 $L_2 \equiv 1/\sqrt{(z-\theta+1)(z-\theta+2)}$, while the gauge fields are $A_\tau = \frac{\mu(z-\theta)\alpha}{\varsigma}$ and $\mathcal{A}_\tau = \frac{\mu(2+z-\theta)\alpha}{\varsigma}$, respectively.

3. Analytical study of shear viscosity

The shear viscosity of the dual boundary theory is related with the retarded Green function via the Kubo formula

$$\eta = -\lim_{\omega\to 0}\frac{\mathbf{Im}G^R_{xy,xy}(\omega)}{\omega}. \tag{22}$$

According to the real-time recipe proposed in [54], the formula of the retarded Green function is

$$G^R_{xy,xy}(\omega) = \frac{1}{16\pi G}\sqrt{-g}g^{uu}h^{x*}_y(u)\partial_u h^x{}_y(u)\mid_{u\to\infty}, \tag{23}$$

where $h^x{}_y(u)e^{-i\omega t}$ is the tensor perturbation, $g_{xy} = \bar{g}_{xy}+h_{xy}$, of the background Eqs. (7)–(11), satisfying the linearized equation of motion

$$u^2 f h^{x''}_y(u) + \left(3uf + uzf - u\theta f + u^2 f'\right)h^{x'}_y(u) + \frac{\omega^2}{u^{2z}r_h^{2z}f}h^x{}_y(u) = 0. \tag{24}$$

In the equation above, the prime denotes to the derivative to u.

3.1. Shear viscosity at zero temperature

We shall calculate the shear viscosity according to the retarded Green function following the matching method proposed in [43]. To this end, we divide the radial axis into inner region with $u-1 = \frac{\alpha\omega}{\varsigma}$ and outer region with $u-1 > \frac{\alpha\omega}{\epsilon}$, and we consider the limit

$$\omega\to 0,\quad \varsigma = \text{finite},\quad \epsilon\to 0,\quad \frac{\alpha\omega}{\epsilon}\to 0. \tag{25}$$

And then, we match the solutions of the inner region and the outer region in the matching region with $\varsigma\to 0$ and $u-1 = \frac{\alpha\omega}{\epsilon}\to 0$. For convenience of notation, we will set $h^x{}_y = \psi$. After introducing

$$\mathfrak{w} = \alpha\omega \text{ and } \zeta = \omega\varsigma, \tag{26}$$

we have $u = 1+\frac{\mathfrak{w}}{\zeta}$ from (20). So the matching region near the horizon means taking a double limit $\zeta\to 0$ and $\mathfrak{w}/\zeta\to 0$. The equation (24) can be rewritten in term of $\mathfrak{w}$ as

$$u^2 f\psi''(u) + \left(3uf + uzf - u\theta f + u^2 f'\right)\psi'(u) + \frac{((z-\theta+1)(z-\theta+2))^2\mathfrak{w}^2}{u^{2z}f}\psi(u) = 0. \tag{27}$$

3.1.1. Solution of inner region

The perturbation mode near the horizon can be expanded in the low frequency limit as

$$\psi_I(\zeta) = \psi_I^{(0)}(\zeta) + \mathfrak{w}\psi_I^{(1)}(\zeta) + \mathfrak{w}^2\psi_I^{(2)}(\zeta) + \cdots. \tag{28}$$

Here the leading term attributes to the near horizon $AdS_2\times\mathcal{R}^2$ geometry. Subsequently, in terms of the coordinate ζ, the leading term of (27) reads as

$$\psi_I^{(0)''}(\zeta) + \psi_I^{(0)}(\zeta) = 0, \tag{29}$$

the general solution of which is

$$\psi_I^{(0)}(\zeta) = a_I^{(0)}\exp(i\zeta) + b_I^{(0)}\exp(-i\zeta). \tag{30}$$

Keeping the regularity in mind, we intend to choose the in-going boundary condition which requires to set $b_I^{(0)} = 0$ to cancel the out-going branch.

Then, near the matching region, i.e., in the limit of $\zeta\to 0$, the in-going result can be expanded as

$$\psi_I^{(0)}(\zeta)\mid_{\zeta\to 0}\simeq a_I^{(0)}(1+i\zeta) = a_I^{(0)}\left[1+\mathcal{G}_R(\mathfrak{w})\frac{1}{u-1}\right] \tag{31}$$

where we have used $\zeta = \frac{\mathfrak{w}}{u-1}$ to express the second equality in the coordinate u. Meanwhile,

$$\mathcal{G}_R(\mathfrak{w}) = i\mathfrak{w} \tag{32}$$

is nothing but the retarded Green function of a zero-charge scalar operator with conformal dimension one in the IR conformal field theory, of which the dual field is $\psi_I^{(0)}$ in the framework of AdS/CFT correspondence. This is explicit because when we rescale ζ into $\zeta/\mathfrak{w}$, equation (29) coincides with the equation of motion for a massless and chargeless scalar field in AdS_2 geometry [43].

Considering equation (27) in term of the order of $\mathfrak{w}$, it's easy to obtain that the solution in the inner region near the matching region has the form

$$\psi_I(u) = a_I^{(0)} + \frac{a_I^{(0)}}{u-1}\mathcal{G}_R(\mathfrak{w}) + \cdots \tag{33}$$

where the dots denote the non-vanishing $\mathfrak{w}$ term, but they vanish in the limit of $\mathfrak{w}\to 0$. Note that the solution of inner region is universal, meaning that it does not depend on the geometrical parameters. Then we will move on to explore the outer region solution in order to match the solutions.

3.1.2. Solution of outer region, matching and the shear viscosity

In this subsection, we shall work out the solution in the outer region, which can be expanded for both at asymptotical boundary and at the matching region so we can match the solution in outer region to that in inner region obtained in the above subsection.

Now, we expand the perturbation field of the outer region in the low $\mathfrak{w}$ limit as

$$\psi_O(u) = \psi_O^{(0)}(u) + \mathfrak{w}\psi_O^{(1)}(u) + \mathfrak{w}^2\psi_O^{(2)}(u) + \cdots. \tag{34}$$

It's straightforward to get the leading order equation of (27)

$$u^2 f\psi_O^{(0)''}(u) + \left(3uf + uzf - u\theta f + u^2 f'\right)\psi_O^{(0)'}(u) = 0, \tag{35}$$

from which we see that this equation only explicitly depends on the value of $\sharp = z-\theta$.

For arbitrary $z-\theta = \sharp$, the general solution of (35) is

$$\psi_O^{(0)}(u) = a_O^{(0)} + \int_1^r e^{\int_1^{K[2]} -\frac{2-\sharp-\sharp^2-(2+2\sharp)K[1]^\sharp+(\sharp^2+3\sharp)K[1]^{2+2\sharp}}{K[1](2+\sharp-(2+2\sharp)K[1]^\sharp+\sharp K[1]^{2+2\sharp})}dK[1]}b_O^{(0)}dK[2], \tag{36}$$

where $K[i]$ $(i = 1, 2)$ is the complete elliptic integral of the first kind. And $a_O^{(0)}$ and $b_O^{(0)}$ are integral constants which will be discussed soon. Although the above solution is not explicit for general $\sharp$, once we give the value of $\sharp$, we can obtain the concrete expression of the solution. Furthermore, we can take its behavior near the matching region and the boundary. Specially, the asymptotical behavior of the above equation has the form

$$\psi_O^{(0)}(u)|_{u\to\infty} = (a_O^{(0)} + C_0 b_O^{(0)}) - \frac{b_O^{(0)}}{(\sharp+2)u^{\sharp+2}} + \cdots \quad (37)$$

with the dots denoting the higher subleading terms than $u^{-(\sharp+2)}$, while near the matching region, the behavior of (36) is

$$\psi_O^{(0)}(u)|_{u\to 1} = -\frac{b_O^{(0)}}{(\sharp+1)(\sharp+2)(u-1)} + a_O^{(0)} + \tilde{C_0} b_O^{(0)} + \cdots \quad (38)$$

where the dots represent the vanishing term in the limit of $\mathfrak{w} \to 0$. Note that C_0 and $\tilde{C_0}$ are some functions of $\sharp$. Matching (33) and (38) will give us the following relations

$$b_O^{(0)} = -(\sharp+1)(\sharp+2)\mathcal{G}_R(\mathfrak{w})a_I^{(0)},$$
$$a_O^{(0)} = \left[1 + (\sharp+1)(\sharp+2)\tilde{C_0}\mathcal{G}_R(\mathfrak{w})\right]a_I^{(0)}. \quad (39)$$

Then putting the above relations in to (37), we can write asymptotic form of $h^x{}_y(u)$ as

$$\begin{aligned}\psi_O(u)|_{u\to\infty} &= a_I^{(0)}\left[1 + (\sharp+1)(\sharp+2)(\tilde{C_0} - C_0)\mathcal{G}_R(\mathfrak{w}) + \cdots\right] \\ &\quad + (\sharp+1)a_I^{(0)}\mathcal{G}_R(\mathfrak{w})[1+\cdots]u^{-(\sharp+2)} + \cdots \\ &= a_I^{(0)}\left[1 + A(\sharp)\mathcal{G}_R(\mathfrak{w}) + \cdots\right] \\ &\quad + (\sharp+1)a_I^{(0)}\mathcal{G}_R(\mathfrak{w})[1+\cdots]u^{-(\sharp+2)} + \cdots \quad (40)\end{aligned}$$

where in the second line we have defined the constant function $A(\sharp) = (\sharp+1)(\sharp+2)(\tilde{C_0} - C_0)$. Then taking account of (22) and (23) and setting $a_I^{(0)} = 1$, we obtain the ratio of shear viscosity to the entropy density is

$$\begin{aligned}\eta/s &= -\lim_{\omega\to 0}\frac{\mathbf{Im}G^R_{xy,xy}(\omega)}{\omega s} \\ &= -\lim_{\omega\to 0}\frac{1}{\omega s}\mathbf{Im}\left[\frac{1}{16\pi G}\sqrt{r_h^{4+2z-4\theta}u^{2+2z-4\theta}}r_h^{\theta}u^{2+\theta}\right. \\ &\quad \left.\times f[-(\sharp+1)(\sharp+2)\mathcal{G}_R(\mathfrak{w})(1+\mathcal{O}(\mathfrak{w}))u^{-(\sharp+3)}]\right]_{u\to\infty} \\ &= \frac{r_h^{2-\theta}}{16\pi G s} = \frac{1}{4\pi}, \quad (41)\end{aligned}$$

where we have substituted $\mathcal{G}_R(\mathfrak{w}) = i\alpha\omega = \frac{i\omega}{(\sharp+1)(\sharp+2)r_h^z}$ in the second line. Note that the result in AdS case with $z=1$ and $\theta=0$ discussed in [44] can be recovered by our case with $\sharp = 1$. So we conclude that in HV background, the KSS bound $\eta/s = \frac{1}{4\pi}$ always hold at zero temperature, independent on the geometrical exponents.

3.2. Shear viscosity at finite temperatures

At finite temperature, we do not have AdS_2 geometry near the horizon because the first oder of the expansion of the redshift now dominants. But we will take some approximation to work out the solutions. In details, we will solve the equation in the near region with $\omega < u\omega \ll 1$ and outer region with $u \gg 1$, and then match the solutions in the near horizon region $u - 1 \ll 1$.[1] We rewrite equation (24) as

$$\psi''(u) + \left(\frac{3+z-\theta}{u} + \frac{f'}{f}\right)\psi'(u) + \frac{\omega^2}{r_h^{2z}u^{2z+2}f^2}\psi(u) = 0. \quad (42)$$

3.2.1. The matching near horizon region

In the matching near region, we have $u - 1 \ll 1$ and $f(u) = f'(r_h)(u-1) + \cdots$ with $f'(r_h) = 4\pi T$, which gives us the leading order of equation as (42) as

$$\psi''(u) + \left(\frac{1}{u-1}\right)\psi'(u) + \frac{\omega^2}{(4\pi T r_h^z)^2(u-1)^2}\psi(u) = 0. \quad (43)$$

The solution to the above equation is $\psi(u) = C_1(u-1)^{-i\omega/4\pi T r_h^z} + C_2(u-1)^{i\omega/4\pi T r_h^z}$. To regularize, we choose the infalling solution by setting $C_2 = 0$. Then in the low frequency limit, the solution behaves as

$$\psi(u) = C_1\left(1 - \frac{i\omega}{4\pi T r_h^z}\log(u-1)\right). \quad (44)$$

3.2.2. The near region

The region with $\omega < u\omega \ll 1$ is our near region, in which the equation becomes

$$\psi''(u) + \left(\frac{3+z-\theta}{u} + \frac{f'}{f}\right)\psi'(u) = 0. \quad (45)$$

Its solution can express as

$$\psi(u) = C_3 + C_4\int\frac{du}{fu^{3+z-\theta}}. \quad (46)$$

Near horizon with $u \to 1$ and $f(u) = 4\pi T(u-1)$, the solution (46) behaves as

$$\psi(u) = C_3 + C_4\int\frac{du}{4\pi T(u-1)} = C_3 + \frac{C_4}{4\pi T}\log(u-1), \quad (47)$$

while, at large radius with $u \gg 1$ and $f(u) \to 1$, it becomes

$$\psi(u) = C_3 + C_4\int\frac{du}{u^{3+z-\theta}} = C_3 - \frac{C_4}{u^{2+z-\theta}}. \quad (48)$$

3.2.3. Outer region, matching solution and the shear viscosity

In the asymptotic of the outer region $u \gg 1$, it is easy to get $f'(u) \to 0$ and $f(u) \to 1$, so the perturbation equation can be simplified as

$$\psi''(u) + \left(\frac{3+z-\theta}{u}\right)\psi'(u) + \frac{\omega^2}{r_h^{2z}u^{2z+2}}\psi(u) = 0. \quad (49)$$

We rewrite the above equation in the coordinate $\mathfrak{u} = 1/u$ and obtain the following solution to the equation of motion

$$\psi = \left(\frac{\omega\mathfrak{u}^z}{zr_h^z}\right)^p\left[C_5 J_p\left(\frac{\omega\mathfrak{u}^z}{zr_h^z}\right) + C_6 Y_p\left(\frac{\omega\mathfrak{u}^z}{zr_h^z}\right)\right], \quad (50)$$

where J and Y are first and second kind of Bessel function with $p = \frac{z-\theta+2}{2z}$, respectively. According to the feature of Bessel function, in the low frequency limit, the leading order the above solution is

$$\psi = \tilde{C_6} + \tilde{C_5}\left(\frac{\omega}{r_h^z}\right)^{2p}\mathfrak{u}^{2zp} = \tilde{C_6} + \tilde{C_5}\left(\frac{\omega}{r_h^z}\right)^{\frac{z-\theta+2}{z}}\frac{1}{u^{z-\theta+2}}. \quad (51)$$

Now we are ready to match the inner and outer solutions in matching region. This can be achieved by identifying (47) with (44), and (48) with (51), which gives us the following relations between the coefficients

[1] Note that in the coordinate r, the near region is $r_h\omega < r\omega \ll 1$ and the outer region is $r > r_h$ while the matching near horizon region is $r - r_h \ll r_h$ [47].

$$\tilde{C_6} = C_3 = C_1, \quad C_4 = -\frac{i\omega}{r_h^z} C_1,$$

$$\tilde{C_5}\left(\frac{\omega}{r_h^z}\right)^{\frac{z-\theta+2}{z}} = \frac{i\omega}{r_h^z} C_1. \tag{52}$$

Thus, the leading order of the asymptotic solution (51) at low frequency limit is

$$\psi = C_1 + C_1\left(\frac{i\omega}{r_h^z}\right)\frac{1}{u^{z-\theta+2}}, \tag{53}$$

where we will consider the normalizability of the solution near the horizon, i.e. we set $C_1 = 1$. Having the above behavior, with the similar algebraic computation using (22) and (23), it is straightforward to obtain the shear viscosity

$$\eta = \frac{r_h^{2-\theta}}{16\pi G}. \tag{54}$$

The shear viscosity is the same as the result (41) at zero temperature, so that the KSS bound is also fulfilled at finite temperature as we expect.

4. Conclusion and discussion

In this letter, via matching method, we have calculated the shear viscosity of a holographic charged non-relativistic effective field theory at both zero and finite temperature, which is dual to a charged HV gravity. We find that the ratio of shear viscosity and the entropy density is always $1/4\pi$ at any temperature as that found in RN-AdS black hole, which satisfies the KSS bound. Our result shows that the domain of universality of the ratio can be enlarged to include holographic HV effective field theory. Here we only consider the perturbations formula $e^{-i\omega t+i\vec{k}\cdot\vec{x}}$ in the limit with momentum $\vec{k}=0$. It is worthwhile to calculate transport coefficients at finite $\vec{k}$. Also, we can calculate the shear viscosity of the other non-relativistic backgrounds, for example, the background with the Schrödinger symmetry [55].

Besides the shear viscosity, another important transport coefficient is the conductivity of the dual field theory. It is addressed in [56] that the electric conductivity at large frequency in three dimensional field theory dual to an AdS geometry approaches to be a constant. So it is of great interest to further investigate the conductivity in Lifshitz and HV gravitational theory. For the shear channel with the radial gauge, the vector perturbations, $g_{ti} = \bar{g}_{ti} + h_{ti}$, $A_i = \bar{A}_i + a_i$, $\mathcal{A}_i = \bar{\mathcal{A}}_i + b_i$ with $i = x, y$, control the conductivity, including electric conductivity, heat conductivity and thermoelectric conductivity of the dual boundary field theory [54]. Due to the $SO(2)$ symmetry in the x–y plane, one can only consider the perturbation in x direction, which satisfy the following equations of motion in linear order

$$\begin{aligned}0 &= u^4 h_t^{x\prime\prime}(u) + (5u^3 - u^3 z - u^3\theta) h_t^{x\prime}(u)\\ &\quad + u^{z+\theta-1}[r_h^{2z-4}(z-\theta)\mu a_x'(u) + r_h^{2\theta-6}(z-\theta+2)\not{\mu} b_x'(u)],\end{aligned} \tag{55}$$

$$\begin{aligned}0 &= u^2 f a_x''(u) + \left(u(3z-\theta-1)f + u^2 f'\right) a_x'(u)\\ &\quad + u^{3-3z+\theta}(z-\theta)\mu r_h^{2-2z} h_t^{x\prime}(u) + \frac{\omega^2}{u^{2z} r_h^{2z} f} a_x(u),\end{aligned} \tag{56}$$

$$\begin{aligned}0 &= u^2 f b_x''(u) + \left(u(z+\theta-3)f + u^2 f'\right) b_x'(u)\\ &\quad + u^{5-z-\theta}(z-\theta+2)\not{\mu} r_h^{2-2z} h_t^{x\prime}(u) + \frac{\omega^2}{u^{2z} r_h^{2z} f} b_x(u),\end{aligned} \tag{57}$$

$$\begin{aligned}0 &= u^{z+3} h_t^{x\prime}(u) + u^{2z+\theta-2}[r_h^{2z-4}(z-\theta)\mu a_x(u)\\ &\quad + r_h^{2\theta-6}(z-\theta+2)\not{\mu} b_x(u)].\end{aligned} \tag{58}$$

Note that the last equation is a constraint equation from the xu-component of linearized Einstein equations.

The AC electric conductivity has been studied in [47,57] in Lifshitz black brane with $\theta = 0$, however, the authors turned off the perturbation b_x and only work with the simplified perturbation equation of a_x. They argued that when the Lifshitz exponent $z > 1$, the AC electric conductivity behaves with a (non-)power scaling in large frequency limit which is analogous to the phenomena found in some disorder realistic materials [58]. Recently, the DC conductivity dual to both a_x and b_x in the hyperscaling violation theory with additional massless axions has been disclosed in [59,60]. Considering both perturbations of the two Maxwell field, a new matrix computational method was proposed. By calculating the mixed DC thermoelectric conductivities, both linear dependent and quadratic dependent of the resistivity on the temperature can be recovered. So it would be very interesting to solve the coupled system (55)–(58) to disclose the properties of various AC conductivity of the system and extended the study to HV theory with momentum relaxation. We will address the results elsewhere in the near future.

Acknowledgments

We are grateful to Xian-Hui Ge, Wei-Jia Li and Zhenhua Zhou for valuable discussions. This work is supported by the Natural Science Foundation of China under Grant Nos. 11705161, 11775036 and 11305018. X. M. Kuang is also supported by Natural Science Foundation of Jiangsu Province under Grant No. BK20170481 and she is also funded by Chilean FONDECYT grant No. 3150006. J. P. Wu is also supported by by Natural Science Foundation of Liaoning Province under Grant No. 201602013.

References

[1] J.M. Maldacena, The large N limit of superconformal field theories and supergravity, Adv. Theor. Math. Phys. 2 (1998) 231, Int. J. Theor. Phys. 38 (1999) 1113.
[2] S.S. Gubser, I.R. Klebanov, A.M. Polyakov, A semiclassical limit of the gauge string correspondence, Nucl. Phys. B 636 (2002) 99.
[3] E. Witten, Anti-de Sitter space and holography, Adv. Theor. Math. Phys. 2 (1998) 253.
[4] X. Dong, S. Harrison, S. Kachru, G. Torroba, H. Wang, Aspects of holography for theories with hyperscaling violation, J. High Energy Phys. 1206 (2012) 041, arXiv:1201.1905 [hep-th].
[5] L. Huijse, S. Sachdev, B. Swingle, Hidden Fermi surfaces in compressible states of gauge-gravity duality, Phys. Rev. B 85 (2012) 035121, arXiv:1112.0573 [cond-mat.str-el].
[6] E. Shaghoulian, Holographic entanglement entropy and Fermi surfaces, J. High Energy Phys. 1205 (2012) 065, arXiv:1112.2702 [hep-th].
[7] M. Alishahiha, E.O. Colgain, H. Yavartanoo, Charged black branes with hyperscaling violating factor, J. High Energy Phys. 1211 (2012) 137, arXiv:1209.3946 [hep-th].
[8] I. Papadimitriou, Hyperscaling violating Lifshitz holography, arXiv:1411.0312 [hep-th].
[9] Z.Y. Fan, H. Lu, Electrically-charged Lifshitz spacetimes, and hyperscaling violations, J. High Energy Phys. 1504 (2015) 139, arXiv:1501.05318 [hep-th].
[10] C. Charmousis, B. Gouteraux, B.S. Kim, E. Kiritsis, R. Meyer, Effective holographic theories for low-temperature condensed matter systems, J. High Energy Phys. 1011 (2010) 151, arXiv:1005.4690 [hep-th].
[11] W. Chemissany, I. Papadimitriou, Lifshitz holography: the whole shebang, J. High Energy Phys. 1501 (2015) 052, arXiv:1408.0795 [hep-th].
[12] A. Lucas, S. Sachdev, Conductivity of weakly disordered strange metals: from conformal to hyperscaling-violating regimes, Nucl. Phys. B 892 (2015) 239, arXiv:1411.3331 [hep-th].
[13] A. Lucas, S. Sachdev, K. Schalm, Scale-invariant hyperscaling-violating holographic theories and the resistivity of strange metals with random-field disorder, Phys. Rev. D 89 (6) (2014) 066018, arXiv:1401.7993 [hep-th].

[14] X.M. Kuang, E. Papantonopoulos, B. Wang, J.P. Wu, Formation of Fermi surfaces and the appearance of liquid phases in holographic theories with hyperscaling violation, J. High Energy Phys. 1411 (2014) 086, arXiv:1409.2945 [hep-th].
[15] X.M. Kuang, E. Papantonopoulos, B. Wang, J.P. Wu, Dynamically generated gap from holography in the charged black brane with hyperscaling violation, J. High Energy Phys. 1504 (2015) 137, arXiv:1411.5627 [hep-th].
[16] J.P. Wu, X.M. Kuang, Scalar boundary conditions in hyperscaling violating geometry, Phys. Lett. B 753 (2016) 34, arXiv:1512.03499 [hep-th].
[17] P. Bueno, P.F. Ramirez, Higher-curvature corrections to holographic entanglement entropy in geometries with hyperscaling violation, J. High Energy Phys. 1412 (2014) 078, arXiv:1408.6380 [hep-th].
[18] L.K. Chen, H. Guo, F.W. Shu, Crystalline geometries from fermionic vortex lattice with hyperscaling violation, arXiv:1511.01370 [hep-th].
[19] Q. Pan, S.J. Zhang, Revisiting holographic superconductors with hyperscaling violation, arXiv:1510.09199 [hep-th].
[20] S.J. Zhang, Q. Pan, E. Abdalla, Holographic superconductor in hyperscaling violation geometry with Maxwell-dilaton coupling, arXiv:1511.01841 [hep-th].
[21] Z. Zhou, J.P. Wu, Y. Ling, DC and Hall conductivity in holographic massive Einstein–Maxwell-Dilaton gravity, J. High Energy Phys. 1508 (2015) 067, arXiv: 1504.00535 [hep-th].
[22] G. Policastro, D.T. Son, A.O. Starinets, The shear viscosity of strongly coupled N=4 supersymmetric Yang–Mills plasma, Phys. Rev. Lett. 87 (2001) 081601, arXiv:hep-th/0104066.
[23] G. Policastro, D.T. Son, A.O. Starinets, From AdS/CFT correspondence to hydrodynamics, J. High Energy Phys. 0209 (2002) 043, arXiv:hep-th/0205052.
[24] K. Maeda, M. Natsuume, T. Okamura, Viscosity of gauge theory plasma with a chemical potential from AdS/CFT, Phys. Rev. D 73 (2006) 066013, arXiv:hep-th/0602010.
[25] R.G. Cai, Y.W. Sun, Shear viscosity from AdS Born–Infeld black holes, J. High Energy Phys. 0809 (2008) 115, arXiv:0807.2377 [hep-th].
[26] P. Benincasa, A. Buchel, Transport properties of N=4 supersymmetric Yang–Mills theory at finite coupling, J. High Energy Phys. 0601 (2006) 103, arXiv:hep-th/0510041.
[27] X.H. Ge, H.Q. Leng, L.Q. Fang, G.H. Yang, Transport coefficients for holographic hydrodynamics at finite energy scale, Adv. High Energy Phys. 2014 (2014) 915312, arXiv:1408.4276 [hep-th].
[28] P. Kovtun, D.T. Son, A.O. Starinets, Holography and hydrodynamics: diffusion on stretched horizons, J. High Energy Phys. 0310 (2003) 064, arXiv:hep-th/0309213.
[29] P. Kovtun, D.T. Son, A.O. Starinets, Viscosity in strongly interacting quantum field theories from black hole physics, Phys. Rev. Lett. 94 (2005) 111601, arXiv: hep-th/0405231.
[30] Y. Kats, P. Petrov, Effect of curvature squared corrections in AdS on the viscosity of the dual gauge theory, J. High Energy Phys. 0901 (2009) 044, arXiv: 0712.0743 [hep-th].
[31] M. Brigante, H. Liu, R.C. Myers, S. Shenker, S. Yaida, Viscosity bound violation in higher derivative gravity, Phys. Rev. D 77 (2008) 126006, arXiv:0712.0805 [hep-th].
[32] M. Brigante, H. Liu, R.C. Myers, S. Shenker, S. Yaida, The viscosity bound and causality violation, Phys. Rev. Lett. 100 (2008) 191601, arXiv:0802.3318 [hep-th].
[33] I.P. Neupane, N. Dadhich, Entropy bound and causality violation in higher curvature gravity, Class. Quantum Gravity 26 (2009) 015013, arXiv:0808.1919 [hep-th].
[34] R.G. Cai, Z.Y. Nie, Y.W. Sun, Shear viscosity from effective couplings of gravitons, Phys. Rev. D 78 (2008) 126007, arXiv:0811.1665 [hep-th].
[35] R.G. Cai, Z.Y. Nie, N. Ohta, Y.W. Sun, Shear viscosity from Gauss–Bonnet gravity with a dilaton coupling, Phys. Rev. D 79 (2009) 066004, arXiv:0901.1421 [hep-th].
[36] X.H. Ge, Y. Matsuo, F.W. Shu, S.J. Sin, T. Tsukioka, Viscosity bound, causality violation and instability with stringy correction and charge, J. High Energy Phys. 0810 (2008) 009, arXiv:0808.2354 [hep-th].
[37] X.H. Ge, S.J. Sin, Shear viscosity, instability and the upper bound of the Gauss–Bonnet coupling constant, J. High Energy Phys. 0905 (2009) 051, arXiv: 0903.2527 [hep-th].
[38] X.H. Ge, S.J. Sin, S.F. Wu, G.H. Yang, Shear viscosity and instability from third order Lovelock gravity, Phys. Rev. D 80 (2009) 104019, arXiv:0905.2675 [hep-th].
[39] A. Bhattacharyya, D. Roychowdhury, Viscosity bound for anisotropic superfluids in higher derivative gravity, J. High Energy Phys. 1503 (2015) 063, arXiv: 1410.3222 [hep-th].
[40] M. Sadeghi, S. Parvizi, Hydrodynamics of a black brane in Gauss–Bonnet massive gravity, arXiv:1507.07183 [hep-th].
[41] A. Rebhan, D. Steineder, Violation of the holographic viscosity bound in a strongly coupled anisotropic plasma, Phys. Rev. Lett. 108 (2012) 021601, arXiv: 1110.6825 [hep-th].
[42] D. Giataganas, Probing strongly coupled anisotropic plasma, J. High Energy Phys. 1207 (2012) 031, arXiv:1202.4436 [hep-th].
[43] T. Faulkner, H. Liu, J. McGreevy, D. Vegh, Emergent quantum criticality, Fermi surfaces, and AdS(2), Phys. Rev. D 83 (2011) 125002, arXiv:0907.2694 [hep-th].
[44] M. Edalati, J.I. Jottar, R.G. Leigh, Transport coefficients at zero temperature from extremal black holes, J. High Energy Phys. 1001 (2010) 018, arXiv:0910.0645 [hep-th].
[45] M. Edalati, J.I. Jottar, R.G. Leigh, Shear modes, criticality and extremal black holes, J. High Energy Phys. 1004 (2010) 075, arXiv:1001.0779 [hep-th].
[46] R.G. Cai, Y. Liu, Y.W. Sun, Transport coefficients from extremal Gauss–Bonnet black holes, J. High Energy Phys. 1004 (2010) 090, arXiv:0910.4705 [hep-th].
[47] J.R. Sun, S.Y. Wu, H.Q. Zhang, Novel features of the transport coefficients in Lifshitz black branes, Phys. Rev. D 87 (2013) 086005, arXiv:1302.5309 [hep-th].
[48] I. Kanitscheider, K. Skenderis, Universal hydrodynamics of non-conformal branes, J. High Energy Phys. 0904 (2009) 062, arXiv:0901.1487 [hep-th].
[49] B. Gouteraux, J. Smolic, M. Smolic, K. Skenderis, M. Taylor, Holography for Einstein–Maxwell-dilaton theories from generalized dimensional reduction, J. High Energy Phys. 1201 (2012) 089, arXiv:1110.2320 [hep-th].
[50] E. Kiritsis, Y. Matsuo, Charge-hyperscaling violating Lifshitz hydrodynamics from black-holes, J. High Energy Phys. 1512 (2015) 076, arXiv:1508.02494 [hep-th].
[51] S. Kulkarni, B.H. Lee, J.H. Oh, C. Park, R. Roychowdhury, Transports in non-conformal holographic fluids, J. High Energy Phys. 1303 (2013) 149, arXiv: 1211.5972 [hep-th].
[52] J. Sadeghi, A. Asadi, Hydrodynamics in a black brane with hyperscaling violation metric background, Can. J. Phys. 92 (12) (2014) 1570, arXiv:1404.5282 [hep-th].
[53] J. Tarrio, S. Vandoren, Black holes and black branes in Lifshitz spacetimes, J. High Energy Phys. 1109 (2011) 017, arXiv:1105.6335 [hep-th].
[54] D.T. Son, A.O. Starinets, Minkowski space correlators in AdS/CFT correspondence: recipe and applications, J. High Energy Phys. 0209 (2002) 042, arXiv: hep-th/0205051.
[55] A. Adams, C.M. Brown, O. DeWolfe, C. Rosen, Charged Schrodinger black holes, Phys. Rev. D 80 (2009) 125018, arXiv:0907.1920 [hep-th].
[56] S.A. Hartnoll, J. Polchinski, E. Silverstein, D. Tong, Towards strange metallic holography, J. High Energy Phys. 1004 (2010) 120, arXiv:0912.1061 [hep-th].
[57] J.R. Sun, S.Y. Wu, H.Q. Zhang, Mimic the optical conductivity in disordered solids via gauge/gravity duality, Phys. Lett. B 729 (2014) 177, arXiv:1306.1517 [hep-th].
[58] J.C. Dyre, T.B. Schroder, Universality of AC conduction in disordered solids, Rev. Mod. Phys. 72 (2000) 873.
[59] X.H. Ge, Y. Tian, S.Y. Wu, S.F. Wu, S.F. Wu, Linear and quadratic in temperature resistivity from holography, J. High Energy Phys. 1611 (2016) 128, arXiv: 1606.07905 [hep-th].
[60] S. Cremonini, H.S. Liu, H. Lu, C.N. Pope, DC conductivities from non-relativistic scaling geometries with momentum dissipation, J. High Energy Phys. 1704 (2017) 009, arXiv:1608.04394 [hep-th].

Permissions

All chapters in this book were first published in PLB, by Elsevier; hereby published with permission under the Creative Commons Attribution License or equivalent. Every chapter published in this book has been scrutinized by our experts. Their significance has been extensively debated. The topics covered herein carry significant findings which will fuel the growth of the discipline. They may even be implemented as practical applications or may be referred to as a beginning point for another development.

The contributors of this book come from diverse backgrounds, making this book a truly international effort. This book will bring forth new frontiers with its revolutionizing research information and detailed analysis of the nascent developments around the world.

We would like to thank all the contributing authors for lending their expertise to make the book truly unique. They have played a crucial role in the development of this book. Without their invaluable contributions this book wouldn't have been possible. They have made vital efforts to compile up to date information on the varied aspects of this subject to make this book a valuable addition to the collection of many professionals and students.

This book was conceptualized with the vision of imparting up-to-date information and advanced data in this field. To ensure the same, a matchless editorial board was set up. Every individual on the board went through rigorous rounds of assessment to prove their worth. After which they invested a large part of their time researching and compiling the most relevant data for our readers.

The editorial board has been involved in producing this book since its inception. They have spent rigorous hours researching and exploring the diverse topics which have resulted in the successful publishing of this book. They have passed on their knowledge of decades through this book. To expedite this challenging task, the publisher supported the team at every step. A small team of assistant editors was also appointed to further simplify the editing procedure and attain best results for the readers.

Apart from the editorial board, the designing team has also invested a significant amount of their time in understanding the subject and creating the most relevant covers. They scrutinized every image to scout for the most suitable representation of the subject and create an appropriate cover for the book.

The publishing team has been an ardent support to the editorial, designing and production team. Their endless efforts to recruit the best for this project, has resulted in the accomplishment of this book. They are a veteran in the field of academics and their pool of knowledge is as vast as their experience in printing. Their expertise and guidance has proved useful at every step. Their uncompromising quality standards have made this book an exceptional effort. Their encouragement from time to time has been an inspiration for everyone.

The publisher and the editorial board hope that this book will prove to be a valuable piece of knowledge for researchers, students, practitioners and scholars across the globe.

List of Contributors

S. Hoseinzadeh and A. Rezaei-Aghdam
Department of Physics, Faculty of Science, Azarbaijan Shahid Madani University, 53714-161, Tabriz, Iran

Andreas Ipp and David Müller
Institut für Theoretische Physik, Technische Universität Wien, Wiedner Hauptstr 8-10, A-1040 Vienna, Austria

Daniel Elander
Departament de Física Quàntica i Astrofísica & Institut de Ciències del Cosmos (ICC), Universitat de Barcelona, Martí Franquès 1, ES-08028, Barcelona, Spain

Maurizio Piai
Department of Physics, College of Science, Swansea University, Singleton Park, Swansea SA2 8PP, UK

M. Kord Zangeneh
Physics Department, Faculty of Science, Shahid Chamran University of Ahvaz 61357-43135, Iran
Research Institute for Astronomy and Astrophysics of Maragha (RIAAM), Maragha, P. O. Box: 55134-441, Iran
Physics Department and Biruni Observatory, Shiraz University, Shiraz 71454, Iran
Center of Astronomy and Astrophysics, Department of Physics and Astronomy, Shanghai Jiao Tong University, Shanghai 200240, China

A. Sheykhi
Research Institute for Astronomy and Astrophysics of Maragha (RIAAM), Maragha, P. O. Box: 55134-441, Iran
Physics Department and Biruni Observatory, Shiraz University, Shiraz 71454, Iran

Z.Y. Tang
Center of Astronomy and Astrophysics, Department of Physics and Astronomy, Shanghai Jiao Tong University, Shanghai 200240, China

B. Wang
Center of Astronomy and Astrophysics, Department of Physics and Astronomy, Shanghai Jiao Tong University, Shanghai 200240, China
Center for Gravitation and Cosmology, College of Physical Science and Technology, Yangzhou University, Yangzhou 225009, China

D. Bazeia, L. Losano and M.A. Marques
Departamento de Física, Universidade Federal da Paraíba, 58051-970 João Pessoa, PB, Brazil

R. Menezes
Departamento de Ciências Exatas, Universidade Federal da Paraíba, 58297-000 Rio Tinto, PB, Brazil
Departamento de Física, Universidade Federal de Campina Grande, 58109-970, Campina Grande, PB, Brazil

Ioannis Dalianis
Physics Division, National Technical University of Athens, 15780 Zografou Campus, Athens, Greece

Fotis Farakos
Dipartimento di Fisica e Astronomia "Galileo Galilei", Università di Padova, Via Marzolo 8, 35131 Padova, Italy
INFN, Sezione di Padova, Via Marzolo 8, 35131 Padova, Italy

Tomasz Romańczukiewicz
Institute of Physics, Jagiellonian University, Kraków, Poland

P.A. Krachkov and A.I. Milstein
Budker Institute of Nuclear Physics, 630090 Novosibirsk, Russia

Á. Ballesteros, I. Gutiérrez-Sagredo and F.J. Herranz
Departamento de Física, Universidad de Burgos, E-09001 Burgos, Spain

G. Gubitosi
Theoretical Physics, Blackett Laboratory, Imperial College, London SW7 2AZ, United Kingdom

N.G. Kelkar and D. Bedoya Fierro
Dept. de Fisica, Universidad de los Andes, Cra. 1E No.18A-10, Santafe de Bogotá, Colombia

David C. Dunbar, John H. Godwin, Guy R. Jehu and Warren B. Perkins
College of Science, Swansea University, Swansea, SA2 8PP, UK

S. Gonzalez-Martin
Departamento de Física Teórica and Instituto de Física Teórica (IFT-UAM/CSIC), Universidad Autónoma de Madrid, Cantoblanco, 28049, Madrid, Spain

C.P. Martin
Universidad Complutense de Madrid (UCM), Departamento de Física Teórica I, Facultad de Ciencias Físicas, Av. Complutense S/N (Ciudad Univ.), 28040 Madrid, Spain

George Georgiou
Institute of Nuclear and Particle Physics, National Center for Scientific Research Demokritos, Ag Paraskevi, GR-15310 Athens, Greece

Konstantinos Sfetsos
Department of Nuclear and Particle Physics, Faculty of Physics, National and Kapodistrian University of Athens, Athens 15784, Greece

Konstantinos Siampos
Albert Einstein Center for Fundamental Physics, Institute for Theoretical Physics, Laboratory for High-Energy Physics, University of Bern, Sidlerstrasse 5, CH3012 Bern, Switzerland

Wen-Cong Gan, Fu-Wen Shu and Meng-He Wu
Department of Physics, Nanchang University, Nanchang, 330031, China
Center for Relativistic Astrophysics and High Energy Physics, Nanchang University, Nanchang 330031, China

Mahdi Kord Zangeneh
Center for Gravitation and Cosmology, College of Physical Science and Technology, Yangzhou University, Yangzhou 225009, China
Research Institute for Astronomy and Astrophysics of Maragha (RIAAM), Maragha, P.O. Box 55134-441, Iran
Physics Department and Biruni Observatory, Shiraz University, Shiraz 71454, Iran
Center for Astronomy and Astrophysics, Department of Physics and Astronomy, Shanghai Jiao Tong University, Shanghai 200240, China

Yen Chin Ong and Bin Wang
Center for Gravitation and Cosmology, College of Physical Science and Technology, Yangzhou University, Yangzhou 225009, China
Center for Astronomy and Astrophysics, Department of Physics and Astronomy, Shanghai Jiao Tong University, Shanghai 200240, China

M. Rahimi, M. Ali-Akbari and M. Lezgi
Department of Physics, Shahid Beheshti University G.C., Evin, Tehran 19839, Iran

Francesco Benini
SISSA, INFN, Sezione di Trieste, via Bonomea 265, 34136 Trieste, Italy
Blackett Laboratory, Imperial College London, London SW7 2AZ, United Kingdom

Kiril Hristov
INRNE, Bulgarian Academy of Sciences, Tsarigradsko Chaussee 72, 1784 Sofia, Bulgaria

Alberto Zaffaroni
Dipartimento di Fisica, Università di Milano-Bicocca, I-20126 Milano, Italy
INFN, sezione di Milano-Bicocca, I-20126 Milano, Italy

Marco Matone
Dipartimento di Fisica e Astronomia "G. Galilei", Istituto Nazionale di Fisica Nucleare, Università di Padova, Via Marzolo, 8-35131 Padova, Italy

R. Bufalo
Departamento de Física, Universidade Federal de Lavras, Caixa Postal 3037, 37200-000 Lavras, MG, Brazil

S. Upadhyay
Centre for Theoretical Studies, Indian Institute of Technology, Kharagpur, Kharagpur 721302,WB, India

Si-Wen Li
Department of Modern Physics, University of Science and Technology of China, Hefei 230026, Anhui, China

Göran Fäldt and Andrzej Kupsc
Division of Nuclear Physics, Department of Physics and Astronomy, Uppsala University, Box516, 75120 Uppsala, Sweden

Adi Armoni and Edwin Ireson
Department of Physics, Swansea University, Swansea SA28PP, United Kingdom

Fabricio M. Ferreira
Departamento de Física, ICE, Universidade Federal de Juiz de Fora, Campus Universitário - Juiz de Fora, 36036-330, MG, Brazil
Instituto Federal de Educação, Ciência e Tecnologia do Sudeste de Minas Gerais, IF Sudeste MG - Juiz de Fora, 36080-001, MG, Brazil

Ilya L. Shapiro
Departamento de Física, ICE, Universidade Federal de Juiz de Fora, Campus Universitário - Juiz de Fora, 36036-330, MG, Brazil
Tomsk State Pedagogical University, Tomsk, 634041, Russia
National Research Tomsk State University, Tomsk, 634050, Russia

M. Gómez-Ramos and A. M. Moro
Departamento de Física Atómica, Molecular y Nuclear, Facultad de Física, Universidad de Sevilla, Apartado 1065, E-41080 Sevilla, Spain

J. Casal
Departamento de Física Atómica, Molecular y Nuclear, Facultad de Física, Universidad de Sevilla, Apartado 1065, E-41080 Sevilla, Spain

European Centre for Theoretical Studies in Nuclear Physics and Related Areas (ECT*) and Fondazione Bruno Kessler, Villa Tambosi, Strada delle Tabarelle 286, I-38123 Villazzano (TN), Italy

Shahar Hod
The Ruppin Academic Center, Emeq Hefer 40250, Israel
The Hadassah Academic College, Jerusalem 91010, Israel

N. Barnea, E. Friedman and A. Gal
Racah Institute of Physics, The Hebrew University, 91904 Jerusalem, Israel

B. Bazak
IPNO, CNRS/IN2P3, Univ. Paris-Sud, Université Paris-Saclay, F-91406 Orsay, France

Andreas Athenodorou
Computation-based Science and Technology Research Center, The Cyprus Institute, 20 Kavafi Str., Nicosia 2121, Cyprus
Department of Physics, University of Cyprus, POB 20537, 1678 Nicosia, Cyprus

Michael Teper
Rudolf Peierls Centre for Theoretical Physics, University of Oxford, 1 Keble Road, Oxford OX1 3NP, UK

Ali Imaanpur and Razieh Zameni
Department of Physics, School of sciences, Tarbiat Modares University, P.O. Box 14155-4838, Tehran, Iran

D.X. Horváth and G. Takács
MTA-BME "Momentum" Statistical Field Theory Research Group, Budafoki út 8, 1111 Budapest, Hungary
Department of Theoretical Physics, Budapest University of Technology and Economics, Budafoki út 8, 1111 Budapest, Hungary

Vivek M.Vyas
Raman Research Institute, Sadashivnagar, Bengaluru, 560 080, India

Prasanta K. Panigrahi
Department of Physical Sciences, Indian Institute of Science Education and Research Kolkata, Mohanpur, 741 246, India

Victor E. Ambruş
Department of Physics, West University of Timi¸soara, Bd. Vasile Pârvan 4, Timi¸soara, 300223, Romania

Paweł Caputa
Center for Gravitational Physics, Yukawa Institute for Theoretical Physics, Kyoto University, Kyoto 606-8502, Japan

Sumit R. Das
Department of Physics and Astronomy, University of Kentucky, Lexington, KY 40506, USA

Masahiro Nozaki
Kadanoff Center for Theoretical Physics, University of Chicago, Chicago, IL 60637, USA

Akio Tomiya
Key Laboratory of Quark & Lepton Physics (MOE) and Institute of Particle Physics, Central China Normal University, Wuhan 430079, China

Yuri V. Kovchegov
Department of Physics, The Ohio State University, Columbus, OH 43210, USA

Daniel Pitonyak
Division of Science, Penn State University-Berks, Reading, PA 19610, USA

Matthew D. Sievert
Theoretical Division, Los Alamos National Laboratory, Los Alamos, NM 87545, USA

Reza Fareghbal
Department of Physics, Shahid Beheshti University, G.C., Evin, Tehran 19839, Iran

Ali Naseh
School of Particles and Accelerators, Institute for Research in Fundamental Sciences (IPM), P.O. Box 19395-5531, Tehran, Iran

Shahin Rouhani
School of Particles and Accelerators, Institute for Research in Fundamental Sciences (IPM), P.O. Box 19395-5531, Tehran, Iran
Department of Physics, Sharif University of Technology, P.O. Box 11365-9161, Tehran, Iran

R.R. Metsaev
Department of Theoretical Physics, P.N. Lebedev Physical Institute, Leninsky prospect 53, Moscow 119991, Russia

Xiao-Mei Kuang
Center for Gravitation and Cosmology, College of Physical Science and Technology, Yangzhou University, Yangzhou 225009, China
Instituto de Física, Pontificia Universidad Católica de Valparaíso, Casilla 4059, Valparaíso, Chile

Jian-Pin Wu
Institute of Gravitation and Cosmology, Department of Physics, School of Mathematics and Physics, Bohai University, Jinzhou 121013, China

Index

A

Ads, 1-2, 4, 8, 20, 23, 25-26, 48, 52-53, 75, 77-80, 83-88, 90-91, 95-96, 125, 135, 159-160, 162-163, 182, 192, 195-198, 204, 209

Anti De Sitter, 52, 95

Atomic Field, 43, 45

Axion, 108, 110-113, 152-157, 162

Axion Mass, 110, 112, 152-153

B

Baryon, 114-116, 118, 130, 146-148, 150, 191

Bekenstein-hawking Entropy, 99

Boost Invariance, 9, 12

Branes, 20, 23-26, 95, 99, 114-116, 118, 133, 160, 163, 209

Btz Black Hole, 25, 77, 79-80, 88, 90

C

C-function, 192, 195-196

Cft, 20, 25, 75, 77-80, 82-84, 90-91, 95-96, 135, 159-160, 163, 166, 182, 192, 195-196, 204, 209

Chern, 8, 13, 27, 31, 99, 113, 115, 118, 171, 174-175

Classical Solutions, 8

Cmera, 77-79, 83

Color Glass Condensate, 9, 13, 191

Conformal Anomaly, 132, 135, 192, 196

Continuous Spin, 197-198, 203

Current, 10, 15, 31, 33, 70-71, 75, 112, 171, 174, 188

Current Confinement, 171, 174

D

D-state Probability, 54, 57

Deformation, 47-48, 50, 52, 70, 75, 79, 159-160

Deuteron, 54, 57-58, 142, 147

Diagrammar, 62

Dilaton, 7-8, 15, 18, 20, 23-26, 118, 162, 209

Dirac Field, 176, 179

Domain Walls, 38, 41-42

E

Effective Action, 70-71, 114, 132, 157, 171, 174, 191

Eft Regulator, 146

Electrodynamics, 26, 46, 108, 112-113

Entanglement Entropy, 78, 82-84, 86, 88, 91, 93, 95, 182-185, 196

Entropy, 25, 78, 82-84, 86, 88, 90-91, 93, 95-96, 99-100, 182-185, 196, 204, 209

Equivalent, 1-2, 12, 40, 50, 70, 75, 79, 96, 99, 102-103, 105-106, 115, 118, 164, 195

F

Fermionic Field, 197-198

Frw, 1, 5, 7-8

G

Gauge Symmetric, 1-2, 4, 8

Gauge Theories, 8, 91, 100, 113, 152-153, 157-158, 175, 190

Gauged Supergravity, 96, 99, 163

Glasma, 9-10, 12-14

Glueball, 114-116, 118, 155, 157-158

Gordon Model, 40, 164-165

Gravity, 1-2, 5, 7-8, 15, 17-20, 25-26, 31, 37, 47, 52-53, 59, 62-64, 66, 69, 77-79, 82-84, 91, 93-96, 99, 114-115, 118, 132, 135, 144-145, 157, 163, 174-175, 192, 195-196, 204, 209

Gravity Model, 1-2, 7-8, 91

H

Hadron, 9, 125, 146

Hamiltonian Densities, 70

Heavy-ion Collisions, 9, 13-14

Helicity Distribution, 187, 191

Holes, 8, 20, 23-26, 83, 95-96, 99-100, 135, 145, 209

Holographic Complexity, 84-85, 87, 90

Holographic Entanglement Entropy, 82, 84, 95

Holographic Superconductors, 26, 85, 90, 159, 163, 209

Holography, 8, 26, 77, 82-85, 88, 95, 130, 135, 182, 192, 196, 204, 209

Hyperbolic, 47-48, 50, 52

Hyperbolic Momenta, 47

K

Kibble, 38, 42, 101-102, 107, 182-186

Klein-gordon Field, 176-177, 179

L

Landau Frame, 176-177, 179

Lifshitz Exponent, 20, 24-25

N

Negative Radiation Pressure, 40

Non-supersymmetric, 18, 159, 163

Nuclei, 9-10, 12-13, 43, 114, 137-138, 141-142, 146-147

P

Particle, 8-10, 13, 15, 43, 45-46, 48, 70, 99, 108, 113-114, 137, 140-141, 147, 152, 157, 164-166, 169, 177, 179, 182

Q

Qcd, 13-15, 95, 108, 112, 114, 118, 130, 157-158, 163, 171, 174, 191

Quantum Quench, 182

Quasi, 20, 23-26, 156, 164

R

Rigidly-rotating Thermal States, 176, 179

S

S-matrix, 59, 62, 75, 166

Scalar Hair, 25, 145

Simons, 8, 13, 27, 31, 99, 113, 115, 118, 171, 174-175, 186

Sitter Algebra, 47

Solitons, 38, 107

State Correspondence, 77-79, 82

String Theory, 1, 5, 7-8, 25, 69, 75-76, 82, 91, 95-96, 106, 114-115, 118, 132, 157, 197, 203

Superfields, 32-33, 36-37, 175

Supergravity, 17, 32-33, 35-37, 82, 96, 99, 114, 118, 159-160, 163

T

Thermal Cft, 77-78

Three-body, 137-139, 141-142, 147, 150

U

Ultra-relativistic Field Theories, 192, 196

V

Veneziano Amplitude, 130

Very Special Relativity, 113

Vortices, 27, 31, 107

W

Wzw, 1, 7-8, 70-71, 75

Y

Yukawa Couplings, 64

Z

Zurek, 38, 182-186